Proceedings of the 6th Asia-Pacific

BIOINFORMATICS

Conference

SERIES ON ADVANCES IN BIOINFORMATICS AND COMPUTATIONAL BIOLOGY

ISSN: 1751-6404

Vol. 1: Proceedings of the 3rd Asia-Pacific Bioinformatics Conference
Eds: Yi-Ping Phoebe Chen and Limsoon Wong

Vol. 2: Information Processing and Living Systems
Eds: Vladimir B. Bajic and Tan Tin Wee

Vol. 3: Proceedings of the 4th Asia-Pacific Bioinformatics Conference
Eds: Tao Jiang, Ueng-Cheng Yang, Yi-Ping Phoebe Chen and Limsoon Wong

Vol. 4: Computational Systems Bioinformatics 2006
Eds: Peter Markstein and Ying Xu
ISSN: 1762-7791

Vol. 5: Proceedings of the 5th Asia-Pacific Bioinformatics Conference
Eds: David Sankoff, Lusheng Wang and Francis Chin

Vol. 6: Proceedings of the 6th Asia-Pacific Bioinformatics Conference
Eds: Alvis Brazma, Satoru Miyano and Tatsuya Akutsu

Series on Advances in Bioinformatics and Computational Biology – Volume 6

Proceedings of the 6th Asia-Pacific

BIOINFORMATICS

Conference

Kyoto, Japan 14 – 17 January 2008

editors

Alvis Brazma
European Bioinformatics Institute, UK

Satoru Miyano
University of Tokyo, Japan

Tatsuya Akutsu
Kyoto University, Japan

ICP

Imperial College Press

Published by

Imperial College Press
57 Shelton Street
Covent Garden
London WC2H 9HE

Distributed by

World Scientific Publishing Co. Pte. Ltd.
5 Toh Tuck Link, Singapore 596224
USA office: 27 Warren Street, Suite 401-402, Hackensack, NJ 07601
UK office: 57 Shelton Street, Covent Garden, London WC2H 9HE

British Library Cataloguing-in-Publication Data
A catalogue record for this book is available from the British Library.

PROCEEDINGS OF THE 6TH ASIA-PACIFIC BIOINFORMATICS CONFERENCE

ISBN-13 978-1-84816-108-5
ISBN-10 1-84816-108-5

Printed in Singapore by World Scientific Printers (S) Pte Ltd

PREFACE

High-throughput sequencing and functional genomics technologies have given us the human genome sequence as well as those of other experimentally, medically and agriculturally important species, and have enabled large-scale genotyping and gene expression profiling of human populations. Databases containing large numbers of sequences, polymorphisms, structures, gene expression profiles of normal and diseased tissues are rapidly being generated for human and model organisms. Databases containing various kinds of biological networks, which include metabolic networks, protein-protein interaction networks and gene regulatory networks, are also being developed. Bioinformatics is thus rapidly growing in importance in the annotation of genomic sequences, in the understanding of the interplay among and between genes and proteins, in the analysis of the genetic variability of species, in the identification of pharmacological targets and in the inference of evolutionary origins, mechanisms and relationships.

The Asia-Pacific Bioinformatics Conference series is an annual forum for exploring research, development, and novel applications of bioinformatics. It brings together researchers, students, professionals, and industrial practitioners for interaction and exchange of knowledge and ideas. The Sixth Asia-Pacific Bioinformatics Conference, APBC 2008, was held in Kyoto, Japan, 14-17 January, 2008.

A total of 100 papers were submitted to APBC 2008. These submissions came from Australia, Belgium, Canada, China, Denmark, France, Germany, Hong Kong, India, Iran, Ireland, Israel, Italy, Japan, Latvia, Netherlands, Norway, Pakistan, Poland, Portugal, Saudi Arabia, Singapore, South Korea, Spain, Switzerland, Taiwan, Turkey, UK, and USA. We assigned each paper to at least three members of the programme committee. Although not all members of the programme committee managed to review all the papers assigned to them, a total of 286 reviews were received, so that there were about three reviews per papers on average. It is to be mentioned that there were at least two reviews for each paper.

A total of 36 papers (36 %) were accepted for presentation and publication in the proceedings of APBC 2008. Based on affiliations of the authors, 3.75 of the accepted papers were from Australia, 2.67 were from Canada, 3.75 were from China, 0.25 were from France, 4.32 were from Germany, 2 were from Hong Kong, 0.5 were from Italy, 3.5 were from Japan, 1 were from Netherland, 2.75 were from Singapore, 0.5 were from Spain, 1.25 were from Switzerland, 1 were from Turkey, 0.18 were from UK and 8.33 were from USA. The topics of the accepted papers cover wide range of bioinformatics, which includes population genetics/SNP/ haplotyping, comparative genetics, evolution and phylogeny, database and data integration, pathways and networks, text mining and data mining, prediction and analysis of RNA and protein structures, gene expression analysis, sequence analysis, and algorithms. In addition to the accepted papers, the scientific programme of APBC 2008 also included three keynote talks, by Andreas Dress, Minoru Kanehisa and Alfonso Valencia, as well as tutorial and poster sessions.

We had a great time in Kyoto, enhancing the interactions between many researchers and practitioners, and reuniting the Asia-Pacific bioinformatics community in the context of an international conference with world-wide participation.

Lastly, we wish to express our gratitude to the authors of the submitted papers, the members of the programme committee and their subreferees, the members of the organizing committee, Phoebe Chen and Limsoon Wang (our liaisons in the APBC steering committee), the keynote speakers and our generous sponsors and supporting organizations, which include Bioinformatics Center (Kyoto University), Human Genome Center (University of Tokyo), The Telecommunications Advancement Foundation, Special Interest Group on Mathematical Modeling and Problem Solving (SIGMPS, IPSJ), Special Interest Group on Bioinformatics (SIGBIO, IPSJ) and Japanese Society for Bioinformatics, for making APBC 2008 a great success.

Alvis Brazma
Satoru Miyano
Tatsuya Akutsu

17 January 2008

APBC2008 ORGANIZATION

Conference Chair

Tatsuya Akutsu, Kyoto University, Japan

Organizing Committee

Tatsuya Akutsu (Chair), Kyoto University, Japan
Susumu Goto, Kyoto University, Japan
Morihiro Hayashida, Kyoto University, Japan
Hiroshi Mamitsuka, Kyoto University, Japan
Satoru Miyano, University of Tokyo, Japan

Steering Committee

Phoebe Chen (Chair), Deakin University, Australia
Sang Yup Lee, KAIST, Korea
Satoru Miyano, University of Tokyo, Japan
Mark Ragan, University of Queensland, Australia
Limsoon Wong, National University of Singapore, Singapore

Program Committee

Alvis Brazma (Co-Chair), European Bioinformatics Institute
Satoru Miyano (Co-Chair), University of Tokyo
Tatsuya Akutsu, Kyoto University
Masanori Arita, University of Tokyo
Kiyoshi Asai, University of Tokyo
Catherine Ball, Stanford University
Vladimir Brusic, Dana-Farber Cancer Institute
Yi-Ping Phoebe Chen, Deakin University
Francis Y.L. Chin, University of Hong Kong
Roderic Guigo, Centre de Regulacio Genomica, Barcelona
Sridhar Hannenhalli, University of Pennsylvania
Wen-Lian Hsu, Academia Sinica
Tao Jiang, University of California, Riverside
Inge Jonassen, Bergen University
Samuel Kaski, Helsinki University of Technology
Sang Yup Lee, KAIST
Jinyan Li, Institute for Infocomm Research
Ming Li, University of Waterloo
Jingchu Luo, Peking University
Bin Ma, University of Western Ontario
Hiroshi Mamitsuka, Kyoto University
Shinichi Morishita, University of Tokyo
Laxmi Parida, IBM, T.J. Watson Research Center
John Quackenbush, Harvard University
Mark Ragan, University of Queensland
Shoba Ranganathan, Macquarie University
Marie-France Sagot, University Claude Bernard Lyon
Yasubumi Sakakibara, Keio University
David Sankoff, University of Ottawa
Thomas Schlitt, King's College London
Paul Spellman, Berkeley National Laboratory
Alfonso Valencia, Spanish National Cancer Research Centre
Jean-Philippe Vert, Ecole des Mines de Paris
Juris Viksna, University of Latvia
Martin Vingron, Max Planck Institute for Molecular Genetics
Lusheng Wang, The City University of Hong Kong
Limsoon Wong, National University of Singapore
Ying Xu, University of Georgia
Ueng Cheng Yang, National Yang Ming University

Byoung-Tak Zhang, Seoul National University
Louxin Zhang, National University of Singapore
Michael Zhang, Cold Spring Harbor Laboratory
Xuegong Zhang, Tsinghua University

Additional Reviewers

Timothy Bailey	Mikael Boden	Jia-Ming Chang
Ching-Tai Chen	Feng Chen	Qingfeng Chen
Kwok Pui Choi	Ivan Gesteira Costa	Larry Croft
Aaron Darling	Stefan Haas	Jingyu Hou
Seiya Imoto	Bo Jiang	Rui Jiang
Hisanori Kiryu	Arto Klami	Leo Lahti
Tingting Li	Xiaowen Liu	Marta Luksza
Scott Mann	Andrew Newman	Janne Nikkilä
Merja Oja	Utz Pape	Jaakko Peltonen
Hugues Richard	Kengo Sato	Petteri Sevon
Teppei Shimamura	Christine Steinhoff	Wanwan Tang
Xuebing Wu	Zhengpeng Wu	Chenghai Xue
Rui Yamaguchi	Tomasz Zemojtel	Xueya Zhou

CONTENTS

RECENT PROGRESS IN PHYLOGENETIC COMBINATORICS

ANDREAS DRESS

CAS-MPG Partner Institute for Computational Biology, Shanghai Institutes for Biological Sciences, Chinese Academy of Sciences, Shanghai 200031, P.R. China and Max-Planck-Institut fuer Mathematik in den Naturwissenschaften, Inselstrasse 22-26, D-04103 Leipzig, Germany

1. Background

Phylogenetic combinatorics deals with the combinatorial aspects of phylogenetic tree reconstruction. A starting point was the following observation:

Given a metric $D : X \times X \longrightarrow \mathbb{R}$ representing the approximative genetic distances between the members of a collection X of taxa, it was shown in Ref. 1 that the following assertions relating to the "object of desire", a phylogenetic-tree X-tree, all are equivalent:

(i) The space "tight span"

$$T_D := \{f \in \mathbb{R}^X : \forall_{x \in X} f(x) = \sup_{y \in X}(D(x,y) - f(y))\}$$

of D is an $\mathbb{R}$-*tree*.

(ii) There exists a tree (V, E) whose vertex set V contains X, and an edge weighting $\ell : E \longrightarrow \mathbb{R}$ that assigns a positive length $\ell(e)$ to each edge e in E, such that D is the restriction of X to the shortest-path metric induced on V.

(iii) There exists a map $w : \mathcal{S}(X) \longrightarrow \mathbb{R}_{\geq 0}$ from the set $\mathcal{S}(X)$ of all bi-partitions or *splits* of X into the set $\mathbb{R}_{\geq 0}$ of non-negative real numbers such that, given any two splits $S = \{A, B\}$ and $S' = \{A', B'\}$ in $\mathcal{S}(X)$ with $w(S), w(S') \neq 0$, at least one of the four intersections $A \cap A', B \cap A'$, $A \cap B'$, and $B \cap B'$ is empty and $D(x,y) = \sum_{S \in \mathcal{S}(X:x \leftrightarrow y)} w(S)$ holds where $\mathcal{S}(X : x \leftrightarrow y)$ denotes the set of splits $S = \{A, B\} \in \mathcal{S}(X)$ that *separate* x and y.

(iv) $D(x,y) + D(u,v) \leq \max\big(D(x,u) + D(y,v), D(x,v) + D(y,u)\big)$ holds for all $x, y, u, v \in X$

Moreover, the metric space T_D actually coincides in this case with the $\mathbb{R}$-tree that is canonically associated with a weighted X-tree (V, E, ℓ).

2. Discussion

This observation suggested to further investigate (1) the tight-span construction and (2) representations of metrics by weighted split systems with more or less specific properties, even if the metric in question would not satisfy the very special properties described above which investigations have, in turn, given rise to a full-fledged research program dealing with many diverse aspects of these two topics (see the list of references below).

In my lecture, I will focus on the rather new developments relating to *block decomposition* and *virtual cut points* of metric spaces reported, respectively, in References 2 and 3 that allow one to canonically decompose any given finite metric space into a sum of pairwise compatible *block metrics*, thus providing a far-reaching generalization of the result recalled above.

References

1. A. Dress, *AiM* **53**, 321 (1984).
2. A. Dress, K. Huber, J. Koolen and V. Moulton, Compatible decompositions and block realizations of finite metric spaces., submitted.
3. A. Dress, K. Huber, J. Koolen, V. Moulton and A. Spillner, A note on the metric cut point and the metric bridge partition problems., submitted.
4. J. Apresjan, *Mashinnyi perevod: prikladnaja lingvistika* **9**, 3 (1966).
5. E. Baake, *Math. Biosci.* **154**, 1 (1998).
6. H. J. Bandelt, *SIAM J. Disc. Math.* **3**, 1 (1990).
7. H. Bandelt and A. Dress, *Bul. Math. Biol.* **51**, 133 (1989).
8. H. Bandelt and A. Dress, *AiM* **92**, 47 (1992).
9. H. Bandelt and A. Dress, *Molecular Phylogenetics and Evolution* **1**, 242 (1992b).
10. H. Bandelt, V. Chepoi, A. Dress and J. Koolen, *Eur. J. Comb.* **27**, 669 (2006).
11. H. Bandelt and M. Steel, *SIAM. J. Disc. Math.* **8**, 517 (1995).
12. J.-P. Barthélemy and A. Guénoche, *Trees and proximity representations* (Wiley, 1991).
13. S. Böcker and A. Dress, *AiAM* **138**, 105 (1998).
14. B. Bowditch, *Notes on Gromov's hyperbolicity criterion for path metric spaces.*, in *Group theory from a geometric viewpoint*, eds. E. Ghys, A. Haefliger and A. Verjovsky (World Scientific, 1991), pp. 64–167.
15. P. Buneman, *The recovery of trees from measures of dissimilarity.*, in *Mathematics in the Archeological and Historical Sciences*, ed. F. H. et al. (Edinburgh University Press, 1971), pp. 387–395.
16. D. Bryant and V. Berry, *AiAM* **27**, 705 (2001).
17. D. Bryant and A. Dress, Linearly independent split systems., Europ. J. Comb., to appear.
18. A. Dress, *AiM* **74**, 163 (1989).
19. A. Dress, *Mathematical hierarchies and biology* , 271 (1996), DIMACS Ser. Discrete Math. Theoret. Comput. Sci., 37, Amer. Math. Soc., Providence, RI, 1997.
20. A. Dress, *AML* **15**, 995 (2002).
21. A. Dress, *Graphs and Metrics.*, in *Encyclopaedia of Genetics, Genomics, Proteomics and Bioinformatics*, ed. A. K. e. a. Shankar Subramaniam (John Wiley and Sons, 2005).
22. A. Dress, *The Category of X-nets.*, in *Networks: From Biology to Theory*, eds. J. Feng, J. Jost and M. Qian (Springer, 2006).

23. A. Dress, Phylogenetic analysis, split systems, and boolean functions., In: *Aspects of Mathematical Modelling*, Roger Hosking and Wolfgang Sprössig eds, Birkhäuser, 2006, to appear.
24. A. Dress, A note on group-valued split and set systems., *Contributions to Discrete Mathematics*, to appear.
25. A. Dress, Split decomposition over an abelian group, part i: Generalities., *AoC*, to appear.
26. A. Dress, Split decomposition over an abelian group, part ii: Group-valued split systems with weakly compatible support., *Discrete Applied Mathematics, Special Issue on Networks in Computational Biology*, to appear.
27. A. Dress, Split decomposition over an abelian group, part iii: Group-valued split systems with compatible support., submitted.
28. A. Dress, B. Holland, K. Huber, J. Koolen, V. Moulton and J. Weyer-Menkhoff, *Discrete Applied Mathematics* **146**, 51 (2005).
29. A. Dress, K. Huber, A. Lesser and V. Moulton, *AoC, Special Volume on Biomathematics* **10**, 63 (2006).
30. A. Dress, K. Huber, J. Koolen and V. Moulton, An algorithm for computing virtual cut points in finite metric spaces (2007).
31. A. Dress, K. Huber, J. Koolen and V. Moulton, Cut points in metric spaces, *AML*, in press.
32. A. Dress, K. Huber and V. Moulton, *AoC* **2**, 299 (1998).
33. A. Dress, K. Huber and V. Moulton, *AoC* **1**, 339 (1997).
34. A. Dress, K. Huber and V. Moulton, *AoC* **4**, 1 (2000).
35. A. Dress, K. Huber and V. Moulton, *AiM* **168**, 1 (2002).
36. A. Dress, K. Huber and V. Moulton, Some uses of the farris transform in mathematics and phylogenetics — a review., AoC, Special Volume on Biomathematics, to appear.
37. A. Dress, J. Koolen and V. Moulton, *AoC* **8**, 463 (2004).
38. A. Dress, V. Moulton and W. Terhalle, *Europ. J. Comb.* **17**, 161 (1996).
39. A. Dress and M. Steel, *AoC, Special Volume on Biomathematics* **10**, 77 (2006).
40. A. Dress and M. Steel, *AoC, Special Volume on Biomathematics* **10**, 77 (2006).
41. A. Dress and M. Steel, Phylogenetic diversity over an abelian group., AoC, Special Volume on Biomathematics, to appear.
42. J. Farris, *On the phylogenetic approach to vertebrate classification.*, in *Major patterns in vertebrate evolution*, eds. M. Hecht, P. Goody and B. Hecht (Plenum Press, 1976).
43. J. Farris, *Sys. Zool.* **28**, 200 (1979).
44. J. Farris, *Sys. Zool.* **28**, 483 (1979).
45. J. Farris, A. Kluge and M. Eckardt, *Sys. Zool.* **19**, 172 (1970).
46. M. Gromov, *Hyperbolic Groups,*, in *Essays in Group Theory, MSRI series, Vol. 8*, ed. S. Gersten (Spring-Verlag, 1988).
47. S. Grünewald, K. Forslund, A. Dress and V. Moulton, *Mol. Biol. Evol.* **24**, 532 (2007).
48. D. Huson, *Bioinformatics* **14**, 68 (1998), http://www-ab.informatik.uni-tuebingen.de/software/splits/welcome_en.html.
49. D. Huson and A. Dress, *ACM Transactions in Computational Biology and Bioinformatics* **1**, 109 (2004).
50. J. Isbell, *Comment. Math. Helv.* **39**, 65 (1964).
51. N. Jardine, *Biometrics* **25**, 609 (1969).
52. P. Lockhart, A. Meyer and D. Penny, *J. Mol. Evol.* **41**, 666 (1995).
53. A. Parker-Rhodes and R. Needham, *Information processing, Proceedings of the International Conference on Information Processing, Paris, 1960* , 321 (1960).
54. C. Semple and M. Steel, *AiAM* **23**, 300 (1999).

55. C. Semple and M. Steel, *Phylogenetics* (Oxford University Press, 2003).
56. M. Steel, *AML* **7**, 19 (1994).

KEGG FOR MEDICAL AND PHARMACEUTICAL APPLICATIONS

MINORU KANEHISA
Bioinformatics Center, Institute for Chemical Research, Kyoto University, Uji, Kyoto 611-0011, Japan
Human Genome Center, Institute of Medical Science, University of Tokyo, Minato-ku, Tokyo 108-8639, Japan

KEGG (http://www.genome.jp/kegg/) is a suite of databases that integrates genomic, chemical, and systemic functional aspects of the biological systems. KEGG provides a reference knowledge base for linking genomes to life through the process of PATHWAY mapping, which is to map, for example, a genomic or transcriptomic content of genes to KEGG reference pathways to infer systemic behaviors of the cell or the organism. In addition, KEGG provides a reference knowledge base for linking genomes to the environment, such as for the analysis of drug-target relationships, through the process of BRITE mapping. KEGG BRITE is an ontology database representing functional hierarchies of various biological objects, including molecules, cells, organisms, diseases, and drugs, as well as relationships among them. The KEGG resource is being expanded to suit the needs for practical applications. KEGG PATHWAY now contains 26 pathway maps for human diseases in four subcategories: neurodegenerative disorders, infectious diseases, metabolic disorders, and cancers. Although such maps will continue to be added, they will never be sufficient to represent our knowledge of molecular mechanisms of diseases because in many cases it is too fragmentary to represent as pathways. KEGG DISEASE is a new addition to the KEGG suite accumulating molecular-level knowledge on diseases represented as lists of genes, drugs, biomarkers, etc. KEGG DRUG now covers all approved drugs in the U.S. and Japan. KEGG DRUG is a structure-based database. Each entry is a unique chemical structure that is linked to standard generic names, and is associated with efficacy and target information as well as drug classifications. Target information is presented in the context of KEGG pathways and drug classifications are part of KEGG BRITE. The generic names are linked to trade names and subsequently to outside resources of package insert information whenever available. This reflects our effort to make KEGG more useful to the general public.

PROTEIN INTERACTIONS EXTRACTED FROM GENOMES AND PAPERS

ALFONSO VALENCIA

Structural and Computational Biology Programme,
Spanish National Cancer Research Centre (CNIO)

To assess the feasibility of extracting protein interactions from text we have recently organized the BIOCREATVE II challenge (http://biocreative.sourceforge.net) in collaboration with the MINT and INTACT databases. The competition was divided in four sub-tasks: a) ranking of publications by their relevance on experimental determination of protein interactions, b) detection of protein interaction partners in text, c) detection of key sentences describing protein interactions, and d) detection of the experimental technique used to determine the interactions. 20 teams participated in the competition that used full text and the information on interactions, sentences and experimental vocabularies provided by the associated databases. The results showed quite promising results and clearly pointed to the main challenges setting the path for future research. Furthermore BioCreative has channel the collaboration of several teams for the creation of the first text mining meta-server (the complete set of BioCreative papers to be published in a special issue of Genome Biology).
Regarding the extraction of information on protein interactions from genomic information along the years my group and others have contributed to the developed of a set of methods based on the concept of concerted evolution between interacting protein families. In a new effort we have recently developed a completely new approach that uses the full power of co-evolution to integrate information from complete collections of protein families.

STRING KERNELS WITH FEATURE SELECTION FOR SVM PROTEIN CLASSIFICATION

WEN-YUN YANG[1] and BAO-LIANG LU[1,2,*]

[1] *Department of Computer Science and Engineering, Shanghai Jiao Tong University*
[2] *Laboratory for Computational Biology, Shanghai Center for System Biomedicine*
Shanghai 200240, China
E-mail: {ywy, bllu}@sjtu.edu.cn

We introduce a general framework for string kernels. This framework can produce various types of kernels, including a number of existing kernels, to be used with support vector machines (SVMs). In this framework, we can select the informative subsequences to reduce the dimensionality of the feature space. We can model the mutations in biological sequences. Finally, we combine contributions of subsequences in a weighted fashion to get the target kernel. In practical computation, we develop a novel tree structure, coupled with a traversal algorithm to speed up the computation. The experimental results on a benchmark SCOP data set show that the kernels produced by our framework outperform the existing spectrum kernels, in both efficiency and ROC50 scores.

Keywords: kernel methods, SVMs, homology detection, feature selection

1. Introduction

Kernel methods and support vector machines (SVMs) have been proved to be highly successful in machine learning and pattern classification fields. In computational biology community, SVMs have also been widely used to yield valuable insights into massive biological data sets. However, since biological data, such as DNA, RNA, and protein sequences, are naturally represented as strings, one needs to convert string format of biological data into a numerical vector, which is the standard input format for SVMs. However, this additional conversion could brings additional computational cost and even unexpected results. Fortunately, this conversion can be avoided by using kernel methods. The key advantage of kernel methods is that they depend only on the inner products of the samples. As a result, we can calculate the inner products directly from the sequences instead of calculating the numerical vectors. In other words, the $n \times n$ matrix of inner products between each two samples is the so-called *kernel* of SVMs. We define the kernels of SVMs directly upon strings, which are also called "string kernels".[1]

The pioneering work on convolution kernels and dynamic alignment kernels for discrete objects, such as strings and trees, was conducted by Haussler[2] and

*To whom correspondence should be addressed.

Watkins,[3] respectively. Thereafter, a number of string kernels have been extensively studied. In general, those kernels take the same idea as the convolution kernels. They all define some kinds of "sub-structures" and employ recursive calculation over all those "sub-structures" to get the kernels. For example, Leslie et al. proposed spectrum string kernels,[1] mismatch string kernels,[4] and a series of inexact matching string kernels,[5] all of which are based on the "sub-structures" called "k-mers"(k-length subsequences). The only difference among those kernels relies on the specific definition for each mapping function. Moreover, Vishwanathan and Smola[6] proposed another type of fast string kernels based on weighted sum for inner products, each of which corresponds to one of the exact matching subsequences. Those above two kinds of string kernels were both applied to a protein classification problem, called remote homology detection. Besides, string kernels have also been successfully applied to natural language processing (NLP) tasks.[7–9]

We introduce a framework to reconstruct string kernels to be used with SVMs. This framework is rather general that the string kernels aforementioned can be regarded as specific instances of it. We also develop a tree data structure and an algorithm for the computation of these string kernels.

2. A string kernel framework

2.1. *Notations*

We begin by introducing some notations. Let $\mathcal{A}$ be the *alphabet* and each element in $\mathcal{A}$ is called *character*. Then we denote the whole *string space* as $P(\mathcal{A}) = \bigcup_k \mathcal{A}^k$, where $\mathcal{A}^k$ denotes the *k-spectrum set* containing all the k-length strings produced by character concatenation from $\mathcal{A}$. At the next step, we make use of *feature groups* to take the biologically mutation effect into account. Each feature group is a subset of the string space, containing certain number of relatively similar strings. Formally, we use $\mathcal{T} = \{T_i \subseteq P(\mathcal{A}) | 1 \leq i \leq m\}$ to denote the set of all the feature groups and $P(\mathcal{T}) = \bigcup_i T_i$ to denote all the strings contained in these feature groups. For each feature group T_i, we use $|T_i|$ to denote its size, and t_{ij} for $j = 1$ to $|T_i|$ to index its elements.

In the following section, "*none of two feature groups are identical*" means that $T_i \neq T_j$ if $i \neq j$ for all i and j. "*All the feature groups cover the set $\mathcal{S}$*" means $\bigcup_i T_i = \mathcal{S}$.

2.2. *Framework definition*

We propose a string kernel framework as follows. First, we define the sub-kernel between strings x and y for each feature group T_i,

$$\begin{aligned} k_{T_i}(x,y) = \mathrm{num}_{T_i}(x) \cdot \mathrm{num}_{T_i}(y) &= \sum_{j=1}^{|T_i|} \mathrm{num}_{t_{ij}}(x) \cdot \sum_{k=1}^{|T_i|} \mathrm{num}_{t_{ik}}(y) \\ &= \sum_{j=1}^{|T_i|} \sum_{k=1}^{|T_i|} \mathrm{num}_{t_{ij}}(x) \cdot \mathrm{num}_{t_{ik}}(y) \qquad (1) \end{aligned}$$

where $\text{num}_{T_i}(x) = \sum_{j=1}^{|T_i|} \text{num}_{t_{ij}}(x)$ counts the total numbers of occurrences of T_i's members in x. Then we combine all the sub-kernels in a weighted fashion to obtain the target kernel, formally,

$$k(x, y) = \sum_{i=1}^{m} w_{T_i} k_{T_i}(x, y) \tag{2}$$

where each w_{T_i} is the weight used to measure the significance for the corresponding feature group T_i. Following this construction framework, we can derive various kinds of string kernels. Several typical string kernel instances are given below as examples:

- Setting $w_{T_i} = 1$ and $|T_i| = 1$ for all $i = 1$ to m. None of two feature groups are identical and all the feature groups cover the k-spectrum set. It yields the *k-spectrum string kernel*.[1]
- Setting $|T_i| = 1$ for all $i = 1$ to m. None of two feature groups are identical and all the feature groups cover the string space $P(\mathcal{A})$. It yields the family of kernels proposed by Vishwanathan and Smola.[6]
- All *the kernels using inexact matching* proposed by Leslie and Kuang[5] can be regarded as specific cases of $|T_i| > 1$.
- If we can customize the members for each feature group T_i, then we will achieves a new family of string kernels which has never been studied.

2.3. *Relations with existing string kernels*

Roughly speaking, existing string kernels can be divided into two categories, kernels using exact matching and using inexact matching. Kernels using exact matching[1,6–8] only take the perfect matching subsequences into account and design optimal algorithms for the computation. However, the kernels using inexact matching can model mismatches, gaps, substitutions and other wildcards. Such kernels are more suitable for biological data. Conceptually, it is clear that the kernels using exact matching are specific instances of the our string kernel framework. Since we can assign only one feature to each feature group then produce those kernels. However practically, we note that the kernels using exact matching have been computed using various optimal algorithms.[6–8]

On the other hand, all the kernels using inexact matching[5] can be constructed equally by feature re-mapping as follows,

$$k(x, y) = \sum_{s} \left(\sum_{\alpha = R^{-1}(s)} \sum_{\beta = R^{-1}(s)} \text{num}_{\alpha}(x) \cdot \text{num}_{\beta}(y) \right) \tag{3}$$

where $R^{-1}(s) = \{s' : R(s', s)\}$ defines the set of substrings that have specific relations with substring s, for example, at most m mismatches and at most g gaps. s is used to enumerate the k-spectrum set $\mathcal{A}^k$. Comparing this definition with Equations (1) and (2), we could immediately find that the kernels using inexact matching can be constructed by $|\mathcal{A}^k|$ feature groups, each of which corresponds to one k-length substring s, containing the set $R^{-1}(s)$. Conceptually, the only difference among all these kernels depends on the specific relation R.

3. Efficient computation

Instead of calculating and storing the feature vectors explicitly, we develop an algorithm based on a novel tree data structure to efficiently compute the kernel matrix, which can be used with the SVM classifier.

3.1. *Tree data structure with leaf links*

This tree data structure shown in Fig. 1 is similar to a suffix tree or mismatch tree used before.[4] The different part is that we add leaf links to generalize the algorithm. The calculation of the kernel matrix can be summarized as follows: firstly we construct the tree based on given feature groups. Note that the tree structure is determined only by the given feature groups. Then we use an essentially sliding window to perform lexical traverse of all the substrings occurring in the data set. As a result, in each leaf we store the number of the leaf substring occurring in each sample string. Finally we calculate the kernel matrix in one traversal for all the leaves of the tree.

3.2. *Leaf traversal algorithm*

The leaves of this tree represent all the substrings occurring in the feature groups, so the number of these leaves is $|P(\mathcal{T})|$. Accordingly, all the leaves are indexed by s_i for $i = 1$ to $|P(\mathcal{T})|$. The tree is organized like a trie: the concatenation of the edge labels from root to leaf interprets the string of the leaf. Unlike the standard tree structure, we add links between two leaves if they are contained in the same feature group T_i (probably not only one). Formally we define the whole set of links as,

$$L = \{l_{ij} | \exists k, s_i \in T_k \wedge s_j \in T_k\}. \tag{4}$$

Then we define the set of leaves, with links to leaf s_i as $L[s_i] = \{j | l_{ij} \in L\}$. For each linked leaf pair, we can define the weight of that link as

$$w(l_{ij}) = \sum_{k: s_i \in T_k \wedge s_j \in T_k} w_{T_k}. \tag{5}$$

In the following part, we use w_{ij} as a shorthand for $w(l_{ij})$. The kernel matrix calculation within the traversal of all the leaves is summarized in Algorithm 3.1.

The correctness of this algorithm follows from the analysis of how many times the term $num_{s_i}(x) \cdot num_{s_j}(y)$ is added up to the kernel value $k(x, y)$. It can be observed from Equations (1) and (2).

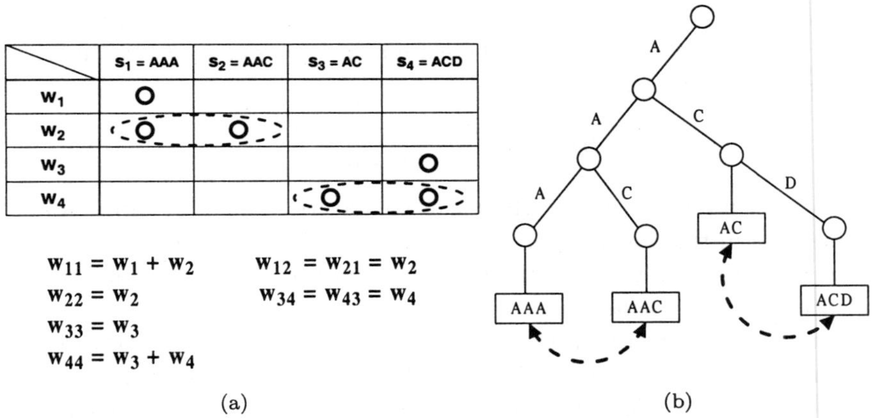

Fig. 1. An example of the tree structure and leaf links: (a) 4 feature groups with weights from w_1 to w_4, respectively; (b) The tree constructed for the given feature groups. Here, a total of 6 links are connected. Note that for clarity, we omit the self links for each leaf node and only draw the leaf links between leaves.

Algorithm 3.1 The calculation of the kernel value $k(x, y)$

$k(x, y) \leftarrow 0$
for all leaf s_i **do**
  for all $j \in L[s_i]$ **do**
    $k(x, y) \leftarrow k(x, y) + w_{ij} \cdot num_{s_i}(x) \cdot num_{s_j}(y)$
  end for
end for

4. Selecting feature groups and weights

The feature group aforementioned is a new concept for string kernels. Immediate extension can also be made for other kinds of machine learning methods. Actually we extend the notion of "feature" to "feature group" to let string kernels be more suitable to biological data. Meanwhile, it makes the construction procedure more flexible to produce various kinds of string kernels. In this section, we will develop several new approaches to demonstrate the effectiveness of the proposed framework.

Existing string kernel methods usually use the whole set of k-length subsequences as the feature set, and treat them equally in the kernel constructions. Unluckily, it leads not only to the loss of discriminative ability of significant subsequences, but also to the increase of computational cost. Apart from those, we start from learning the distribution of subsequences. Then we extracts statistically significant subsequences or groups of subsequences, which are then combined in a weighted fashion to reconstruct the string kernels.

To simplify this discussion, we restrict ourselves to two-class classification problems. Without loss of generalization, we explain our methods by using the following

BW criterion, which is based on the ratio of between-class scatter to within-class scatter. However, we also note that there are many types of statistical metrics that can be used in our proposed method.

$$BW(s) = \frac{|m^+(s) - m^-(s)|^2}{\sigma^+(s) + \sigma^-(s)} \tag{6}$$

where $m^+(s)$ and $\sigma^+(s)$ denote the mean composition and standard variance for subsequence s in the positive class, respectively, and $m^-(s)$ and $\sigma^-(s)$ are for the negative class. Usually, the numerator is called *between-class scatter* and the divisor is called *within-class scatter.*

To measure the statistical significance of a feature group, we also extend the definition of $BW(s)$ in Equation (6) to $BW(T_i)$, just by naturally defining the number of occurrences of feature group T_i as the sum of those of its members.

By using our framework, we propose two kinds of new string kernels in the following sections. Essentially, one is the reduced version of k-spectrum string kernel, and the other is the reduced version of (k, m)-mismatch string kernel.

4.1. *Reduction of spectrum string kernel*

We reconstruct the spectrum string kernels in two respects, the number of feature groups and the weights. Corresponding to the spectrum string kernel definition in Section 2, the number of feature groups is denoted by $|\mathcal{A}^k|$ and the weights are denoted by w_{T_i} for $i = 1$ to $|\mathcal{A}^k|$. For sake of computational efficiency and performance, we try to reduce feature groups $|\mathcal{A}^k|$ using two thresholds, minimum occurrence O_{min} and minimum score BW_{min}. Since we assume that the subsequences with low occurrences are either non-informative for discrimination or not influential in global performance. Similarly, the subsequences with low BW scores are also regarded with low discriminative ability.

For a proof of concept, we simply use the power of BW score, $w_{T_i} = [BW(T_i)]^\lambda$ to weight each of the feature groups, where the exponent λ is a parameter used to control the scale of weights.

4.2. *Statistically selecting feature groups*

How to choose the most discriminative feature groups and weights is at least as hard as the feature selection problem, which has 2^n subsets to be tested. This is clear since we can regard the feature selection as a specific case of feature group selection. Hence, we do not have an optimal solution for it. As an alternative approach, we propose a heuristic method to construct feature groups, each of which contains multiple members.

This method can be summarized as two steps: selecting the base subsequences s and then using a greedy expansion. The greedy expansion is an iterative process. At each iteration, the subsequence s' that lets $R(s', s)$ hold and maximize the $BW(T_i)$ score among the candidate subsequences, is selected into the feature group. This

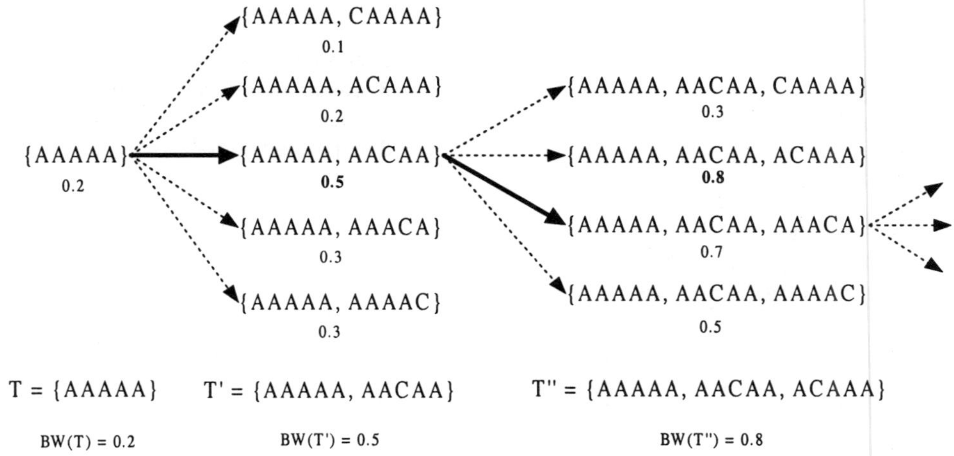

Fig. 2. An example of the greedy expansion in (5, 1) mismatch set.

process ends when no such s' is found. We give a simple example in Fig. 2. In this figure, for simplicity, we assume that the alphabet contains two letters, 'A' and 'C'. At the first iteration, AACAA is selected into the feature group, since it increases BW score more than other candidates. Then ACAAA is selected. Finally this greedy expansion terminates when there are no any features that let the BW score increase.

5. Experiment

We report the experiments on a benchmark SCOP data set (SCOP version 1.37) designed by Jaakkola et al.,[10] which is widely used to evaluate the methods for remote homology detection of protein sequences.[1,4–6] The data set[a] consists of 33 families, each of which has four sets of protein sequences, namely positive training and test sets, and negative training and test sets. The target family serves as the positive test set. The positive training set is chosen from the remaining families in the same superfamily. The negative training and test sets are chosen from the folds outside the fold of the target family.

We use ROC_{50} score[11] to evaluate the performance of homology detection. The ROC_{50} score is the area under the receiver operating characteristic curve (the plot of true positives as a function of false positives) up to the first 50 false positives. A score of one indicates perfect separation of positives from negatives, whereas a score of zero indicates that none of the top 50 sequences selected by the algorithm is positives. This ROC_{50} score is the most standard way to evaluate the performance of remote homology detection methods in computational biology.[1,6,11]

---

[a]Data is available at www.cse.ucsc.edu/research/compbio/discriminative

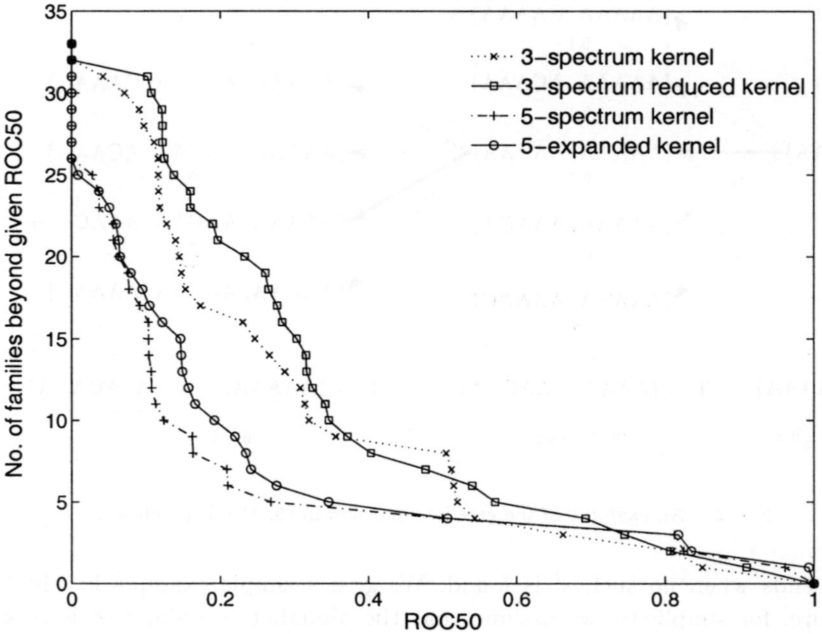

Fig. 3. Comparison of four kinds of kernels.

Table 1. The numbers of used subsequences in four kernels.

3-spectrum	3-spectrum reduced (mean/ ± SD)	5-spectrum	5-expanded (mean/ ± SD)
8000	2706/ ± 865	3.2×10^6	44926/ ± 20508

We give a performance overview in Fig. 3 for the four kinds of kernels. Table 1 shows the number of used subsequence for each kernel. The 3-spectrum and 5-spectrum kernels are the existing methods developed by Leslie et al.[1] We reduce the 3-spectrum kernel according to reduction techniques of spectrum kernels (see Section 4). The experimental result shows that better performance could be obtained even with much fewer 3-length subsequences, about 33.4% of the 3-spectrum set. This result strongly suggests that only a small portion of k-spectrum features could hold the discriminative information for remote homology. We would like to note that it is possible to further reduce the number of subsequences with comparative performance, providing that a more powerful feature selection technique is used.

We compare the kernels based on greedy expansion called 5-expanded kernel (see Fig. 2) with the existing 5-spectrum kernel. Our 5-expanded kernel can also be

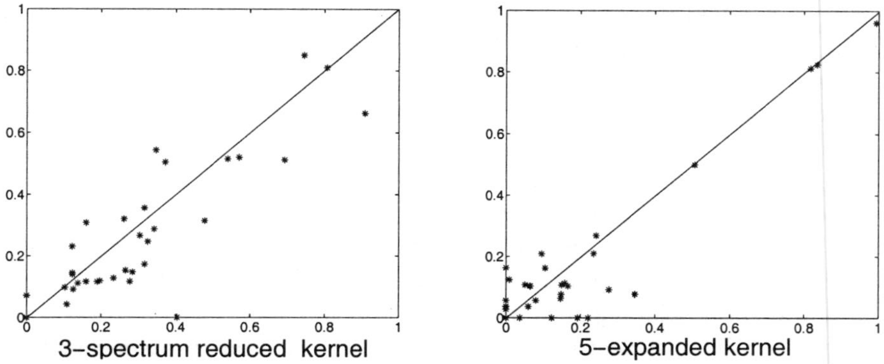

(a) (b)

Fig. 4. Family-by-family comparison of spectrum string kernels and their reduced versions. Here, the coordinates of each point are the ROC_{50} scores for one SCOP family, corresponding to the two labeld kernels, respectively

regarded as a reduced version of $(5,1)$-mismatch string kernel, since we reduce the 5-spectrum set and the members of each $R^{-1}(s)$. From the experimental result, we can observe that this kind of greedy expansion leads to a slight improvement upon 5-spectrum kernel. But our method uses only about 1.4% of 5-spectrum set, which is a significant feature reduction.

We should note that the $(5,1)$-mismatch kernel proposed by Leslie et al.[4] performs comparably with 3-spectrum kernel. On one hand, it means that our reduction of each $R^{-1}(s)$ leads to the performance decline compared with $(5,1)$-mismatch kernel. On the other hand, we obtain computational efficiency by reducing the feature number as a compensation.

We give in Fig. 4 a family-by-family comparison between the existing spectrum string kernels and our methods. It is clear that our methods perform slightly better than the existing spectrum kernels, especially for relatively hard-to-recognize families. This result suggests that carefully selected subsequences benefit hard detection tasks. However, for easy-to-recognize families, it seems always relatively easy to recognize no matter which kinds of features are used.

We select O_{min} from $\{5, 10, 20, 50\}$, BW_{min} from $\{0.5, 0.8, 1\}$, and λ from $\{1, 2, 4, 8\}$, respectively. Then the best results are reported. The 3-reduced kernel is obtained by using $O_{min} = 20$, $BW_{min} = 0.5$, and $\lambda = 2$. The 5-expanded kernel is constructed by using greedy expansion (see Fig. 2) with parameters $O_{min} = 5$, $BW_{min} = 0.8$, and $\lambda = 1$.

6. Discussion and future work

In this research work, we have proposed a general framework for string kernels, coupled with a general algorithm to naturally combine string kernels with feature selection techniques. This framework is applicable to almost all the kernel-based methods in biological sequence analysis. We make experiments on a benchmark SCOP data set for protein homology detection. The experimental results demonstrate that a large number of features can be reduced without any performance reduction, but conversely with improvement. We believe that this kind of string kernels, in conjunction with SVMs, will offer a more flexible and extendable approach to other protein classification problems.

For the further research, we plan to apply these string kernels to the prediction of protein subcellular locations and other biological problems. Meanwhile, we are still interested in developing new approaches to combining of feature selection and string kernels. We hope eventually this method could facilitate protein classification problems with both effectiveness and efficiency.

Acknowledgments

The authors thank James Kwok and Bo Yuan for their valuable comments and suggestions. They also thank National Institute of Information and Communications Technology, Japan, for their support with computational resources. This research is partially supported by the National Natural Science Foundation of China via the grant NSFC 60473040.

References

1. C. Leslie, E. Eleazar and W. S. Noble, The spectrum kernel: a string kernel for SVM protein classification, in *Proceedings of the Pacific Symposium on Biocomputing*, 2002.
2. D. Haussler, *Convolution kernels on discrete structures*, tech. rep., UC Santa Cruz (1999).
3. C. Watkins, *Dynamic alignment kernels*, tech. rep., UL Royal Holloway (1999).
4. C. Leslie, E. Eskin, J. Weston and W. S. Noble, Mismatch string kernels for svm protein classification, in *Advances in Neural Information Processing Systems 15*, (MIT Press, Cambridge, MA, 2003) pp. 1417–1424.
5. C. Leslie and R. Kuang, *Journal of Machine Learning Research* **5**, 1435 (2004).
6. S. Vishwanathan and A. J. Smola, Fast kernels for string and tree matching, in *Advances in Neural Information Processing Systems 15*, (MIT Press, Cambridge, MA, 2003) pp. 569–576.
7. H. Lodhi, J. Shawe-Taylor, N. Cristianini and C. Watkins, Text classification using string kernels, in *Advances in Neural Information Processing Systems 13*, (MIT Press, Cambridge, MA, 2001) pp. 563–569.
8. M. Collins and N. Duffy, Convolution kernels for natural language, in *Advances in Neural Information Processing Systems 14*, (MIT Press, Cambridge, MA, 2002) pp. 625–632.
9. J. Suzuki and H. Isozaki, Sequence and tree kernels with statistical feature mining, in *Advances in Neural Information Processing Systems 18*, (MIT Press, Cambridge, MA, 2006) pp. 1321–1328.
10. T. Jaakkola, M. Diekhans and D. Haussler, *Journal of Computational Biology* **7**, 95 (2000).
11. M. Gribskov and N. L. Robinson, *Computeres and Chemistry* **20**, 25 (1996).

PREDICTING NUCLEOLAR PROTEINS USING SUPPORT-VECTOR MACHINES

MIKAEL BODÉN

ARC Centre of Excellence in Bioinformatics, Institute for Molecular Bioscience, and School of Information Technology and Electrical Engineering
The University of Queensland, QLD 4072, Australia
E-mail: m.boden@uq.edu.au

The intra-nuclear organisation of proteins is based on possibly transient interactions with morphologically defined compartments like the nucleolus. The fluidity of trafficking challenges the development of models that accurately identify compartment membership for novel proteins. A growing inventory of nucleolar proteins is here used to train a support-vector machine to recognise sequence features that allow the automatic assignment of compartment membership. We explore a range of sequence-kernels and find that while some success is achieved with a profile-based local alignment kernel, the problem is ill-suited to a standard compartment-classification approach.

Keywords: Nucleolus, support-vector machine, intra-nuclear protein localisation, kernel

1. Introduction

By virtue of its architecture, the cell nucleus not only encloses the genetic material but also controls its expression. Recent discoveries have exposed morphologically well-defined compartments with which proteins and RNA associate.[1,2] This paper uses emerging experimental data to develop a basic predictive model of intra-nuclear protein association.

Similar to cytoplasmic organelles, intra-nuclear compartments seem to specialize in particular functions (like ribosomal RNA synthesis, spliceosome recycling and chromatin remodeling). However, intra-nuclear compartments are not membrane-bound and thus employ different principles to sustain their functional integrity. Indeed, compartments are in perpetual flux, with some proteins and RNA stably associated and others just transiently binding before they move on to another compartment. Proteins and RNA are trafficked by non-directed, passive diffusion and association with a compartment is based on molecular interactions with its residents.[1,2] The largest compartment inside the nucleus is the nucleolus. With functions primarily related to ribosomal biogenesis, the nucleolus is conveniently located at sites of ribosomal genes. Apart from being involved in producing ribosomes, examples of nucleolar functions include the maturation of tRNA and snRNA of the spliceosome, pre-assembly of the signal recognition particle and the sequestration of several important regulatory proteins.[3]

Recent efforts using mass spectrometry have resulted in the identification of a substantial number of nucleolar proteins in human cells.[4] With the view that proteins are only transiently associated with one or more compartments, we ask if we can build a classifier that is able to distinguish between proteins with nucleolar association from those without. Specifically, a growing protein inventory is leveraged using state-of-the-art machine learning algorithms–support-vector machines equipped with sequence kernels. This paper develops an appropriate data set, and a sequence data-driven model. The model is evaluated on its ability to capture in terms of sequence features the possibly loose association of proteins with the nucleolus.

2. Background

Analysis has shown that there seems to be no single feature that allows the automatic sorting of proteins into nuclear compartments.[5] Several characteristics, like iso-electric point, molecular weight, and amino acid and domain composition may need to be used in conjunction to accurately assign their compartmental association.[5] The nucleolus has the largest number of known proteins, but there appears to be few generic motifs shared by its residents, the so-called DEAD-box helicase and the WD40 repeat being two notable exceptions each occurring in about 6% of known members.[5]

Using the Nuclear Protein Database,[6] Lei and Dai[7,8] developed a predictor using machine learning of six different nuclear compartments including the nucleolus. Multi-compartmental proteins were removed from the data set (prior to training) to avoid the ambiguous presentation of data to a classifier. In their most refined model, there is a Gene Ontology (GO) module which relies on the identification of GO terms of the protein and its homologs (via a BLAST search). Additionally, a separate support-vector machine is trained to map the sequence to one of the six classes. Notably, inclusion of the GO term module elevates overall performance considerably (the correlation coefficient for nucleolus improves from 0.37 to 0.66). However, the GO terms (a) include specific annotations of localisation and (b) need to be known in advance.

Hinsby et al.[3] devised a system from which novel nucleolar proteins could be semi-automatically identified. By cross-checking protein-protein interactions involving known nuclear proteins with mass spectrometry data of the nucleolus, they identified prioritised nucleolar protein complexes and subsequently eleven novel nucleolar proteins (by targeted search for 55 candidates in the raw mass spectrometry data). The approach indicates the potential of assigning intra-nuclear compartment membership in terms of interactions with residents rather than possibly elusive compartment-unifying features.

3. Methods

3.1. *Data set*

We re-use the data set of Hinsby et al.,[3] sourced primarily from the Nucleolar Proteome Database (NOPdb[9]), then adding the eleven novel proteins from Hinsby et al.'s study, resulting in 879 human nucleolus-localised proteins. We further performed redundancy reduction using BlastClust ensuring that only 30% sequence similarity was present in the remaining set of 767 positives. This set consists of proteins which are either stable or transient residents of the nucleolus. Importantly, they could also be present in other locations to varying degrees.

Preliminary investigations which did not employ a negative training set were unsuccessful. More specifically, we used one-class support-vector machines to generate a decision function that included only all positives. Test performance on known negatives clearly indicated the need for pursuing a full discriminative approach. Thus, a negative, non-nucleolar protein set was devised from two sources: the Nuclear Protein Databank[6] and UniProt R51–restricted to mammalian proteins. NPD-extracted proteins had one or more intra-nuclear compartments assigned, not including the nucleolus. UniProt proteins were similarly required to have a non-ambiguous nuclear subcellular localisation with further intra-nuclear association explictly stated, not including the nucleolus. We further cleaned the negative set by removing all proteins that were in the original positive set (or homologs thereof). Finally, to prevent over-estimation of test accuracy, the negative set was reduced so that the remaining sequences had less than 30% similarity.

The final negative 359-sequence set thus represents nuclear proteins with no experimentally confirmed association with the nucleolus. However, due to the inherent fluidity of nuclear proteins, the negative set may still contain proteins that are transiting through the nucleolus. It should be noted that the final data sets differ from the sets used by Lei and Dai who removed any protein not exclusively associated with one of the six compartments. Additionally, 35 nucleolar proteins were found in the original 879-set that were incorrectly assigned exclusively to a non-nucleolar compartment in their study.

3.2. *Model*

Support-vector machines (SVMs[10]) are trained to discriminate between positive and negative samples, i.e. to generate a decision function

$$f(\mathbf{x}) = \sum_{i=1}^{n} y_i \alpha_i K(\mathbf{x}_i, \mathbf{x}) + b \tag{1}$$

where $y_i \in \{-1, +1\}$ is the target class for sample $i \in \{1, ..., n\}$, $\mathbf{x}_i$ is the ith sample, α_i is the ith Lagrange multiplier and b is a threshold. All multipliers and the threshold are tuned by training the SVM.

To determine the Lagrange multipliers, Platt's sequential minimal optimization[11] with convergence improvements[12] is used. Note that only multipliers directly associated with samples on the margin separating positives from negatives are non-zero (these samples are known as the support-vectors). Models based on support-vector machines have previously garnered success for classifying cytoplasmic protein compartmentalisation.[13–16]

Due to the graded membership of intra-nuclear compartments, the SVM output is converted to a probabilistic output, using a sigmoid function

$$p(\mathbf{x}) = \frac{1}{1 + e^{A \cdot f(\mathbf{x}) + B}} \tag{2}$$

where A and B are estimated by minimizing the negative log-likelihood from training samples.[17] The training data assigned to the model is divided internally so that approximately 4/5 is used for tuning the support-vector machine, and 1/5 for tuning the sigmoid function.

A number of sequence-based kernels have been developed recently, primarily targeted to protein classification problems. We evaluate the performance of the Spectrum kernel,[18] the Mismatch kernel,[19] Wildcard kernel,[19] the Local Alignment (LA) kernel[20] and a profile-based Local Alignment kernel, each replacing $K(\cdot,\cdot)$ in Equation 1.

We refer the reader to the literature for detailed information regarding the kernels. Essentially, spectrum-based kernels (including the Mismatch and Wildcard kernels) are based on the sharing of short sequence seqments (of length k, with provision of minor differences, m is the allowed number of "mismatches" in the Mismatch kernel, x is the number of "wildcard" symbols in the Wildcard kernel).[19]

The Local Alignment kernel compares two sequences by exploring their alignments.[20] We explore some details of the Local Alignment kernel to describe the only novel kernel in this paper–the Profile Local Alignment kernel.

An alignment between two sequences is quantified using an amino acid substitution matrix, S, and a gap penalty setting, g. A further parameter, β, controls the contribution of non-optimal alignments to the final score. Let $\Pi(\mathbf{x_1}, \mathbf{x_2})$ be the set of all possible alignments between sequences $\mathbf{x_1}$ and $\mathbf{x_2}$. The kernel can be expressed in terms of alignment-specific scores, $\varsigma_{S,g}$ (for details of this function see[20]).

$$K_\beta^{LA}(\mathbf{x_1}, \mathbf{x_2}) = \sum_{\pi \in \Pi(\mathbf{x_1}, \mathbf{x_2})} exp(\beta \varsigma_{S,g}(\mathbf{x_1}, \mathbf{x_2}, \pi)) \tag{3}$$

When the Local Alignment kernel is used herein, S is the BLOSUM62 matrix.

Evidence is mounting that so-called position-specific substitution matrices (PSSMs; a.k.a. "profiles") disclose important evolutionary information tied to each residue.[21,22] We adapt the alignment-specific function, ς, in the Local Alignment kernel to use such substitution scores generated by PSI-Blast (max three iterations, E-value threshold is 0.001, using Genbank's non-redundant protein set) in place of the generic substitution matrix, S. Specifically, we define the substitution score as the average of the PSSM-entries for the two sequences (where the entry coordinates are determined from the sequence position of one sequence and the symbol of the other).

Table 1. Accuracy of classification for different kernel settings when the output cut-off is set to 0.5. Mean correlation coefficient on test data in 10-fold crossvalidation, repeated 10 times, is shown (1.0 indicates ideal agreement, 0.0 indicates chance agreement with target data). The standard deviation is provided for each configuration after ±.

Kernel	Parameters	Correlation coefficient
Spectrum	$k = 3$	0.340 ± 0.016
Wildcard	$k = 3,\ x = 1$	0.391 ± 0.012
Wildcard	$k = 4,\ x = 1$	0.388 ± 0.013
Mismatch	$k = 3,\ m = 1$	0.382 ± 0.015
Mismatch	$k = 4,\ m = 1$	0.420 ± 0.017
Local Alignment	$\beta = 0.1$	0.399 ± 0.012
Profile Local Alignment	$\beta = 0.1$	0.447 ± 0.017

4. Results

Models are trained and tested using 10-fold crossvalidation. Essentially, the available data is first partitioned into ten evenly sized sub-sets. Second, ten models are trained on 9 of the ten sub-sets, each sub-set combination chosen so that it is unique. Third, each of the ten models is tested only on their respective remaining sub-set. Note that no model is trained on any of their test samples, and each of the original samples is used as a test sample by exactly one model. Finally, the test results are collated and the whole crossvalidation procedure is repeated ten times to establish variance in prediction accuracy.

All kernels are normalised, i.e. kernel values are adjusted such that the diagonal of the kernel matrix is 1.0. Due to substantive computation requirements, only a few kernel parameters were trialled but care was exercised to explore the configurations most successful in the literature.

Support-vector machines require the manual setting of regularisation parameters (C-values). Preliminary parameter-sweeps with two C-values (one for the positive and one for the negative set) identified that when they exceed 1.0 the support-vector machine generalised stably for all kernels. C-values were thus fixed at 1.0 throughout.

We use the correlation coefficient (CC) between experimentally confirmed association with the nucleolus and the prediction to illustrate the accuracy.

$$CC = \frac{tp \cdot tn - fp \cdot fn}{\sqrt{(tp + fn)(tp + fp)(tn + fp)(tn + fn)}} \tag{4}$$

where tp, tn, fp and fn is the number of true positives, true negatives, false positives and false negatives, respectively.

The classification of proteins as nucleolar-associated (or not) reached 77% accuracy on our data set with a SVM equipped by the Profile Local Alignment kernel. This corresponds to a correlation of $CC = 0.447$ (± 0.017) between observed and predicted nucleolar association. All classification results when using the default output cut-off at 0.5 are presented in Table 1.

To further illustrate the accuracy we generated ROC curves for the SVMs with the Profile Local Alignment kernel and the Mismatch kernel (see Figure 1). That is, by varying the threshold which needs to be exceeded by the probabilistic output, the sensitivity and specificity of the model is monitored.

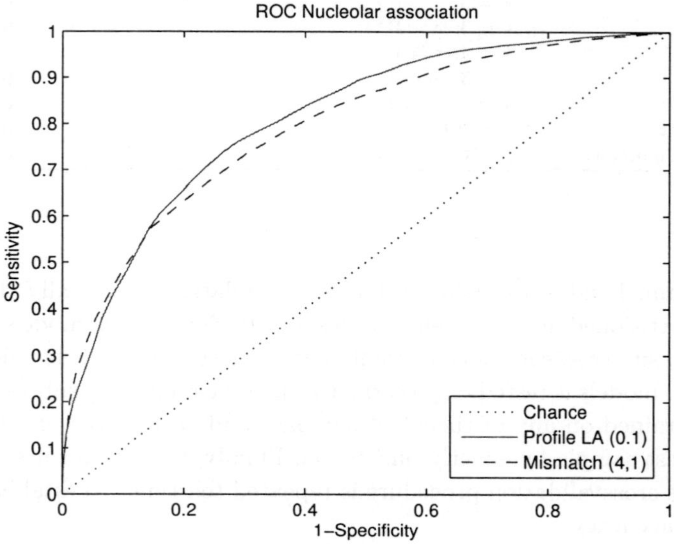

Fig. 1. ROC curves illustrating the change in sensitivity as a function of specificity. The area under the ROC is 0.811 for the Profile LA kernel ($\beta = 0.1$) and 0.794 for the Mismatch kernel ($k = 4$, $m = 1$). Maximum correlation coefficient 0.451 of the Profile LA SVM is seen at an output threshold of 0.66 (sensitivity=0.71, specificity=0.76). Sensitivity is defined as $tp/(tp + fn)$ and specificity as $tn/(tn + fp)$.

The probabilistic output has the potential of indicating the certainty of the prediction. We computed the mean output for the four classification outcomes using a 0.5 cut-off (again over 10 runs using our best configuration, i.e. over $(767+359)\cdot 10$ test samples). (a) A true positive is 0.81 ($\pm$0.12), (b) a false positive is 0.71 ($\pm$0.12), (c) a true negative is 0.26 ($\pm$0.13) and (d) a false negative is 0.34 ($\pm$0.12). Hence, it is reasonable to regard a prediction closer to the cut-off as uncertain.

In the absence of known motifs clearly identifying nucleolar association, we attempted to characterise the basis of generalisation of the best predictive model by qualifying the mistakes made by it.

Over all ten runs, we collated all proteins mistakenly predicted to be nucleolar. These false positives were divided into their location as assigned by the Nuclear Protein Database[6] and as used as training data by Lei and Dai.[7] Available assignments are shown in Table 2. The reader is reminded that this data set has limited coverage, thus we present ratios based on available data. The mistakes are seemingly distributed evenly between alternative intra-nuclear locations. Noteworthy, we discovered one protein (O95347) that was consistently misclassified as nucleolar. O95347 is indeed nucleolar according to NPD but associated with Chromatin in UniProt.

Table 2. Number of proteins falsely classified as nucleolar and their location according to the Nuclear Protein Database as used by Lei and Dai. Average counts (of 359 possible) are shown over 10 repeats of 10-fold crossvalidation tests. The "absolute" percentage of a mistaken location refers to the location-count over the total number of false positives. The "relative" percentage refers to the location-count relative the number of proteins known in each location in Lei and Dai's data set (assuming the distribution of proteins is uniform).

Location	Proteins (count)	% (absolute)	% (relative)
Chromatin	26.4	15	21
Lamina	30.4	17	27
Nucleolus	1.0	1	0
Nucleoplasm	25.4	15	17
PML	14.8	8	19
Speckles	17.9	10	16
Unknown	58.6	34	

We similarly collated all proteins that were incorrectly predicted to not associate with the nucleolus. The false negatives were cross-checked by identifying their function according to the Nucleolar Proteome Database.[9] Hence, the tabulation seen in Table 3 illustrates functions commonly confused with alternative locations. Not surprising, beside the "unknowns", at the top of the list there are functions that relate to alternative compartments rather than being uniquely nucleolar, e.g speckles are associated with both splicing and transcription related factors[2] and the nuclear lamina consists mainly of filament proteins, lamins.

On average a model in one fold of a cross-validation run is trained on about 1000 samples. Of these, about 600 were usually selected to be support-vectors, ultimately defining the model's decision boundary. To further qualify the nature of subscribed generalisation, about 10% of all support-vectors of one model were analysed using a kernelised hierarchical cluster analysis (using normalised Profile Local Alignment kernel and average-linkage). The cluster dendrogram is shown in Figure 2. Each support-vector is labelled with its target label (Pos=Nucleolar or Neg=Other locations), function as determined from the Nucleolar Proteome Database or location as used by Lei and Dai. Proteins without functional annotation or location were excluded. Functional groups are visible (e.g. splicing/transcription, chromatin, lamina/cytoskeleton) further indicating that generalisation is based on protein function rather than intra-nuclear location.

5. Conclusion

We develop a model that is able to predict nucleolar association of proteins from their sequence. A support-vector machine fitted with a profile-based adaptation of

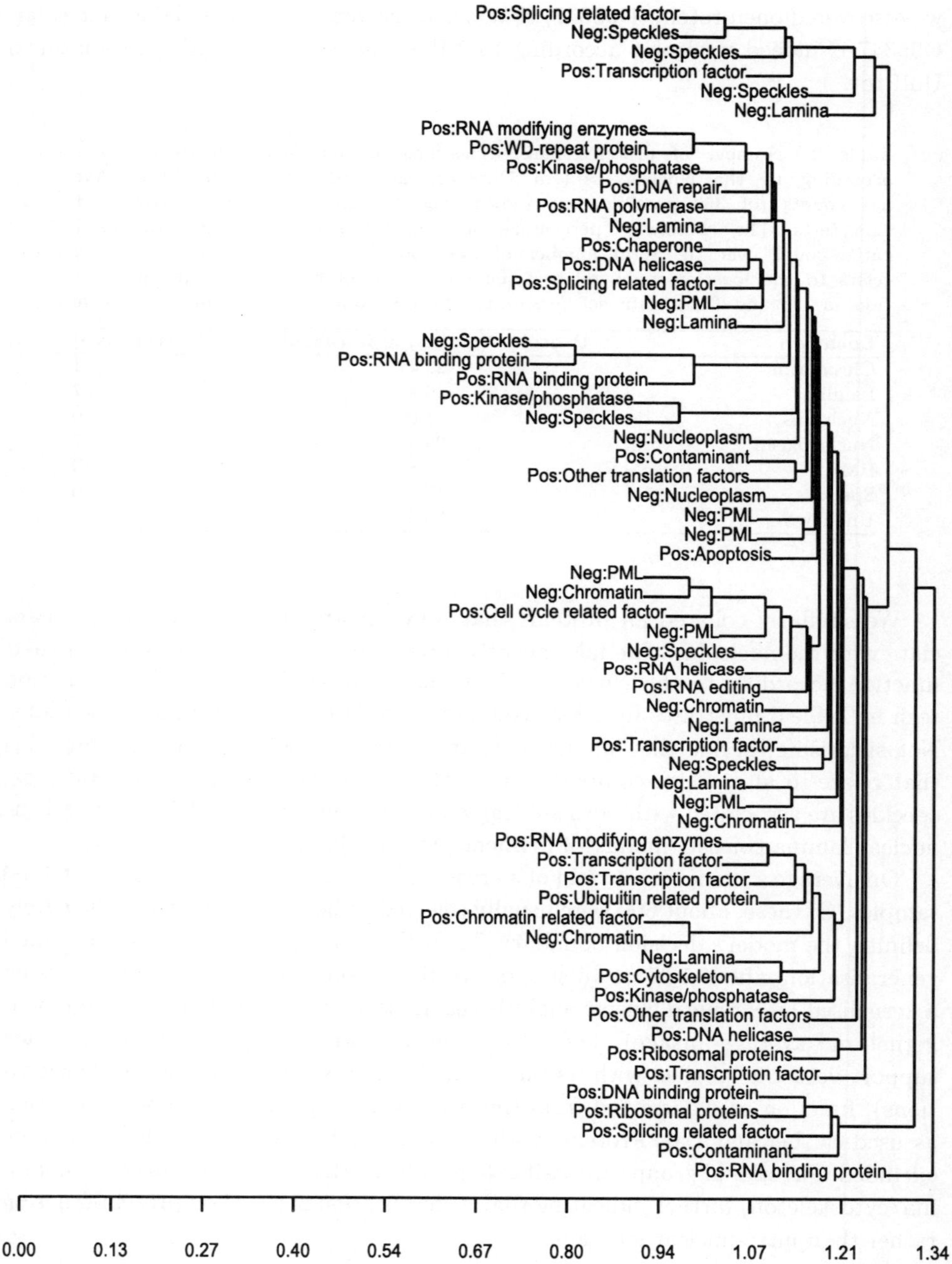

Fig. 2. A cluster dendrogram illustrating the organisation of the support-vectors implemented by one arbitrarily selected model. The distance in feature-space is indicated by the horisontal axis. Several clusters exemplify the mixing of location but coherence of function, e.g. on top nucleolar splicing and transcription factors are mixed with speckles-native proteins (the primary site for such functions).

Table 3. Number of proteins falsely predicted as non-nucleolar and their function according to the Nucleolar Proteome Database. Average counts (out of 767 positives) are shown over 10 repeats of 10-fold crossvalidation tests.

Function	Proteins (count)
Function unknown	49.9
Cell cycle related factor	4.7
Transcription factor	3.9
Splicing related factor	3.8
Ubiquitin related protein	2.2
DNA binding protein	1.8
Lamina	1.8
Kinase/phosphatase	1.7
WD-repeat protein	1.7
Contaminant	1.6
RNA binding protein	1.5
RNA modifying enzymes	1.4
p53 activating	1.3
DNA repair	1.0
Intermediate filaments	1.0
RNA polymerase	1.0
Chromatin related factor	0.5
Chaperone	0.4
Other translation factors	0.4
DNA methyltransferase	0.1
Exonuclease mRNA	0.1

the Local Alignment kernel and a probabilistic output achieves a correlation coefficient of about 0.45 (or 77% on our specific data set). It is difficult to directly compare this result with Lei and Dai's work since their ensemble predictor distinguishes between six classes as well as using differently scoped training and test data. Their SVM-only model has a lower correlation coefficient, but their GO term model (which requires the prior identification of such terms, some of which are explicitly concerned with location) exceeds the accuracy presented herein.

Compartmentalisation of proteins inside the nucleus is fluid and categorically discriminating between such compartments may thus be objectionable. To alleviate issues with multiple localisations, positive data used for model-tuning did not exclude proteins for which additional compartments were known. Moreover, the model presented here incorporates a probabilistic output which allows graded membership to be reflected.

Analysis shows that false positive predictions are drawn evenly from other intranuclear compartments. Conversely, nucleolar proteins not recognised as such are sometimes involved in functions also associated with alternative locations, suggesting that generalisation is based on functional features. Compartment-specific features are thus largely eluding an approach that has garnered success for cytoplasmic localisation, suggesting that to combat intra-nuclear trafficking we may need to reconsider model designs.

Acknowledgments

This work was supported by the ARC Centre of Complex Systems. Lynne Davis and Johnson Shih both contributed to this paper by implementing some of the kernels.

References

1. T. Misteli, *Science* **291**, 843 (2001).
2. K. E. Handwerger and J. G. Gall, *Trends in Cell Biology* **16**, 19 (2006).
3. A. M. Hinsby, L. Kiemer, E. O. Karlberg, K. Lage, A. Fausboll, A. S. Juncker, J. S. Andersen, M. Mann and S. Brunak, *Molecular Cell* **22**, 285 (2006).
4. J. S. Andersen, Y. W. Lam, A. K. Leung, S. E. Ong, S. E. Lyon, A. I. Lamond and M. Mann, *Nature* **433**, 77 (2005).
5. W. Bickmore and H. Sutherland, *The EMBO Journal* **21**, 1248 (2002).
6. G. Dellaire, R. Farrall and W. Bickmore, *Nucl. Acids Res.* **31**, 328 (2003).
7. Z. Lei and Y. Dai, *BMC Bioinformatics* **6**, p. 291 (2005).
8. Z. Lei and Y. Dai, *BMC Bioinformatics* **7**, p. 491 (2006).
9. A. K. L. Leung, L. Trinkle-Mulcahy, Y. W. Lam, J. S. Andersen, M. Mann and A. I. Lamond, *Nucleic Acids Research* **34**, D218 (2006).
10. V. Vapnik, *Statistical Learning Theory* (Wiley, New York, 1998).
11. J. Platt, Fast training of support vector machines using sequential minimal optimization, in *Advances in Kernel Methods–Suport Vector Learning*, eds. B. Schölkopf, C. J. C. Burgess and A. J. Smola (MIT Press, Cambridge, MA, 1999) pp. 185–208.
12. S. S. Keerthi, S. K. Shevade, C. Bhattacharyya and K. R. K. Murthy, *Neural Computation* **13**, 637 (2001).
13. V. Atalay and R. Cetin-Atalay, *Bioinformatics* **21**, 1429 (2005).
14. D. Sarda, G. Chua, K.-B. Li and A. Krishnan, *BMC Bioinformatics* **6**, p. 152 (2005).
15. A. Garg, M. Bhasin and G. P. S. Raghava, *J. Biol. Chem.* **280**, 14427 (2005).
16. A. Pierleoni, P. L. Martelli, P. Fariselli and R. Casadio, *Bioinformatics* **22**, e408 (2006).
17. J. Platt, Probabilistic outputs for support vector machines and comparison to regularized likelihood methods, in *Advances in Large Margin Classifiers*, eds. A. Smola, P. Bartlett, B. Schölkopf and D. Schuurmans (MIT Press, Cambridge, MA, 2000)
18. C. Leslie, E. Eskin and W. S. Grundy, The spectrum kernel: A string kernel for svm protein classification, in *Proceedings of the Pacific Symposium on Biocomputing*, eds. R. B. Altman, A. K. Dunker, L. Hunter, K. Lauerdale and T. E. Klein (World Scientific, 2002).
19. C. Leslie and R. Kuang, *Journal of Machine Learning Research* **5**, 1435 (2004).
20. H. Saigo, J.-P. Vert, N. Ueda and T. Akutsu, *Bioinformatics* **20**, 1682 (2004).
21. H. Rangwala and G. Karypis, *Bioinformatics* **21**, 4239 (2005).
22. R. Kuang, E. Ie, K. Wang, K. Wang, M. Siddiqi, Y. Freund and C. Leslie, *Journal of Bioinformatics and Computational Biology* **3**, 527 (2005).

SUPERVISED ENSEMBLES OF PREDICTION METHODS FOR SUBCELLULAR LOCALIZATION

JOHANNES AßFALG, JING GONG, HANS-PETER KRIEGEL,
ALEXEY PRYAKHIN, TIANDI WEI and ARTHUR ZIMEK

Institute for Informatics, Ludwig-Maximilians-Universität München, Germany
www: `http://www.dbs.ifi.lmu.de`
E-mail: `{assfalg,gongj,kriegel,pryakhin,tiandi,zimek}@dbs.ifi.lmu.de`

In the past decade, many automated prediction methods for the subcellular localization of proteins have been proposed, utilizing a wide range of principles and learning approaches. Based on an experimental evaluation of different methods and on their theoretical properties, we propose to combine a well balanced set of existing approaches to new, ensemble-based prediction methods. The experimental evaluation shows our ensembles to improve substantially over the underlying base methods.

1. Introduction

In cells, different regions have different functionalities. Certain functionalities are performed by specific proteins. To function properly, a protein must be localized in the proper region of a cell. Co-translational or post-translational transport of proteins into specific subcellular localizations is therefore a highly regulated and complex cellular process. Knowing of the subcellular localization of a protein helps to annotate its possible interaction partners and functionalities.

Starting in the mid-nineties of the last century, until now a plethora of automated prediction methods for the subcellular localization of proteins has emerged. These methods are based on different sources of information like the amino acid composition of the protein, specific sorting signals or targeting sequences contained in the protein sequence, or homology search in databases of proteins with known localization. Furthermore, hybrid methods combine the different sources of information often in a very specialized way. Besides different sources of information, prediction methods differ in the employed learning algorithms (like naive Bayes and Bayes networks, k-nearest neighbor methods, support vector machines (SVM), and neural networks). Due to their different sources of information, prediction methods differ widely in their coverage of different localizations. For example, methods based on targeting sequences generally have a low coverage of only a few localizations. Methods based on amino acid composition vary considerably in their coverage. The coverage of a method is also directly related to the available classes in the data sets used for training of the corresponding method. As most prediction methods are trained and evaluated on data sets suitable to their requirements in coverage, it is a hard task to compare different methods w.r.t. their performance.[9]

In this paper, we survey shortly prominent methods for prediction of subcellular localization of proteins, particularly considering their different properties (Section 2). Based on a diverse selection of the best methods, we propose combined methods using a well balanced set of prediction methods as new ensemble-methods (Section 3). Section 4 presents the evaluation of selected localization prediction methods in comparison to our new ensemble methods. Finally, Section 5 concludes the paper.

2. Survey on Prominent Prediction Methods for Subcellular Localization

For our evaluation of localization prediction methods, we confined the selection to those that are available (excluding methods like NNPSL[24] or fuzzy_loc[16]), and that focus on eukaryotic localization prediction (excluding methods like PSORT-B[11] or PSLPred[3]). In the following, we survey prominent examples from these methods, choosing representatives for the different sources of information the methods are based upon.

2.1. *Amino Acid Composition*

Predicting the subcellular localization based on amino acid composition was suggested by Nakashima and Nishikawa.[22] They presented a method to discriminate between intracellular and extracellular proteins using the amino acid composition. In the following years, a number of approaches using the amino acid composition was proposed.

SubLoc[15] uses one-versus-rest support vector machines (SVM) to predict the localization. No additional information aside from the amino acid composition (like, e.g., dipeptide composition) is used for the prediction. In contrast to SubLoc, PLOC[23] additionally considers the dipeptide composition and the gapped amino acid composition aside from the standard amino acid composition. Like SubLoc, this method employs one-versus-rest SVMs. By using pairs of peptides the authors take more sequence-order information than SubLoc into account. The gapped pair composition corresponds to periodic occurrences of certain amino acids in the sequence. Similar to PLOC, CELLO[17] incorporates several kinds of compositions, including single, dipeptide, and partitioned amino acid compositions. Furthermore, compositions based on physicochemical properties of the amino acids were derived. These features are again used as input for one-versus-rest SVMs.

2.2. *Sorting Signals*

One of the earliest works trying to identify a certain location based on protein sorting signals was already presented in 1986.[27] Most of the methods based on sorting signals are very specialized. For example, Mitoprot[5] predicts only mitochondrial proteins, SignalP[2] predicts only proteins of the secretory pathway. More general methods in this category are iPSORT[1] and Predotar.[25] The comparison of these two methods is especially interesting because they use very different computational approaches: iPSORT uses simple and interpretable rules based on protein sequence features. These features are derived from the so-called amino acid index, a categorization of amino acids based on different kinds of properties. iPSORT uses N-terminal sorting signal sequences. Predotar considers N-terminal sorting signals as well and processes the input information with a feed forward neural network. As an output value, this method yields probability values for the presence of a certain localization sequence rather than an abstract score.

2.3. *Homology*

Prominent methods based on homology search are PredictNLS[6] and PA_SUB.[19] PredictNLS is also based on sorting signals, as it is trained on a data set of experimentally confirmed nuclear localization signal (NLS) sequences. This data set is extended by homology search. Nevertheless, NLSPred is specialized on recognizing nuclear proteins. PA_SUB is purely based on PSI-BLAST homology search using database annotations from homologous proteins. In many cases, homology search is very accurate. However, the result will be arbitrary if no homologous protein with localization annotation is available. The combination of homology search with other methods is a common way to overcome this shortcoming.

2.4. *Hybrid Methods*

As in PredictNLS, most of the methods using homology search combine this technique with some other sources of information. In this category, great effort was already spent to develop refined combinations of information and methods. One often finds series of related approaches from certain groups like the PSORT series (PSORT,[21] PSORT-II,[20] PSORT-B,[10,11] and WoLFPSORT[14]) or ESLPred,[4] HSLPred,[12] and PSLPred.[3] The PSORT-B approaches and PSLPred are specialized for bacteria. PSORT is one of the earliest methods at all, based on amino acid composition, N-terminal targeting sequence information, and motifs. Like iPSORT, it is based on a set of rules. PSORT-II uses a k-NN approach. WoLF-PSORT uses a feature selection procedure and incorporates new features, based on new sequence data, simultaneously increasing the coverage of localizations and organisms. ES-LPred uses an SVM approach, combining amino acid composition, dipeptide composition, overall physicochemical properties, and PSI-BLAST scores. The extensions HSLPred and PSLPred focus on human and prokaryotic proteins, respectively. MITOPRED[13] uses Pfam domains and amino acid composition, and is specialized for mitochondrial proteins. MultiLoc[18] traines SVMs based on N-terminal targeting sequences, sequence motifs, and amino acid composition.

3. Ensemble Methods

In preliminary tests on our data set, the accuracy of all compared methods was not as high as reported in their original literature for other data sets, meaning our data set can be considered as not too easy. Furthermore, there were sequences with certain localizations always wrongly predicted by some methods, e.g. there was no protein with localization vacuole within fungi group predicted positively although there were 68 vacuole proteins in this group. Some other methods could predict more accurately for these proteins while they might be incapable of accurate prediction of other localizations. In other words, each method has its own advantages and disadvantages. These findings motivate the idea to combine some of these methods.

3.1. *Theory*

Combining several self-contained predicting algorithms to an ensemble to yield a better performance in terms of accuracy than any of the base predictors, is backed by a sound theoretical background.[7,8,26] In short, a predictive algorithm can suffer from several limitations such as statistical variance, computational variance, and a strong bias. *Statistical variance* describes the phenomenon that different prediction models result in equally good performance on training data. Choosing arbitrarily one of the models can then result in deteriorated performance on new data. Voting among equally good classifiers can reduce this risk. *Computational variance* refers to the fact, that computing the truly optimal model is usually intractable and hence any classifier tries to overcome computational restrictions by some heuristics. These heuristics, in turn, can lead to local optima in the training phase. Obviously, trying several times reduces the risk of choosing the wrong local optimum. A restriction of the space of hypotheses a predictive algorithm may create is refered to as *bias* of the algorithm. Usually, the bias allows for learning an abstraction and is, thus, a necessary condition of learning a hypothesis instead of learning by heart the examples of the training data (the latter resulting in random performance on new data). However, a strong bias may also hinder the representation of a good model of the true laws of nature one would like to learn. A weighted sum of hypotheses may then expand the space of possible models.

To improve over several self-contained classifiers by building an ensemble of those classifiers requires the base algorithms being accurate (i.e., at least better than random) and diverse (i.e., making different errors on new instances). It is easy to understand why these two conditions are necessary and also sufficient. If several individual classifiers are not diverse, then all of them will be wrong whenever one of them is wrong. Thus nothing is gained by voting over wrong predictions. On the other hand, if the errors made by the classifiers were uncorrelated, more individual classifiers may be correct while some individual classifiers are wrong. Therefore, a majority vote by an ensemble of these classifiers may be also correct. More formally, suppose an ensemble consisting of k hypotheses, and the error rate of each hypothesis is equal to a certain $p < 0.5$ (assuming a dichotomous problem), though independently. The ensemble will be wrong, if more than $k/2$ of the ensemble members are wrong. Thus the overall error rate $\bar{p}$ of the ensemble is given by the area under the binomial distribution, where $k \geq \lceil k/2 \rceil$, that is for at least $\lceil k/2 \rceil$ hypotheses being wrong: $\bar{p}(k,p) = \sum_{i=\lceil k/2 \rceil}^{k} \binom{k}{i} p^i (1-p)^{k-i}$. The overall error-rate is rapidly decreasing for an increasing number of ensemble members.

3.2. *Selection of Base Methods for Ensembles*

Comparing several methods based on amino acid compositions we found an increase of accuracy by adding more sequence-order information. CELLO behaved best no matter for which taxonomy group because it used the most sequence-order information: single amino acid composition, dipeptide composition, n-peptide composition, and even physicochemical properties of amino acids in the updated version that we used. In contrast, PLOC which used only amino acid composition and dipeptide composition had more false predictions than CELLO, but it was more accurate than SubLoc which used only single amino acid composition. In comparison, the methods based on detecting N-terminal sorting signals performed better than expected, although they have to handle missing N-terminal sorting signals. Of the hybrid methods the two newest, WoLFPSORT (2006) and MultiLoc (2006), had similar prediction ability and their accuracy is higher than that of the others in this category.

Based on the results of our preliminary experimental comparisons and the criteria of usability, reliability, efficiency, coverage, and, for theoretical reasons, as discussed above, diversity in the underlying methods and sources of information, we chose the following methods to build an ensemble: From the methods based on amino acid composition SubLoc was excluded because of its too simple foundation and its lower rank during the preliminary tests. In addition, both PLOC and CELLO use the single amino acid composition too and predict more accurately than SubLoc. iPSORT and Predotar as prominent examples of methods based on sorting signals had similar prediction ability in our preliminary tests but use quite different algorithms, so both of them were chosen for the combination. PA_SUB is a purely homology-based method. The data set used for generating PA_SUB consists of virtually all Swiss-Prot[29] entries that provide a localization annotation. As we evaluate the considered methods and our combination of methods on an up-to-date data set also compiled from Swiss-Prot, we exclude PA_SUB from the experiments, as it is highly overfitted to the data set. Usually, as discussed above, homology-based approaches are combined with other approaches. From the hybrid methods only the method PSORT II was excluded, because we use its extension WoLFPSORT which is more accurate and has a larger taxonomy coverage than PSORT II. HSLPred is used for the human proteins. Although its localization coverage is very narrow, it is still very sensitive for the three localizations within its coverage. Finally we chose 7 methods for the plant, animal and fungi groups and 8 methods for the human group to construct an ensemble method: PLOC, CELLO, iPSORT, Predotar, WoLFPSORT, MultiLoc, ESLPred, and, for human proteins, HSLPred.

3.3. *Ensemble Method Based on a Voting Schema*

Despite a clear theoretical background for ensemble learning in general, the combination of localization prediction methods is not trivial due to the wide range of localization and taxonomic coverage. Imagining a prediction method as a function from some feature space to some class space, the base learners map the proteins into different class spaces. Thus, for unifying the prediction methods, the class spaces must be unified first. The unified class space should contain the classes supported by most of the methods (resulting in the set of ten localization classes as described above). Methods that are unable to predict some of the classes contained in the unified class space must be treated especially. Furthermore, some methods (PLOC, CELLO, WoLFPSORT, and MultiLoc) predict exactly one localization for a query protein while others (iPSORT, Predotar, ESLPred, and HSLPred) predict a range of possible localizations. We define therefore a voting schema as follows: Methods in the first group give their vote to one certain localization at a time if the predicted localization belongs to the 10 localizations in our data set. Otherwise their vote is blanked out. Methods

Table 1. Ranks of different classification methods for the considered taxonomic groups.

Taxonomic group	CELLO	ESLPred	HSLPred	iPSORT	MultiLoc	PA_SUB
Animal	2	10	—	3	6	1
Fungi	4	9	—	1	7	3
Human	4	10	7	2	6	1
Plant	3	2	—	9	8	1

Taxonomic group	PLOC	Predotar	PSORT II	SubLoc	WoLFPSORT
Animal	4	5	8	9	7
Fungi	5	2	8	10	6
Human	5	3	9	11	8
Plant	4	7	—	6	5

in the second group may give their vote to several localizations at a time. If a classifier maps the proteins into a class space containing some of the ten classes and a class 'unknown', a prediction for class 'unknown' can be mapped to the set of the remaining classes. However, if a classifier cannot decide between some classes, this will not mean automatically that the protein belongs to the set of unknown classes. For example, if there is no sorting signal being detected by iPSORT or Predotar, we cannot say that this protein is not localized in chloroplast, mitochondrion, or the secretory pathway, because the N-terminal sequence of this protein may be not complete. In this case, iPSORT and Predotar will give up on voting.

Based on the votes of all base classifiers, we derive a vector s of scorings for the localizations, where for localization i the score s_i is computed as follows:

$$s_i = \sum_{j=1}^{N} (v_j \cdot (N - rank_j + 1) ,$$

where N is the number of methods used by the ensemble method, $rank_j$ is the rank in accuracy of method j according to our preliminary tests, and $v_j = 1$ if method j votes for localization i (allowing voting for multiple localizations), otherwise $v_j = 0$. This ensemble is therefore built based on prior knowledge concerning the performance of the base classifiers. We also tried a voting without explicitly ranking the votes of the base classifiers, but the results were not acceptable. The ranks we used for the evaluation can be found in Table 1.

3.4. *Ensemble Method Based on Decision Trees*

As requiring prior knowledge to construct a voting schema is not satisfying, we chose to derive the voting schema by decision trees, trained on the predictions of the single base methods and the correct localization classes. Decision trees combine the benefits of generally good accuracy and interpretable models, i.e. the derived voting schema provides further information regarding the performance of the underlying methods on different localization classes. For example, the decision tree for the taxonomic group "plant" learns a rule like *If CELLO predicts class 6 and WoLFPSORT predicts class 4, then class 4 is correct.* We trained decision trees using J48 of WEKA[28] for each taxonomic group.

Table 2. Covered subcellular localizations and corresponding keywords in SWISS-PROT.

ID	Subcellular localization	Keywords in SWISS-PROT	ID	Subcellular localization	Keywords in SWISS-PROT
1	Chloroplast	Chloroplast	8	Peroxisome	Peroxisome, Peroxisomal
2	Cytoplasm	Cytoplasm(ic)			Microsome, Microsomal
3	ER	Endoplasmic reticulum			Glyoxysome, Glyoxysomal
4	Golgi apparatus	Golgi			Glycosome, Glycosomal
5	Lysosome	Lysosome, Lysosomal	9	Extracellular	Extracellular
6	Mitochondrion	Mitochondrion, Mitochondrial			Secreted
7	Nucleus	Nucleus, Nuclear	10	Vacuole	Vacuole, Vacuolar

4. Evaluation

Although more and more prediction methods for subcellular localization have been developed, several limitations exist. First, the coverage of predicted localizations, which ranges from just a few localizations to all possible localizations. While e.g., SubLoc predicts only 4 localizations, PLOC is able to predict 12 localizations. Second, most existing methods were trained by a limited number of sequences from a specific taxonomic category of organisms, so the methods differ in their taxonomic coverage. The third aspect is the so-called sequence coverage, which is the number of sequences the different approaches learn from. Nonetheless, many newly developed methods still use the data set created by Reinhardt and Hubbard in 1998.[24] Thus, we decided to compile an up-to-date data set based on Swiss-Prot.[29] In order to compare methods differing widely in many aspects, we restricted the data set to 10 localization classes which are commonly accepted by most of the methods. These localization classes are listed in Table 2. This selection accommodates most of the available and rather general methods. For methods with a narrower localization coverage we used their reliability indices and assigned query sequences with lower reliability indices to the class "unknown". While their coverage is narrower, these methods often exceed others in their performance for the covered localization classes.

Based on Swiss-Prot (release 53.0), we at first selected all eukaryotic proteins with a unique subcellular localization annotation, where the localization annotation was one of the 10 localization classes listed in Table 2. Then, all proteins with a sequence length smaller than 60 amino acids were removed, as this is the required minimal sequence length for Predotar, the method with the largest required minimal length. Finally we kept only those proteins whose localization annotation was experimentally confirmed and belonged to one of the taxonomic groups "plant", "fungi", "human", or "animal". As the golgi group of plants was too small (7 entries), we complemented this group with 28 proteins whose localization information was not confirmed experimentally. This yielded 4 subsets corresponding to the 4 taxonomic groups. Table 3 lists the final number of proteins for each taxonomic group and each localization class.

Both the ensemble methods as well as the single base classifiers were evaluated by 10-fold cross-validations on our data set. The results are illustrated in Figures 1 and 2. Figure 1 shows the total accuracy. The simple weighted voting schema ("Voting") performs slightly better than the base classifiers. The decision tree ensembles ("DT-Ensemble") clearly outperform all other methods (including the voting schema). The most prominent improvement

Table 3. Number of proteins for different taxonomic groups and localization classes.

		Plant	Fungi	Animal	Human	Total
1	Chloroplast	3425	0	0	0	3425
2	Cytoplasm	470	578	1394	511	2953
3	ER	66	170	391	164	791
4	Golgi	35	55	78	55	223
5	Lysosome	0	0	102	56	158
6	Mitochondrion	370	632	1341	347	2690
7	Nucleus	308	899	2221	1094	4522
8	Peroxisome	50	85	181	72	388
9	Extracellular	149	199	596	4723	5667
10	Vacuole	35	68	0	0	103
	Total	4908	2686	10431	2895	20920

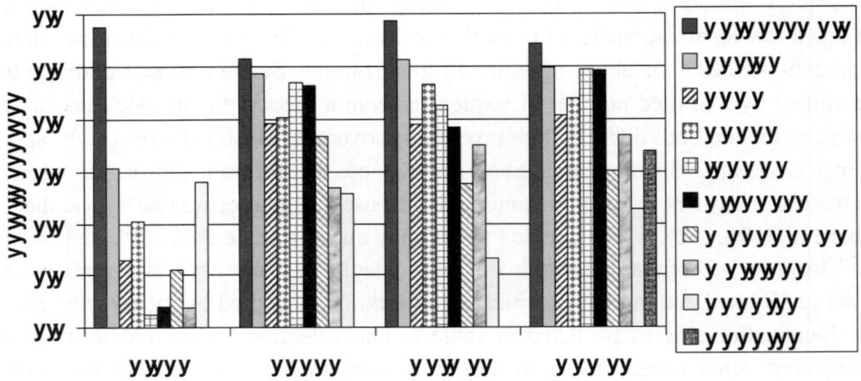

Fig. 1. Comparison between single and ensemble classification methods: Total Accuracy, i.e., the overall percentage of correctly predicted instances.

can be seen in the plant group, were the other methods mostly perform rather weak (at best, ESLPred reaches an accuracy of just below 60%), while the accuracy of the decision tree ensemble is well above 80%.

Most methods perform comparably well in terms of specificity (cf. Figure 2). Again, in the plant group the improvement of both ensemble methods is most prominent. In the remaining taxonomic groups the best base classifiers already reach almost 100%. Thus, no significant improvement can be expected. However, the ensemble methods perform as well as the best base classifiers. The decision tree ensembles even slightly improve over the already very good values.

All our methods are available via a webinterface at `http://www.dbs.ifi.lmu.de/research/locpred/ensemble/`.

5. Conclusions

In this paper, we shortly surveyed some prominent prediction methods for subcellular localization of proteins. The spectrum of underlying information (as amino acid composition,

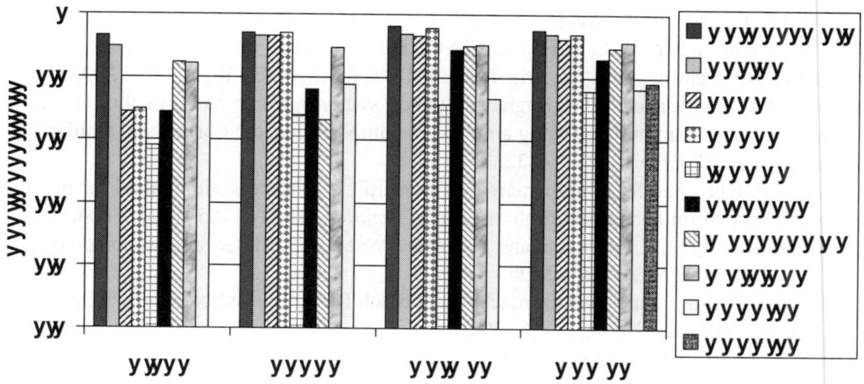

Fig. 2. Comparison between single and ensemble classification methods: Average Specificity, i.e., the percentage averaged over all localization classes to correctly exclude an instance from the corresponding class.

sorting signals, and homology search) makes these methods ideally diverse to expect an ensemble composed of these methods to improve considerably in terms of accuracy. We developed two ensemble methods: First, a simple voting scheme using the votes of the base learners weighted according to their average performance (based on prior knowledge), second, decision trees trained on the prediction values of the base methods (thus learning the weight of the methods on the fly and allowing for a more complex weighting). Both ensembles are shown to improve over the base classifiers in most cases. The decision tree ensemble can even said to outperform the remaining methods.

References

1. H. Bannai, Y. Tamada, O. Maruyama, K. Nakai, and S. Miyano. Extensive feature detection of n-terminal protein sorting signals. *Bioinformatics*, 18(2):298–305, 2002.
2. J.-D. Bendtsen, H. Nielsen, G. von Heijne, and S. Brunak. Improved prediction of signal peptides: SignalP 3.0. *J. Mol. Biol.*, 340(4):783–795, 2004.
3. M. Bhasin, A. Garg, and G.-P.-S. Raghava. PSLpred: prediction of subcellular localization of bacterial proteins. *Bioinformatics*, 21(10):2522–2524, 2005.
4. M. Bhasin and G. P. S. Raghava. ESLpred: SVM-based method for subcellular localization of eukaryotic proteins using dipeptide composition and PSI-BLAST. *Nucleic Acids Res.*, 32(Web Server Issue):W414–W419, 2004.
5. M.-G. Claros and P. Vincens. Computational method to predict mitochondrially imported proteins and their targeting sequences. *Eur. J. Biochem.*, 241(3):779–786, 1996.
6. M. Cokol, R. Nair, and B. Rost. Finding nuclear localization signals. *EMBO Rep.*, 1(5):411–415, 2000.
7. T. G. Dietterich. Ensemble methods in machine learning. In *Proc. MCS*, 2000.
8. T. G. Dietterich. Ensemble learning. In M. A. Arbib, editor, *The Handbook of Brain Theory and Neural Networks*, pages 405–408. MIT Press, second edition, 2003.
9. P. Dönnes and A. Höglund. Predicting protein subcellular localization: Past, present, and future. *Geno. Prot. Bioinfo.*, 2(4):209–215, 2004.
10. J. L. Gardy, M. R. Laird, F. Chen, S. Rey, C. J. Walsh, M. Ester, and F. S. L. Brinkman. PSORTb v.2.0: expanded prediction of bacterial protein subcellular localization and insights gained from comparative proteome analysis. *Bioinformatics*, 21(5):617–623, 2005.

11. J. L. Gardy, C. Spencer, K. Wang, M. Ester, G. E. Tusnady, I. Simon, S. Hua, K. deFays, C. Lambert, K. Nakai, and F. S. L. Brinkman. PSORT-B: improving protein subcellular localization prediction for gram-negative bacteria. *Nucleic Acids Res.*, 31(13):3613–3617, 2003.
12. A. Garg, M. Bhasin, and G. P. S. Raghava. Support vector machine-based method for subcellular localization of human proteins using amino acid compositions, their order, and similarity search. *J. Biol. Chem.*, 280(15):14427–14432, 2005.
13. C. Guda, E. Fahy, and S. Subramaniam. MITOPRED: a genome-scale method for prediction of nucleus-encoded mitochondrial proteins. *Bioinformatics*, 20(11):1785–1794, 2004.
14. P. Horton, K.-J. Park, T. Obayashi, and K. Nakai. Protein subcellular localization prediction with WoLF PSORT. In *Proc. APBC*, 2006.
15. S. Hua and Z. Sun. Support vector machine approach for protein subcellular localization prediction. *Bioinformatics*, 17(8):721–728, 2001.
16. Y. Huang and Y. Li. Prediction of protein subcellular locations using fuzzy k-NN method. *Bioinformatics*, 20(1):21–28, 2004.
17. J.-K. Hwang, C.-J. Lin, and C.-S. Yu. Predicting subcellular localization of proteins for gram-negative bacteria by support vector machines based on n-peptide compositions. *Protein Science*, 13:1402–1406, 2004.
18. A. Höglund, P. Dönnes, T. Blum, H.-W. Adolph, and O. Kohlbacher. MultiLoc: prediction of protein subcellular localization using N-terminal targeting sequences, sequence motifs and amino acid composition. *Bioinformatics*, 22(10):1158–1165, 2006.
19. Z. Lu, D. Szafron, R. Greiner, P. Lu, D.S. Wishart, B. Poulin, J. Anvik, C. Macdonell, and R. Eisner. Predicting subcellular localization of proteins using machine-learned classifiers. *Bioinformatics*, 20(4):547–556, 2004.
20. K. Nakai and P. Horton. PSORT: a program for detecting sorting signals in proteins and predicting their subcellular localization. *Trends Biochem. Sci.*, 24(1):34–36, 1999.
21. K. Nakai and M. Kanehisa. A knowledge base for predicting protein localization sites in eukaryotic cells. *Genomics*, 14(4):897–911, 1992.
22. H. Nakashima and K. Nishikawa. Discrimination of intracellular and extracellular proteins using amino acid composition and residue-pair frequencies. *J. Mol. Biol.*, 238(1):54–61, 1994.
23. K.-J. Park and M. Kanehisa. Prediction of protein subcellular locations by support vector machines using compositions of amino acids and amino acid pairs. *Bioinformatics*, 19(13):1656–1663, 2003.
24. A. Reinhardt and T. Hubbart. Using neural networks for prediction of the subcellular location of proteins. *Nucleic Acids Res.*, 26:2230–2236, 1998.
25. I. Small, N. Peeters, F. Legeai, and C. Lurin. Predotar: A tool for rapidly screening proteomes for n-terminal targeting sequences. *Proteomics*, 4(6):1581–1590, 2004.
26. G. Valentini and F. Masulli. Ensembles of learning machines. In *Proc. Neural Nets WIRN*, 2002.
27. G. von Heijne. A new method for predicting signal sequence cleavage sites. *Nucleic Acids Res.*, 14(11):4683–4690, 1986.
28. I. H. Witten and E. Frank. *Data Mining: Practical machine learning tools and techniques*. Morgan Kaufmann, San Francisco, 2nd edition, 2005.
29. C.H. Wu, R. Apweiler, A. Bairoch, D.A. Natale, W.C. Barker, B. Boeckmann, S. Ferro, E. Gasteiger, H. Huang, R. Lopez, M. Magrane, M.J. Martin, R. Mazumder, C. O'Donovan, N. Redaschi, and B. Suzek. The universal protein resource (UniProt): an expanding universe of protein information. *Nucleic Acids Res.*, 34:D187–D191, 2006.

CHEMICAL COMPOUND CLASSIFICATION WITH AUTOMATICALLY MINED STRUCTURE PATTERNS

[1]AARON M. SMALTER, [1]J. HUAN and [2]GERALD H. LUSHINGTON

[1]*Department of Electrical Engineering and Computer Science*
[2]*Molecular Graphics and Modeling Laboratory*
University of Kansas, Lawrence, KS 66045, USA
E-mail: {asmalter, jhuan, glushington}@ku.edu

In this paper we propose new methods of chemical structure classification based on the integration of graph database mining from data mining and graph kernel functions from machine learning. In our method, we first identify a set of general graph patterns in chemical structure data. These patterns are then used to augment a graph kernel function that calculates the pairwise similarity between molecules. The obtained similarity matrix is used as input to classify chemical compounds via a kernel machines such as the support vector machine (SVM). Our results indicate that the use of a pattern-based approach to graph similarity yields performance profiles comparable to, and sometimes exceeding that of the existing state-of-the-art approaches. In addition, the identification of highly discriminative patterns for activity classification provides evidence that our methods can make generalizations about a compound's function given its chemical structure. While we evaluated our methods on molecular structures, these methods are designed to operate on general graph data and hence could easily be applied to other domains in bioinformatics.

1. Introduction

The development of accurate models for chemical activity prediction has a range of applications. They are especially useful in the screening of potential drug candidates, currently a difficult and expensive process that can benefit enormously from accurate *in silico* methods. These models have proved difficult to design, due to the complex nature of most biological classification problems. For example, the toxicity of a particular chemical compound is determined by a large variety of factors, as there are innumerable ways that a foreign chemical might interfere with an organism, and the situation is further complicated by the possibility that a benign chemical may be broken down into toxic metabolites *in vivo*. Clearly, there is no single set of chemical features that can be easily applied to to all problems in all situations, and therefore the ability to isolate problem-specific chemical features from broader data collections is a critical issue.

Here we address the problem of identifying structure characteristics that link a chemical compound to its function by integrating graph database mining methods from the data mining community with graph kernel functions from the machine learning community. Graphs are powerful mathematical structures and have been widely used in bioinformatics and other research [1], and are ubiquitous in the representation of chemical compounds. In our method, we identify frequently occurring subgraphs from a group of chemical structures represented as graphs, and define a graph similarity measure based on the obtained subgraphs. We then build a model to predict the function of a chemical structure based on the previously generated similarity measures.

Traditional approaches to graph similarity rely on the comparison of compounds using a variety of molecular attributes known *a priori* to be involved in the activity of interest. Such methods are problem-specific, however, and provide little assistance when the relevant descriptors are not known in advance. Additionally, these methods lack the ability to provide explanatory information regarding what structural features contribute to the observed chemical activity. Our proposed method alleviates both of these issues through the mining and analysis of structural patterns present in the data in order to identify highly discriminating patterns, which then augment a graph kernel function that computes molecular similarity.

We have applied our methods to three chemical structure-activity benchmarks: predictive toxicology, human intestinal absorption, and virtual screening. Our results indicate that the use of a pattern-based approach to graph similarity yields performance profiles comparable to, and sometimes exceeding that of previous non-pattern-based approaches. In addition, the presence and identification of highly discriminative patterns for chemical activity classification provides evidence that our methods can make generalizations about a compound's function given its chemical structure.

The rest of the paper is organized in the following way. In Section 2, we present an overview of related work on graph kernels and frequent subgraph mining. In Section 3, we present background information about graph representation of chemical structures. In Section 4, we present the algorithmic details of the work and in Section 5, we present our empirical study of the proposed algorithm using several chemical structure benchmarks. We conclude our paper with a short discussion about pros and cons of our proposed methods.

2. Related Work

The term kernel function refers to an operation for computing the inner product between two vectors in a feature space, thus avoiding the explicit computation of coordinates in that feature space. Graph kernel functions are simply kernel functions that have been defined to compute the similarity between two graph structures. In recent years a variety of graph kernel functions have been developed, with promising results as described by Ralaviola et al[2]. Here we review the two methods that are most similar to ours. The first compares graphs using random, linear substructures; and the second is based on matching and aligning the vertices of two graphs. We also review the technique used to identify substructure patterns in our proposed method.

2.1. *Marginalized and Optimal Assignment Graph Kernels*

The work of Kashima et al.[3] is based on the use of shared label sequences in the computation of graph kernels. Their marginalized graph kernel uses a Markov model to randomly generate walks of a labeled graph. The random walks are created using a transition probability matrix combined with a walk termination probability. These collections of random walks are then compared and the number of shared sequences is used to determine the overall similarity between two molecules.

The optimal assignment kernel, described by Frölich et al[4], differs significantly from the marginalized graph kernel. This kernel function first computes the similarity between all vertices in one graph and all vertices in another. The similarity between the two graphs is then computed by finding the maximal weighted bipartite graph between the two sets of vertices, called the optimal assignment. The authors investigate an extension of this method whereby certain structure patterns defined *a priori* by expert knowledge, are collapsed into single vertices, and this reduced graph is used as input to the optimal assignment kernel.

2.2. *Frequent Subgraph Mining*

Frequent subgraph mining is a technique used to enumerate graph substructures that occur in a graph database with at least some specified frequency. This minimum frequency threshold is termed the *support threshold* by the data mining community. After limiting returned subgraphs by frequency, we can further constrain the types we find by setting upper and lower limits on the number of vertices they can contain. In this paper, we use the FFSM algorithm[11], for fast computation of frequent subgraphs. Figure 1, adopted from[11], shows an example of this frequent subgraph enumeration. Some work has been done by Deshpande et al.[5] toward the use of these frequent substructures in the classification of chemical compounds with promising results.

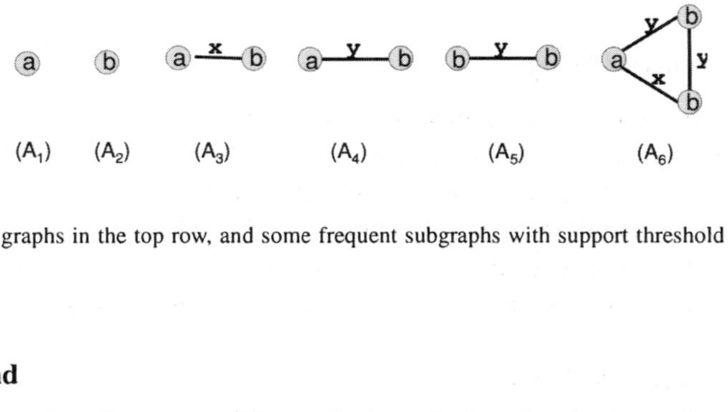

Figure 1. Set of graphs in the top row, and some frequent subgraphs with support threshold 2/3 in the bottom row.

3. Background

Before we proceed to discuss specific methods and other details, let us first provide some general background information regarding both chemical structures and graph mining.

3.1. *Chemical Structure*

Chemical compounds are well-defined structures that are easily encapsulated by a graph representation. Compounds are composed of a number of atoms which are represented as vertices in a graph, and a number of bonds between atoms represented as edges in the graph. Vertices are labeled with the atom element type, and edges are labeled with the bond type. The edges in the graph are undirected, since there is no directionality associated with chemical bonds. Figure 2 shows an example chemical structure.

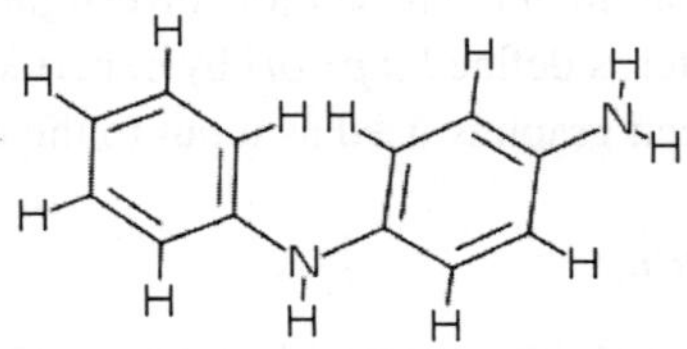

Figure 2. An example chemical structure from the PTC data set. Unlabeled vertices are assumed to be carbon C.

4. Algorithm Design

The following sections outline the algorithm that drives our experimental method. In short, we measure the similarity of graph structures whose vertices and edges have been labeled with various descriptors. These descriptors represent physical and chemical information such as atom and bond types. They are also used to represent the membership of atoms in specific structure patterns that have been mined from the data. To compute the similarity of two graphs, the vertices of one graph are aligned with the vertices of the second graph, such that the total overall similarity is maximized with respect to all possible alignments. Vertex similarity is measured by comparing vertex descriptors, and is computed recursively so that when comparing two vertices, we also compare the neighbors of those vertices, and their neighbors, etc.

4.1. *Structure Pattern Mining*

The frequent subgraph mining problem can be phrased as such: given a set of labeled graphs, the support of an arbitrary subgraph is the fraction of all graphs in the set that contain that subgraph. A subgraph is frequent if its support meets a certain minimum threshold. The goal is to enumerate all the frequent, connected subgraphs in a graph database. The extraction of important subgraph patterns can be controlled by selecting the proper frequency threshold, as well as other parameters such as size and density of subgraph patterns.

4.2. *Optimal Assignment Kernel*

The optimal assignment kernel function computes the similarity between two graph structures. This similarity computation is accomplished by first representing the two sets graph vertices as a bipartite graph, and then finding the set of weighted edges assigning every vertex in one graph to a vertex in the other. The edge weights are calculated via a recursive vertex similarity function. We present the equations describing this algorithm in detail, as discussed by Frölich et al[4]. The top-level equation describing the similarity of two molecular graphs is:

$$k_A(M_1, M_2) := max_\pi \sum_{h=1}^{m} k_{nei}(v_{\pi(h)}, v_h) \tag{1}$$

Where π denotes a permutation of a subset of graph vertices, and m is the number of vertices in the smaller graph. This is needed since we want to assign all vertices of the smaller graph to vertices in the large graph. The k_{nei} function, which calculates the similarity between two vertices using their local neighbors, is given as follows:

$$k_{nei}(v_1, v_2) := k_v(v_1, v_2) + R_0(v_1, v_2) + S_{nei}(v_1, v_2) \tag{2}$$

$$S_{nei}(v_1, v_2) := \sum_{l=1}^{L} \gamma(l) R_l(v_1, v_2) \tag{3}$$

The functions k_v and k_e compute the similarity between vertices (atoms) and edges (bonds), respectively. These functions could take a variety of forms, but in the OA kernel they are RBF functions between vectors of vertex/edge labels.

The $\gamma(l)$ term is a decay parameter that weights the similarity of neighbors according to their distance from the original vertex. The l parameter controls the topological distance within which to consider neighbors of vertices. The R_l equation, which recursively computes the similarity between two specific vertices is given by the following equation:

$$R_l(v_1, v_2) = \frac{1}{|v_1||v_2|} \sum_{i,j} R_{l-1}(n_i(v_1), n_j(v_2)) \tag{4}$$

Where $|v|$ is the number of neighbors of vertex v, and $n_k(v)$ is the set of neighbors of v. The base case for this equation is R_0, defined by:

$$R_0(v_1, v_2) := \frac{1}{|v_1|} max_\pi \sum_{i=1}^{|v_2|} (k_v(a, b) | k_e(x, y)) \tag{5}$$

$$a = n_{\pi(i)}(v_1),\ b = n_i(v_2) \tag{6}$$

$$x = v_1 \rightarrow n_{\pi(i)}(v_1),\ y = v_2 \rightarrow n_i(v_2) \tag{7}$$

The notation $v \rightarrow n_i(v)$ refers to the edge connecting vertex v with the ith neighboring vertex. The functions k_v and k_e are used to compare vertex and edge descriptors, by counting the total number of descriptor matches.

4.3. *Reduced Graph Representation*

One way in which to utilize the structure patterns that are mined from the graph data is to collapse the specific subgraphs into single vertices in the original graph. This technique is explored by Frölich et al.[4] with moderate results, although they use predefined structure patterns, so called pharmacophores, identified *a priori* with the help of expert knowledge. Our method ushers these predefined patterns in favor of the structure patterns generated via frequent subgraph mining.

The use of a reduced graph representation does have some advantages. First, by collapsing substructures, we can compare an entire set of vertices at once, reducing the graph complexity and marginally decreasing computation time. Second, by changing the substructure size we can adjust the resolution at which graph structures are compared. The disadvantage of a reduced graph representation is that substructures can only be compared directly to other substructures, and cannot align partial structure matches. As utilized in Frölich et al.[4], this is not as much of a burden since they have defined the best patterns *a priori* using expert knowledge. In our case, however, this is a significant downside, as we have no *a priori* knowledge to guide our pattern generation and we wish to retain as much structural information as possible.

4.4. *Pattern-based Descriptors*

The loss of partial substructure alignment following the use of a reduced graph representation motivated us to find another way of integrating this pattern-based information. Instead of collapsing graph substructures, we simple annotate vertices with additional descriptor labels indicating the vertex's membership in the structure patterns that were previously mined. These pattern-based descriptors are calculated for each vertex and are used by the optimal assignment kernel in the same way that other vertex descriptors are handled. In this way we are able to capture substructure information in the graph vertices without needing to alter the original graph structure.

5. Experimental Study

We conducted classification experiments on five different biological activity data sets, and measured support vector machine (SVM) classifier prediction accuracy for several different feature generation methods. The data sets and classification methods are described in more detail in the following subsections, along with the associated results. Figure 3 gives a graphical overview of the process.

We performed all of our experiments on a desktop computer with a 3Ghz Pertium 4 processor and 1 GB of RAM. Generating a set of frequent subgraphs is very quick, generally a few seconds. Optimal assignment requires significantly more computation time, but not intractable, at less than half an hour for the largest data set.

5.1. *Data Sets*

We have selected five data sets used in various problem areas to evaluate our classifier performance. The Predictive Toxicology Challenge data set, discussed by Helma et al[6], contains a set of chemical compounds classified according to their toxicity in male rats (PTC-MR), female rats (PTC-FR), male mice (PTC-MM), and female mice (PTC-FM).

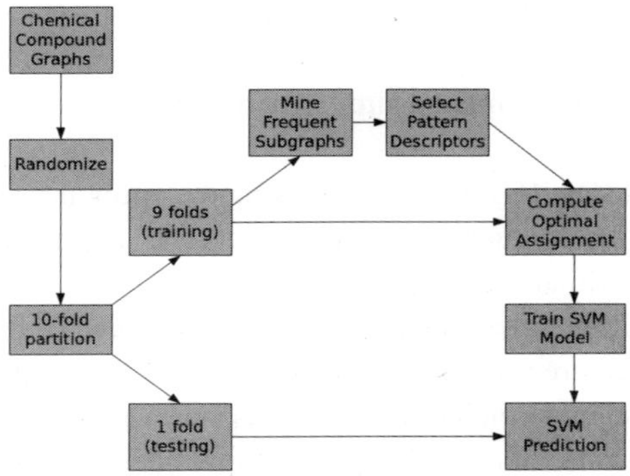

Figure 3. Experimental workflow for a single cross-validation trial.

The Human Intestinal Absorption (HIA) data set (Wessel et al.[7]) contains chemical compounds classified by intestinal absorption activity. We also included two different virtual screening data sets (VS-1,VS-2) used to predict various binding inhibitors from Fontaine et al.[8] and Jorissen et al[9]. The final data set (MD) is from Patterson et al[10], and was used to validate certain molecule descriptors. Various statistics for these data sets can be found in Table 1.

Table 1. Data set statistics.

Dataset	Number of Compounds	Number of Positives	Number of Negatives	Average Compound Size
HIA	86	47	39	22.45
MD	310	148	162	10.38
VS-1	435	279	156	59.81
VS-2	1071	125	946	39.93
PTC-MR	344	152	192	25.56
PTC-MM	336	129	207	25.05
PTC-FR	351	121	230	26.08
PTC-FM	349	143	206	25.25

5.2. *Methods*

We evaluated the performance of the SVM classifier when trained using several different feature sets. The first set of features (FSM) consists only of frequent subgraphs. Those subgraphs are mined using the FFSM software [11] with minimum subgraph frequency of 50%. Each chemical compound is represented by a binary vector with length equal to the number of mined subgraphs. Each subgraph is mapped to a specific vector index, and if a chemical compound contains a subgraph then the bit at the corresponding index is set to one, otherwise it is set to zero.

The second feature set (OA) consists of the similarity values computed by the optimal assignment kernel, as proposed by Frölich et al[4]. Each compound is represented as a real-valued vector containing the computed similarity between it and all other molecules in the data set.

The third feature set (OARG) is computed using the optimal assignment kernel as well, except that we embed the frequent subgraph patterns as a reduced graph representation before computing the optimal assignment. The reduced graph representation is described by Frölich et al.[4] as well, but they use *a priori* patterns instead of frequently mined ones.

Finally, the fourth feature set (OAPD) also consists of the subgraph patterns combined with the optimal assignment kernel, however in this case we do not derive a reduced graph, and instead annotate vertices in a graph with additional descriptors indicating its membership in specific subgraph patterns.

In our experiments, we used the support vector machine (SVM) classifier in order to generate activity predictions. The use of SVM has recently become quite popular for a variety of biological machine learning applications because of its efficiency and ability to operate on high-dimensional data sets. We used the SMO SVM classifier implemented by Platt[13] and included in the Weka data-mining software package by Witten et al[14]. The SVM parameters were fixed, and we used a linear kernel with C = 1. Classifier performance was averaged over a ten-fold cross-validation set.

We perform some feature selection in order to identify the most discriminating frequent patterns. Using a simple statistical formula, known as the Pearson correlation coefficient (PCC), we measure the correlation between a set of feature samples (in our case, the occurrences of a particular subgraph in each of the data samples) and the corresponding class labels. Frequent patterns are ranking according to correlation strength, and the top patterns are selected.

5.3. *Results*

Table 2 contains results reporting the average and standard deviation of the prediction accuracy over the 10 cross-validation trials. With the table, we have the following observations.

Table 2. Average and standard deviation of 10-fold cross-validation accuracy for each data set.

Dataset	Method			
	FSM	OA	OARG	OAPD
HIA	57.36 ±19.11	63.33 ±20.82	62.92 ±22.56	65.28 ±15.44
MD	68.39 ±7.26	70.00 ±6.28	69.35 ±6.5	70.32 ±5.65
VS-1	60.00 ±5.23	64.14 ±3.07	62.07 ±4.06	63.91 ±4.37
VS-2	90.29 ±2.3	94.96 ±1.88	93.18 ±2.68	94.77 ±2.17
PTC-FM	54.16 ±5.82	61.35 ±9.53	59.03 ±6.46	59.29 ±8.86
PTC-FR	63.28 ±5.32	60.10 ±9.21	64.68 ±3.96	64.39 ±3.6
PTC-MM	60.45 ±3.87	62.16 ±6.43	62.75 ±7.69	63.05 ±5.24
PTC-MR	58.42 ±4.43	56.41 ±6	54.07 ±7.52	60.76 ±7.32

First, we notice that OAPD (and OARG) outperforms FSM methods in all of the tried data sets except one (FSM is better than OARG on the PTC-MR data set). This results indicate that if we use frequent subgraph alone without using the optimal alignment kernel, we do not have a good classifier. Although the conclusion is generally true, interestingly, we found that for the PTC-MR data set, the FSM method outperforms both the OA and OARG methods, while the OAPD method outperforms FSM. This seems to suggest that important information is encoded in the frequent subgraphs, and is being lost in the OARG, but is still preserved in the OAPD method.

Second, we notice that OAPD (or OARG) method outperforms the original OA method in 5 of the tried 8 data sets: HIA, MD, PTC-FR, PTC-MM, PTC-MR. OAPD has a very close performance to that of OA in the rest of the three data sets. The results indicate that our OAPD method provides good performance for diverse data sets which involve tasks such as predicting chemical's toxicology, predicting human intestinal absorption of chemicals, and virtual screening of drugs.

Table 3. Top five highest ranked frequent subgraph patterns for each data set, expressed as SMARTS strings that encode a specific subgraph.

HIA	MD	VS-1	VS-2
[NH3+]C(C)C	C(=CC)(C)S	C(=CC=C)C=C	C(=CCC)C
C(=C)(C)C	C(=CC=CC)(C)S	C(=CC)CNC	C=CCC
C(=CC)(C)C	C(=C)(C=CC=C)S	C(=C)CNC	[NH2+](CC=C)CC
C(=CC)(C=C)C	C(=CCC)C=C	CC(=CC)N	[NH2+](CCC)CC
C(=CC=C)(C=C)C	C(=CS)C=C	CNCC=CC	[NH3+]CC(=CC)C

PTC-MR	PTC-MM	PTC-FR	PTC-FM
[NH2+]C(=C)C=C	[NH3+]CC	[NH2+]C(=CC)C=C	OCC=C
[NH2+]C=CC	c1ccccc1	[NH2+]C(=C)C=C	C(=CC)C(=C)C
[NH3+]CC	C(=CC)C(=C)C	[NH3+]CC	CCC=CC
CC=C	C(=CC=C)C	CC=C	C(=C)(C)C
C(CC)C	C(=C)C(=C)C	C(CC)C	c1ccccc1

In addition to outperforming the previous methods, our new method also reports the specific subgraph patterns that were mined from the training data and used to augment the optimal assignment kernel function. By identifying highly discriminating patterns, our method can offer additional insight into the structural features that contribute to a compound's chemical function. Table 3 contains the five highest ranked (using Pearson correlation coefficient) subgraph patterns for each data set, expressed as SMARTS strings that encode the specific pattern. Many of the patterns in all sets denote various carbon chains (C(CC)C, C=CC, etc.), however there seem to be some unique patterns as well. The MD data set contains carbon chain patterns with some sulfur atoms mixed in, while the VS-1 data set has carbon chains with nitrogen mixed in. The [NH2+] and [NH3+] patterns appear to be important in the VS-2 data set, as well as some of the PTC data sets.

6. Conclusions

Graph structures are a powerful and expressive representation for chemical compounds. In this paper we present a new method, termed OAPD, for computing the similarity of chemical compounds, based on the use of an optimal assignment graph kernel function augmented with pattern-based descriptors that have been mined from a set of molecular graphs. Our experimental study demonstrate that our OAPD method integrates the structural alignment capabilities of the existing optimal alignment kernel method with the substructure discovery capabilities of the frequent subgraph mining method and delivers better performance in most of the tried benchmarks. In the future, we plan to involve domain experts to evaluate the performance of our algorithm, including the prediction accuracy and the capability of identifying structure important features, in diverse chemical structure data sets.

Acknowledgments

This work has been supported by the Kansas IDeA Network for Biomedical Research Excellence (NIH/NCRR award #P20 RR016475) and the KU Center of Excellence for Chemical Methodology and Library Development (NIH/NIGM award #P50 GM069663)

References

1. M. E. J. Newman. The structure and function of complex networks. *SCIAM Rev.*, 45(2):167-256, 2003.
2. L. Ravaliola, S. J. Swamidass, H. Saigo. Graph Kernels for Chemical Informatics. *Neural Networks*, 18(8):1093-1110, September 2005.
3. H. Kashima, K. Tsuda, A. Inokuchi. Marginalized Kernels Between Labeled Graphs. *Proc. of the Twentieth Int. Conf. on Machine Learning (ICML-03)*, 2003.
4. H. Fröohlich, J. Wegner, F. Sieker, A. Zell. Kernel Functions for Attriubted Molecular Graphs - A new Similarity-Based Approach to ADME Prediction in Classification. *QSAR & Combinatorial Science*, 25(4):317-326, 2006.
5. M. Deshpande, M. Kuramochi, G. Karypis. Frequent Sub-Structure-Based Approaches for Classifying Chemical Compounds. *IEEE Transactions on Knowledge and Data Engineering*, 17(8):1036-1050, August 2005.
6. C. Helma, R. King, S. Kramer. The predictive toxicology challenge 2000-2001. *Bioinformatics*, 17(1):107-108, 2001.
7. M. Wessel, P. Jurs, J. Tolan, S. Muskal. Prediction of Human Intestinal Absorption of Drug Compounds from Molecular Structure. *J. Chem. Inf. Comput. Sci.*, 38(4):726-735, 1998.
8. F. Fontaine, M. Pastor, I. Zamora, and F. Sanz. Anchor-GRIND: Filling the Gap between Standard 3D QSAR and the GRid-INdependent Descriptors. *J. Med. Chem.*, 48(7):2687-2694, 2005.
9. R. Jorissen and M. Gilson. Virtual Screening of Molecular Databases Using a Support Vector Machine. *J. Chem. Inf. Model.*, 45(3):549-561, 2005.
10. D. Patterson, R. Cramer, A. Ferguson, R. Clark, L. Weinberger. Neighbourhood Behaviour: A Useful Concept for Validation of "Molecular Diversity" Descriptors. *J. Med. Chem.*, 39:3049-3059, 1996.
11. J. Huan, W. Wang, J. Prins. Efficient Mining of Frequent Subgraphs in the Presence of Isomorphism. *Proc. of the 3rd IEEE Int. Conf. on Data Mining (ICDM-03)*, 549-552, 2003.
12. V. Vapnik. *Statistical Learning Theory*. John Wiley, New York, NY, 1998.
13. J. Platt. Fast Training of Support Vector Machines using Sequential Minimal Optimization. *Advances in Kernel Methods - Support Vector Learning*, MIT Press, Cambridge, MA, 1998.
14. I. Witten, E. Frank. *Data Mining: Practical machine learning tools and techniques*, 2nd Edition. Morgan Kaufmann, San Francisco, CA, 2005.

STRUCTURE-APPROXIMATING DESIGN OF STABLE PROTEINS IN 2D HP MODEL FORTIFIED BY CYSTEINE MONOMERS

ALIREZA HADJ KHODABAKHSHI, JÁN MAŇUCH, ARASH RAFIEY and ARVIND GUPTA

School of Computing Science
8888 University Drive, Simon Fraser University
Burnaby, BC, V5A 1S6, Canada
E-mail: alireza@cs.sfu.ca, jmanuch@sfu.ca, arafieyh@cs.sfu.ca, arvind@mitacs.cs

The inverse protein folding problem is that of designing an amino acid sequence which has a prescribed native protein fold. This problem arises in drug design where a particular structure is necessary to ensure proper protein-protein interactions. The input to the inverse protein folding problem is a shape and the goal is to design a protein sequence with a unique native fold that closely approximates the input shape. Gupta *et al.*[1] introduced a design in the 2D HP model of Dill that can be used to approximate any given (2D) shape. They conjectured that the protein sequences of their design are stable but only proved the stability for an infinite class of very basic structures. The HP model divides amino acids to two groups: hydrophobic (H) and polar (P), and considers only hydrophobic interactions between neighboring H amino in the energy formula. Another significant force acting during the protein folding are sulfide (SS) bridges between two cysteine amino acids. In this paper, we will enrich the HP model by adding cysteines as the third group of amino acids. A cysteine monomer acts as an H amino acid, but in addition two neighboring cysteines can form a bridge to further reduce the energy of the fold. We call our model the HPC model. We consider a subclass of linear structures designed in Gupta *et al.*[1] which is rich enough to approximate (although more coarsely) any given structure. We refine the structures for the HPC model by setting approximately a half of H amino acids to cysteine ones. We conjecture that these structures are stable under the HPC model and prove it under an additional assumption that non-cysteine amino acids act as cysteine ones, i.e., they tend to form their own bridges to reduce the energy. In the proof we will make an efficient use of a computational tool 2DHPSolver which significantly speeds up the progress in the technical part of the proof. This is a preliminary work, and we believe that the same techniques can be used to prove this result without the artificial assumption about non-cysteine H monomers.

Keywords: HP model; protein stability; protein design; 2D square lattice; cysteine.

1. Introduction

It has long been known that protein interactions depend on their native three-dimensional fold and understanding the processes and determining these folds is a long standing problem in molecular biology. Naturally occurring proteins fold so as to minimize total free energy. However, it is not known how a protein can choose the minimum energy fold amongst all possible folds.[2]

Many forces act on the protein which contribute to changes in free energy including hydrogen bonding, van der Waals interactions, intrinsic propensities, ion pairing, disulfide bridges and hydrophobic interactions. Of these, the most significant is hydrophobic interaction.[3] This led Dill to introduce the *Hydrophobic-Polar model*.[4] Here the 20 amino acids from which proteins are formed are replaced by two types of monomers: hydrophobic (H or '1') or polar (P or '0') depending on their affinity to water. To simplify the problem, the protein is laid out on vertices of a lattice with each monomer occupying exactly one vertex and neighboring monomers occupy neighboring vertices. The free energy is minimized when the maximum number of hydrophobic monomers are adjacent in the lattice. Therefore, the "native" folds are those with the maximum number of such HH contacts. Even though the HP model is the simplest model of the protein folding process, computationally it is an NP-hard problem for both the two-dimensional[5] and the three-dimensional[6] square lattices.

In many applications such as drug design, we are interested in the complement problem to protein folding: *inverse protein folding* or *protein design*. The *inverse protein folding problem* involves starting with a prescribed target fold or structure and designing an amino acid sequence whose native fold is the target (positive design). A major challenge in designing proteins that attain a specific native fold is to avoid proteins that have multiple native folds (negative design). We say that a protein is *stable* if its native fold is unique. In Gupta *et al.*,[1] a design in the 2D HP model that can be used to approximate any given (2D) shape was introduced and it was shown that approximated structures are native for designed proteins (positive design). It was conjectured that the protein sequences of their designed structures are also stable but only proved for an infinite class of very basic structures (arbitrary long "I" and "L" shapes), as well as computationally tested for over 48,000 structures (including all with up to 9 tiles). Design of stable proteins of arbitrary lengths in the HP model was also studied by Aichholzer *et al.*[7] (for 2D square lattice) and by Li *et al.*[8] (for 2D triangular lattice), motivated by a popular paper of Brian Hayes.[9] In this paper we aim to show stability for a subclass of the structures introduced by Gupta *et al.*[1] which is still rich enough to approximate (although more coarsely) any target shape.

In natural proteins, sulfide bridges between two cysteine monomers play an important role in improving stability of the protein structure.[10] We believe that enriching the HP model with the third type of monomers, cysteines, and incorporating sulfide bridges between two cysteines into energy model results in a model with even more stable designs. This added level of stability can help in proving formally that the designed proteins are indeed stable. We call this new model, the HPC model (hydrophobic-polar-cysteine). The cysteine monomers act as hydrophobic, but in addition two neighboring cysteines can form a bridge to further reduce the energy of the fold.

The class of structures which we use is a subset of *linear structures* introduced by Gupta *et al.*[1] They are formed by a sequence of "plus" shape tiles, cf. Figure 1(a), connected by overlapping two pairs of polar monomers (each coming from a different

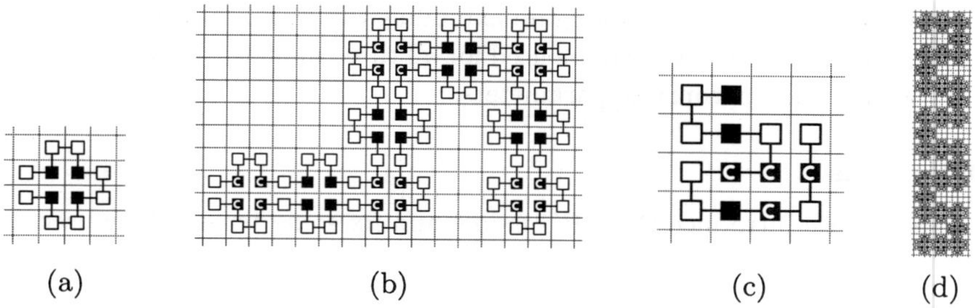

Fig. 1. (a) The basic building tile for constructible structures: black squares represent hydrophobic and white polar monomers. The lines between boxes represent the peptide bonds between consecutive monomers in the protein string. (b) An example of snake structure. The bending tiles use cysteines (black squares marked with C). (c) Example of energy calculation of a fold in HPC model. There are 5 contacts between hydrophobic monomers, thus the contact energy is -5. There are three potential sulfide bridges sharing a common vertex, hence only one can be used in the maximum matching. Thus the sulfide bridge energy is -2 and the total energy is -7.

tile). The structures are linear which means that every tile except the first and the last is attached to exactly two other tiles. In addition, we assume that the sequence of tiles has to change direction ("bend") in every odd tile. The hydrophobic monomers of these "bending" tiles are set to be cysteines, and all other hydrophobic monomers are non-cysteines, cf. Figure 1(b). We call these structures the *snake structures*. Note that approximately 40% of all monomers in snaked structures are hydrophobic and half of those are cysteines. Thus approximately 20% of all monomers are cysteines. Although, the most of naturally occurring proteins have much smaller frequency of cysteines, there are some with the same or even higher ratios: 1EZG (antifreeze protein from the beetle[11]) with 19.5% ratio of cysteines and the protein isolated from the chorion of the domesticated silkmoth[12] with 30% ratio.

Note that the snake structures can still approximate any given shape, although more coarsely than the linear structures. The idea of approximating a given shape with a linear structure is to draw a non-intersecting curve consisting of horizontal and vertical line segments. Each line segment is a linear chain of basic tiles depicted in Figure 1(a). At first glance, the snake structures seem more restricted than linear structures, as the line segments they use are very short and have the same size (3 tiles long). However, one can simulate arbitrary long line segments with snake structures forming a zig-zag pattern, cf. Figure 1(d).

We conjecture that the proteins for the snake structures are stable in the HPC model and that this can be proved using the techniques presented in this paper. These techniques are (i) the case analysis (also used in Gupta *et al.*[1]) and (ii) the induction on diagonals. Furthermore, to increase the power of the case analysis technique, we developed a program called "2DHPSolver" for semi-automatic proving of hypothesis about the folds of proteins of the designed structures. In this preliminary paper, we demonstrate the power of our techniques by showing that all snake structures are stable in the "strong" HPC model. The strong HPC model adds an artificial assumption that non-cysteine monomers form bridges as well to minimize the energy. We are currently working on extending our proof for the "proper" HPC model. Note that 2DHPSolver can be used for all three models: HP, HPC and strong HPC by setting the appropriate parameters.

2. Definitions

In this section we introduce the HPC model and fix some terminology used in the paper.

2.1. *Hydrophobic-polar-cysteine (HPC) model*

Proteins are chains of monomers where each monomer is either hydrophobic or polar. Furthermore, we will distinguish two types of hydrophobic monomers: cysteines which can form sulfide bridges to decrease the energy of the fold and non-cysteines. We can represent a protein chain as a string $p = p_1 p_2 \dots p_{|p|}$ in $\{0,1,2\}^*$, where "0" represents a polar monomer, "1" a hydrophobic non-cysteine monomer and "2" a cysteine monomer.

The proteins are folded onto the regular lattice. A *fold* of a protein p is embedding of a path of length n into lattice, i.e., vertices of the path are mapped into distinct lattice vertices and two consecutive vertices of the path are mapped to lattice vertices connected by an edge (a peptide bond). In this paper we use the 2D square lattice.

A protein will fold into a fold with the minimum free energy, also called a *native fold.* In the HP model only hydrophobic interactions between two adjacent hydrophobic monomers which are not consecutive in the protein sequence (*contacts*) are considered in the energy model, with each contact contributing with -1 to the total energy. In addition, in the HPC model, two adjacent non-consecutive cysteines can form a sulfide bridge contributing with -2 to the total energy. (Note that the results in the paper are independent on the exact value of the energy of sulfide bridge, as long as it is negative, and therefore we did not research on determination of the correct value for this energy.) However, each cysteine can be involved in at most one sulfide bridge. More formally, any two adjacent non-consecutive hydrophobic monomers (cysteine or non-cysteine) form a contact and the contact energy is equal to -1 times the number of contacts; and any two adjacent non-consecutive cysteines form a *potential* sulfide bridge and the sulfide-bridge energy is equal to -2 times the number of matches in the maximum matching in the graph of potential sulfide bridges. The total energy is equal to the sum of the contact and sulfide bridge energies. For example, the energy of the fold in Figure 1(c) is $(-5) + (-2) = -7$. Note that there might be several native folds for a given protein. A protein with a unique native fold is called *stable* protein.

2.2. *Snake structures*

In Gupta *et al.*,[1] a wide class of 2D structures, called *constructible structures*, was introduced. They are formed by a sequence of "plus" shape tiles, cf. Figure 1(a), connected by overlapping two pairs of polar monomers (each coming from different tile). It was conjectured that these structures are stable and proved for two very simple subclasses of the linear structures, namely for L_0 and L_1 structures. The L_0 and L_1 structures consist of an arbitrary large sequence of tiles in the shape of a straight line and the letter L, respectively. Note that although L_1 structures are still quite simple, the proof of their stability involves analysis of a large number of cases.

In this paper, we consider a rich subclass of constructible structures. The structures in the subclass are *linear* which means that every tile t_i except the first t_1 and the last t_n is attached to exactly two other tiles t_{i-1} and t_{i+1} (and the first and the last ones are attached to only one tile, t_2 and t_{n-1}, respectively). In addition, we assume that the sequence of tiles has to change direction ("bend") in every odd tile. The hydrophobic monomers of these "bending" tiles are set to be cysteines, and all other hydrophobic monomers are non-cysteines, cf. Figure 1(b). We call these structures the *snake structures* and their proteins the *snake proteins*.

2.3. *The strong HPC model*

We conjecture that the snake proteins are stable in the HPC model, and furthermore that it can be proved with techniques presented in this paper. As a preliminary result, we present the proof that the snake proteins are stable in the artificial *strong* HPC model. In this model, the energy function consists of three parts (first two are the same as in the HPC model): (i) the bond energy, (ii) the sulfide bridge energy and (iii) non-cysteine bridge energy. The last part is equal to -2 times the number of matches (pairings) in the maximum matching of the graph of potential non-cysteine bridges, where there is a potential non-cysteine bridge between any two non-consecutive adjacent non-cysteine hydrophobic monomers. Thus, the fold in Figure 1(c) had energy -9 in the strong HPC model. This energy model can be interpreted as follows: we assume that we have two types of cysteine-like hydrophobic monomers each forming bridges, but no bridges are possible between "cysteines" of different types. Furthermore, in our design we only use cysteine-like hydrophobic monomers (in bending tiles we use the first type, in non-bending tiles the second type).

3. Proof techniques

In this section we review some basic proof techniques used in this paper.

3.1. *Saturated folds*

The proteins used by Gupta *et al.*[1] in the HP model and the snake proteins in HPC or strong HPC models have a special property. The energy of their native folds is the smallest possible with respect to the numbers of hydrophobic cysteine and non-cysteine monomers contained in the proteins. We call such folds *saturated.* In saturated folds all parts of energy function produce minimum possible values. This means: (i) every hydrophobic monomer (cysteine or non-cysteine) has two contacts with other monomers; (ii) there is a sulfide bridge matching containing all or all but one cysteine monomers; and (iii) in the strong HPC model, there is a non-cysteine bridge matching containing all or all but one non-cysteine monomers. Obviously, a saturated fold of a protein must be native, and furthermore, if there is a saturated fold of a protein, then all native folds of this protein must be saturated.

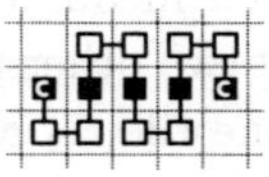

Fig. 2. Forbidden configuration in saturated fold under the strong HPC model

To illustrate the main difference between the HPC and the strong HPC models, consider a part of the fold in Figure 2 and assume that the number of non-cysteine hydrophobic monomers in the whole fold is even. In the HPC model, it is possible to extend the configuration in the figure to a complete saturated fold, while in the strong HPC model, this is not possible, as the non-cysteine hydrophobic monomers will never form a complete matching. Thus, the power of strong HPC model is in ability to faster eliminate a lot of cases, for instance, cases containing a configuration depicted in Figure 2, while in the HPC model the same proof will require a much deeper case analysis.

3.2. *2DHPSolver: a semi-automatic prover*

2DHPSolver is a tool for proving the uniqueness of a protein design in 2D square lattice under the HP, HPC or strong HPC models. 2DHPSolver is not specifically designed to analyze the snake structures or even the constructible structures. It can be used to prove the stability of any 2D HP design based on the induction on the boundaries. It starts with an initial configuration (initial field) which is given as the input to the program. In each iteration, one of the fields is replaced by all possible extensions at one point in the field specified by user. Note that in displayed fields red 1 represents a cysteine monomer, blue 1 a non-cysteine monomer and finally, uncolored 1 is hydrophobic monomer, but it is not known whether it is cysteine or not.

These extensions are one of the following type:

- extending a path (of consecutive monomers in the protein string);
- extending a 1-path (of a chain of hydrophobic monomers connected with contacts);
- coloring an uncolored H monomer.

There are 6 ways to extend a path, 3 ways to extend a one-path and 2 ways to color an uncolored H monomer. For each of these possibilities, 2DHPSolver creates a new field which is then checked to see if it violates the rules of the design. Those which do not violate the design rules will replace the original field.

However, this approach will result in producing too many fields, which makes it hard for the user to keep track of. Therefore, 2DHPSolver contains utilities to assist in automatically finding an extending sequence for a field which leads to either no valid configurations, in which case the field is automatically removed, or to only one valid configuration, in which case the field is replaced by the new more completed configuration. This process is referred to as a *self-extension*. The time required for searching for such extending sequence depends on the depth of the search, which can be specified by user through two parameters "depth" and "max-extensions". Thus, leaving the whole process of proving to 2DHPSolver by setting the parameters to high values is not practical as it could take enormous amount of time. Instead, one should set parameters to moderate values and use intuition in choosing the next extension point when 2DHPSolver is unable to automatically find self-extending sequences. Note that these parameters can be changed at any time during the use of the program by the user.

2DHPSolver is developed using C++ and its source code is freely available to all users under the GNU Public Licence (GLP). For more information on 2DHP-Solver and to obtain a copy of the source codes please visit `http://www.sfu.ca/~ahadjkho/2dhpsolver/`.

4. Stability of the snake structures

In this section we prove that the protein of any snake structure is stable. Let S be a snake structure (fold), p its protein and let F be an arbitrary native (i.e., saturated) fold of p.

Define a path in F as a sequence of vertices such that no vertex appears twice and any pair of consecutive vertices in the path are connected by peptide bonds. A cycle is a path whose start and end vertices are connected by a peptide bond.
For $i \in \{0, 1, 2\}$, an i-vertex in the fold F is a lattice vertex (square) containing a monomer i. For instance, a square containing a cysteine monomer in F is called a 2-vertex. An H-vertex is a vertex which is either 1-vertex or 2-vertex. Define a 1-path in F to be a sequence of H-vertices such that each H-vertex appears once and any pair of consecutive ones form an HH contact. A 1-cycle in F is a 1-path whose first and last vertices form an HH contact. A 1-cycle of length 4 is called a core in F.

A core c is called *monochromatic* if all its H-vertices are either cysteines or non-cysteines. Let c_1 and c_2 be two cores in F. We say, c_1 and c_2 are adjacent if there is a path of length 2 or 3 between an H-vertex of c_1 and an H-vertex of c_2. We say c_1 and c_2 are correctly aligned if they are adjacent in one of the forms in Figure 3.

In what follows we prove that every H-vertex in F belongs to a monochromatic core and the cores are correctly aligned.

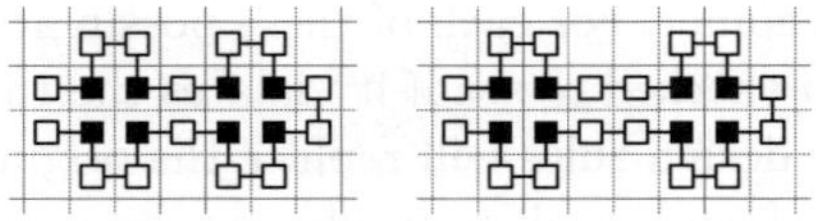

Fig. 3. Correctly aligned cores.

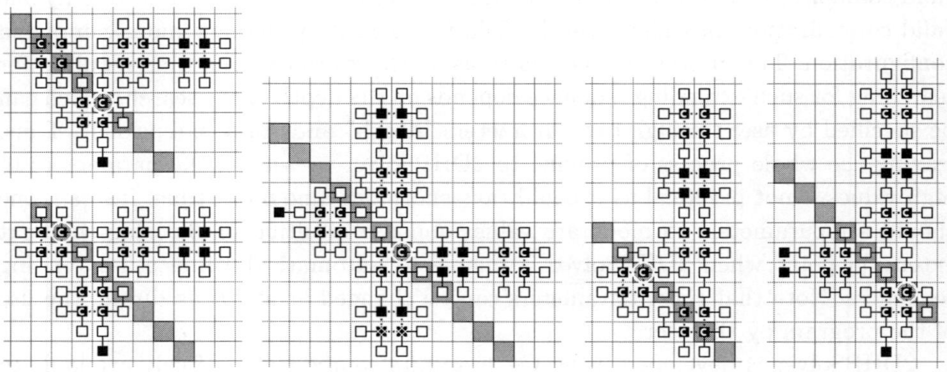

Fig. 4. Configurations with misaligned cores. The circled cysteine monomer is the one used as the starting point in induction proof by 2DHPSolver. The hatched black square depict hydrophobic monomers for which it was not yet determined whether they are cysteines or non-cysteines.

Lemma 4.1. *Every* H*-vertex in F belongs to a monochromatic core and all the cores are either correctly aligned or there is only one occurrence of one of the configurations depicted in Figure 4, in which 3 cores are not correctly aligned while others are correctly aligned.*

Proof. For any integer i, let SW_i be the set of lattice vertices $\{[x, y]; x+y = i\}$. Let m be the maximum number such that SW_i, $i < m$ does not contain any H-vertex, i.e., SW_m is a boundary of diagonal rectangle enclosing all H-vertices.

We start by proving the following claim.

Claim 4.1. *If there is an* H*-vertex w on SW_i then*

(1) w is on a monochromatic core c; and
(2) if c is adjacent to core c' which has a H*-vertex, on SW_j, $j < i$, then either c and c' are correctly aligned or one of the configurations depicted in Figure 4 occurs.*
(3) if c is adjacent to core c' which has a H*-vertex, on SW_j, $j > i$, then either c and c' are correctly aligned or one of the configurations depicted in Figure 4 occurs.*

Proof. We prove the (1) and (2) by induction on i. Note that one can prove (1) and (3) in a similar way.

For the base case, assume that w is an H-vertex on SW_m. It is enough to show that w is in a monochromatic core (case (1)). Since w lies on the boundary, this can be easily proved by short case analysis or by 2DHPSolver.

Now suppose $i > m$. Suppose none of the configuration in Figure 4 happens. By induction hypothesis, the part of the fold F that lies between SW_m and SW_{i-1} contains only correctly aligned monochromatic cores. We prove that any H-vertex w located on SW_i is on a monochromatic core c and if c is adjacent to a core c' which has a 1-vertex on $SW_{k'}$ for some $k' < i$ then c is correctly aligned to c'.

We show that if (1) and (2) does not happen for w then we see a subsequence in F which is not in p. This is done by enumerative case analysis of all possible extensions of this configuration and showing that each branch will end in a configuration has a subsequence not in p.

This process requires the analysis of many configurations which is very hard and time consuming to do manually. Therefore, we used 2DHPSolver to assist in analyzing the resulting configurations. The program generated proof of this step of the induction can be found on our website at `http://www.sfu.ca/~ahadjkho/2dhpsolver/`. Please be advised that this is a PDF document containing 2707 pages and 16543 images.

One can see that in all of the configurations depicted in Figure 4, there are 3 cysteine cores c, c' and c'' which are adjacent pairwise and contain two occurrences of the subsequence $es = (020)^4$. The subsequence es occurs exactly twice in S_n and that is in t_1 and t_n. □

Analogously the SE_i is the set of vertices $\{[x,y]; x - y = i\}$ of the lattice. We have a similar claim for an H-vertex on SE_i. In each of the configurations in Figure 4 subsequence es occurs twice. Combining the two claims completes the proof of the lemma.

Theorem 4.1. *Every H-vertex in F belongs to a monochromatic core and all the cores are correctly aligned.*

Proof. By Lemma 4.1, every H-vertex is on a core. Consider a graph G defined as follows. For every core c of F, let x_c be a vertex in G. Furthermore, two vertices x_c, and $x_{c'}$ are connected in G if and only if cores c and c' are adjacent in F. We show that G is acyclic. For the contrary, let C be a cycle in G. If all the cores corresponded to vertices of C in F are correctly aligned we get a closed subsequence of Q which is not the entire Q. Thus C contains vertex x_c which c is one of the core shown in Figure 4. Each core c in Figure 4 is adjacent to at least three other cores in F. Therefore vertex x_c has degree at least three in G. If C is of length more than three then C contains only two of the three cores in Figure 4 and all other cores of F corresponded to C' are correctly aligned. However again we get a close subsequence of Q which is not the entire Q. Thus C has only three vertices, since x_c is of degree 2 and there is only one cycle in G, there is one vertex of degree 1. Now we have three occurrence of $(020)^4$ in F, a contradiction. Therefore G is acyclic. Similarly G

has no vertex of degree more than 2 as otherwise there would be three occurrences of $(020)^4$ in F. Thus all the cores are correctly aligned and each core is adjacent to at most two other cores, except the first and the last one. Note that since there is no vertex of degree 3 in G, every core in F is adjacent to other cores in a way that cores in S are connected. Now the first core c_1 in F (c_1 is adjacent to exactly one core) is correspond to t_1 of S. By continuing the sequence of p in core c_i of F and t_i of S for $i > 1$ we see that F has the same structure as S. Thus F is unique. □

5. Conclusions

In this paper we have enriched the HP model of Dill with the third type of amino acids, cysteines, and a new interaction acting between monomers, disulfide bridges. We consider a robust subclass of constructible structures introduced by Gupta *et al.*[1] able to approximate any given shape, and refine these structures for the new HP-cysteine model. We believe that introduction of cysteine monomers into structure design improves the stability of designed structures which in turn helps in proving the stability. To formally prove that the considered structures are stable, it is necessary to consider an enormous number of cases. For that reason, we have developed semi-automated prover 2DHPSolver. Using 2DHPSolver we are able to prove stability under one additional assumption on the HPC model. We are currently working on the proof of stability without this assumption.

We conjecture that use of cysteines in the design of proteins might help to improve their stability. To verify this, we would like to extend our results to 3D lattice models and test them using existing protein folding software.

References

1. A. Gupta, J. Maňuch and L. Stacho, *Journal of Computational Biology* **12**, 1328 (2005).
2. K. A. Dill, S. Bromberg, K. Yue, K. M. Fiebig, D. P. Yee, P. D. Thomas and H. S. Chan, *Protein Science* **4**, 561 (1995).
3. K. A. Dill, *Biochemistry* **29**, 7133 (1990).
4. K. A. Dill, *Biochemistry* **24**, 1501 (1985).
5. P. Crescenzi, D. Goldman, C. Papadimitriou, A. Piccolboni and M. Yannakakis, On the complexity of protein folding, in *Proc. of STOC'98*, 1998.
6. B. Berger and T. Leighton, *J. Comp. Biol.* **5**, 27 (1998).
7. O. Aichholzer, D. Bremner, E. Demaine, H. Meijer, V. Sacristán and M. Soss, *Computational Geometry: Theory and Applications* **25**, 139 (2003).
8. Z. Li, X. Zhang and L. Chen, *Appl. Bioinformatics* **4**, 105 (2005).
9. B. Hayes, *American Scientist* **86**, 216 (1998).
10. R. Jaenicke, *Eur. J. Biochem.* **202**, 715 (1991).
11. Y. Liou, A. Tocilj, P. Davies and Z. Jia, *Nature* **406**, 322 (2000).
12. G. C. Rodakis and F. C. Kafatos, *Proc. Natl. Acad. Sci. USA* **79**, 3551 (1982).

DISCRIMINATION OF NATIVE FOLDS USING NETWORK PROPERTIES OF PROTEIN STRUCTURES

ALPER KÜÇÜKURAL

Faculty of Engineering and Natural Sciences, Sabanci University, Orhanli, Tuzla, Istanbul, Turkey

O. UĞUR SEZERMAN

Faculty of Engineering and Natural Sciences, Sabanci University, Orhanli, Tuzla, Istanbul, Turkey

AYTÜL ERÇİL

Faculty of Engineering and Natural Sciences, Sabanci University, Orhanli, Tuzla, Istanbul, Turkey

Graph theoretic properties of proteins can be used to perceive the differences between correctly folded proteins and well designed decoy sets. 3D protein structures of proteins are represented with graphs. We used two different graph representations: Delaunay tessellations of proteins and contact map graphs. Graph theoretic properties for both graph types showed high classification accuracy for protein discrimination. Fisher, linear, quadratic, neural network, and support vector classifiers were used for the classification of the protein structures. The best classifier accuracy was over 98%. Results showed that characteristic features of graph theoretic properties can be used in the detection of native folds.

1 Introduction

Proteins are the major players responsible for almost all the functions within the cell. Protein function, moreover, is mainly determined by its structure. Several experimental methods already exist to obtain the protein structure, such as x-ray crystallography and NMR. All of these methods, however, have their limitations: they are neither cost nor labor effective. Therefore, an imminent need arises for computational methods that determine protein structure which will reveal clues about the mechanism of its function. Determining the rules governing protein function will enable us to design proteins for specific function and types of interactions. [1] This course of action has vast application areas ranging from the environmental to the pharmaceutical industries. Additionally, these designed proteins should have native like protein properties to perform their function without destabilizing under physiological conditions.

There are several methods developed to find the three dimensional structure of proteins. Since these models are created by computer programs their overall structural properties may differ from those of native proteins. There is a need for distinguishing near native like structures (accurate models) from those that do not show native like structural properties. This paper aims to define a function that can distinguish the native protein structures from artificially generated non native like protein structures. The proposed function can also be used in the protein folding problem as well as domain recognition and structural alignment of proteins.

2 Methods

The evaluation function consists of two parts: the network properties of the graphs obtained from the proteins and the contact potentials. Graphs are employed to solve many problems in protein structure analysis as a representation method. [2, 3] Protein structure can be converted into a graph where the nodes represent the C_α atoms of the residues and the links between them represent interactions (or contacts) between these residues.

The two most commonly used representations of 3D structures of proteins in graph theory are contact maps and Delaunay tessellated graphs [4, 5]. Both graphs can be represented as an N×N matrix S for a protein which has N residues. If the residues are in contact the $S_{ij}=1$, otherwise $S_{ij}=0$ [6, 7]. Contact definition differs for both graphs. In contact map, if the distance between C_α atoms of residues i and j is smaller than a cut-off value then they are considered to be in contact. Several distances ranging from 6.5 A° to 8 A° have been used in the literature. 6.8 A° has been found to be a good definition of a contact between residues, therefore in our work we used 6.8 A° as the contact cut off value [5].

On the other hand, Delaunay tessellated graphs consist of partitions produced between a set of points. A point is represented by an atom position in the protein for each residue. This atom position can be chosen as α carbon, β carbon or the center of mass of the side chain. There is a certain way to connect these points by edges so as to have Delaunay simplices which form non-overlapping tetrahedrals [4]. A Delaunay tessellated graph includes the neighborhood (contact) information of these Delaunay simplices. In this work, we used Qhull program to derive the Delaunay tessellated graph of our proteins using the alpha carbon atoms as simplices [8, 21].

Several network properties of the graphs are employed to distinguish the graphs of native proteins from those obtained from artificially created near native conformations, called decoy sets. The first network property is the degree or connectivity k which is the number of edges incident of a vertex i [4]. The average degree of a protein structure is calculated by the mean of the degree distribution of the graph. If the average degree is high, this points out to a globular structure where many residues establish many contacts with each other. Unfolded proteins would have very low average degree value. Natural proteins folds are compact, and measures using the compactness of the proteins can distinguish the native folds from those of artificially generated decoy set. The second graph property is the second connectivity which is calculated by the sum of the contacts of each neighbor of a node. The second connectivity is a measure we defined that also shows the compactness of the graph. If the structure is composed of small compact domains rather than one globular structure, the structure would have high average degree but low second connectivity numbers. The attractiveness of this value is its ability to distinguish such structures.

The third graph property is the clustering coefficient which measures how well the neighbors are connected to each other, thus forming a network of contacts (clique). The clustering coefficient C for each node is calculated by

$$C_n = \frac{2E_n}{k(k-1)}$$

where En is the actual edges between the neighbors of the residue n and k is the degree. If all the neighbors of a node i are connected to each other, then they form a tight clique and the C_i value becomes 1. The clustering coefficient of the graph C is the average of all the C_n values [4, 9].

Graph properties can only capture overall structural properties of the proteins but do not measure physiochemical interactions between the atoms that are in contact in the folded form. The second part of the evaluation function uses contact potentials to capture the favorability of physicochemical interactions between the contacting residues of the folded protein. Contact potentials are statistical potentials that are calculated from experimentally known 3D structures of proteins which calculate the frequencies of occurrences of all possible contacts and convert them into energy values so that frequently occurring contacts have favorable contact scores. This method is an approximation to actual physico-chemical potentials but they have been shown to work as target energy functions on the protein folding problem [7, 8, 12, 13].

In this study, the average contact potential scores were calculated using contact potential matrix by Jernigan *et. al.* [10]. There are other contact potential matrices that are widely used as well [11], since they are highly correlated with each other, we found it sufficient to use Jernigan matrix to see the discriminative power of contact potentials in our problem. The degree, clustering coefficient, second connectivity and their moments along with Jernigan potential scores are employed as dimensions of the classification methods. Using the average values causes loss of information on the distribution of each variable; therefore we used moments to better capture the distributions of all the features.

Several classification methods are used to find out whether the graph theoretic properties can discriminate the native proteins while determining which graph representation and data classification method yields the best results.

3 Background and Related Works

Several attempts have been made to define a function to distinguish native folds from incorrectly folded proteins. In early studies, Novotny *et. al.* looked at various concepts such as solvent-exposed side-chain non-polar surface, number of buried ionizable groups, and empirical free energy functions that incorporate solvent effects for ability to discriminate between native folds and those misfolded ones in 1988 [25]. Vajda *et. al.*

used combination of hydrophobic folding energy and the internal energy of proteins which showed importance of relaxation of bond lengths and angles contributing to the internal energy terms in detection of native folds [2, 22].

McConkey *et. al.* have used contact potentials as well to distinguish native proteins. They calculated the contacts from Voronoi tessellated graphs of the native proteins and the decoy sets. They assumed a normal distribution of contact energy values and calculated the z scores to show if the native protein has a very high z-score compared to z-score of the decoy structures (or the contact energy of the native structure ranks high compared to decoy structures created for that structure). The scoring function can effectively distinguish 90% of the native structures on several decoy sets created from native protein structures [14].

Another scoring function derived by Wang *et. al.* is based on calculating distances (RMSD) between all the Cα atoms in native proteins and other conformations in given decoy sets. They show their function distinguish better than other functions depending on the quality of the decoy sets [15].

Beside the knowledge based potentials, approximate free energy potentials are also used to discriminate native proteins by Gatchel *et. al.* [15]. In their approach they defined a free energy potential that combines molecular mechanics potentials with empirical solvation and entropic terms. Their free energy potential's discrimination power improved when the internal energy of the structure was added to the solvation energy. [16]

The hydrophobic effect on protein folding and its importance to discrimination of proteins is also stated by Fain *et. al.* Their approach is based on discovering optimal hydrophobic potentials for this specific problem, by using different optimization methods. [17]

Using graph properties to distinguish native folds was first done by Taylor *et. al.* They state that using degree, clustering coefficient, and the average path length information can help distinguish native proteins. They determine a short list based on these properties. The natives' appearance in the short list indicates that these properties can distinguish the native like structures. Of 43 structures set in which they worked, the native was placed in the short list in 27 of them. [4]

All of the previous works do not treat the problem as a classification problem; they only check whether the native structure ranks high according to their scoring scheme. Several classification and clustering methods such as neural network based approaches and support vector machines have been widely used in other successful applications related to protein structure. The success of the classification depends on the features that are used to discriminate the classes [7, 18, 19].

In this paper we use combination of contact potentials (to capture the physicochemical interactions between the contacting residues that are formed upon folding) and network properties of the graph (which shows compactness of the structure). Using these values as the feature vectors, we used several classification methods to distinguish native and decoy protein classes.

4 Dataset

The first data set employed in the experiments, which is from PISCES database[20], has 1364 non-homologous proteins, and their resolution < 2.2Å, crystallographic R factor < 0.23, and maximum pair wise sequence identity < 30%. The second data set consists of 1364 artificially generated and well designed decoy set; the third one is 101 artificially generated straight helices. Decoy sets are generated by randomly locating C_α atoms at about 3.83A° distance while avoiding the self-intersection of C_α atoms and keeping the globular structure approximately at the same size and shape of an average protein [4]. Further details of decoy set generation stage can be found in the article of Wang *et. al.* [26].

The feature values in the data set possessed large variations in some cases. Therefore, to see the impact of outliers in classification accuracy, we performed a simple outlier analysis technique based on the elimination of all the values that are three standard deviations away from the mean for the given data set. Approximately 9% of the data was eliminated for each dataset.

5 Results

Average degree, clustering coefficient, second connectivity are used as structural features. Besides the averages for the properties, moments of the probability distributions were calculated for each property such as standard deviation, skewness and kurtosis of the distributions whereas skewness measures the asymmetry of the distribution and kurtosis measures the "peakedness" of the distribution. Average Jernigan potential scores are given as sequence dependent energy features. These features are supplied as input vector to several classification methods in PRTools [19]. We first tested which graph representation method is more suitable for the given problem. The results from Delaunay tessellated graphs and contact map results are given in Table 1. The contact map had much better prediction accuracy since it captures actual compactness information of the protein structure. In some cases, tessellated graphs may represent the distant residues as if they are in close contact; this representation may be the reason for the difference in classification accuracy.

We randomly selected half of the data five times and performed a five fold cross validation on each data set to reduce to run time for the classifiers especially for the support vector classifier. The classification accuracy and two standard deviation neighborhood of these values are shown in the tables.

Table 1. indicates that the best classification accuracy was obtained from normal density based quadratic classifier (qdc) [19]. Even though some of the other classifiers performed very close to the qdc, we proceeded to focus on qdc for the rest of the paper. Table 1. also shows that outlier analysis improved the results by a minimum of 1 % independent of the classification method used.

We optimized the SVM results using kernel parameters (σ) and regularization parameters (C) for each of the kernel function separately. Changing the regularization parameter (C) did not affect classification error rates. After parameter optimization the best results from SVM were obtained when the polynomial kernel was used with while σ was 2.

Table 1. Classification accuracy table using all the features including the moment values

Classifier	Contact Maps After OA	Contact Maps Before OA	Delaunay Tes. After OA	Delaunay Tes. Before OA
Support vector class.	98.02%± 0.44	96.47%± 0.93	94.78%± 1.62	93.56%± 1.12
Norm. dens. based linear	98.72%± 0.53	97.12%± 1.02	94.85%± 1.67	93.41%± 0.94
Norm. dens. based quad.	98.87%± 0.49	98.08%± 1.32	94.81%± 1.20	92.91%± 0.52
Binary decision tree	95.61%± 1.97	94.04%± 1.88	85.77%± 2.01	82.23%± 4.17
Quadratic classifier	98.54%± 0.71	98.11%± 0.88	94.97%± 1.13	93.51%± 0.74
Linear perceptron	95.28%± 1.56	93.98%± 1.13	50.46%±10.81	54.46%± 8.53
Random neural network	96.76%± 0.76	95.40%± 1.72	88.81%± 2.27	86.10%± 2.13
k-nearest neighbor (k=3)	97.67%± 1.26	95.93%± 0.98	85.06%± 0.82	83.95%± 2.32
Parzen classifier	97.04%± 0.86	95.25%± 1.12	85.89%± 2.43	84.51%± 2.94
Parzen density based	98.59%± 0.56	97.12%± 1.77	88.62%± 3.08	86.66%± 2.71
Naive Bayes classifier	96.24%± 1.77	95.17%± 1.11	87.70%± 2.14	82.99%± 1.92
Normal densities based	96.86%± 1.67	96.35%± 1.56	89.88%± 1.37	86.04%± 2.39
Subspace classifier	93.85%± 2.96	93.93%± 1.56	85.52%± 2.82	82.18%± 1.24
Scaled nearest mean	96.26%± 1.22	96.41%± 1.36	89.20%± 1.23	86.35%± 1.37
Nearest mean	83.84%± 2.35	84.23%± 3.02	74.78%±10.72	69.39%±17.02

Different combinations of features are used in normal density based quadratic classifier to discover the effect of these features on classification accuracy and some of the results are summarized in Table 2. When we use degree, clustering coefficient, second connectivity, and contact potential score together, classification accuracy is close to 99%. Even without contact potential score, the method had 98.13% (kCS) prediction accuracy using only the graph properties after outlier analysis. Use of Jernigan contact potentials only decreased the classification accuracy drastically to 51.77%.

Table 2. Classification accuracy rates for different combination of properties with moments. (k: Degree. C: Clustering coefficient. S: Second Connectivity. . J: Profile Score from Jernigan *et. al.*. OA: Outlier Analysis)

	Contact Maps After OA	Contact Maps Before OA	Delaunay Tes. After OA	Delaunay Tes. Before OA
kCSJ	98.87%± 0.25	98.08%± 0.66	94.81%± 0.60	92.91%± 0.26
CSJ	98.95%± 0.28	97.82%± 0.41	94.60%± 1.18	91.13%± 1.06
SJ	98.15%± 0.25	98.22%± 0.16	89.53%± 0.93	88.36%± 0.48
kC	98.72%± 0.17	97.26%± 0.34	94.72%± 0.32	92.01%± 0.86
k	96.74%± 0.41	96.27%± 0.74	88.68%± 1.21	87.23%± 0.90
kCS	98.13%± 0.60	97.60%± 0.10	94.19%± 1.26	92.12%± 1.17
kS	96.93%± 0.81	95.73%± 0.86	90.43%± 0.74	87.80%± 1.08
J	51.77%± 0.23	48.53%± 0.62	47.71%± 0.84	44.45%± 1.12

Structural properties have more discriminating power, using the degree (k) distribution only we could accurately classify the native and non native structures with 96.74% accuracy. Addition of second connectivity information did not improve the accuracy much. Cliquishness (C) along with degree (k) distribution improved the classification accuracy to 98.72%. Using only the degree and the second connectivity resulted in 96.93% classification accuracy.

6 Conclusion and Discussion

The difference of this study from previous studies can be summarized in four points:

- Using contact maps to derive the structural properties of the proteins yielded much better results than tessellated graphs.
- Combining structural and physicochemical features distinguished the native folds.
- Graph properties have much more discriminative power than the contact potentials.
- Representing the problem as a classification problem, testing the success rate of several classification methods, and building an optimized predictor that can predict native folds about 99 % accuracy.

Classification using the contact potentials only resulted in 51% five fold cross validation accuracy using the quadratic classifier. Thus it is apparent that the structural features are necessary for accurate prediction. As can be seen from the results additional contribution to the prediction accuracy from contact potentials was assumed at less than 1%. Even the non native structures can create favorable interactions between contacting residues so the contact potentials alone are not sufficient to distinguish native structures.

Important structural features were the degree and the clustering coefficient. The second connectivity did not contribute much to the classification accuracy since it is highly correlated to the degree. Previous works focused on the eligibility of different kinds of potentials in discrimination of native folds; this work indicates that structural properties are more important features and, furthermore, these properties can be employed for other problems related to protein structure. This work also shows that contact map provides a better representation of protein structure.

One drawback of our method is all the features that are used in a way capture different aspects of compactness of the protein structure. Our function might fail when trying to identify natively unfolded proteins from random generated counterparts. Since an important feature in the discrimination process is compactness of structure, the method would rule out disordered regions as decoy sets, even though this disorder is a characteristic feature of native states and is functional as well (eg: calcineurin) Such proteins constitute a small subset of all the known protein structures and out of the scope of the proposed work. In addition to this, if decoy sets are generated from naturally unfolded proteins, the native proteins would have more contacts than the artificially generated structures of these native proteins and therefore these naturally unfolded proteins could be captured by our function [23]. This needs to be explored further in a future study.

Another application of our function is to distinguish bad models from good ones (computer generated structures) for protein structure prediction competitions (CASP) [24]. As a preliminary study, we tested the method on CASP VI data set of 59 proteins and 28956 model predictions. Our method correctly assigned 58 proteins as native and 6118 model structures as non native. The predicted non native structures had more than 12 A° root mean square deviation (rmsd) from the crystal structure. The non native structures assigned as native had much smaller rmsd to the corresponding crystal structures. This shows that the graph properties can easily filter out the bad models. We are currently working on finding a function using graph properties that can measure closeness of the prediction to the crystal structure on CASP VII data sets and compare it with other ranking methods.

References

1. Baker. D.: Prediction and design of macromolecular structures and interactions. *Philos. Trans. R. Soc. Lond. B. Biol. Sci.* **361** (2006) 459-463
2. Strogatz S.H.: Exploring Complex Networks. Nature **410** (2001) 268-276
3. Albert. R. and Barabasi. A.-L.: Statistical mechanics of complex networks. *Rev. Mod. Phys.* **74** (2002) 47–97
4. Taylor T. Vaisman I.I.: Graph theoretic properties of networks formed by the Delaunay tessellation of protein structures. *Phys. Rev. E. Stat. Nonlin. Soft. Matter Phys.* **73** (2006) 041925
5. A. R. Atilgan. P. Akan. C. Baysal: Small-World Communication of Residues and Significance for Protein Dynamics. *Biophys. J.* **86** (2004) 85-91
6. Vendruscolo. M.. E. Kussel. and E. Domany: Recovery of Protein Structure from Contact Maps. *Structure Fold. Des.* **2** (1997) 295-306.
7. Fariselli. P. and R. Casadio: A Neural Network Based predictor of Residue Contacts in Proteins. *Protein Eng.* **9** (1996) 941-948.
8. Soyer. A.. J. Chomiller. J.-P. Mornon. R. Jullien. and J.-F. Sadoc: Voronoi Tesselation Reveals the Condensed Matter Character of Folded Proteins. *Phys. Rev. Lett.* **85** (2000) 3532-3535.
9. Vendruscolo. M.. N. V. Dokholyan. E. Paci. and M. Karplus: Small-world view of the amino acids that play a key role in protein folding. *Phys. Rev.* **65** (2002) 061910
10. Miyazawa. S.. and R. L. Jernigan: Residue-residue potentials with a favorable contact pair term and an unfavorable high packing density term. for simulation and threading. *J. Mol. Biol.* **256** (1996) 623–644
11. Liang. J. and K.A. Dill: Are proteins Well-Packed? *Biophys. J.* **81** (2001) 751-766
12. Lazaridis. T. and Karplus. M.: Effective energy functions for protein structure prediction. *Curr. Opin. Struct. Biol.* **10** (2000) 139–145
13. Bonneau. R. and Baker. D.: Ab Initio Protein Structure Prediction: Progress and Prospects. *Annu. Rev. Biophys. Biomol. Struct.* **30** (2001) 173–189
14. McConkey. B.J.. Sobolev. V.. and Eldman. M.: Discrimination of native protein structures using atom–atom contact scoring. *Proc. Natl. Acad. Sci.* **100** (2003) 3215–3220
15. Wang K.. Fain B.. Levitt M.. Samudrala R.: Improved protein structure selection using decoy-dependent discriminatory functions. *BMC Struct. Biol.* **4** (2004) 296
16. Gatchell D. Dennis S. and Vajda S.: Discrimination of near-native protein structures from misfolded models by empirical free energy functions. *Proteins* **41** (2000) 518-534
17. Fain B.. Xia H. And Levitt M.: Design of an optimal Chebyshev-expanded discrimination function for globular proteins. *Protein Sci.* **11** (2002) 2010–2021

18. Zhao. Y.. Karypis. G.: Prediction of contact maps using support vector machines. Proceedings of the IEEE Symposium on BioInformatics and BioEngineering. *IEEE Computer Society* (2003) 26- 33
19. Ferdi van der Heijden. Robert P.W. Duin. Dick de Ridder and David M.J. Tax. John Wiley & Sons: Classification. parameter estimation and state estimation - an engineering approach using Matlab. ISBN 0470090138 (2004)
20. G. Wang and R. L. Dunbrack. Jr.: PISCES: a protein sequence culling server. *Bioinformatics* **19** (2003) 1589-1591
21. C. B. Barber. D. P. Dobkin. and H. Huhdanpaa: The quickhull algorithm for convex hulls. *ACM Trans. Math. Softw.* **22** (1996) 469-483
22. Vajda S, Jafri MS, Sezerman OU, DeLisi C.: Necessary conditions for avoiding incorrect polypeptide folds in conformational search by energy minimization. *Biopolymers* **33** (1993)173-192
23. Uversky VN, Gillespie JR, and Fink AL. Why are "natively unfolded" proteins unstructured under physiologic conditions? *Proteins* **2000**; 41:415-427
24. P.E. Bourne, CASP and CAFASP experiments and their findings, *Methods Biochem Anal* **44** (2003), pp. 501–507.
25. Novotny J, Rashin AA, Bruccoleri RE. 1988. Criteria that discriminate between native proteins and incorrectly folded models. *Proteins Struct. Fund. Genet.* **4**:19-30.
26. B. Park and M. Levitt, Energy Functions that Discriminate X-ray and Near-native Folds from Well-constructed Decoys, *J. Mol. Biol.* **258**, 367, 1996.

INTERACTING AMINO ACID PREFERENCES OF 3D PATTERN PAIRS AT THE BINDING SITES OF TRANSIENT AND OBLIGATE PROTEIN COMPLEXES

SURYANI LUKMAN and KELVIN SIM

Institute for Infocomm Research,
21 Heng Mui Keng Terrace, Singapore 119613
E-mail: slukman, shsim@i2r.a-star.edu.sg

JINYAN LI

School of Computer Engineering,
Nanyang Technological University, Singapore 639798
E-mail: JYLi@ntu.edu.sg

YI-PING PHOEBE CHEN

Faculty of Science and Technology,
Deakin University, Australia
E-mail: phoebe@deakin.edu.au

To assess the physico-chemical characteristics of protein-protein interactions, protein sequences and overall structural folds have been analyzed previously. To highlight this, discovery and examination of amino acid patterns at the binding sites defined by structural proximity in 3-dimensional (3D) space are essential. In this paper, we investigate the interacting preferences of *3D pattern pairs* discovered separately in transient and obligate protein complexes. These 3D pattern pairs are not necessarily sequence-consecutive, but each residue in two groups of amino acids from two proteins in a complex is within certain Å threshold to most residues in the other group. We develop an algorithm called *AA-pairs* by which every pair of interacting proteins is represented as a bipartite graph, and it discovers all *maximal quasi-bicliques* from every bipartite graph to form our 3D pattern pairs. From 112 and 2533 highly conserved 3D pattern pairs discovered in the transient and obligate complexes respectively, we observe that Ala and Leu is the highest occuring amino acid in interacting 3D patterns of transient (20.91%) and obligate (33.82%) complexes respectively. From the study on the dipeptide composition on each side of interacting 3D pattern pairs, dipeptides Ala-Ala and Ala-Leu are popular in 3D patterns of both transient and obligate complexes. The interactions between amino acids with large hydrophobicity difference are present more in the transient than in the obligate complexes. On contrary, in obligate complexes, interactions between hydrophobic residues account for the top 5 most occuring amino acid pairings.

Keywords: bipartite graph; amino acid preferences; pattern pairs; transient complexes; obligate complexes

1. Introduction

Amino acid interactions are fundamental for protein-protein interactions. These protein-protein interactions are important in all facets of life, from metabolism to disease fighting. Since different amino acids possess distinct functional groups, a preliminary step towards understanding protein-protein interactions is to analyze the pairing preferences of amino acids at the binding sites of distinct protein complexes. Previous studies to determine interacting amino acid preferences are of various conclusions. Some studies report that the amino acid composition of interfaces of distinct protein complexes are similar,[7,9] whereas others report significant differences.[5,10,11] Furthermore, some groups discovered that polar and charged residues are the major contributors in protein-protein interactions,[14,18] whereas other reported that hydrophobic interactions are favoured.[19] Therefore, we introduce the concept of interacting 3D pattern pairs, defined by spatial proximity, to understand the interacting preferences of conserved amino acids involved in distinct protein complexes. We focus our study on two types of protein complexes: transient and obligate.

Individual proteins are capable of adopting their native fold as monomers and they may interact transiently upon a molecular stimulus, to fulfill a particular function and dissociate after that. This type of protein complexes is termed as transient complex. Another type of protein complexes is obligate complex, in which the protein chains remain bounded to each other throughout their functional lifetime. These proteins may not fold and function properly when they are unbound. A classical example of obligate complex is the interaction between the β- and γ- subunits of heterotrimeric G proteins, whereas the α-subunit forms a transient interaction with the β- and γ-subunits. Since the transient and obligate interactions are characterized by distinct physico-chemical characteristics,[16] it is crucial to distinguish between these two kinds of interactions when analyzing the interacting preferences of amino acid pattern pairs.

We propose a graph-based approach to discover interacting 3D pattern pairs efficiently. We represent a pair of interacting protein chains using a *bipartite* graph[6] based on their residues' 3D-coordinate distance information. We discover *maximal quasi-biclique* subgraphs[15] from every bipartite graph. We then mine across the maximal quasi-bicliques to obtain quasi-bicliques that are significantly frequent and large, and each of them corresponds to a 3D pattern pair. We choose to discover maximal quasi-bicliques, instead of the classical *maximal bicliques*[6] because maximal quasi-bicliques are more tolerant to missing data.

2. *AAPAIRS*: Our algorithm to discover 3D pattern pairs

We present an algorithm to discover **A**mino **A**cid 3D pattern **pairs**, called *AApairs* (Algorithm 2.1). A preprocessing step of AApairs is to classify a pair of interacting polypeptide chains into one of three different classes: crystal packing, transient or obligate complex. Crystal packing is excluded from our further consideration. The first step of our *AApairs* algorithm finds a special type of subgraph—maximal quasi-biclique[15]—from a pair of interacting polypeptide chains which we represent as a bipartite graph in our implementation. At the second step, *closed patterns*[8,13] across the maximal quasi-biclique subgraphs are detected. If a pair of closed patterns can form a quasi-biclique subgraph, and the pair occurs frequently in many pairs of interacting proteins, then we call such a pair a 3D pattern pair, which is of our interest. Our experiments were conducted in Windows XP environment, using Intel Xeon CPU 3.4GHz with 2GB RAM. *AApairs* was implemented in C++.

2.1. *Classifying pairs of interacting proteins into obligate or transient complexes*

This preprocessing step deals with all X-ray crystallographic protein structures having resolution better than 2.5Å from the Protein Data Bank (PDB, `http://www.rcsb.org`). We do not consider Nuclear Magnetic Resonance (NMR)-determined protein structures and any of the nucleic acids in PDB. We consider only PDB entries with two or more polypeptide chains. Given such an entry, we use the NOXclass method,[20] a support-vector machine prediction method, to remove biologically-irrelevant crystal packing interaction between any two polypeptides, and then to classify those remaining biologically-relevant interactions into either transient or obligate interactions. Using interface properties such as interface area, ratio of interface area to protein surface area, and amino acid composition of the interface as input, NOXclass was reported to be highly accurate, achieving an accuracy of 91.8% for the classification of those interactions.[20]

2.2. *AApair (step one): Discovering maximal quasi-biclique subgraphs from transient and obligate complexes*

We represent a pair of transient or obligate polypeptide chains as an undirected bipartite graph. An undirected bipartite graph G is a graph consisting of two disjoint vertex sets V_1 and V_2 satisfying the condition that there is no edge between any two vertices within V_1 or within V_2. Such a graph is usually denoted as $G = (V_1, V_2, E)$ where E is the set of edges of G. A polypetide chain can be mathematically represented as a set of amino acid residues (with location information). Thus, to transform a pair of polypeptide chains into a bipartite graph, we just represent every residue as a vertex, and we assign an edge between a residue x_1 in one chain and a residue x_2 in the other chain if and only if there exists at least a pair of atoms between x_1 and x_2 whose distance is less than a threshold. In this study, we use a threshold of 5.0Å.[17]

After constructing a bipartite graph $G = (V_1, V_2, E)$ representing a pair of interacting polypetide chains, we discover the complete set of maximal quasi-biclique subgraphs from G. A *quasi-biclique* H of G is a subgraph consisting of two sets of vertices $X_1 \subseteq V_1$ and $X_2 \subseteq V_2$ such that every vertex in X_i, $i = 1, 2$, is adjacent to at least $|X_j| - \epsilon$, $j \neq i$, vertices in X_j. The tolerance rate ϵ is a small integer number, e.g. 1 or 2, defined by users. A quasi-biclique subgraph H is maximal in G if there is no other quasi-biclique in G that contains H. We use our $CompleteQB$ algorithm[15] to discover maximal quasi-bicliques. The $CompleteQB$ algorithm has another user input parameter ms—the minimum number of vertices in each side of a maximal quasi-biclique. That is, only those maximal quasi-bicliques, whose vertex set size is at least ms, are enumerated. Therefore, by mining maximal quasi-biclique subgraphs, we can discover pairs of closely interacting residues from a pair of interacting polypeptide chains. We note that the residues in one side of a maximal quasi-biclique are not necessarily consecutive in one chain.

The above procedure is performed for all possible pairs of interacting (transient or obligate) polypetide chains within a PDB entry. Thus, after going through all PDB entries, we can obtain many maximal quasi-bicliques representing pairs of closely interacting residues.

2.3. *AApair (step two): Identifying significant 3D pattern pairs*

As some maximal quasi-bicliques may occur in interacting polypetide chains by chance, we identify those quasi-bicliques that occur in PDB entries with a high frequency. Let m be the number of all pairs of interacting polypetide chains from all PDB entries used in this study, and let CHAINPAIRS represents all these pairs. We then denote CHAINPAIRS as $\{chainPair^{(i)} \mid i = 1, 2, \cdots, m\}$, where $chainPair^{(i)} = (C_1^{(i)}, C_2^{(i)})$, $C_1^{(i)}$ or $C_2^{(i)}$ represents the set of amino acid residues in one of the two chains. Let n be the number of all maximal quasi-bicliques discovered from $\{chainPair^{(i)} \mid i = 1, 2, \cdots, m\}$, after transforming every $chainPair^{(i)} = (C_1^{(i)}, C_2^{(i)})$ into a bipartite graph $G^{(i)} = (V_1^{(i)}, V_2^{(i)}, E^{(i)})$. A maximal quasi-biclique is denoted as $H^{(j)} = (X_1^{(j)}, X_2^{(j)})$, $j = 1, 2, \cdots, n$. We then discover frequent closed patterns from $segmentDB$ to construct our desired patterns, where

$$segmentDB = \{X_1^{(1)}, X_2^{(1)}, X_1^{(2)}, X_2^{(2)}, \cdots, X_1^{(n)}, X_2^{(n)}\}$$

A closed pattern is the maximal pattern among patterns that occur in the same set of objects. For example, abc is the closed pattern occurring in $abcde$ and $abcfg$, but ab is not. Suppose there are k number of closed patterns of $segmentDB$, denoted as $\{P_1, P_2, \cdots, P_k\}$.We then pairwise $P_1, P_2, \cdots$, and P_k, and for every pair (P_u, P_v), we go through $\{chainPair^{(i)} \mid i = 1, 2, \cdots, m\}$ to count the number of them containing (P_u, P_v). If the number exceeds a pre-defined threshold sup, then (P_u, P_v) is a significant interacting 3D pattern pair.

Formally,

Definition 2.1. (Interacting 3D pattern pair). A pair of closed patterns P and Q forms a 3D pattern pair (P, Q) if and only if

- $|P| \geq ms$ and $|Q| \geq ms$, as specified in Step one;
- The occurrence number in CHAINPAIRS exceeds sup, as specified in Step two;
- P and Q can form a quasi-biclique.

3. Results

From 17,630 X-ray crystallographic protein complexes data in the Protein Data Bank (PDB, `http://www.rcsb.org`), we collect 4,661 transient and 7,985 obligate interactions. Only polypeptide chains containing >30 amino acids are considered in our analysis. Similar sequences at 90% identity were removed. To ensure that our results are supported with experimental evidence, we consider only interactions between two polypeptide chains found in a single PDB entry.

Our *AApair* algorithm allows users to set three parameter values prior to discovering interacting 3D pattern pairs. The three parameters are: the minimum number of residues in one side of pattern pairs (ms), the minimum occurence in the pairs of interacting proteins (sup), and the error tolerance rate (ϵ) of the maximal quasi-bicliques.

By varying the three parameters, we obtain different numbers of 3D pattern pairs. We observe that ms set to 3 or 4 is ideal because the average number of residues in one side

Algorithm 2.1 Algorithm $AApairs$

Input:

Set of pairs of interacting polypeptide chains, $ppDB$
ms is the minimum size threshold; ϵ is the error tolerance rate;
sup is the minimum occurrence;

Description:

use $NOXclass$ to classify all pairs of interacting polypeptide chains in $ppDB$. Let the set of qualified pairs of interacting polypeptide chains be CHAINPAIRS $= \{chainPair^{(i)}, i = 1, 2, \ldots, m\}$;
convert $\{chainPair^{(i)}, i = 1, 2, \ldots, m\}$ into a set of bipartite graphs $\{G^{(i)}, i = 1, 2, \ldots, m\}$;
use $CompleteQB$ to mine maximal quasi-biclique subgraphs from every $G^{(i)}$. Let the set of maximal quasi-biclique subgraphs be $\{H^{(j)}, j = 1, 2, \ldots, n\}$, where $H^{(j)} = (X_1^{(j)}, X_2^{(j)})$;
$segmentDB = \{X_1^{(1)}, X_2^{(1)}, X_1^{(2)}, X_2^{(2)}, \cdots, X_1^{(n)}, X_2^{(n)}\}$
use $MineLMBC$ on $segmentDB$ to mine a set of closed patterns $\{P_1, P_2, \cdots, P_k\}$;
for all $P_i, P_j \in \{P_1, P_2, \cdots, P_k\}$ **do**
  $count = 0$;
  for all $chainPair^{(x)} \in$ CHAINPAIRS **do**
    if P_i and $P_j \in chainPair^{(x)}$ **then** $count$++;
  if $count \geq sup$ **then** (P_i, P_j) forms a 3D pattern pair;
output all 3D pattern pairs;

of the pattern pairs mined by maximal quasi-bicliques are 3.60 and 5.14 for transient and obligate complexes respectively. When ϵ is set to 1, ms set to 3, sup set to 100, we discover 112 and 2,533 3D pattern pairs from the transient and obligate interaction datasets respectively. When ϵ is set to 0, while the two other parameters remain the same, only one 3D pattern pair is discovered each from the transient and obligate datasets.

The introduction of error tolerance rate in maximal-quasi bicliques, allows reasonable numbers of 3D pattern pairs to be found across different sup settings, up to sup=100. This also solves the challenges faced by maximal biclique technique in which very few or even zero 3D pattern pairs can be found when sup is high. Mining maximal quasi-bicliques is appropriate as not all existing structural data are complete. Even in a complete data of protein complexes, we cannot expect that the inter-protein residual interactions have a perfect all-versus-all relationship as represented by maximal biclique subgraphs. In addition, the ϵ parameter also accommodates the rapidly growing structural data.

The statistic of the numbers of 3D pattern pairs discovered by varying the parameters is reported in `http://research.i2r.a-star.edu.sg/AApairs`. A note of caution for setting the parameters is essential as parameters set too low (e.g. sup) may result in very big numbers of 3D pattern pairs.

3.1. *Amino acid distributions on each side of 3D pattern pairs*

We consider average amino acid composition in percentage for the complete database in release 53.0 of UniProtKB/Swiss-Prot (`http://cn.expasy.org/sprot/relnotes/`). If the percentage of a particular amino acid in our 3D pattern pairs is much greater than its percentage in Swiss-Prot database, the amino acid is likely to play important role in protein-protein interactions. Amino acids such as Leu, Ala, and Gly have high compositions in protein sequences, whereas amino acids such as Cys and Trp have low compositions. We study the amino acid distributions in 3D patterns found within both transient and obligate complexes and compare them with the amino acid distributions in protein sequences (Figure 1). Leu is the highest occuring amino acid in interacting 3D patterns of obligate complexes, accounting for 33.82% as compared to 9.66% in protein sequences. Ala, a less hydrophobic amino acid than Leu, is the highest and second highest occuring amino acid in interacting 3D patterns of transient (20.91%) and obligate complexes (16.78%) respectively. Ala was reported to have high α helix-forming tendency and two α helices can wrap around one another to form a coiled-coil as a mechanism for protein-protein interaction.[1] Though Ala side chain is very non-reactive, it can play a role in substrate recognition or specificity.[3]

Except for Ala whose presence in 3D patterns of transient complexes shows a significant increase, the frequencies of hydrophobic amino acids (Ile, Val, Leu, Phe, and Met) in 3D patterns of transient complexes are generally lower than those in overall protein sequences. On the other hand, more polar amino acids, especially the charged ones (His, Glu, Asp, Lys, Arg) occur less frequently in interacting 3D patterns of obligate complexes. Only 0.06% of amino acids in 3D patterns of obligate complexes are His. For the other two positively-charged residues, Arg occurs more often than Lys does in the 3D patttern pairs of both transient (6.06% versus 0.89%) and obligate (2.16% versus 0.33%) complexes. This is in agreement with previous studies.[12]

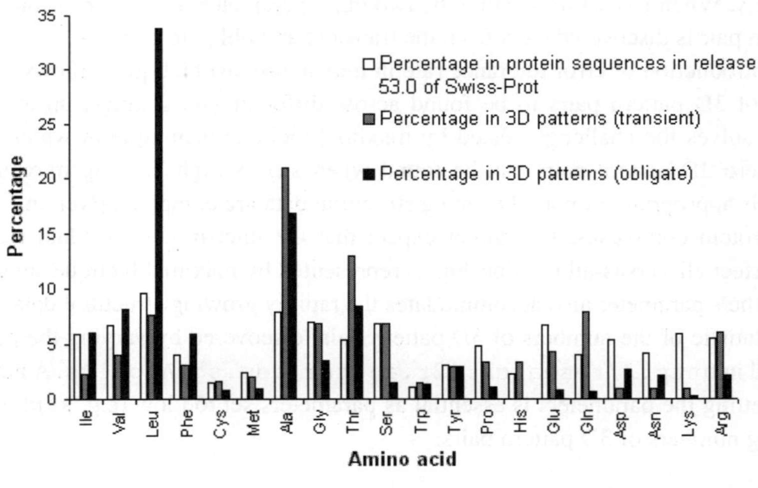

Fig. 1. The comparison of amino acid distributions among protein sequences, 3D patterns in transient and those in obligate protein complexes. The amino acid residues are ordered according to their hydrophobicity, with Ile as the most hydrophobic and Arg as the least hydrophobic.

We also calculate the frequencies of all 210 possible dipeptides in each side of interacting 3D pattern pairs of both transient and obligate complexes. As we exclude sequence-consecutiveness, we consider dipeptide AT and TA as the same, and any occurence of TA is added to the number of occurence of AT. We present the top 10 dipeptides to compare the

3D patterns on the interfaces of the two distinct protein-protein interactions (Table 1). Only AA and AL are within the top 10 dipeptides of 3D patterns of both transient and obligate complexes. The transient dipeptides contain more combinations of hydrophobic and polar amino acids (AT, AQ, GT, AS, GS, and AE) whereas such combinations are less observed in the obligate dipeptides, in which only LT and DL are highly present.

Table 1. The list of the top 10 highest occuring dipeptides in 3D patterns of transient and obligate protein-protein interactions.

Dipeptide	Transient (%)	Dipeptide	Obligate (%)
AT	6.63	LL	9.59
AQ	6.51	AL	8.84
AA	4.25	LT	7.73
TT	4.19	LV	6.34
GT	4.04	FL	6.26
AL	3.71	IL	5.74
AS	2.89	AA	4.20
GS	2.20	AV	3.91
AE	2.13	DL	3.16
AG	1.90	GL	2.38

Table 2. The distributions of dipeptides comprising charged residues in 3D patterns of transient and obligate complexes.

Dipeptide containing residues of same charges	Transient (%)	Obligate (%)	Dipeptide containing residues of opposite charges	Transient (%)	Obligate (%)
DD	0	0	DH	0	1.37
DE	0	1.58	DK	0	1.38
EE	18.47	0	DR	15.36	23.31
HH	0	0.69	EH	3.50	4.19
HK	13.45	0	EK	11.05	0.70
HR	28.09	1.97	ER	6.51	11.52
KK	0	0			
KR	3.56	2.76			
RR	0	50.53			

Note: The percentage of each dipeptide containing charged side chains is calculated from the 3D patterns found separately in the transient and obligate complexes.

The clustering of neighboring polar amino acid side chains can alter their reactivity.[1] For example, if a number of negatively charged side chains are forced together against their mutual repulsion in a particular site by the way of protein folding, the affinity of this site for a positively charged ion is greatly increases. We investigate if such phenomenon is present in our 3D pattern pairs (Table 2). We observe that 50.53% of contributing charged dipeptides in 3D patterns of obligate complexes is RR, which seems to support the hypothesis of clustering of same charged amino acids. However, caution has to be exercised while interpreting this result as Arg is one of the three abundant hotspot residues in binding energy for protein interfaces.[4] To support this, DR and ER dipeptides are also highly present in the 3D patterns of obligate complexes.

In 3D patterns of transient complexes, dipeptides HR and EE are highly present. The high percentage of EE in 3D patterns of transient complexes compared to those in obligate ones, suggests the role of EE as characteristic dipeptide of 3D patterns in transient complexes.

3.2. *Amino acid pairing preferences of 3D pattern pairs*

To reveal residue pairing preferences across various protein-protein interactions, we study the pairing preferences of amino acids in our 3D pattern pairs (Figure 2). In transient complexes, the interactions between residues of large hydrophobic difference are observed more often than those in obligate complexes. The pairings between Gly and Thr, Ala and Ser, and Ala and Glu (4.85%, 4.47%, and 3.90% respectively) are among the top 10 pairing preferences of interacting 3D pattern pairs of transient complexes. On contrary, no pairings between Gly and Thr, Ala and Glu, and only 1.14% of pairing is observed for Ala and Ser.

In obligate complexes, there is a high occurrence of interactions between identical amino acids. The interactions between all identical amino acids, especially the hydrophobic residues, account for 29.48% of all possible residual interactions. Interactions between hydrophobic residues, such as Ile, Val, Leu, and Ala, occur much more often than those between polar residues. In particular, interactions involving Leu are highly common. The pairings between Leu and Leu, Ala and Leu, Leu and Thr, Leu and Val, which are 14.6%, 6.92%, 6.56% and 4.99% respectively, are among the top 5 pairing preferences of interacting 3D pattern pairs of obligate complexes.

Covalent interactions such as the Cys-Cys disulphide bonds are also observed, though they are uncommon. Only six Cys-Cys pairings are present in 3D pattern pairs of obligate complexes. The sulphur atoms of the two Cys residues from the interacting proteins form a disulphide bond if they are at most 2.0Å apart. Since disulphide bonds are more rigid and stable than ionic and van der Waals interactions, it is not surprising to detect such interactions only in 3D pattern pairs of obligate complexes, but not in those of transient ones.

4. Discussion

To draw veritable observations and knowledge from the available structural data, it is essential to analyze as many protein-protein interactions as possible. Our study addresses this requirement by analyzing 12,646 interactions, which is further classified into either transient or obligate interactions. Inclusion of large dataset allows for distinguishing transient and obligate complexes, which cannot be achieved by most studies using small datasets.[12]

3D pattern pairs can be used as building blocks for the model building of protein complexes in crystallography. They are also applicable to automated structure prediction and determination of protein complexes based on crystallography data. Furthermore, 3D pattern pairs can facilitate the incremental acquisition and indexing of structural data of protein complexes into knowledge bases, which can be organized based on substructural similar-

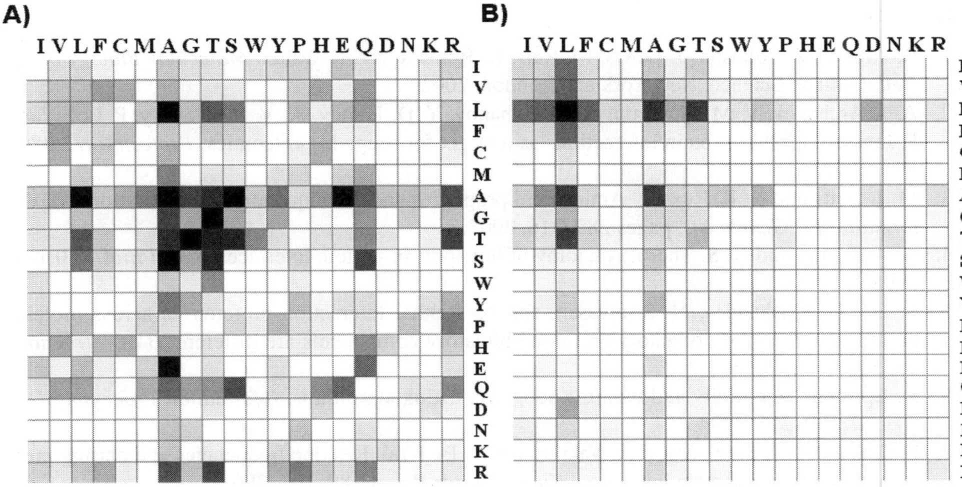

Fig. 2. The pairing preferences of amino acids of the 3D pattern pairs discovered in A) transient, and B) obligate complexes. The darkness of each cell corresponds to the normalized frequency of amino acid interactions: the darker it is, the more frequently the interaction occurs. The amino acid residues are ordered according to their hydrophobicity, with I (Ile) as the most hydrophobic and R (Arg) as the least hydrophobic.

ity. From this study, we observe the high occurence of interactions between hydrophobic amino acids in obligate complexes. These hydrophobic interface resemble domain interfaces or the protein core. As obligate complexes may not fold and function properly when their proteins are unbound, it is sensible to consider obligate complexes as an extension of protein folding.[2] On the other hand, the high-occuring pairings between amino acids of large hydrophobicity difference in transient complexes suggest a kind of interactions that are less permanent than obligate ones. This supports that transient complexes share some similarities with active sites of enzymes, but they are more conservative than active sites of enzymes.[2]

Although 3D pattern pairs are not sufficient to predict a complete structure of protein complexes, there are differences between the interacting amino acid preferences in 3D pattern pairs of transient and those of obligate complexes, which are useful in understanding the difference between transient and obligate complexes. This will be useful in large-scale structural proteomics initiatives, especially for assemblies of protein complexes, in which the physico-chemical characterization is incomplete. In brief, using maximal quasi-bicliques allows us the flexibility of having a wider range of parameter settings to obtain 3D pattern pairs.

Although there are only 20 possible types of amino acid in most binding sites of protein-protein interactions, there are many more variations that can occur through subsequent modification. This necessistates the inclusion of post-translational modification information in the future analysis of binding sites of distinct protein-protein interactions.

References

1. B. Alberts, A. Johnson, J. Lewis, M. Raff, K. Roberts, and P. Walter. *Molecular Biology of the Cell*. Garland Science, New York and London, 2002.
2. A. I. Archakov, V. M. Govorun, A. V. Dubanov, Y. D. Ivanov, A. V. Veselovsky, P. Lewi, and P. Janssen. Protein-protein interactions as a target for drugs in proteomics. *Proteomics*, 3:380–391, 2003.
3. M. J. Betts and R. B. Russell. Amino acid properties and consequences of substitutions. *Bioinformatics for Geneticists*, pages 289–316, 2003.
4. A. A. Bogan and K. S. Thorn. Anatomy of hot spots in protein interfaces. *J Mol Biol*, 280:1–9, 1998.
5. S. De, O. Krishnadev, N. Srinivasan, and N. Rekha. Interaction preferences across protein-protein interfaces of obligatory and non-obligatory components are different. *BMC Structural Biology*, 5:15, 2005.
6. D. Eppstein. Arboricity and bipartite subgraph listing algorithms. *Information Processing Letters*, 51:207–211, 1994.
7. F. Glaser, D. M. Steinberg, I. A. Vakser, and N. Ben-Tal. Residue frequencies and pairing preferences at protein-protein interfaces. *Nucleic Acids Res*, 43:89–102, 2001.
8. G. Grahne and J. Zhu. Fast algorithms for frequent itemset mining using fp-trees. *IEEE Transactions on Knowledge and Data Engineering*, 17(10):1347–1362, October 2005.
9. S. Jones, A. Marin, and J. M. Thornton. Protein domain interfaces: characterization and comparison with oligomeric protein interfaces. *Protein Engineering*, 13:77–82, 2000.
10. S. Jones and J. Thornton. Analysis of protein-protein interaction sites using surface patches. *J Mol Biol*, 272:121–32, 1997.
11. L. Lo Conte, C. Chothia, and J. Janin. The atomic structure of protein-protein recognition sites. *J Mol Biol*, 285(2):177–198, 1999.
12. Y. Ofran and B. Rost. Analysing six types of protein-protein interfaces. *J Mol Biol*, 325(2):377–387, 2003.
13. N. Pasquier, Y. Bastide, R. Taouil, and L. Lakhal. Discovering frequent closed itemsets for association rules. *ICDT*, pages 398–416, 1999.
14. F. B. Sheinerman, R. Norel, and B. Honig. Electrostatic aspects of protein-protein interactions. *Curr Opin Struct Biol*, 10:153–159, 2000.
15. K. Sim, J. Li, V. Gopalkrishnan, and G. Liu. Mining maximal quasi-bicliques to co-cluster stocks and financial ratios for value investor. *ICDM*, pages 1059–1063, 2006.
16. E. Sprinzak, Y. Altuvia, and H. Margalit. Characterization and prediction of protein-protein interactions within and between complexes. *PNAS*, 103:14718–14723, 2006.
17. C. J. Tsai, S. L. Lin, H. J. Wolfson, and R. Nussinov. A dataset of protein-protein interfaces generated with a sequence-order-independent comparison technique. *J Mol Biol*, 260:604–620, 1996.
18. D. Xu, S. Lin, and R. Nussinov. Protein binding versus protein folding: the role of hydrophilic bridges in protein associations. *J Mol Biol*, 265:68–84, 1997.
19. H.-X. Zhou and Y. Shan. Prediction of protein interaction sites from sequence profile and residue neighbor list. *Proteins Structure Function and Genetics*, 44:336 – 343, 2001.
20. H. Zhu, F. S. Domingues, I. Sommer, and T. Lengauer. Noxclass: prediction of protein-protein interaction types. *BMC Bioinformatics*, 7:27, 2006.

STRUCTURAL DESCRIPTORS OF PROTEIN-PROTEIN BINDING SITES

OLIVER SANDER*, FRANCISCO S. DOMINGUES, HONGBO ZHU,
THOMAS LENGAUER and INGOLF SOMMER

Max-Planck-Institute for Informatics,
Stuhlsatzenhausweg 85
66123 Saarbrücken, Germany
E-mail: osander@mpi-sb.mpg.de

Structural bioinformatics provides new tools for investigating protein-protein interactions at the molecular level. We present two types of structural descriptors for efficiently representing and comparing protein-protein binding sites and surface patches. The descriptors are based on distributions of distances between five types of functional atoms, thereby capturing the spatial arrangement of physicochemical properties in 3D space. Experiments with the method are performed on two tasks: (1) detection of binding sites with known similarity from homologous proteins, and (2) scanning of the surfaces of two non-homologous proteins for similar regions.

Keywords: Structural Descriptor; Protein-Protein Interactions; Binding Sites.

1. Introduction

Throughout the life of a cell protein-protein interactions (PPI) are the driving force behind many molecular functions and cellular activities. Protein-protein interactions also drive many of the processes related to diseases, such as host-pathogen interactions including the immune response to invading pathogens and disease-related protein misfolding and aggregation.

Experimental high-throughput techniques such as yeast-two-hybrid screens, tandem affinity purification and co-immunoprecipitation can afford a comprehensive view of the network of protein-protein interactions inside a cell. The wealth of interaction data that are generated with these methods is further increased by predicting interactions computationally based on homology. However, these data suffer from severe limitations. First, experimentally derived interactions show inaccuracies,[31] which are then propagated by homology-based annotation. The discrepancies between different experimental and predicted data sets are considerable.[21] Second, binary interaction data lack molecular details about where and when, in which relative positioning, and how strongly proteins interact. This information is vital for assessing the effect of mutations on binding sites and for the development of inhibitors of PPIs.[2]

The spectrum of questions and methods in the field of protein-protein interactions is

*to whom correspondence should be addressed

wide. Appropriate classification of interactions e.g., as permanent or transient interactions,[17,18,37] is of high relevance in the cellular context but requires an in-depth study of structural features. In their paper from 2006,[1] Aloy and Russell proclaim structural analysis of protein interactions, interfaces, and binding sites as a crucial step towards understanding interactions at a system level. Temporal dynamics of interactions, spatial organizations of assemblies, locations of interactions and types of interactions need to be understood to place single interactors in their cellular or systems context.

In this work we study the similarities between binding sites of proteins. Global similarity of two proteins is neither necessary, nor sufficient for sharing similar binding partners. On the one hand, proteins from the same family can exhibit different binding specificities due to subtle changes in their binding sites. On the other hand, mimicking of binding sites enables two proteins with different global folds to bind to the same partner, such as the viral M3 protein imitating a chemokine homodimeric binding site[11] or the mimicking of CD4 by scorpion toxin.[12]

To study these phenomena purely data-driven analysis as well as similarity-based methods have been applied. The idea underlying data-driven analysis is that if complexes AB and A'C involving the domains A, A', B, and C were observed, A and A' being from the same homologous family, an alignment of A and A' can be used to analyze whether B and C are binding at equivalent binding sites to A and A', respectively. This indicates that the binding sites of B and C are likely to share some properties, as they are able to bind to similar partners. Henschel *et al.*[11] studied similar binding sites by extracting them from known complex structures using this concept.

In contrast to this data-driven analysis, similarity-based methods have been used to detect similarities between binding sites, despite global dissimilarity between the respective proteins. Similarity-based methods use either combinatorial matching procedures to find common properties or "structural descriptors" to capture the essential characteristics of a binding site. We use the notion "structural descriptor" here to mean an abstract representation allowing for efficient comparison, in contrast to methods like geometric hashing or clique search on correspondence graphs, which use simpler representations, but more complex combinatorial matching procedures.

Geometric hashing[35] and other combinatorial matching techniques have been applied numerous times for the comparison of binding sites (i.e., the binding residues of one interaction partner) as well as protein interfaces (i.e., the binding residues from both interaction partners).[24,26] Keskin *et al.*[15] identified similar binding behaviour by structural alignment of interface residues. In contrast to the data-driven approach by Henschel *et al.*[11] outlined above, this procedure requires less data, but relies heavily on the structural alignment method, making it difficult to differentiate between differences in binding sites and methodological artefacts. For the comparison of enzymatic active sites the software packages TESS[32] and JESS[3] were developed by the Thornton group. The Klebe group developed clique search and geometric hashing approaches for the comparison of small-ligand binding sites.[23,33]

The concept of using structural descriptors for representing functional sites or structural arrangements has been described previously. Stahl *et al.*[29] have used distance-based descriptions for comparing active sites of enzymes based on chemical and geometric properties. Subsequently the sites were clustered and visualized using a self-organizing-map (SOM) approach. For the analysis of protein-protein interaction interfaces, Mintseris and Weng[16] have proposed atomic contact vectors which consist of contact counts derived from thresholded distance matrices. Distributions of atomic distances have been used successfully in structure comparison.[6,7] In protein structure prediction, distributions of distances have been applied in the form of knowledge-based potentials for evaluating the fit of a sequence to a structure.[28] Bock *et al.*[4,5] use spin-image representations to represent the arrangement of neighboring residues around a residue of interest. Via *et al.*[30] provide a recent review of methods for detecting of protein surface similarities. Several of the methods, such as distance distributions and spin-image representations, stem from the computer vision research field.[19,25]

In our recent work[22] we demonstrated the applicability of structural descriptors to the specific task of predicting HIV coreceptor usage based on properties of the viral V3 loop region. Here, we examine their applicability to the more general task of binding site comparison.

We propose a method for representing and comparing protein-protein binding sites. The structural descriptor is based on distributions of pairwise distances between functional atoms. Thereby, the descriptor encodes the spatial arrangement of those physico-chemical properties in a vector of fixed length.

We evaluate two modes of analysis: (1) using the structural descriptor to describe a whole binding site, i.e. the set of all residues binding to the partner in a protein-protein interaction, and (2) to describe a set of surface residues as defined by a sphere of radius r around the $C\alpha$-atom of a given central residue. The first mode can be used for comparing predefined protein-protein binding sites, whereas the second mode can be used to scan the surfaces of two proteins for similarities, if the binding patches are not known *a priori*.

This article is organized as follows. In Section 2 we describe the details of the distance-based descriptor and methodological variants and the nearest-neighbor prediction protocol. In Section 3 we present results of the performance evaluation on a data set of protein kinases and a case study on scanning protein surfaces for similar patches.

2. Comparison of protein binding sites and surface patches

Here, we introduce two variants of structural descriptors SD_{bsite} and SD_{patches}. SD_{bsite} describes the spatial arrangement of physico-chemical properties for given set of residues for a predefined binding site. In contrast, SD_{patches} provides a representation for several small surface patches and computes a combined match score. See Figure 1 for a schematic overview of the two methods.

2.1. *Structural descriptors of protein-protein binding sites*

The structural descriptor SD_{bsite} takes a set R of binding site residues as input and encodes their relative positioning in three-dimensional space. Residues losing more than 1 Å^2 of solvent accessible surface area upon complexation with the binding partner[14] are defined

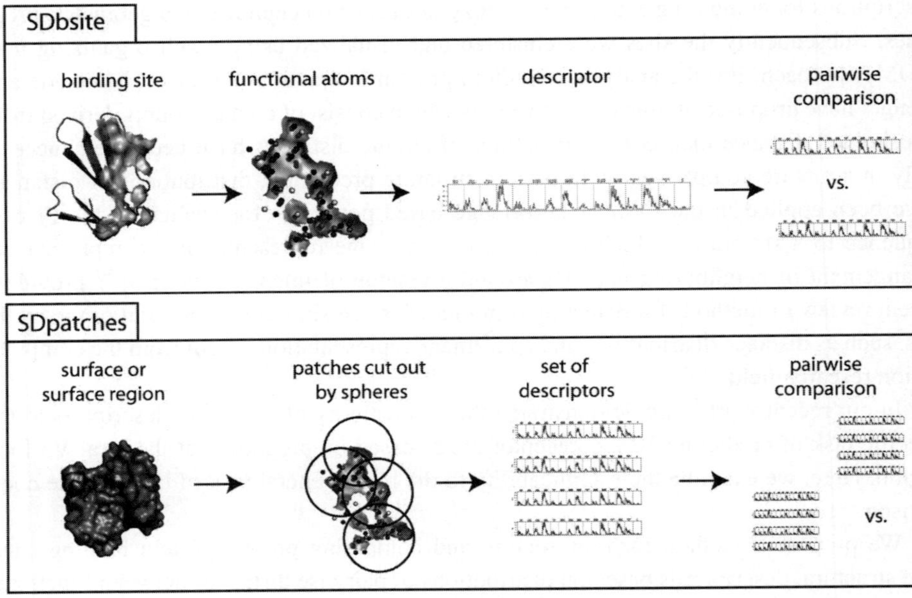

Fig. 1. Schematic overview of the structural descriptors $SD_{\mathtt{bsite}}$ and $SD_{\mathtt{patches}}$

as binding site residues. The solvent-accessible surface areas for the single domains as well as their complexes are computed using NACCESS.[13]

Following Schmitt *et al.*[23] we represent the side chains using five functional atom types, namely, hydrogen-bond donor, acceptor, ambivalent donor/acceptor, aliphatic, or aromatic ring. Amino acids R, N, Q, K, and W are classified as donors. Acceptors are N, D, Q, and E. Ambivalent donor/acceptors comprise H, S, T, and Y. As aliphatic amino acids we consider A, R, C, I, L, K, M, P, T, and V. Pi-stacking centers are H, F, W, and Y. Pseudo-atoms for donors, acceptors, and ambivalent donor/acceptor interaction centers are placed at the respective nitrogen or oxygen atoms. For aliphatic and aromatic interaction centers all involved atom positions were averaged per residue to compute a pseudo-atom. We used the unweighted average of carbons to determine the center of aliphatic side chains, and do not consider backbone atoms as pi-stacking interaction centers.

To derive the structural descriptor, the spatial arrangement of these functional pseudo-atoms is encoded by distance distributions. For each of the 15 combinations of functional atom types (i.e. donor-donor, donor-acceptor, etc.) pairwise Euclidean distances between the respective pseudo-atoms in the residue set R are calculated. Note that the number of these distances depends on the number of pseudo-atoms in the two respective groups. From these distance matrices we derive distance distributions using a kernel density estimate with a Gaussian kernel and a smoothing kernel bandwidth of 1 Å. The density estimates are then discretized by uniform sampling at intervals of 1 Å from 1Å to 10Å, resulting in a 15 (distance distributions for atom type combinations) times 10 (sample points) dimensional vector. The resulting vector is used as a structural descriptor for a given set of binding site residues R. Distance distributions are representations of protein structure invariant under translation and rotation. The smoothing kernel bandwidth as well as the sampling intervals for the distance-based descriptors have been set to values based on empirical observations. Variations within a reasonable range did not result in significant changes of the performance.

2.2. *Comparison and retrieval of structural descriptors*

The structural descriptor is a vector of fixed length. The length only depends on the parameters of the method, not on the size or number of residues of the binding site to be described. By using a vectorial representation of a binding site or surface patch multivariate analysis and statistical learning techniques can directly be applied to the descriptors. Here we use simple nearest-neighbor classification, but in principle, kernel-based discriminative methods can be applied directly.

A wide variety of distance functions, Minkowski norms like the Euclidean or Manhattan metric, information theoretic measures like the Kullback-Leibler distance or Jensen-Shannon divergence, or other statistical approaches like χ^2-test, dot-products, or cosine distance can be used to compare two descriptor vectors. On the tasks and data sets studied here, the cosine and Euclidean measures provide very good performance. While on the rather small data sets used here for evaluation we used pairwise distance computations to determine nearest neighbors, spatial indexing methods like kd-trees[8] can be used to speed up the retrieval of nearest neighbors from a massive set of hundred thousands or millions of descriptors.

2.3. *Structural descriptors of protein surface patches*

While the SD_{bsite} descriptor relies on a predefined set of residues, here for SD_{patches}, we drop this prerequisite. In contrast to predefined protein-protein binding sites the comparison of two proteins for similar surface regions does not provide a defined set of residues to be described by the structural descriptor. SD_{patches} describes the surface of a protein or parts of it by a set of patches. Each patch is composed of the residues within a sphere of radius r around the Cα atom of a given central residue. In the current implementation we use one sphere per surface residue. A multi-resolution approach can be implemented by using spheres of different radii and combining the matches appropriately in the subsequent p-value computation.

Each surface patch is represented by a distance distribution like with SD_{bsite}. Thus the comparison of two protein surfaces turns into the comparison of two sets of descriptors. From the raw descriptor match scores described above, we compute p-values. This is done by generating a background distribution of similarity scores of unrelated pairs of descriptors. For efficient lookup of p-values the cumulative distribution function of the top 5% scores in the distribution is smoothed by a cubic spline with 4 knots and fitted by a piecewise linear function. p-values above a threshold of 5% are set to 1, to avoid the accumulation of spurious similarities.

To compare two sets of patches, each patch in the first set receives the score of the

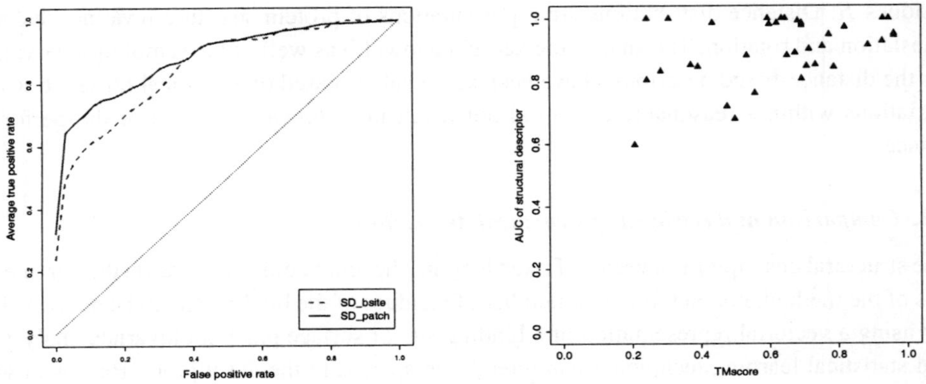

(a) (b)

Fig. 2. Retrieval of similar binding sites: (a) ROC curve, (b) comparison of AUC performance of the structural descriptor against TMscore

respective best hit in the second set and the p-values of all hits get accumulated by multiplication (assuming statistical independence). To avoid numerical instabilities $-log_{10}(p)$ scores are computed and accumulated by summation.

3. Experiments & Evaluation

Experiments are performed on a set of binding sites from Pkinases and their respective binding partners. In addition to the quantitative evaluation, a case study on an instance of viral mimicking is presented.

3.1. *Retrieving Similar Binding Sites - Kinases*

We analyzed and evaluated the structural binding site descriptors SD_{bsite} and SD_{patches} on a set of protein kinase binding sites and the binding sites of their respective partners. This data sets consists of binding sites derived from domain interfaces from 25 Pkinase complexes comprising 50 binding sites. For the selection of these binding sites we used the SCOPPI database.[34] SCOPPI provides an extensive data set of domain-domain interactions for all SCOP domains in the protein data bank PDB. In addition to the preprocessed list of pairwise interactions, SCOPPI supplies a categorization of binding sites into face types. The binding sites of all domains within a specific family are compared on the basis of how many residues they share that are matched in an alignment of the two protein families. Based on this criterion strongly overlapping binding sites on equivalent regions of the domain surface are classified into the same face type.

Complexes with redundant entries (i.e., using the same binding faces in both interactions) were removed if they exhibited a sequence identity level of at least 90% with already included complexes. From the resulting set of 50 binding sites we removed one of the two binding sites in each symmetric homo-dimeric complex. Due to symmetry, these pairs are highly similar and would be trivial to find in the subsequent matching experiment. The resulting data set consists of a set of 38 binding sites. Each of these binding sites is labelled with a four-tuple: SCOP family of the domain of the binding site, SCOPPI face type of the binding site, SCOP family of the binding partner, face type of the partnering binding site.

In the retrieval experiment we aimed at efficiently recovering similar binding sites, as defined by the label described above. Both compared methods, SD_{bsite} and SD_{patches}, use a set of binding site residues as input. While SD_{bsite} represents the binding site globally, SD_{patches} represents the binding site by a set of smaller local patches, as described in Section 2.1. In order to assess the predictive performance of the structural descriptors we performed leave-one-out cross-validation. Evaluation of predictive performance was done using ROCR.[27]

The measure used for evaluation of predictive performance is the area under the ROC curve (AUC). The AUC is calculated by adding the area of trapezoid strips under the ROC curve. This is equal to the value of the Wilcoxon-Mann-Whitney test statistic and also to the probability that the classifier will score a randomly drawn positive sample higher than a randomly drawn negative sample.[10] A ROC curve is computed for each sample in the data set, quantifying, how well similar sites are being retrieved.

In Figure 2 (a) a vertically averaged ROC curve is shown for each of the two descriptors SD_{bsite} and SD_{patches}. SD_{patches} clearly outperforms SD_{bsite} on the retrieval task of the 38 kinase binding sites. This is due to the sensitivity of SD_{bsite} to small changes in the binding site definitions. For example, augmenting a highly similar binding site by a small terminal tail changes the descriptor considerably. The AUC values (one per query binding site) have a mean of 0.9078 and a median of 0.9364 for SD_{bsite} and a mean of 0.9236 and a median of 1.0000 for SD_{patches}. Thus, for at least half of the 38 binding sites SD_{patches} is able to perform a perfect classification, i.e. all similar binding sites from the same class are ranked above binding sites from other classes.

While the AUC quantifies the overall ability to rank samples with the same label higher than samples with another label, the accuracy at the top-rank or in the top-k ranks quantifies the fraction of training samples for which a similar site could be detected. SD_{bsite} is able to find for 68.42%, 73.68%, and 81.58% of the 38 binding sites a similar binding site with the same label at the top-rank, in the top-3, and in the top-5, respectively. SD_{patches} finds for 71.05%, 81.58%, and 89.47% of the 38 binding sites a similar binding site with the same label at the top-rank, in the top-3, and in the top-5, respectively.

Figure 2 (b) shows a scatter plot of the AUC of SD_{bsite} per binding site versus the TMscore[36] between that binding site and its closest binding site from the same family. TMscore is the structural similarity measure provided by the TM-align program. Although this program performs the structural alignment respecting sequence ordering, it can be applied here, as the binding sites labelled to be similar are from the same families. It can be observed that the variability in performance of SD_{bsite} depends on the TMscore: a high TMscore implies a high AUC performance, whereas lower TMscores might result in worse AUCs. This means that SD_{bsite} performs very well on binding sites with a structurally similar closest hit. With decreasing similarity of the best hit the worst-case AUC performance decreases linearly, but for some dissimilar binding sites good performance is still possible.

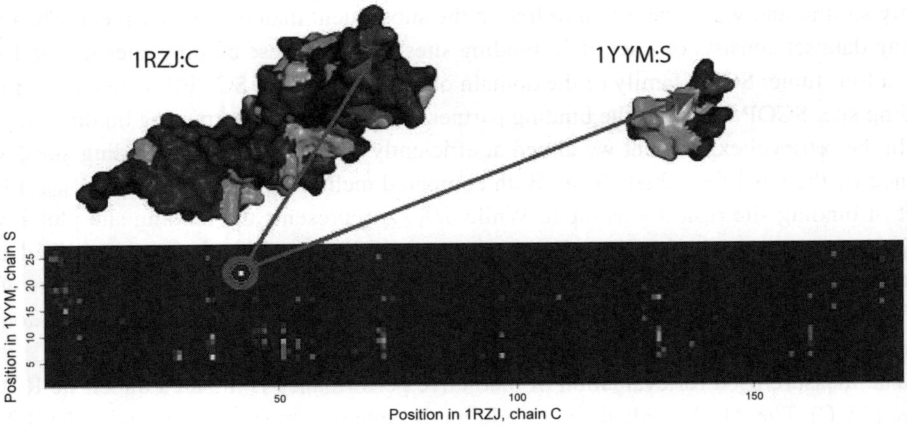

Fig. 3. Using $SD_{\mathtt{patches}}$ to compare CD4 (1RZJ:C) with its mimicking scorpion toxin protein (1YYM:S)

3.2. *Scanning for similar surface patches*

Huang *et al.*[12] analyze the mimicking of CD4 by a small scorpion toxin fold. The scorpion toxin is a 31-amino-acid protein consisting of two beta strands and an α-helix, held together by disulphide bonds. It has been designed to mimick the binding site of CD4 to the viral protein gp120.

We use the structural descriptor $SD_{\mathtt{patches}}$ to compare the surface of CD4 against the surface of the mimicking scorpion toxin. Figure 3 shows the pairwise similarities between patches in CD4 (1RZJ:C) and the scorpion toxin (1YYM:S). Patches are colored by significance of their similarity. The most similar red surface patches in both proteins show a p-value of $10^{-4.223}$ for similarity assessed with $SD_{\mathtt{patches}}$ and, in fact, they correspond to the loops mimicking each other. The matrix in Figure 3 shows the pairwise similarity p-values for all patches in both proteins, ordered along the sequences of both proteins. The highest match is highlighted. The second highest similarity (p-value of $10^{-2.742}$) is clearly less pronounced. The structural descriptor $SD_{\mathtt{patches}}$ is able to pick the binding site from the scorpion toxin mimicking the CD4 binding site, despite the global dissimilarities of the two proteins.

4. Conclusion and Outlook

The proposed structural descriptor is an efficient and accurate method for describing binding sites and surface patches. The major remaining problem is the evaluation, as data with annotation is scarce. There is no clear notion of "non-trivial" similarity that should be detected by methods focusing on local similarity of proteins. Even if the proteins have the same functions or bind to the same partner at the same respective site, it is not guaranteed that they share some detectable similarities. Further directions are (1) relating local similarities of surface patches to protein function, and (2) comparison of the described approach against other descriptor based methods like ACVs, spin-image representations and combinatorial matching approaches like geometric hashing or clique search.

Acknowledgement

We would like to thank Christoph Winter for providing a flatfile version of the SCOPPI database. Analysis of the results and prediction was performed using the statistical language R[20] with the package ROCR.[27] Protein structure visualizations were created using PyMOL.[9]

References

1. Patrick Aloy and Robert B Russell. Structural systems biology: modelling protein interactions. *Nat Rev Mol Cell Biol*, 7(3):188–197, Mar 2006.
2. Michelle R Arkin and James A Wells. Small-molecule inhibitors of protein-protein interactions: progressing towards the dream. *Nat Rev Drug Discov*, 3(4):301–317, Apr 2004.
3. Jonathan A Barker and Janet M Thornton. An algorithm for constraint-based structural template matching: application to 3D templates with statistical analysis. *Bioinformatics*, 19(13):1644–9, Sep 2003.
4. Mary Ellen Bock, Guido M. Cortelazzo, Carlo Ferrari, and Concettina Guerra. Identifying similar surface patches on proteins using a spin-image surface representation. In A. Apostolico, M. Crochemore, and K. Park, editors, *CPM 2005, LNCS 3537*, pages 417–428, Heidelberg, 2005. Springer-Verlag Berlin.
5. Mary Ellen Bock, Claudio Garutti, and Concettina Guerra. Discovery of similar regions on protein surfaces. *Journal of Computational Biology*, 14(3):285–299, 2007.
6. Stefan Canzar and Jan Remy. Shape distributions and protein similarity. In *Proceedings of the German Conference on Bioinformatics (GCB '06)*, pages 1–10, 2006.
7. Oliviero Carugo and Sandor Pongor. Protein fold similarity estimated by a probabilistic approach based on C(alpha)-C(alpha) distance comparison. *J Mol Biol*, 315(4):887–898, 2002.
8. Mark de Berg, Marc de Kreveld, and Mark Overmars. *Computational Geometry. Algorithms and Applications*. Springer, Berlin, 2000.
9. Warren L. DeLano. The PyMOL molecular graphics system, 2002. DeLano Scientific. San Carlos, CA, USA. http://www.pymol.org.
10. Tom Fawcett. An introduction to ROC analysis. *Pattern Recognition Letters*, 27(8):861–874, 2006.
11. Andreas Henschel, Wan Kyu Kim, and Michael Schroeder. Equivalent binding sites reveal convergently evolved interaction motifs. *Bioinformatics*, 22(5):550–555, Mar 2006.
12. Chih-Chin Huang, Francois Stricher, Loic Martin, Julie M Decker, Shahzad Majeed, Philippe Barthe, Wayne A Hendrickson, James Robinson, Christian Roumestand, Joseph Sodroski, Richard Wyatt, George M Shaw, Claudio Vita, and Peter D Kwong. Scorpion-toxin mimics of CD4 in complex with human immunodeficiency virus gp120 crystal structures, molecular mimicry, and neutralization breadth. *Structure*, 13(5):755–768, May 2005.
13. Simon J. Hubbard and Janet M. Thornton. Naccess. Computer Program, Department of Biochemistry and Molecular Biology, University College London., 1993.
14. Susan Jones and Janet M. Thornton. Principles of protein-protein interactions. *Proc Natl Acad Sci U S A*, 93(1):13–20, Jan 1996.
15. Ozlem Keskin and Ruth Nussinov. Similar binding sites and different partners: implications to shared proteins in cellular pathways. *Structure*, 15(3):341–354, Mar 2007.
16. Julian Mintseris and Zhiping Weng. Atomic contact vectors in protein-protein recognition. *Proteins*, 53(3):629–639, 2003.

17. Irene M A Nooren and Janet M Thornton. Diversity of protein-protein interactions. *EMBO J*, 22(14):3486–92, Jul 2003.
18. Yanay Ofran and Burkhard Rost. Analysing six types of protein-protein interfaces. *J Mol Biol*, 325(2):377–87, Jan 2003.
19. Robert Osada, Thomas Funkhouser, Bernard Chazelle, and David Dobkin. Shape distributions. *ACM Transactions on Graphics (TOG)*, 21(4):807–832, October 2002.
20. R Development Core Team. *R: A language and environment for statistical computing*. R Foundation for Statistical Computing, Vienna, Austria, 2005. ISBN 3-900051-07-0.
21. Fidel Ramirez, Andreas Schlicker, Yassen Assenov, Thomas Lengauer, and Mario Albrecht. Computational analysis of human protein interaction networks. *Proteomics*, 7(15):2541–52, 2007.
22. Oliver Sander, Tobias Sing, Ingolf Sommer, Andrew J Low, Peter K Cheung, P. Richard Harrigan, Thomas Lengauer, and Francisco S Domingues. Structural descriptors of gp120 V3 loop for the prediction of HIV-1 coreceptor usage. *PLoS Comput Biol*, 3(3):e58, Mar 2007.
23. Stefan Schmitt, Daniel Kuhn, and Gerhard Klebe. A new method to detect related function among proteins independent of sequence and fold homology. *J Mol Biol*, 323(2):387–406, 2002.
24. Maxim Shatsky, Alexandra Shulman-Peleg, Ruth Nussinov, and Haim J. Wolfson. *Recognition of Binding Patterns Common to a Set of Protein Structures*, volume 3500, chapter Lecture Notes in Computer Science, pages 440 – 455. Springer-Verlag GmbH, May 2005.
25. Philip Shilane, Patrick Min, Michael Kazhdan, and Thomas Funkhouser. The Princeton shape benchmark. In *Shape Modeling International, Genova, Italy*, 2004.
26. Alexandra Shulman-Peleg, Ruth Nussinov, and Haim J. Wolfson. SiteEngines: recognition and comparison of binding sites and protein-protein interfaces. *Nucleic Acids Research*, 33:W337–W341, 2005.
27. Tobias Sing, Oliver Sander, Niko Beerenwinkel, and Thomas Lengauer. ROCR: visualizing classifier performance in R. *Bioinformatics*, 21(20):3940–3941, 2005.
28. Manfred J. Sippl. Knowledge-based potentials for proteins. *Curr Opin Struct Biol*, 5(2):229–235, 1995.
29. Martin Stahl, Chiara Taroni, and Gisbert Schneider. Mapping of protein surface cavities and prediction of enzyme class by a self-organizing neural network. *Protein Eng*, 13(2):83–88, 2000.
30. Allegra Via, Fabrizio Ferre, Barbara Brannetti, and Manuela Helmer-Citterich. Protein surface similarities: a survey of methods to describe and compare protein surfaces. *Cell Mol Life Sci*, 57(13-14):1970–1977, Dec 2000.
31. Christian von Mering, Roland Krause, Berend Snel, Michael Cornell, Stephen G Oliver, Stanley Fields, and Peer Bork. Comparative assessment of large-scale data sets of protein-protein interactions. *Nature*, 417(6887):399–403, May 2002.
32. Andrew C. Wallace, Neera Borkakoti, and Janet M. Thornton. TESS: a geometric hashing algorithm for deriving 3D coordinate templates for searching structural databases. Application to enzyme active sites. *Protein Sci*, 6(11):2308–23, Nov 1997.
33. Nils Weskamp, Daniel Kuhn, Eyke Hüllermeier, and Gerhard Klebe. Efficient similarity search in protein structure databases by k-clique hashing. *Bioinformatics*, 20(10):1522–6, Jul 2004.
34. Christof Winter, Andreas Henschel, Wan Kyu Kim, and Michael Schroeder. Scoppi: a structural classification of protein-protein interfaces. *Nucleic Acids Res*, 34(Database issue):D310–D314, Jan 2006.
35. Haim J. Wolfson and Isidore Rigoutsos. Geometric hashing: an overview. *Computational Science and Engineering, IEEE*, 4(4):10–21, 1997.
36. Yang Zhang and Jeffrey Skolnick. TM-align: a protein structure alignment algorithm based on the TM-score. *Nucleic Acids Res*, 33(7):2302–2309, 2005.
37. Hongbo Zhu, Francisco S Domingues, Ingolf Sommer, and Thomas Lengauer. NOXclass: prediction of protein-protein interaction types. *BMC Bioinformatics*, 7:27, 2006.

A MEMORY EFFICIENT ALGORITHM FOR STRUCTURAL ALIGNMENT OF RNAs WITH EMBEDDED SIMPLE PSEUDOKNOTS

THOMAS WONG, Y. S. CHIU, TAK-WAH LAM and S. M. YIU

Department of Computer Science, The University of Hong Kong
Pokfulam Road, Hong Kong

In this paper, we consider the problem of structural alignment of a target RNA sequence of length n and a query RNA sequence of length m with known secondary structure that may contain embedded simple pseduoknots. The best known algorithm for solving this problem (Dost et al. [13]) runs in O(mn4) time with space complexity of O(mn3), which requires too much memory making it infeasible for comparing ncRNAs (non-coding RNAs) with length several hundreds or more. We propose a memory efficient algorithm to solve the same problem. We reduce the space complexity to O(mn2 + n3) while maintaining the same time complexity of Dost et al.'s algorithm. Experimental reslts show that our algorithm is feasible for comparing ncRNAs of length more than 500.

Availability: The source code of our program is available upon request.

1. Introduction

A non-coding RNA (ncRNA) is a RNA molecule which is not translated into a protein. It is a general belief that ncRNAs are involved in many cellular functions. The number of ncRNAs within the genome was underestimated before, but recently some databases reveal over 30,000 ncRNAs [1] and more than 500 ncRNA families[2]. Large discoveries of ncRNAs and families show the possibilities that ncRNAs may be as diverse as protein molecules [3]. Identifying these ncRNAs becomes an important problem.

It is known that the secondary structure of an ncRNA molecule usually plays an important role in its biological functions. Some researches attempted to identify ncRNAs by considering the stability of secondary structures formed by the substrings of a given genome [15]. However, this method is not effective because a random sequence with high GC composition also allows an energetically favorable secondary structure [8]. A more promising direction is comparative approach which makes use of the idea that if a substring of genome from which a RNA is transcribed with similar sequence and structure to a known ncRNA, then this genome region is likely to be an ncRNA gene whose corresponding ncRNA is in the same family of the known ncRNA. Thus, to locate ncRNAs in a genome, we can use a known ncRNA as a query and searches along the genome for substrings with similar sequence and structure to the query. The key of this approach is to compute the structural alignment between a query sequence with known structure and a target sequence with unknown structure. The alignment score represents their sequence and structural similarity. RSEARCH [9], FASTR [10], and a recent tool INFERNAL [11] for Rfam are using this approach.

However, all of these tools do not support pseudoknots. Given two base-pairs at positions (i, j) and (i', j'), where $i < j$ and $i' < j'$, pseudoknots are base-pairs either $i < i' < j < j'$ or $i' < i < j' < j$. In some studies, secondary structures including pseudoknots are found involved in some functions such as telomerase [5], catalytic functions [6], and self-splicing introns [7]. The presence of pseudoknots makes the problem computationally harder, so finding ncRNA genes with secondary structure including pseudoknots are limited. Usually the large time complexity and considerable memory required for these algorithms make it impractical to search long pseudoknotted ncRNA along the genome. Among over 500 known ncRNA families in Rfam, only 24 families that are in pseudoknotted structure exist. The small number may reflect the uncommon situation of pseudoknotted ncRNA, but it may also reflect the difficulty of finding pseudoknotted ncRNAs due to the limitation of existing tools.

Matsui *et al.* [12] developed a method of computing the structural alignment to support a pseudoknot structure. They used a pseudoknot definition that a secondary structure has m-clossing property if and only if there exists m base pairs in which any two of them are crossing each other. For 2-crossing pseudoknots, their algorithm runs in $O(mn^5)$ with space complexity of $O(mn^4)$ where m is the length of the query sequence and n is the length of the searching sequence. The large time and space complexity makes the method infeasible for practical use. Pseudoknots can exist within another pseudoknot forming recursive pseudoknots. Since known ncRNA families are found to have a simpler structure (with only a single level of recursion), called embedded simple pseudoknots. Some focuses on this simpler structure. Dost *et al.* [13] developed a tool called PAL using dynamic programming approach that supports secondary structures with embedded simple pseudoknots. By restricting their supporting structure to be a subset of the structures having 2-crossing properties, their dynamic programming algorithm runs faster and uses less memory with time complexity of $O(mn^4)$ and space complexity of $O(mn^3)$.

However, their algorithm is still not feasible for long RNA sequences due to the extensive memory required. For example, for the pseduknotted ncRNA family RF00024 (found in the database Rfam), the average length of the members is about 548. It is estimated that performing a pair-wise structural alignment for members in this family using PAL requires at least 10GB memory. Therefore, the tool becomes impractical for ncRNA families with members of length several hundreds or more. In this paper, we proposed a memory-efficient algorithm for solving the same structural alignment problem with space complexity reduced to $O(mn^2 + n^3)$ while maintaining the same time complexity of $O(mn^4)$.

2. Definitions

Let $A = a_1a_2...a_m$ be a RNA sequence and M be the secondary structure of A. M is represented as a set of base pairs (a_i, a_j), $1 \le i < j \le m$. Let $M_{x,y} \subseteq M$ be the set of base pairs in the subsequence $a_xa_{x+1}...a_y$, $1 \le x < y \le m$. $M_{x,y} = \{(a_i, a_j) \in M \mid x \le i < j \le y\}$.

$M_{x,y}$ is a regular structure if there does not exist two pairs (i, j), $(k, l) \in M_{x,y}$ such that $I < k < j < l$ or $k < i < l < j$. Note that an empty set is considered as a regular structure.

$M_{x,y}$ is a simple pseudoknot if $\exists\, x < x_1, x_2 < y$ such that

1. each $(i, j) \in M_{x,y}$ satisfies either $x \le i < x_1 \le j < x_2$ or $x_1 \le i < x_2 \le j \le y$; and,
2. M_L and M_R are both regular where $M_L = \{(i, j) \in M_{x,y} \mid x \le i < x_1 \le j < x_2\}$ and $M_R = \{(i, j) \in M_{x,y} \mid x_1 \le i < x_2 \le j \le y\}$.

An embedded simple pseudoknot structure is defined as follows [13].

M is an embedded simple pseudoknot structure if $\exists\, 1 \le x_1 < y_1 < ... < x_s < y_s \le m$ such that

1. $M_{xi,yi}$, for $1 \le i \le s$, is a simple pseudoknot structure; and,
2. $\left(M - \bigcup_{1 \le i \le s} M_{xi,yi}\right)$ is a regular structure

Note that simple pseudoknot structure is a subset of embedded simple pseudoknot structure. In this paper, our method is designed for ncRNAs with embedded simple pseudoknot structures.

3. Algorithm

3.1. *Structural alignment*

Let $S[1...m]$ be a query sequence with known secondary structure M, and $T[1...n]$ be a target sequence with unknown secondary structure. S and T are sequences from the character set {A, C, G, U}. A structural alignment between S and T can be represented by a pair of sequences $S'[1...r]$ and $T'[1...r]$ where $r \ge m, n$. S' is from S and T' is from T with spaces inserted in between the characters to make both sequences of the same length. A space cannot appear in the same position of S' and T'. The score of the alignment, which determines the sequence and structure similarity between S' and T', is defined as follows.

$$score = \sum_{i=1}^{r} \gamma(S'[i], T'[i]) + \sum_{i,j \text{ s.t. } (\eta(i),\eta(j)) \in M \text{ and } S'[i],S'[j],T'[i],T'[j] \ne '\_'} \delta(S'[i], S'[j], T'[i], T'[j])$$

where $\eta(i)$ is the corresponding position in S according to the position i in S'; $\gamma(t_1, t_2)$ and $\delta(x_1, y_1, x_2, y_2)$, where $t_1,t_2 \in$ {A,C,G,U, '_'} and $x_1,x_2,y_1,y_2 \in$ {A,C,G,U}, are the score for sequence similarity and the score for structural similarity respectively. The calculation of structural alignment score is not restricted to any kind of secondary structure. It works in the same way for pseudoknot structure. The objective is to find an alignment such that the corresponding score is maximized. Higher score represents higher similarity between the two sequences according to their sequences and structures. Also, if the score is high, then the alignment can reasonably reveal the secondary structure of the target sequence.

3.2. *Structural alignment for simple pseudoknot*

Consider a length-m subsequence $S[i_0 \ldots k_0]$ with a simple pseudoknot structure, there exists $i_0 \leq x_1$, $x_2 \leq k_0$ such that the simple pseudoknot structure M can be divided into two regular structures M_L and M_R as mentioned in Section 2. A subpseudoknot $P(i,j,k)$ of S is defined as the union of subinterval $[i_0..i]$ and $[j..k]$, where $i_0 \leq i < x_1$, $x_1 \leq j < x_2$, $x_2 \leq k \leq k_0$, as shown in Figure 1. Let $B[p,q,r,i,j,k]$ be the optimal alignment score between $P(i,j,k)$ of S and $P(p,q,r)$ of another length-n subsequence $T[p_0 \ldots r_0]$ whose structure is unknown.

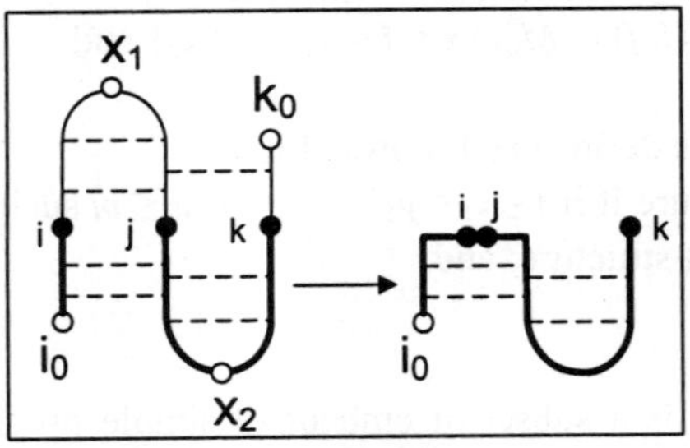

Figure 1: Subpseudoknot P(i,j,k) = $[i_0..i][j..k]$, where $i_0 \leq i < x_1$, $x_1 \leq j < x_2$, $x_2 \leq k \leq k_0$

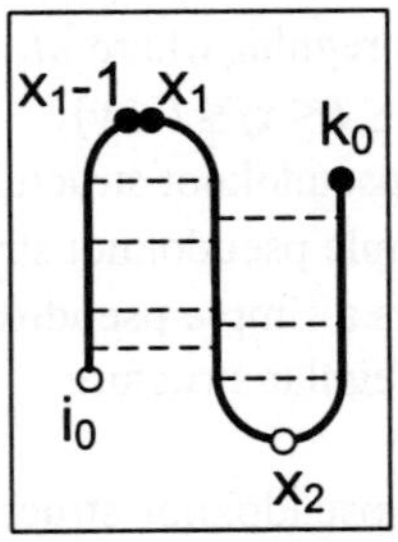

Figure 2: $P(x_1\text{-}1,x_1,k_0)$ represents the whole pseudoknot structure of $S[i_0 \ldots k_0]$

Consider the case of $(i, j) \in M_L$, the following recurrence equation [13] gives the value for B. The case for $(i, j) \in M_R$ is similar.

$$B[p,q,r,i,j,k] = \max \begin{cases} \text{// match} \\ B[p-1,q+1,r,i-1,j+1,k] + \delta(T[p],T[q],S[i],S[j]) \\ \qquad + \gamma(T[p],S[i]) + \gamma(T[q],S[j]) \\ \text{// delete} \\ B[p-1,q,r,i-1,j+1,k] + \gamma(T[p],S[i]) + \gamma('\_',S[j]) \\ B[p,q+1,r,i-1,j+1,k] + \gamma('\_',S[i]) + \gamma(T[q],S[j]) \\ B[p,q,r,i-1,j+1,k] + \gamma('\_',S[i]) + \gamma('\_',S[j]) \\ \text{// insert} \\ B[p-1,q,r,i,j,k] + \gamma(T[p],'\_') \\ B[p,q-1,r,i,j,k] + \gamma(T[q],'\_') \\ B[p,q,r-1,i,j,k] + \gamma(T[r],'\_') \end{cases}$$

As shown in Figure 2, $P(x_1\text{-}1,x_1,k_0)$ represents the whole pseudoknot structure of $S[i_0 \ldots k_0]$. Therefore, the score of an optimal alignment between $S[i_0,k_0]$ and $T[p_0,r_0]$ is $\max_{p_0 \leq p \leq r_0} B[p-1,p,r_0,x_1-1,x_1,k_0]$. Note the changes of indices of (i,j,k) in the recursive calculation. The value of i decreases from x_1-1 to i_0, the value of j increases from x_1 to x_2-1 and the value of k deceases from k_0 to x_2. That sequence of triples can be first built from a simple-pseudoknot structure in linear time [13]. The triple $(x_1\text{-}1,x_1,k_0)$ is chosen as the first item (called *root*). The sequence of triples ensures each nucleotide is touched by at least one triple, and every base pair is reached by one and only one triple (i.e. for each base pair $B(x,y)$, there must exist one and only one triple (i,j,k) such that $i=x$ and $j=y$ if $B \in M_L$ or $j=x$ and $k=y$ if $B \in M_R$). The number of triples in the sequences is $O(m)$, where m is the length of the query sequence S.

By using this sequence of triples, the function B can be rewritten as $B[p,q,r,v]$ where v is a triple in the sequence. The recurrence relationship for the case of $v \in M_L$ can be modified as follows.

$$B[p,q,r,v] = \max \begin{cases} // \textit{ match} \\ B[p-1,q+1,r,next(v)] + \delta(T[p],T[q],S[v_i],S[v_j]) \\ \qquad + \gamma(T[p],S[v_i]) + \gamma(T[q],S[v_j]) \\ // \textit{ delete} \\ B[p-1,q,r,next(v)] + \gamma(T[p],S[v_i]) + \gamma('\_',S[v_j]) \\ B[p,q+1,r,next(v)] + \gamma('\_',S[v_i]) + \gamma(T[q],S[v_j]) \\ B[p,q,r,next(v)] + \gamma('\_',S[v_i]) + \gamma('\_',S[v_j]) \\ // \textit{ insert} \\ B[p-1,q,r,v] + \gamma(T[p],'\_') \\ B[p,q-1,r,v] + \gamma(T[q],'\_') \\ B[p,q,r-1,v] + \gamma(T[r],'\_') \end{cases}$$

where v_i,v_j is the i value and j value of the triple v respectively. Next(v) represents the next triple after v in the sequence of triples. The cases for $v \in M_R$, $v \notin M_L$ or M_R are similar [13]. Since $p_0 \le p < q < r \le r_0$ and the number of $v = O(m)$, then both time and space complexities are $O(mn^3)$. The score of an optimal alignment between $S[i_0,k_0]$ and $T[p_0,\mathbf{r}_0]$ is $\max_{p_0 \le p \le r_0} B[p-1,p,r_0,root]$.

3.3. *Our memory-efficient algorithm*

A simple-pseudoknot has two interesting features. Firstly, as shown in the Figure 3a, the reversal of a simple pseudoknot is also a simple pseudoknot. The subpseudoknot $P(i,j,k)$ becomes the upper region including (i,j,k) in the reverse structure. If we consider the alignment between a reverse query sequence and a reverse target sequence, the previous algorithm should also work, but the order of B calculation will be in reversed order according to the reverse sequence of triples in which the root becomes (x_2+1,x_2, i_0) instead of (x_1-1,x_1,k_0).

Secondly, as shown in the Figure 3b, a simple pseudoknot can be separated into two simple-pseudoknots according to a triple (i,j,k): the upper region including (i,j,k) and the lower region excluding (i,j,k). This indicates that the alignment problem between a pair of sequences can be divided into two alignment problems between two pairs of shorter

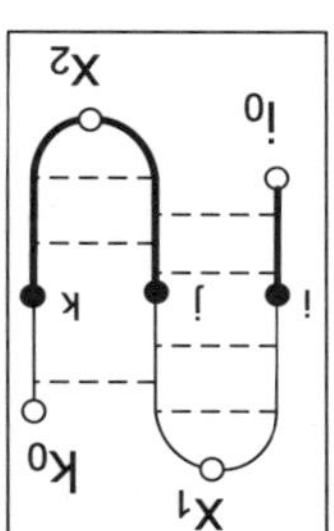

Figure 3a: Reverse of a simple-pseudoknot

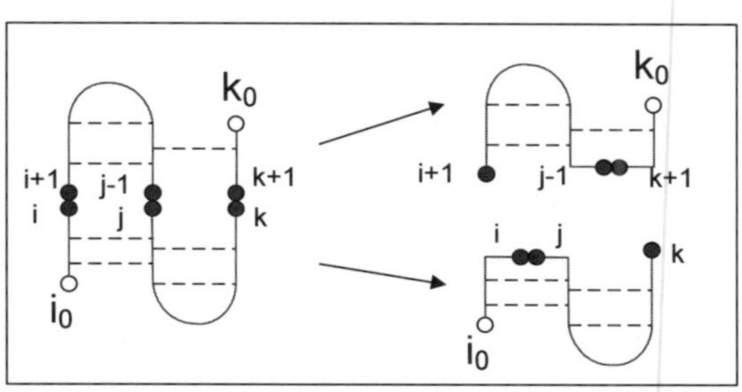

Figure 3b: a simple-pseudoknot can be separated into two simple-pseudoknots

sequences. Based on these two features and inspired by Hirschberg's algorithm [14], we derive a method which can reduce the memory consumption of structural alignment algorithm for simple pseudoknots from $O(mn^3)$ to $O(n^3)$, while maintaining the same time complexity of $O(mn^3)$.

For the sake of simplicity, we rename the indices of $S[i_0,k_0]$ as $S[1\ldots m]$ and $T[p_0,r_0]$ as $T[1\ldots n]$. We first show a *score-only* algorithm is to compute $B[p,q,r,root]$ where $1 \le p<q<r \le n$. B is calculated triple by triple, from the last triple in the triple sequence up to root. At the end of every iteration, we discard all values of B calculated in the previous iterations and only keep those calculated in this iteration for the B calculation in the next iteration.

<u>score-only (V, T)</u>

```
Initialize B_prev
for v = last-item to root in the sequence of triples V
        for all p,q,r where 1≤p<q<r≤n
                compute B[p,q,r] with respect to current value of v and B_prev
        Swap the pointer B_prev with B
return B_prev
```

The score-only algorithm can compute $B[p,q,r,root]$ where $1 \le p<q<r \le n$ in $O(mn^3)$ time and requires $O(n^3)$ memory space. The score-only algorithm can only give the alignment scores.

Let $S^R[1\ldots m]$ be the reversal of a query sequence S (i.e. $S^R[i] = S[m-i+1]$ for $1 \le i \le m$). Let v^r be the corresponding triple v of S for S^R and let v_i,v_j,v_k be the positions i,j,k of triple v, respectively. $v^r=((v^r)_i,(v^r)_j,(v^r)_k)=(m-v_k+1,m-v_j+1,m-v_i+1)$. As illustrated in Figure 4, the subpseudoknot $P(v^r)$ of S^R is the upper region including the triple v of S. The union of the subpseudoknot $P(v_i-1,v_j+1,v_k-1)$ of S and the subpseudoknot $P(v^r)$ of S^R is the whole pseudoknot of S. If a triple (p,q,r) of T is mapped to v of S in the optimal alignment between S and T, then the optimal alignment score between S and T = the optimal alignment score between $P(v_i-1,v_j+1,v_k-1)$ of S and $P(p-1,q+1,r-1)$ of T + the optimal alignment score between $P(v^r)$ of S^R and $P((p,q,r)^r)$ of T^R.

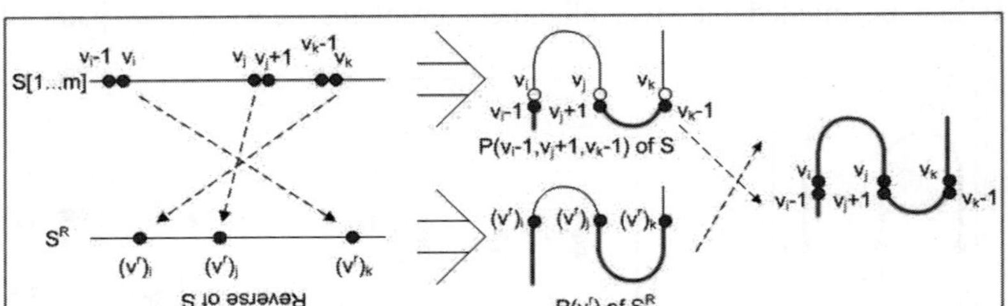

Figure 4: Union of $P(v_i-1,v_j+1,v_k-1)$ of S and $P(v^r)$ of S^R is the whole pseudoknot of S.

The following is a divide-and-conquer approach to get the alignment between $S[1..m]$ and $T[1..n]$.

1. Select the middle triple w of S and prepare the triple sequences for the reverse of the upper region V_{upper}^{R} and for the lower region V_{lower}
   a. Build the sequence of triples $V[1\ldots m_v]$ for S.
   b. Let $w = V\left[\dfrac{m_v}{2}\right]$, the middle triple in the sequence.
   c. As in Figure 3b, partition S into two subseudoknots: upper region including w and lower region excluding w (i.e. $S_{\text{upper}} = [w_i,w_j]\cup[w_k,m]$, $S_{\text{lower}}=[1,w_i\text{-}1] \cup[w_i+1,w_k\text{-}1]$)
   d. Reverse sequence S_{upper} to obtain S_{upper}^{R}
   e. Build the triple sequence V_{upper}^{R} and V_{lower} for S_{upper}^{R} and S_{lower} respectively. Note that the root of V_{upper}^{R} is corresponding to w^r on S and the root of V_{lower} is $(w_i\text{-}1,w_j+1,w_k\text{-}1)$.
2. Find p_h,q_h,r_h such that triple (p_h,q_h,r_h) of T is mapped to triple w of S in the optimal alignment between S and T.
   a. Use score-only(V_{lower},T) to compute $B[p,q,r]$ $\forall$ $1\leq p<q<r\leq n$ for alignment between T and V_{lower}.
   b. Use score-only(V_{upper}^{R}, T^R) to compute $B^R[p,q,r]$ $\forall$ $1\leq p<q<r\leq n$ for alignment between T^R and V_{upper}^{R}.
   c. Find out p_h,q_h,r_h such that $B[p_h\text{-}1,q_h+1,r_h\text{-}1]+ B^R[p',q',r']$ is maximum where $p'= n\text{-}r_h+1,q'=n\text{-}q_h+1,r'=n\text{-}p_h+1$.
3. As in Figure 3b, partition T into two subseudoknots. T_{upper} : upper region including (p_h,q_h,r_h) and T_{lower} : lower region excluding (p_h,q_h,r_h).
4. Recursively find out the optimal alignment of S_{upper} and T_{upper}.
5. Recursively find out the optimal alignment of S_{lower} and T_{lower}.
6. Combine results from Steps 4 and 5 to obtain the optimal alignment.

Lemma 1. The above procedure runs in $O(mn^3)$ time with space complexity of $O(n^3)$.

Proof.

To analyze the total time complexity required for the whole procedure, let $K[m,n]$ be the total time required to align S of length m and T of length n. Both Steps 1 and 6 require $O(m)$ time. Step 2 takes $O(mn^3)$ time. For Step 4, it takes $K[m/2,n_1]$ time where, n_1=length of T_{upper} and for Step 5, it takes $K[m/2,n_2]$ time, n_2=length of T_{lower}.

$$
\begin{aligned}
&K[m,n]\\
&= O(mn^3) + K\left[\frac{m}{2},n_1\right] + K\left[\frac{m}{2},n_2\right]\\
&= O(mn^3) + O(\frac{m}{2}n_1^{\,3}) + O(\frac{m}{2}n_2^{\,3}) + K\left[\frac{m}{4},n_{11}\right] + K\left[\frac{m}{4},n_{12}\right] + K\left[\frac{m}{4},n_{21}\right] + K\left[\frac{m}{4},n_{22}\right]
\end{aligned}
$$

Since $(n_1+n_2)=n$, $(n_{11}+n_{12}+n_{21}+n_{22})=n$, and so on, and the fact that $\left(\sum_i n_i\right)^3 \geq \sum_i (n_i)^3$,

therefore, $n_1^3+n_2^3 \leq n^3$, $n_{11}^3+n_{12}^3+n_{21}^3+n_{22}^3 \leq n^3$ and so on

$$K[m,n] = O(mn^3) + O(\frac{m}{2}n^3) + O(\frac{m}{4}n^3) + ... = O(mn^3)$$

Thus, the whole procedure requires $O(mn^3)$ time. Since the memory space for the table in the score-only algorithm can be reused during the recursion, the total space required is $O(n^3)$.

3.4 *Structural alignment for embedded simple pseudoknot*

When considering the structural alignment for embedded-simple-pseudoknot, we extend the algorithm in [13] in order to apply the above procedure. We first binarize the query RNA converting it into a binary tree structure [13]. Each node represents a pair of nucleotides (which may not be a base-pair). Node *A* is a descendant of node *B* if pair *B* is inside pair *A*. A node would have two children if the region bounded by the pair can be partitioned into two embedded simple pseudoknot regions. If the pair bounds a simple pseudoknot region, then the node will be indicated as a simple pseudoknot and has no child but the corresponding triple sequence is formed and attached under the node. The tree ensures that each nucleotide is touched by at least one node, and every base pair is reached by one and only one node. No two pairs represented by two nodes are crossing each other. Let $A[i,j,v]$ be the score of optimal alignment between a target sequence $T[i...j]$ and the subtree rooted at the node v (v_i,v_j), which is also the subinterval (v_i, v_j) of the query sequence. The following shows the algorithm for embedded simple pseudoknot alignment.

<u>ALIGN(*S*[1..*m*],*T*[1..*n*])</u>

```
Binarizing the query S to obtain the binary tree M'.
for i = n-1 downto 1
        for all nodes v = (v_i,v_j) in M' (from leaves to root)
                if v is a simple pseudoknot
                        let V be the triple sequence of v
                                // return a set of optimal simple pseudoknot alignment scores
                                // between V and T[i,j] where i+1≤j≤n
                        C[i+1…n] = score-only-SP(V,T,i)
                for j = i+1 to n
                        if v is NIL
                                then A[i,j,NIL]=A[i,j-1,NIL]+γ(T[j],'_')
                        if v is a pseudoknot
                                then A[i,j,v]=C[j]
                        if v∈ M'
```

13.
$$A[i,j,v]=\max\begin{cases} //\ match \\ A[i+1,j-1,child(v)+\delta(S[v_i],S[v_j],T[i],T[j]) \\ \qquad\qquad +\gamma(S[v_i],T[i])+\gamma(S[v_j],T[j]) \\ //\ insertion \\ A[i,j-1,v]+\gamma('\_',T[j]) \\ A[i+1,j,v]+\gamma('\_',T[j]) \\ //\ deletion \\ A[i+1,j,child(v)]+\gamma(S[v_i],T[i])+\gamma(S[v_j],'\_') \\ A[i,j-1,child(v)]+\gamma(S[v_i],'\_')+\gamma(S[v_j],T[j]) \\ A[i,j,child(v)]+\gamma(S[v_i],'\_')+\gamma(S[v_j],'\_') \end{cases}$$

14. if $v \notin M'$ and has less than 2 children, A can be computed similarily as in Step 13 [10].

15. if $v \notin M'$ and has 2 children

16. $$A[i,j,v]=\max_{i\le k\le j+1}\{A[i,k-1,left\_child(v)]+A[k,j,right\_child(v)]\}$$

The function score-only-SP(V, $T[1\ldots n]$, p_0) is to compute a set of optimal simple pseudoknot alignment scores between V and $T[p_0,j]$ where $p_0+1\le j\le n$.

<u>score-only-SP(V, $T[1\ldots n]$, p_0)</u>

```
Initialize B_prev
for v = last-item to root in the triple sequence V
        for all p,q,r such that p0≤p<q<r≤n
                compute B[p,q,r] with respect to current v and B_prev
        B_prev ↔ B
Initialize C
for k = p0+2 to n
        for j = p0+1 to k-1
                if C[k]<B_prev[j-1,j,k]
                        C[k]=B_prev[j-1,j,k]
return C[p0+1…n]
```

The total time and space complexity for ALIGN() procedure (with score-only-SP()) is $O(mn^4)$ and $O(mn^2+n^3)$ respectively. After running the ALIGN() procedure, although the optimal alignments between S and T for the pseudoknotted regions are still unknown, we know the locations of the regions on T. Then, for each pseudoknotted region of T mapped to a pseudoknotted region of S, the previous divide-and-conquer procedure can be used to obtain the corresponding alignments. The time and space complexities for this are $O(mn^4)$ and $O(n^3)$, respectively. Therefore, we have the following lemma.

Lemma 2. The overall time and space complexities required for aligning $S[1..m]$ and $T[1..n]$ with embedded simple pseudoknots are $O(mn^4)$ and $O(mn^2+n^3)$, respectively.

4. Experimental Results

We implemented the memory efficient algorithm in C++. Since PAL [13] program is not available, we also implement their method for comparison on the performance. We selected ten RNA families which have embedded simple pseudoknot structures for the experiment. The sequence and structural information of all seed members in the family were downloaded from Rfam database [2]. For each family, we randomly picked one of the seed members as a query sequence and aligned it with the other members one by one.

In the experiment, we found that the time required for both algorithms are almost the same, however the difference between the memory consumption is large, especially for the families with long sequences. Our memory consumption is less than theirs in all ten families. For the families with short sequence length, say less than 70, their algorithm does not need much memory. However, their memory consumption increases dramatically for the families with long sequence length compared with ours. Table 1 shows the comparison of memory usage between our space-efficient algorithm and their algorithm for the families with sequence length greater than 70. For the family Telomerase-vert, their algorithm cannot be executed in our server because it consumes more than 4G memory. We estimated that the actual memory consumption would be more than 10G.

Table 1: Comparison on memory usage between our space-efficient algorithm and their algorithm for the families with sequence length greater than 70. *Mp* is the number of triples for the pseudoknotted region and *N* is the number of members.

RNA Family	Rfam Id	N	Ave. length	Mp	Ave. memory usage in our algorithm (*M*)	Ave. memory usage in their algorithm (*M*)
Corona_FSE	RF00507	23	82	53	5.2	30
Tymo_tRNA-like	RF00233	28	83	12	6.4	10
Parecho CRE	RF00499	5	112	21	11.0	31
IFN-gamma	RF00259	5	168	73	23.0	282
Telomerase-vert	RF00024	37	548	100	839.0	>4,000

5. Discussion and Conclusions

Since we are mainly interested in the RNA sequences (or substrings in the given genome) that have a high score with the given query sequence, we expect that the distances between the bases in two mapped base pairs will not differ a lot. In practice, we can impose an upper bound Δ on the difference of the lengths between the mapped base pairs in order to decrease the time complexity of the algorithm.

The algorithms in this paper are only designed for the embedded simple pseudoknot structures. We have scanned through the existing families and found that there exist other pseudoknot structures. Developing time and memory efficient algorithms for other pseudoknot structures would be essential and helpful for discovery of new members for these pseudoknotted ncRNA families.

Acknowledgments

The project is partially supported by Seed Funding Programme for Basic Research of the University of Hong Kong (200611159001).

References

1. Noncoding RNA database http://biobases.ibch.poznan.pl/ncRNA/
2. Griffiths-Jones S, Bateman A, Marshall M, Khanna A, Eddy SR (2003) Rfam: an RNA family database. NucleicAcid Research, 31(1):439–441, Jan. 2003. http://www.sanger.ac.uk/Software/Rfam/
3. Eddy S (2001) Non-coding RNA genes and the modern RNA world. Nature Reviews in Genetics 2, 919-929
4. Rietveld K, Van Poelgeest R, Pleij CW, Van Boom JH, Bosch L (1082) The tRNA-like structure at the 3'terminus of turnip yellow mosaic virus RNA. Differences and similarities with canonical tRNA. Nucleic Acids Res 10: 1929-1946
5. Hen J, Greider CW (2005) Functional analysis of the pseudoknot structure in human telomerase RNA. PNAS, 102(23):8080-8085
6. Dam E, Pleij K, Draper D (1992) Structural and functional aspects of RNA pseudoknots. Biochemistry, 31(47):11665-11676
7. Adams PL, Stahley MR, Kosek AB, Wang J, Strobel SA (2004) Crystal structure of a self-splicing group I intron with both exons. Nature 430: 45-50
8. Rivas E and Eddy S (2000) Secondary structure alone is generally not statistically significant for the detection of noncoding RNAs. Bioinformatics, 16(7):583-605
9. Klein R and Eddy S (2003) Rsearch: Finding homologs of single structured rna sequences. BMC Bioinformatics, 4(1):44
10. Zhang S, B Hass, E Eskin, V Bafna. (2005) Searching genomes for noncoding RNA using FastR, IEEE/ACM Transactions on Computational Biology and Bioinformatics, 2(4) October-December 2005.
11. Nawrocki EP, Eddy SR (2007) Query-Dependent Banding (QDB) for Faster RNA Similarity Searches, PLoS Comput. Biol., 3:e56
12. Matsui H, Sato K, Sakakibara Y (2005) Pair Stochastic Tree Adjoining Grammars for Aligning and Predicting Pseudoknot RNA Structures. Bioinformatics 21 2611-2617
13. Dost B, Han B, Zhang S, Bafna V (2006) Structural Alignment of Pseudoknotted RNA. RECOMB 2006, LNBI 3909, 43-158
14. Hirschberg DS (1975) *A linear space algorithm for computing maximal common subsequences.* Comm. A.C.M. **18**(6) 341-343
15. Le SY, Chen JH, Maizel J (1990) *Structure and Methods: Human Genome Initiative and DNA Recombination*, volume 1, chapter Efficient searches for unusual folding regions in RNA sequences, 127-130

A NOVEL METHOD FOR REDUCING COMPUTATIONAL COMPLEXITY OF WHOLE GENOME SEQUENCE ALIGNMENT

RYUICHIRO NAKATO[1]

E-mail: rnakato@genome.ist.i.kyoto-u.ac.jp

OSAMU GOTOH[1,2]

E-mail: o.gotoh@i.kyoto-u.ac.jp

[1]*Department of intelligence Science and Technology, Graduate School of informatics, Kyoto University, Yoshida-Honmachi, Sakyo-ku, Kyoto-shi, Kyoto 606-8501, Japan*

[2]*National Institute of Advanced Industrial Science and Technology, Computational Biology Research Center, 2-42 Aomi, Koto-ku, Tokyo 135-0064, Japan*

Genomic sequence alignment is a powerful tool for finding common subsequence patterns shared by the input sequences and identifying evolutionary relationships between the species. However, the running time and space requirement of genome alignment have often been very extensive. In this research, we propose a novel algorithm called Coarse-Grained AlignmenT (CGAT) algorithm, for reducing computational complexity necessary for cross-species whole genome sequence alignment. The CGAT first divides the input sequences into "blocks" with a fixed length and aligns these blocks to each other. The generated block-level alignment is then refined at the nucleotide level. This two-step procedure can drastically reduce the overall computational time and space necessary for an alignment. In this paper, we show the effectiveness of the proposed algorithm by applying it to whole genome sequences of several bacteria.

Keywords: Genome Alignment; Multiple Alignment; Sequence Analysis; Comparative Genomics.

1. Introduction

With the rapid increase in genomic sequence data available in recent years, there is a great demand for alignment programs that can allow direct comparison of the DNA sequences of entire genomes. However, whole genome sequence alignment is a difficult problem in the points of time and space complexity. Optimal pairwise alignment using Dynamic Programming (DP) requires $O(L^2)$ time and $O(L)$ space, where L is the length of an input sequence.[1] As the length of an entire bacterial genome usually exceeds 1Mb, application of full-blown DP is impractical, therefore, it is necessary to devise more efficient methods.

There are several existing algorithms for pairwise genomic sequence alignment. These algorithms generally apply fast word-search algorithms, such as suffix tree, suffix array, and look-up table, to extract high scoring pairs (HSPs) of subsequences from the input genome sequences. The HSPs are then chained to conform to coherent alignment.[2–6] If necessary, the chained HSPs may serve as anchor points to the subsequences between which are aligned by a standard DP algorithm. In this report, we propose a novel algorithm for pairwise alignment named Coarse-Grained AlignmenT (CGAT) algorithm. We developed a preliminary version of computer program, $Cgaln$, that implements the proposed algorithm. Comparison of the results of $Cgaln$ with those of $Blastz$[3] indicated that $Cgaln$ is as sensitive as $Blastz$ while considerably more specific than $Blastz$, when appropriate parameters are given. The block-level local alignments are generated in a very short period of time, and the overall computation speed was an order of magnitude faster than that of $Blastz$ with the default setting.

2. Method

2.1. *Outline*

Figure 1 shows the flow of CGAT. CGAT divides the input sequences into "blocks" with a fixed length. These blocks are taken as "elements" to be aligned. The similarity between two blocks, each from the two input sequences, is evaluated by frequency of words (k-mers) commonly found in the blocks. Similar methods based on word counts have been used for rapid estimation of the degree of similarity between two protein sequences.[7,8] For block-level alignment, we apply the Smith-Waterman local alignment algorithm[9] modified so that sub-optimal similarities are also reported.[10] The nucleotide-level alignment is conducted upon the restricted regions included in the block-level alignment found in the first stage. For the nucleotide-level alignment, we adopt a seed-extension strategy widely used in homology search programs such as $Blast$[2,3] and $PatternHunter$.[4]

2.2. *Block-level alignment*

Let's denote the given input genome sequences G_a and G_b. Let L_a and L_b be the lengths of G_a, G_b, respectively, and m_a and m_b be the numbers of blocks in G_a and G_b, respectively. Thus $m_a = \lceil L_a/J \rceil$ and $m_b = \lceil L_a/J \rceil$, where J is the length of a block. Let b_x^a be the x-th block of G_a and b_y^b be the y-th block of G_b. The measure of similarity between b_x^a and b_y^b is denoted by $M_{x,y}$. We evaluate $M_{x,y}$ by the frequency of words commonly found in both b_x^a and b_y^b, where a word is a contiguous or discrete series of nucleotides of length k (k-mer). (In the discrete case, the value for k refers to the "weight", i.e. the number of positions where nucleotide match is examined.[4]) Thus,

$$M_{x,y} = \sum_i (c - \log p_i^a p_i^b)\delta_a(k_i^x)\delta_b(k_i^y), \tag{1}$$

where the summation is taken for all k-mers, and $\delta_a(k_i^x) = 1$ if k_i is present in b_x^a, otherwise $\delta_a(k_i^x) = 0$. The same notation applies to $\delta_b(k_i^y)$ as well. p_i^a and p_i^b are the probabilities that the word k_i appears in b_x^a and b_y^b, respectively, assuming its random distribution along the entire genome. Thus,

$$p_i^a = n_{k_i}^a J/L_a \simeq n_{k_i}^a/m_a, \tag{2}$$

$$p_i^b = n_{k_i}^b J/L_b \simeq n_{k_i}^b/m_b, \tag{3}$$

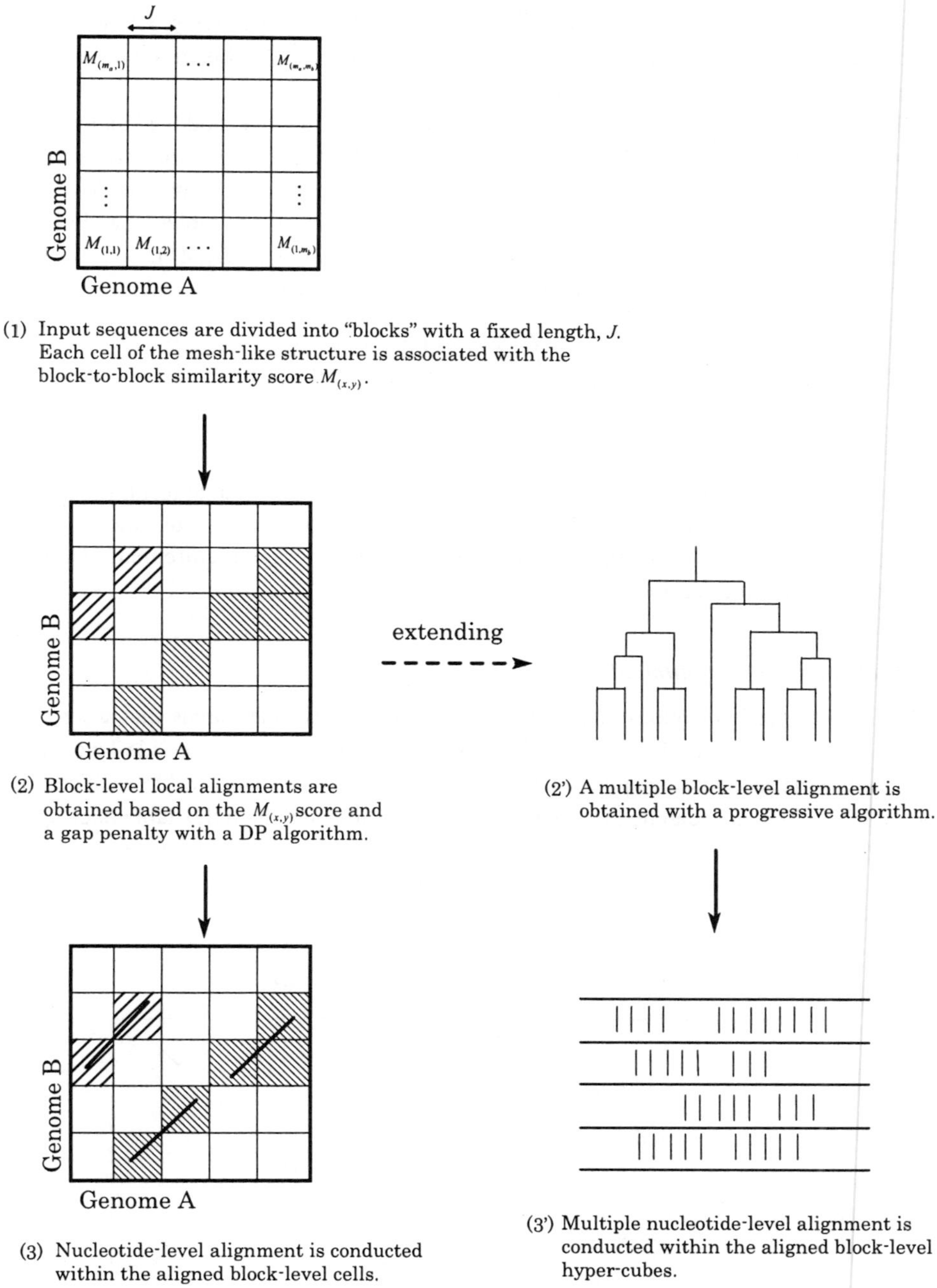

Fig. 1. The flow of CGAT algorithm. (1)-(2)-(3) is a pairwise alignment flow, and (1)-(2)-(2')-(3') is a multiple alignment flow (future work).

where $n^a_{k_i}$ and $n^b_{k_i}$ are the total numbers of k_i in G_a and G_a, respectively. The term c is a constant that may be estimated with some evolutionary model. At this moment, however, we treated c as an adjustable parameter.

The block-level local alignment uses two tables, the "word table" and the "index table." The word table stores the number of occurrences of each word in a genome, $n^a_{k_i}$ and $n^b_{k_i}$, whereas the index table stores the list of positions where a particular k-mer resides. These tables are made only once for each genomic sequence. Using these tables, the similarity measure matrix, $M_{x,y}$ ($x = 1..m_a, y = 1..m_b$), is obtained in $O(L_a L_b / 4^k)$.

The block-level alignment is conducted using DP as follows:

$$F_{x,y} = \max \begin{pmatrix} F_{x-1,y-1} + M_{x,y} \\ F_{x-1,y} + M_{x,y} - d \\ F_{x,y-1} + M_{x,y} - d \\ 0 \end{pmatrix}, \quad (4)$$

where d is the gap penalty. Equation (4) is based on Smith-Waterman algorithm.[9] For obtaining the optimal and suboptimal locally best matched alignments, we use the algorithm presented by Gotoh.[10] This method can greatly reduce the storage requirement, while the computational time remains $O(m_a m_b)$.

2.3. *Nucleotide-level alignment*

$Cgaln$ applies the nucleotide-level alignment within the restricted areas that were composed of cells included in the block-level local alignments. For the nucleotide-level alignment, we use the seed finding approach like $Blast$[2,3] or $PatternHunter$.[4] In each cell, the seed matches (hits) are searched for by using the word table and the index table once again. Figure 2 shows the nucleotide-level alignment within a cell. A group of hits are integrated into one larger matching segment if the hits are closer to each other than a threshold with

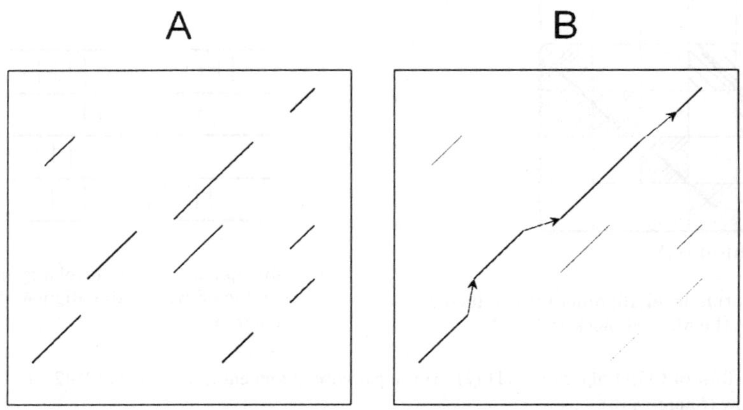

Fig. 2. The nucleotide-level alignment in a cell. (A) Finding seed matches and integrating into HSPs. (B) Computing a maximal-scoring ordered subset of HSPs, and chaining them.

no gap (laid on the same diagonal in the dot matrix). We define such a gap-less matching segment as a high scoring pair (HSP). Next, $Cgaln$ computes a maximal-scoring ordered subset of HSPs, and the HSPs are chained to one global alignment within each block-level local alignment. This step can eliminate the noise such as repeats.

2.4. *Computational complexity*

Here we describe the computational complexity of CGAT (see Table 1.) The block-level alignment phase takes space for the word table, the index table, and the DP matrix as major components. The word table requires $O(4^k)$ space, where k is the size of a word. Both the index table and the DP algorithm require $O(\min(m_a, m_b))$ space. However, we also use the index table at the nucleotide-level alignment, and hence the space requirement is formally $O(\min(L_a, L_b))$. Then, the space requirement for the block-level alignment is $O(4^k + \min(L_a, L_b))$.

Table 1. The computational complexity of CGAT algorithm

		Time		Space
Block-level alignment	making two tables	$O(\min(L_a, L_b))$	the word table	$O(4^k)$
	matrix scoring	$O(L_a L_b / 4^k)$	the index table	$O(\min(m_a, m_b))$ $(O(\min(L_a, L_b)))$
	the DP alignment	$O(m_a m_b)$	the DP matrix	$O(\min(m_a, m_b))$
	total	$O(m_a m_b + \min(L_a, L_b))$	total	$O(4^k + \min(m_a, m_b))$ $(O(4^k + \min(L_a, L_b)))$
Nucleotide-level alignment	generating HSPs	$O(JN_c)$	the two tables	$O(\min(L_a, L_b))$
	chaining HSPs	$O(\overline{n_h}^2 N_l)$	the HSPs	$O(\overline{n_h}^2)$
	total	$O(JN_c + \overline{n_h}^2 N_l)$	total	$O(\min(L_a, L_b) + \overline{n_h}^2)$

The time complexity is $O(\min(L_a, L_b))$ for making the two tables, $O(L_a L_b/4^k)$ for preparing the similarity matrix, and $O(m_a m_b)$ for DP alignment. As we choose k such that $\min(L_a, L_b)/4^k$ is not much greater than 1, the computational complexity is $O(m_a m_b + \min(L_a, L_b))$.

The nucleotide-level alignment phase takes $O(JN_c)$ time for generating HSPs, and $O(\overline{n_h}^2 N_l)$ time for chaining, where N_c is the number of cells included in the block-level local alignments, N_l is the number of the block-level local alignments, and $\overline{n_h}$ is the average number of HSPs included in each block-level local alignment. Then, the time requirement for the nucleotide-level alignment is $O(JN_c + \overline{n_h}^2 N_l)$. The space requirement is $O(\min(L_a, L_b))$ for the two tables and $O(\overline{n_h}^2)$ for chaining HSPs, and the total is $O(\min(L_a, L_b) + \overline{n_h}^2)$.

2.5. *Preparation of data*

Whole genome sequences of *E. coli* CFT073 (5,231,428 bp), *E. coli* O157 (5,498,450 bp), and *S. dysenteriae* (4,369,232 bp) were obtained from DDBJ.[a] Before applying the genomic sequences to alignment programs, we masked repetitive sequences by *WindowMasker*[16] with default parameters. As *WindowMasker* can mask genome sequences without a library of known repetitive elements, it is suitable for a comparative genome analysis.

All experiments were performed on a 2.0 GHz ($\times 2$) Xeon dual core PC with 4 Gbyte memory.

3. Results

3.1. *Comparison of accuracy by dotplots*

We compare the accuracy and computational time of *Cgaln* with *Blastz*.[3] *Blastz* is a pairwise alignment tool for long genomic DNA sequences, and it is used as an internal engine of several multiple genomic sequence alignment tools such as *MultiPipMaker*,[13] *TBA*,[14] *MultiZ*,[14] and Choi's algorithm.[15]

We obtained the global view of the results of *Cgaln* and *Blastz* by the dotplot outputs (Figure 3) for two kinds of pairwise alignments: (A) *E. coli* CFT073 vs. *E. coli* O157, and (B) *E. coli* CFT073 vs. *S. dysenteriae*. The *Cgaln* results were generated by gnuplot[b] whereas the *Blastz* results were generated by *PipMaker*.[c] We examined *Blastz* with two sets of parameter values; with the default parameter set and with a tuned parameter set (T=2, C=2). The option "T=2" disallows transitions, which speeds up computation but slightly reduces sensitivity. The option "C=2" directs "chain and extend", which contributes to reduction in noise.

We did not consider segmental inversion in comparison of the two *E. coli* strains, because the tight evolutionary relationship between the two sequences precludes such a possibility. In the case of cross-species comparison, we did consider the possibility of inversions. We adjusted the value for parameter c for each case of comparison, but the other parameters were unchanged.

3.2. *Comparison of computational time and memory*

Table 2 summarizes the actual computation time and memory used in our experiments. *Blastz* with the default parameters took nearly 200 s for either intra- or inter-species comparison. The computation time was considerably reduced with the tuned parameter set (T=2 and C=2). However, *Cgaln* runs faster than *Blastz* even with this tuned parameter set, spending only 40 s and 14 s (including inversions) for the intra- and inter-species comparisons, respectively. Of these total computation times, 7 s and 11 s were consumed

[a]http://www.ddbj.nig.ac.jp/
[b]http://www.gnuplot.info/
[c]http://pipmaker.bx.psu.edu/pipmaker/

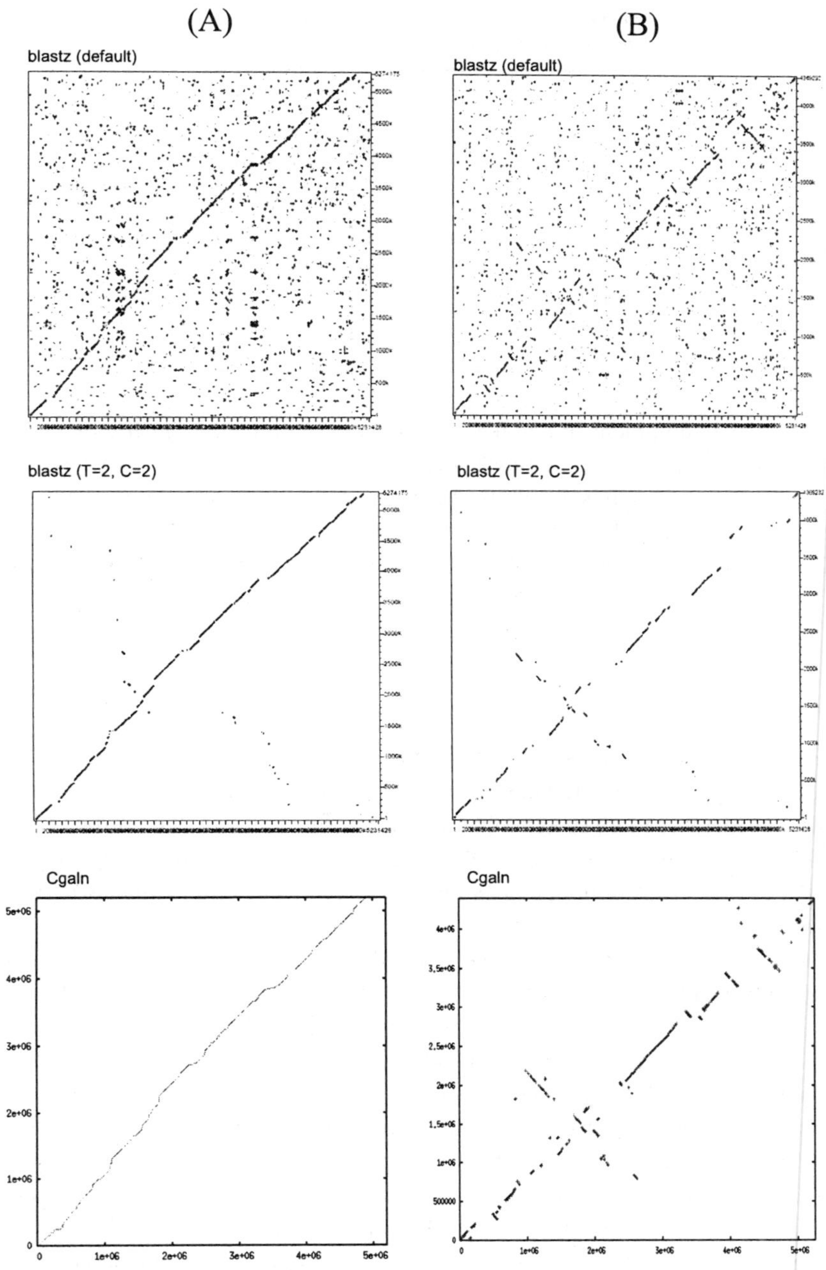

Fig. 3. (A) The dotplot outputs of the alignment between *E. coli* CFT073 and *E. coli* O157. (B) The dotplot outputs of the alignment between *E. coli* CFT073 and *S. dysenteriae*. Top: alignment by *Blastz* with the default parameter set. Middle: alignment by *Blastz* with a tuned parameter set (T=2, C=2). Bottom: alignment by *Cgaln*. In each alignment, the abscissa indicates *E. coli* CFT073 and the ordinate indicates the counterpart. *Cgaln* did not consider the inversions in (A), but considered the inversions in (B).

Table 2. Comparison of computational time and memory used by *Blastz* and *Cgaln*.

	time (s)	memory (MB)
(*E. coli* CFT073 - *E. coli* O157)		
Blastz (default)	224	222
Blastz (T=2, C=2)	51	197
Cgaln	40	155
(*E. coli* CFT073 - *S. dysenteriae*)		
Blastz (default)	192	201
Blastz (T=2, C=2)	36	179
Cgaln	14	143

by the block-level alignment. When the inversions were omitted, *Cgaln* took only 7 s for overall alignment and 5 s for the block-level alignment between *E. coli* CFT073 and *S. dysenteriae*. *Cgaln* requires a slightly smaller memory size than *Blastz*. This is reasonable because *Cgaln* uses 11 mer (11 match positions out of 18 word width) while *Blastz* uses 12 mer (12 match positions out of 19 word width) to index discrete words in their default settings, respectively. A larger k-mer generally increases both speed and memory consumption.

4. Discussion

Comparison of the results presented in Figure 3 indicates that *Cgaln* is as sensitive as *Blastz*, when appropriate parameters are given. Moreover, the results also indicate that *Cgaln* is considerably more specific than *Blastz* as illustrated by the drastic reduction in the level of noise. Although the noise level of *Blastz* output is appreciably attenuated by application of the "C=2" option, *Cgaln* appears to generate better outputs with respect to S/N ratios.

Because the performance of *Cgaln* strongly depends on the outcome of the block-level alignment, a proper choice of parameter values at this level (e.g c, d, and J) is essential for the overall accuracy of *Cgaln*. Although we currently determine these values in an *ad hoc* manner, it would be desired to develop a method for finding a suitable set of the parameter values automatically. More quantitative evaluation of the performance of *Cgaln* with more examples, in comparison with those of *Blastz* and other aligners, remains as a future task.

For the nucleotide-level alignment, we adopted a seed-extension strategy used in homology search programs such as *Blast*[2,3] and *PatternHunter*.[4] In view of sensitivity, this scheme can be improved by adding a recursive step which searches for the seed matches with progressively smaller k-mers in the inter-HSPs regions. The overall computational speed and memory requirement of *Cgaln* was superior to that of *Blastz*. This result suggests that *Cgaln* may be used for alignment of longer sequences such as entire mammalian chromosomes. In fact, we have already proved that the CGAT algorithm is successfully applied to all-by-all comparison of human and mouse chromosomes. However, it still requires prohibitively large memory to be executed on an ordinary computer. Further improvement in the algorithm would be necessary to reduce time and memory requirements.

The very short time consumed by the block-level alignment also suggests the capability of CGAT to be extended to the fast multiple genomic sequence alignment. For this purpose, it is necessary to solve the problems of how to adapt the block-level alignment to progressive or iterative algorithms, and how to treat the rearrangement such as inversions. These problems will be tackled in future work.

Acknowledgments

The authors would like to thank Drs. T. Yada and N. Ichinose for valuable discussions. This work was partly supported by a Grant-in-Aid for Scientific Research on Priority Areas "Comparative Genomics" from the Ministry of Education, Culture, Sports, Science and Technology of Japan.

References

1. Myers, E. and Miller, W. Optimal alignments in linear space. *Computer Applications in the Biosciences(CABIOS)*, Vol.4, No.1, pp. 11–17, 1988.
2. Altschul, S.F., Gish, W., Miller, W., Myers, E.W. and Lipman, D.J. Basic local alignment search tool. *Journal of Molecular Biology*, Vol.215, No.3, pp. 403–410, 1990.
3. Schwartz, S., Kent, W.J., Smit, A., Zheng, Z., Baertsch, R., Hardison, R. C., Haussler, D. and Miller, W. Human-mouse alignments with BLASTZ. *Genome Research*, Vol.13, No.1, pp. 103–107, 2003.
4. Ma, B., Tromp, J. and Li, M. PatternHunter: faster and more sensitive homology search. *Bioinformatics*, Vol.18, No.3, pp. 440–445 2002.
5. Delcher, A.L., Kasif, S., Fleischmann, R.D., Peterson, J., White, O. and Salzberg, S.L. Alignment of whole genomes. *Nucleic Acids Research*, Vol.27, No.11, pp. 2369–2376, 1999.
6. Brudno, M., Steinkamp, R. and Morgenstern, B. The CHAOS/DIALIGN WWW server for multiple alignment of genomic sequences. *Nucleic Acids Research*, Vol.32(Web Server issue), pp. w41–w44, 2004.
7. Edgar, R.C. Local homology recognition and distance measures in linear time using compressed amino acid alphabets. *Nucleic Acids Research*, Vol.32, No.1, pp. 380–385, 2004.
8. Jones, D.T., Taylor, W.R. and Thornton, J.M. The rapid generation of mutation data matrices from protein sequences. *Computer Applications in the Biosciences(CABIOS)*, Vol.8, No.3, pp. 275–282, 1992.
9. Smith, T.F. and Waterman, M.S. Identification of common molecular subsequences. *Journal of Molecular Biology*, Vol.147, No.1, pp. 195–197, 1981.
10. Gotoh, O. Pattern matching of biological sequences with limited storage. *Computer Applications in the Biosciences(CABIOS)*, Vol.3, No.1, pp. 17–20, 1987.
11. Brudno, M., Do, C.B., Cooper, G. M., Kim, M. F., Davydov, E., NISC Comparative Sequencing Program, Green, E. D., Sidow, A. and Batzoglou, S. LAGAN and Multi-LAGAN: Efficient Tools for Large-Scale Multiple Alignment of Genomic DNA. *Genome Research*, Vol.13, No.4, pp. 721–731, 2003.
12. Bray N. and Pachter L. MAVID: Constrained Ancestral Alignment of Multiple Sequences. *Genome Research*, Vol.14, No.4, pp. 693–699, 2004.
13. Schwartz, S., Elnitski, L., Li, M., Weirauch, M., Riemer, C., Smit, A., NISC Comparative Sequencing Program, Green, E.D., Hardison, R.C. and Miller, W. MultiPipMaker and supporting

tools: alignments and analysis of multiple genomic DNA sequences. *Nucleic Acids Research*, Vol.31, No.13, pp. 3518–3524, 2003.

14. Blanchette, M., Kent, W. J., Riemer, C., Elnitski, L., Smit, A. F. A., Roskin, K.M., Baertsch, R., Rosenbloom, K., Clawson, H., Green, E.D., Haussler, D. and Miller, W. Aligning Multiple Genomic Sequences With the Threaded Blockset Aligner. *Genome Research*, Vol.14, No.4, pp. 708–715, 2004.
15. Choi J., Choi, K., Cho, H. and Kim, S. Multiple Genome Alignment by Clustering Pairwise Matches. *Lecture Notes in Computer Science*, Vol.3388, pp. 30–41, 2005.
16. Morgulis, A., Gertz, E.M., Schaffer, A.A. and Agarwala, R. WindowMasker: window-based masker for sequenced genomes. *Bioinformatics*, Vol.22, No.2, pp. 134–141 (2006).

*f*RMSDAlign: PROTEIN SEQUENCE ALIGNMENT USING PREDICTED LOCAL STRUCTURE INFORMATION FOR PAIRS WITH LOW SEQUENCE IDENTITY

HUZEFA RANGWALA and GEORGE KARYPIS

Department of Computer Science and Engineering
University of Minnesota, Minneapolis, Minnesota 55414
Email: rangwala@cs.umn.edu

As the sequence identity between a pair of proteins decreases, alignment strategies that are based on sequence and/or sequence profiles become progressively less effective in identifying the correct structural correspondence between residue pairs. This significantly reduces the ability of comparative modeling-based approaches to build accurate structural models. Incorporating into the alignment process predicted information about the local structure of the protein holds the promise of significantly improving the alignment quality of distant proteins. This paper studies the impact on the alignment quality of a new class of predicted local structural features that measure how well fixed-length backbone fragments centered around each residue-pair align with each other. It presents a comprehensive experimental evaluation comparing these new features against existing state-of-the-art approaches utilizing profile-based and predicted secondary-structure information. It shows that for protein pairs with low sequence similarity (less than 12% sequence identity) the new structural features alone or in conjunction with profile-based information lead to alignments that are considerably better than those obtained by previous schemes.

1. Introduction

Over the years a wide range of comparative modeling-based methods[23,25,28] have been developed for predicting the structure of a protein (target) from its amino acid sequence. The central idea behind these techniques is to align the sequence of the target protein to one or more template proteins and then construct the target's structure from the structure of the template(s) using the alignment(s) as reference.

The overall performance of comparative modeling approaches[30] depends on how well the alignment, constructed by considering sequence and sequence-derived information, agrees with the structure-based alignment between the target and the template proteins. This can be quite challenging, as two proteins can have high structural similarity even though there exists very little sequence identity between them. This led to the development of sophisticated profile-based methods and scoring functions[1,5,8,10,19,31] that allowed high-quality alignments between protein pairs whose sequence identities are as low as 20%. However, these profile-based methods become less effective for protein pairs with lower similarities. As a result, researchers are increasingly relying on alignment scoring methods that also incorporate various predicted structural information such as secondary structure, backbone angles, and protein blocks.[7,12,13,20]

Recently we developed machine-learning methods[21] that can accurately estimate the root mean squared deviation (RMSD) value of a pair of equal-length protein fragments (i.e., contiguous backbone segments) by considering only sequence and sequence-derived information. Our interest in solving this problem is motivated by the operational characteristics of various dynamic-programming-based[18,29] protein structure alignment methods like CE[27] and MUSTANG[14] that score the aligned residues by computing the RMSD value of the optimal superimposition of the two fixed-length fragments centered around each residue. Thus, by being able to accurately predict the RMSD values of all these fragment-pairs from the protein sequence alone, we can enable the target-template alignment algorithms to use the same information as that used by the structure alignment methods.

In this paper we focus on studying the extent to which the predicted fragment-level RMSD (fRMSD) values can actually lead to alignment improvements. Specifically, we study and evaluate various alignment scoring schemes that use information derived from sequence profiles, predicted secondary structure, predicted fRMSD values, and their combinations. Results on two different datasets show that scoring schemes using the predicted fRMSD values alone and/or in combination with scores derived from sequence profiles lead to better alignments than those obtained by current state-of-the-art schemes that utilize sequence profiles and predicted secondary structure information, especially for sequence pairs having less than 12% sequence identity. In addition, we present two methods based on seeded alignments and iterative sampling that significantly reduce the number of fRMSD values that need to be predicted, without a significant loss in the overall alignment accuracy. This significantly reduces the computational requirements of the proposed alignment strategies.

The rest of the paper is organized as follows. Section 2 provides key definitions and notations used throughout the paper. Section 3 describes the datasets and the various computational tools used in this paper. Section 4 describes the scoring schemes used in our study and the various optimizations that we developed. Section 5 presents a comprehensive experimental evaluation of the methods developed. Finally, Section 6 summarizes the work and provides some concluding remarks.

2. Definitions and Notations

Throughout the paper we will use X and Y to denote proteins, x_i to denote the ith residue of X, and $\pi(x_i, y_j)$ to denote the *residue-pair* formed by residues x_i and y_j.

Given a protein X of length n and a user-specified parameter v, we define $v\texttt{frag}(x_i)$ to be the $(2v + 1)$-length contiguous substructure of X centered at position i $(v < i \leq n - v)$. These substructures are commonly referred to as fragments.[14,27] Given a residue-pair $\pi(x_i, y_j)$, we define $f\texttt{RMSD}(x_i, y_j)$ to be the *structural similarity score* between $v\texttt{frag}(x_i)$ and $v\texttt{frag}(y_j)$. This score is computed as the root mean square deviation between the pair of substructures after optimal superimposition. Finally, we define the *fRMSD estimation* problem as that of estimating the $f\texttt{RMSD}(x_i, y_j)$ score for a given residue-pair $\pi(x_i, y_j)$ by considering only information derived from the amino acid sequence of X and Y.

3. Materials

3.1. *Datasets*

We evaluate the accuracy of the alignment schemes on two datasets. The first dataset, referred to as the ce_ref dataset, was used in a previous study to assess the performance of different profile-profile scoring functions for aligning protein sequences.[5] The ce_ref dataset consists of 581 alignment pairs having high structural similarity but low sequence identity ($\leq$ 30%). The gold standard reference alignment was curated from a consensus of two structure alignment programs: FSSP[11] and CE.[27] The second dataset, referred to as the mus_ref dataset, was derived from the SCOP 1.57 database.[17] This dataset consists of 190 protein pairs with an average sequence identity of 9.6%. Mustang[14] was used to generate the gold standard reference alignments.

To better analyze the performance of the different alignment methods, we segmented each dataset based on the pairwise sequence identities of the proteins that they contain. We segmented the ce_ref dataset into four groups, of sequence identities in the range of 6-12%, 12-18%, 18-24%, and 24-30% that contained 15, 140, 203, and 223 pairs of sequences, respectively. We segmented the mus_ref dataset into three groups, of sequence identities in the range of 0-6%, 6-12%, and 12-30% that contained 76, 67, and 47 pairs of sequences, respectively. Note that the three groups of the mus_ref are highly correlated with the bottom three levels of the SCOP hierarchy, with most pairs in the first group belonging to the same fold but different superfamily, most pairs in the second group belonging to the same superfamily but different family, and most pairs in the third group belonging to the same family.

3.2. *Evaluation Methodology*

We evaluate the quality of the various alignment schemes by comparing the differences between the generated candidate alignment and the reference alignment generated from structural alignment programs.[5,6,24] As a measure of alignment quality, we use the Cline Shift score (CS)[2] to compare the reference alignments with the candidate alignments. The CS score is designed to penalize both under- and over-alignment and crediting the parts of the generated alignment that may be shifted by a few positions relative to the reference alignment.[2,5,22] The CS score ranges from a small negative value to 1.0, and is symmetric in nature. We also assessed the performance on the standard Modeler's (precision) and Developer's (recall) score,[24] but found similar trends to the CS score and hence do not report the results here.

3.3. *Profile Generation*

The profile[1] of a sequence X of length n is represented by two $n \times 20$ matrices, namely the position-specific scoring matrix $\mathcal{P}_X$ and the position-specific *frequency* matrix $\mathcal{F}_X$. These profiles capture evolutionary information for a sequence. The $\mathcal{F}_X(i)$ and $\mathcal{P}_X(i)$ are the ith column of X's position-specific scoring and frequency matrices. For our study, the profile matrices $\mathcal{P}$ and $\mathcal{F}$ were generated using PSI-BLAST[1] with the following parameters: `blastpgp -j 5 -e 0.01 -h 0.01`. The PSI-BLAST was performed against NCBI's nr database that was downloaded in November of 2004 and contained 2,171,938 sequences.

3.4. *Secondary Structure Prediction*

For a sequence X of length n we predict the secondary structure and generate a position-specific secondary structure matrix $\mathcal{S}_X$ of length $n \times 3$. The (i, j) entry of this matrix represents the strength of the amino acid residue at position i to be in state j, where $j \in (0, 1, 2)$ corresponds to the three secondary structure elements: alpha helices (H), beta strands (E), and coil regions (C). We use the state-of-the-art secondary structure prediction server YASSPP[13] (default parameters) to generate the $\mathcal{S}$ matrix. The values of the $\mathcal{S}$ matrix are the output of the three one-versus-rest SVM classifiers trained for each of the secondary structure elements.

3.5. *f* RMSD *Estimation*

To estimate the fRMSD scores for a residue-pair $\pi(x_i, y_j)$ we used the recently developed fRMSDPredprogram[21][a]. The fRMSDPredprogram uses an ϵ-SVR learning methodology to estimates the fRMSD score of a reside-pair $\pi(x_i, y_j)$ by taking into account the profile and the predicted secondary structure of a fixed-length window around the x_i and y_j residues. The ϵ-SVR estimation technique deploys a novel second-order pairwise exponential kernel function which shows superior results in comparison to the radial basis kernel function.

The ϵ-SVR implementation used the publicly available support vector machine tool SVM^{light}[26] which has an efficient ϵ-SVR implementation. We used the defaults for regularization and regression tube width parameters. The fRMSDPredprogram was trained on a dataset consisting of 1117 protein pairs derived from the SCOP 1.57 database. This training dataset was used in previous studies,[19,21] and no two protein domains in the dataset shared greater than 75% sequence identity. For each protein pair in the train dataset we use the standard Smith-Waterman[29] algorithm to generate the residue-pairs for which we compute the fRMSD score by considering fragment lengths of seven.

3.6. *Gap Modeling and Shift Parameters*

For all the different scoring schemes, we use a local alignment framework with an affine gap model, and a zero-shift parameter[31] to maintain the necessary requirements for a good optimal alignment.[9] We optimize the gap modeling parameters (gap opening (*go*), gap extension (*ge*)), the zero shift value (*zs*), and weights on the individual scoring matrices for integrating them to obtain the highest quality alignments for each of the schemes. Having optimized the alignment parameters on the ce_ref dataset, we keep the alignment parameters unchanged for evaluation on the mus_ref dataset.

---

[a]This work can be found at http://bioinfo.cs.umn.edu/supplements/fRMSDPred/

4. Methods

4.1. *Scoring Schemes*

We use the standard Smith-Waterman based local alignment[29] algorithm in our methods. The different alignment schemes vary in the computation of the position-to-position similarity scores between residue-pairs.

4.1.1. *Profile-Profile Scoring Scheme*

Many different profile-profile scoring functions[15,16,31] have been developed for determining the similarity between a pair of profile columns (i.e., residue-pairs). We use one of the best performing profile-profile scoring functions called PICASSO,[10,16] which computes the similarity between the ith position of protein's X profile and the jth position of the protein's Y profile as $\langle \mathcal{F}_X(i), \mathcal{P}_Y(j)\rangle + \langle \mathcal{F}_Y(j), \mathcal{P}_X(i)\rangle$. The operator $\langle , \rangle$ denotes a dot-product operation. We will refer to this scoring scheme as *prof.*

4.1.2. *Predicted Secondary Structure-based Scoring Scheme*

For a given residue-pair $\pi(x_i, y_j)$ we compute the similarity score based on the predicted secondary structure information as a dot-product of the ith row of $\mathcal{S}_X$ and the jth row of $\mathcal{S}_Y$, i.e., $\langle \mathcal{S}_X(i), \mathcal{S}_Y(j)\rangle$. This approach of incorporating secondary structure information along with profiles, has been shown to significantly improve the alignment quality.[20] We will refer to this scoring scheme as *ss*.

4.1.3. *fRMSD-based Scoring Scheme*

For a given residue-pair $\pi(x_i, y_j)$, we use the fRMSDPred program[21] to estimate its $f\text{RMSD}(x_i, y_j)$ score. Since this score is actually a distance, we convert it into a similarity score using the transformation: $\log(\alpha / f\text{RMSD}(x_i, y_j))$. This transformation assigns positive values to residue-pairs $\pi(x_i, y_j)$ having an estimated fRMSD score that is less than α. For the purposes of this study the α parameter was set to one, because we observed that the residue-pairs $\pi(x_i, y_j)$ with $f\text{RMSD}(x_i, y_j)$ score of less than one are more likely to be structurally aligned. We will refer to this scoring scheme as *frmsd*.

4.2. *Combination Schemes*

Besides the above scoring schemes, we also investigated their combinations. We used a weighted combination of the profile-based, predicted secondary, and fRMSD-based scoring schemes to compute a similarity score for a residue pair $\pi(x_i, y_j)$. In this approach the similarity score for a residue-pair $\pi(x_i, y_j)$, using the *prof* and *frmsd* scoring schemes is given by

$$\frac{w * prof(i,j)}{maxP} + \frac{(1-w) * frmsd(i,j)}{maxF}, \qquad (1)$$

where $prof(i,j)$ and $frmsd(i,j)$ represent the PICASSO and fRMSD scores for $\pi(x_i, y_j)$, respectively. The value $maxP$ ($maxF$) is the maximum absolute value of all *prof*-based (*frmsd*-based) residue-pair scores between the sequences and is used to normalize the different scores prior to addition. The parameter w defines the weighting for different parts of the scoring function after normalization. The optimal weight parameter w, was determined by varying w from 0.0 to 1.0 with increments of 0.1. This parameter was optimized for the ce_ref dataset, and the same value was then used for the mus_ref dataset.

A similar approach is used to combine *prof* with *ss* and *frmsd* with *ss*. In case of the *frmsd +prof +ss* there are two weight parameters that need to be optimized.

We will denote the various combination schemes by just adding their individual components, e.g., *frmsd +prof* will refer to the scheme that uses the scores obtained by *frmsd* and *prof*.

4.3. *Speedup Optimization*

For a residue-pair, we can compute the PICASSO- and secondary structure-based scores using two and one dot-product operations, respectively. In comparison, the fRMSD score needs $|SV|$ dot-product operations, where $|SV|$ is the number of support vectors determined by the ϵ-SVR optimization method. Hence, the *frmsd* alignment scheme has a complexity of at least $O(|SV|)$, which is significantly higher than that of the *prof* and *ss* alignment schemes. To reduce these computational requirements we developed two heuristic alignment methods that require the estimation of only a fraction of the total number of residue pairs.

4.3.1. *Seeded Alignment*

The first method combines the banded alignment approach and the seed alignment technique[9] and is performed in three steps. In the first step, we generate an initial alignment, referred to as the *seed alignment*, using the Smith-Waterman algorithm and the *prof +ss* scoring scheme. In the second step, we estimate the fRMSD scores for all residue-pairs within a fixed number of residues from the seed alignment, i.e., a *band* around the seed alignment. Finally, in the third step, we compute the optimal local alignment in the restricted band around the initial seed alignment. The computed *frmsd* alignment lies within a fixed band around the *prof +ss* alignment and will be effective if the original *frmsd* alignment and the *prof +ss* alignments are not very far away from each other. The complexity of this method can be controled by selecting bands of different sizes. We refer to this method as the seeded alignment technique.

4.3.2. *Iterative Sampling Alignment*

The second method employs an iterative sampling procedure to optimize the speed of the *frmsd* alignment. The basic idea is fairly similar to the seeded alignment. At iteration i, we estimate 1 out of R_i fRMSD scores in the dynamic-programming matrix for those residue-pairs that lie within the banded region of size K_i around the seed alignment generated in step $i-1$. K_i and R_i denote the band size and the sampling rate at iteration i, respectively. Using the estimated fRMSD scores, an alignment is produced at step i which serves as the seed alignment for step $i+1$. The band size is reduced by half, whereas the sampling rate is doubled at each step (i.e., R_i will be halved), effectively increasing the number of points in the dynamic-programming matrix to be estimated within a confined band. The first iteration can be assumed to have the initial seed as the main diagonal with a band size covering the entire dynamic-programming matrix.

5. Results

We performed a comprehensive study to evaluate the accuracy of the alignments obtained by the scoring scheme derived from the estimated *frmsd* values against those obtained by the *prof* and *ss* scoring schemes and their combinations. These results are summarized in Figures 1 and 2, which show the accuracy performance of the different scoring schemes on the ce_ref and mus_ref datasets, respectively. The alignment accuracy is assessed using the average CS scores across the entire dataset and at the different pairwise sequence identity segments. To better illustrate the differences between the schemes, the results are presented relative to the CS score obtained by the *prof* alignment and are shown on a $\log_2$ scale.

Analyzing the performance of the different scoring schemes we see that most of those that utilize predicted information about the protein structure (*ss*, *frmsd*, and combinations involving them and *prof*) lead to substantial improvements over the *prof* scoring scheme for the low sequence identity segments. However, the relative advantage of these schemes somewhat diminishes for the segments that have higher pairwise sequence identities. In fact, in the case of the 12%–30% segment for mus_ref, most of these schemes perform worse than *prof*. This result is not surprising, and confirms our earlier discussion in Section 1.

Comparing the *ss* and *frmsd* scoring schemes, we see that the latter achieves consistently and substantially better performance across the two datasets and sequence identity segments. For instance, for the first segment of ce_ref (sequence identities in the range of

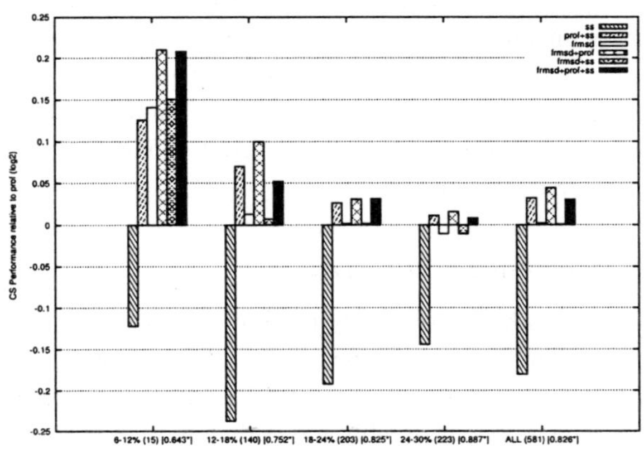

Fig. 1. Relative CS Scores on the ce_ref dataset. For each segment we display the range of percent sequence identity, the number of pairs in the segment, and the average CS score of the baseline *prof* scheme.

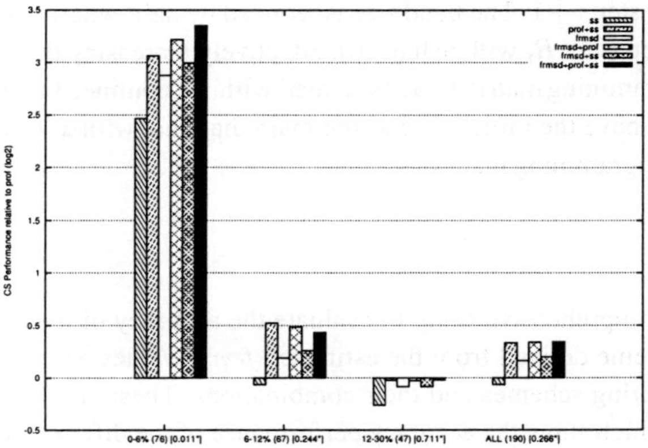

Fig. 2. Relative CS Scores on the mus_ref dataset. For each segment we display the range of percent sequence identity, the number of pairs in the segment, and the average CS score of the baseline *prof* scheme.

6%–12%), *frmsd*'s CS score is 20% higher than that achieved by the *ss* scoring scheme. In the first segment of mus_ref dataset (sequence identity in the range of 0%–6%), *frmsd*'s CS score is 33% higher than achieved by the *ss* scoring scheme, and is 19% higher for the second segment (sequence identity in the range of 6%–12%).

Comparing most of the schemes based on *frmsd* and its combinations with the other scoring schemes we see that for the segments with low sequence identities they achieve the best results. Among them, the *frmsd +prof* scheme achieves the best results for ce_ref, whereas the *frmsd +prof +ss* performs the best for mus_ref. For the first segments of ce_ref and mus_ref, both of these schemes perform 6.1% and 27.8% better than *prof +ss*, respectively, which is the best non-*frmsd*-based scheme. Moreover, for many of these segments, the performance achieved by *frmsd* alone is comparable to that achieved by the *prof +ss* scheme. Also, comparing the results obtained by *frmsd* and *frmsd +ss* we see that by adding information about the predicted secondary structure the performance does improve. In the case of the segments with somewhat higher sequence identities, the relative advantage of *frmsd +prof* diminishes and becomes comparable to *prof +ss*.

Finally, comparing the overall performance of the various schemes on the ce_ref and mus_ref datasets, we see that *frmsd +prof* is the overall winner as it performs the best for ce_ref and similar to the best for mus_ref.

5.1. *Comparison to Other Alignment Schemes*

Since the ce_ref dataset has been previously used to evaluate the performance of various scoring schemes we can directly compare the results obtained here with those presented in.[5] In particular, according to that study, the best PSI-BLAST-profile based scheme achieved a CS score of 0.805, which is considerably lower than the CS scores of 0.854 and 0.845 obtained by the *frmsd +prof* and *prof +ss*, respectively.

Also, to ensure that the CS scores achieved by our schemes on the mus_ref dataset are reasonable, we compared them against the CS scores obtained by the state-of-the-art

CONTRALIGN[3] and ProbCons[4] schemes. These schemes were run locally using the default parameters. CONTRALIGN and ProbCons achieved average CS scores of 0.197 and 0.174 across the 190 alignments, respectively. In comparison the *frmsd* scheme achieved an average CS score of 0.299, whereas *frmsd +prof* achieved an average CS score of 0.337.

5.2. *Optimization Performance*

We also performed a sequence of experiments to evaluate the extent to which the two runtime optimization methods discussed in Section 4.3 can reduce the number of positions whose fRMSD needs to be estimated while still leading to high-quality alignments. These results are shown in Figure 3, which shows the CS scores obtained by the *frmsd* scoring scheme on the ce_ref dataset as a function of the percentage of the residue-pairs whose fRMSD scores were actually estimated. Also, the figure shows the average CS score achieved by the original (not sampled) *frmsd* scheme.

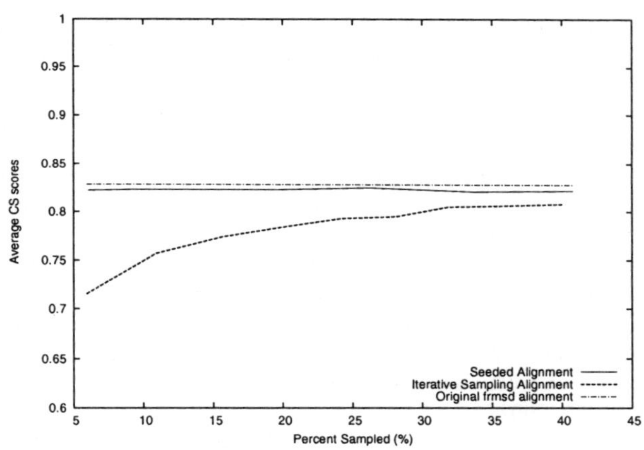

Fig. 3. Speedup using the Seeding and Sampling Alignment Procedure on the ce_ref dataset.

These results show that both the seeded and iterative sampling procedures generate alignments close to the alignment generated from the original complete scheme. The average CS scores of the seeded and iterative sampling alignment by computing just 6% of the original *frmsd* matrix is 0.822 and 0.715, respectively. The average CS score of the original *frmsd* scheme is 0.828. Hence, we get competitive scores by our sampling procedures for almost a 20 fold speedup. The seeded based technique shows better performance compared to the iterative sampling technique.

6. Conclusion

In this paper we evaluated the effectiveness of using estimated fRMSD scores to aid in the alignment of protein sequences. Our results showed that the structural information encoded in these estimated scores are substantially better than the corresponding information in predicted secondary structures and when coupled with existing state-of-the-art profile scoring schemes, they lead to considerable improvements in aligning protein pairs with very low sequence identities.

This approach of estimating the fragment-level RMSD is of similar spirit to learning a profile-profile scoring function to differentiate related and unrelated residue pairs using artificial neural networks.[19]

Acknowledgment

This work was supported by NSF EIA-9986042, ACI-0133464, IIS-0431135, NIH RLM008713A, the Digital Technology Center and the Minnesota Supercomputing Institute at the University of Minnesota.

References

1. S. F. Altschul, L. T. Madden, A. A. Schffer, J. Zhang, Z. Zhang, W. Miller, and D. J. Lipman. Gapped blast and psi-blast: a new generation of protein database search programs. *Nucleic Acids Research*, 25(17):3389–402, 1997.
2. M. Cline, R. Hughey, and K. Karplus. Predicting reliable regions in protein sequence alignments. *Bioinformatics*, 18:306–314, 2002.
3. C. B. Do, S. S. Gross, and S. Batzoglou. Contralign: Discriminative training for protein sequence alignment. *In Proceedings of the Tenth Annual International Conference on Computational Molecular Biology (RECOMB)*, 2006.
4. C. B. Do, M. S. P. Mahabashyam, and S. Batzoglou. Probcons: Probabilistic consistency-based multiple sequence alignment. *Genome Research*, 15:330–340, 2005.
5. R. Edgar and K. Sjolander. A comparison of scoring functions for protein sequence profile alignment. *BIOINFORMATICS*, 20(8):1301–1308, 2004.
6. A. Elofsson. A study on protein sequence alignment quality. *PROTEINS:Structure, Function and Genetics*, 46:330–339, 2002.
7. C. Etchebest, C. Benros, S. Hazout, and A. G. deBrevern. A structural alphabet for local protein structures: improved prediction methods. *Proteins: Structure, Function, and Bioinformatics*, 59(4):810–827, 2005.
8. M. Gribskov and N. Robinson. Use of receiver operating characteristic (roc) analysis to evaluate sequence matching. *Computational Chemistry*, 20:25–33, 1996.
9. Dan Gusfield. *Algorithms on Strings, Trees, and Sequences: Computer Science and Computational Biology*. Cambridge University Press, New York, NY, 1997.
10. A. Heger and L. Holm. Picasso:generating a covering set of protein family profiles. *Bioinformatics*, 17(3):272–279, 2001.
11. L. Holm and C. Sander. Mapping the protein universe. *Science*, 273(5275):595–602, 1996.
12. D. T. Jones, W. R. Taylor, and J. M. Thorton. A new approach to protein fold recognition. *Nature*, 358:86–89, 1992.
13. George Karypis. Yasspp: Better kernels and coding schemes lead to improvements in svm-based secondary structure prediction. *Proteins: Structure, Function and Bioinformatics*, 64(3):575–586, 2006.
14. A. S. Konagurthu, J. C. Whisstock, P. J. Stuckey, and A. M. Lesk. Mustang: a multiple structural alignment algorithm. *Proteins: Structure, Function, and Bioinformatics*, 64(3):559–574, 2006.
15. M. Marti-Renom, M. Madhusudhan, and A. Sali. Alignment of protein sequences by their profiles. *Protein Science*, 13:1071–1087, 2004.

16. D. Mittelman, R. Sadreyev, and N. Grishin. Probabilistic scoring measures for profile-profile comparison yield more accurate short seed alignments. *Bioinformatics*, 19(12):1531–1539, 2003.
17. A. G. Murzin, S. E. Brenner, T. Hubbard, and C. Chothia. Scop: a structural classification of proteins database for the investigation of sequences and structures. *Journal of Molecular Biology*, 247:536–540, 1995.
18. S. B. Needleman and C. D. Wunsch. A general method applicable to the search for similarities in the amino acid sequence of two proteins. *Journal of Molecular Biology*, 48:443–453, 1970.
19. T. Ohlson and A. Elofsson. Profnet, a method to derive profile-profile alignment scoring functions that improves the alignments of distantly related proteins. *BMC Bioinformatics*, 6(253), 2005.
20. J. Qiu and R. Elber. Ssaln: An alignment algorithm using structure-dependent substitution matrices and gap penalties learned from structurally aligned protein pairs. *Proteins: Structure, Function, and Bioinformatics*, 62(4):881–891, 2006.
21. H. Rangwala and G. Karypis. frmsdpred: Predicting local rmsd between structural fragments using sequence information. In *To Appear in Proceedings of the 2007 International Conference on Computational Systems Bioinformatics*, 2007.
22. H. Rangwala and G. Karypis. Incremental window-based protein sequence alignment algorithms. *Bioinformatics*, 23(2):e17–23, 2007.
23. R. Sanchez and A. Sali. Advances in comparative protein-structure modelling. *Current Opinion in Structural Biology*, 7(2):206–214, 1997.
24. J. M. Sauder, J. W. Arthur, and R. L. Dunbrack. Large-scale comparison of protein sequence alignments with structural alignments. *Proteins*, 40:6–22, 2000.
25. T. Schwede, J. Kopp, N. Guex, and M. C. Peltsch. Swiss-model: An automated protein homology-modeling server. *Nucleic Acids Research*, 31(13):3381–3385, 2003.
26. B. Schlkopf, C. Burges, and A. Smola, editors. *Making large-Scale SVM Learning Practical.* Advances in Kernel Methods - Support Vector Learning. MIT Press, 1999.
27. I. Shindyalov and P. E. Bourne. Protein structure alignment by incremental combinatorial extension (ce) of the optimal path. *Protein Engineering*, 11:739–747, 1998.
28. J. Skolnick and D. Kihara. Defrosting the frozen approximation: Prospector–a new approach to threading. *Proteins: Structure, Function and Genetics*, 42(3):319–331, 2001.
29. T. F. Smith and M. S. Waterman. Identification of common molecular subsequences. *Journal of Molecular Biology*, 147:195–197, 1981.
30. C. Venclovas and M. Margelevicius. Comparative modeling in casp6 using consensus approach to template selection, sequence-structure alignment, and structure assessment. *Proteins: Structure, Function, and Bioinformatics*, 7:99–105, 2005.
31. G. Wang and R. L. Dunbrack JR. Scoring profile-to-profile sequence alignments. *Protein Science*, 13:1612–1626, 2004.

RUN PROBABILITY OF HIGH-ORDER SEED PATTERNS AND ITS APPLICATIONS TO FINDING GOOD TRANSITION SEEDS

JIALIANG YANG* and LOUXIN ZHANG

Department of Mathematics, National University of Singapore,
Singapore 117543, Singapore
**E-mail: g0306107@nus.edu.sg*

Transition seeds exhibit a good tradeoff between sensitivity and specificity for homology search in both coding and non-coding regions. But, identifying good transition seeds is extremely hard. We study the hit probability of high-order seed patterns. Based on our theoretical results, we propose an efficient method for ranking transition seeds for seed design.

Keywords: High-order seed pattern, transition Seed, sensitivity analysis, run statistics.

1. Introduction

Biomolecular sequence comparison is one of the most important tasks in bioinformatics in the study of molecular evolution, genomics and molecular medicine. As a result, many sequence comparison programs have been developed to meet the challenge of the rapid increase in size of sequence databases.

The seed alignment is the dominant technique for homology search and genomic sequence alignment. Such a technique was first implemented in BLASTN program.[1] In BLASTN, a local alignment is found by first identifying exact matches of eleven contiguous residues between the two input sequences, called *seed hits*, and then extending them on either side for approximate matches by dynamic programming. The resulting alignments are scored for acceptance. In recent years, more general patterns of conservation have been proposed as seeds for sequence alignment.[5,8,14,20,24] Different seeds are also used as anchor point in whole-genome and multiple sequence alignments.[3,6,11]

Good spaced seeds improve tremendously the sensitivity of seed alignment while keeping speed unchanged.[20] Hence, seed design is an important aspect of seeded alignment. Identifying good seeds relies on efficient computation of seed sensitivity. Dynamic programming[13] and recurrence[10] methods were proposed for computing the sensitivity of the spaced seeds on a simple i.i.d. ungapped alignment model. It has been shown that computing the sensitivity of a spaced seed is NP-hard.[18] Hence, efficient heuristic methods were also developed for identifying good spaced seeds.[7,9,10,15,23,27,29] Algorithms for multi-seed design were also developed.[16,19,21,26,28]

Transition and transversion were first incorporated into seed design in BLASTZ.[24] This leads to the study of the transition seeds that contain fixed match and transition positions. Transition seeds exhibit a good tradeoff between sensitivity and specificity for homology search in both coding and non-coding regions.[27,31] However, identifying good transition seeds is a difficult task. Here we study the run probabilities of high-order seed-like patterns, which include spaced seeds and transition seeds as special cases. We generalize the theoretic study of spaced seeds[30] to high-order seed-like patterns. Using these results, we propose an efficient method for ranking transition seeds for the purpose of seed design.

Due to the space limit, we omit the proofs of the theorems stated in this extended abstract. The reader is referred to the full version of this paper for all proofs.

2. Seeds, Sensitivity and Specificity

2.1. *Spaced seeds*

A **(basic) spaced seed** is defined as a list of indices $\{i_1, i_2, \ldots, i_w\}$ with $i_1 = 0$. It can also be specified by a string $1 *^{i_2-1} 1 *^{i_3-i_2-1} 1 \ldots 1 *^{i_w-i_{w-1}-1} 1$ over alphabet $\{1, *\}$. Two sequences S_1 and S_2 exhibit a seed match at positions x and y if, for $1 \le k \le w$, $S_1[x+i_k] = S_2[y+i_k]$. The number of match positions w is defined to be the weight of the seed; the span of the checked positions, $i_w + 1$, is called the length of the seed.

A **transition spaced seed** is defined as a pair of disjoint lists of indices:

$$M = \{i_1, i_2, \ldots, i_{m'}\}, \quad Z = \{j_1, j_2, \ldots, j_{t'}\}$$

with $i_1 = 0$ or $j_1 = 0$. Two sequences S_1 and S_2 exhibit a match of the transition seed at positions x and y if, for $1 \le k \le m'$, $S_1[x+i_k] = S_2[y+i_k]$ and, for $1 \le k \le t'$, $S_1[x+j_k] = S_2[y+j_k]$, or two residues $S_1[x+j_k]$ and $S_2[y+j_k]$ are both pyrimidines or purines. The positions in M are called match positions; m' is defined to be the match weight of the seed. The positions in Z are called transition positions; t' is defined to be the transition weight of the seed. The length of the seed is defined to be $\max\{i_{m'}, j_{t'}\}+1$. Equivalently, we specify a transition seed of length L_Q by a string of length L_Q over alphabet $\{1, \#, *\}$ in which 1s represent match positions, #s transition positions, and $*$s other so-called 'don't care' positions.

2.2. *Seed sensitivity and specificity*

The sensitivity of a seed is the probability that a biologically meaningful alignment contains a match to the seed. The biological meaningful alignments are usually given through a probabilistic model on nucleotides. Here, we restrict ourselves to the Bernoulli or zero-th order Markov ungapped alignment model. We assume the pair of residues in each position is independently and identically generated from $\{A, G, C, T\} \times \{A, G, C, T\}$. By using 1s and 0s to represent matches and mismatches in the ungapped alignment between two sequences X and Y, a seed match can be viewed as an occurrence of the spaced seed in a binary sequence. Therefore, in the Bernoulli sequence model, the sensitivity of a spaced seed is defined to be the hit probability of a spaced seed pattern in a binary random sequence of a fixed length L (which is 64 by default).

Similar to basic spaced seeds, each match to a transition seed in an ungapped alignment can be viewed as an occurrence of the seed in the corresponding sequence over $\{1, 2, 3\}$, where we use 1s, 2s, 3s to represent matches, mismatches and transversions respectively.

A seed's specificity is defined to be one minus the probability that the seed match occurs in an alignment between two unrelated random sequence by chance. Therefore, the specificity is also a kind of hit probability in a probabilistic alignment model.

3. High-order Seed Patterns and Their Run Probabilities

Motivated by analyzing seed sensitivity and specificity, we study the run probabilities of sequence patterns of a special type in this section.

Let $\Sigma = \{b_1, b_2, \ldots, b_m\}$. An order-$t$ pattern $\mathcal{P}$ consists of a sequence Q of length L_Q on an alphabet $\Sigma' = \{a_1, a_2, \ldots, a_t\}$ and an ordered list of subsets $\{\Sigma_1, \Sigma_2, \ldots, \Sigma_t\}$ such that $Q[1] \neq a_t$, $Q[L_Q] \neq a_t$, and $\Sigma_1 \subset \Sigma_2 \subset \ldots \subset \Sigma_t = \Sigma$. We say the pattern Q to hit a sequence S on Σ at position k if, for $1 \leq i \leq L_Q$, the following condition is satisfied: if $Q[i] = a_j$ for some j, then, $S[i + k - L_Q] \in \Sigma_j$.

Example 3.1. (1)A basic spaced seed π is an order-2 pattern with a sequence over $\{1, *\}$ and the subset list: $\{1\}$, $\{0, 1\}$. (2) A transition seed π_t is an order-3 pattern with a sequence over $\{1, \#, *\}$ and ordered subset list: $\{1\}$, $\{1, 2\}$, $\{1, 2, 3\}$.

We study the hit probability of an order-t pattern in the Bernoulli random sequence on alphabet $\Sigma = \{b_1, b_2, \cdots, b_m\}$, in which a letter b_i is generated with probability p_i at each position and $\sum_{1 \leq i \leq m} p_i = 1$. We use $\mathcal{M}(\Sigma, p_1, p_2, \ldots, p_m)$ to denote this Bernoulli sequence model.

For an order-t pattern Q and a Bernoulli random sequence S, we use q_n to denote the hit probability that the pattern Q hits S before or at position n; we also use f_n to denote the probability that Q first hits S at position n. Let the length of the pattern Q be L_Q. Obviously,

$$f_i = 0, 1 \leq i \leq L_Q - 1,$$

$$f_{L_Q} = p_1^{w_1}(p_1 + p_2)^{w_2} \cdots (\sum_{i=1}^{m-1} p_i)^{w_{m-1}},$$

where w_i is the number of occurrences of the letter a_i in the pattern, and

$$q_n = \sum_{1 \leq i \leq n} f_i. \tag{1}$$

Let $\Sigma = \{b_1, b_2, \cdots, b_m\}$. Given an order-$t$ pattern $\mathcal{P}$ with sequence Q and ordered list of subsets of $\Sigma : \{\Sigma_1, \Sigma_2, \ldots, \Sigma_t\}$ and a sequence S on Σ. By encoding the letters in $\Sigma_i - \Sigma_{i-1}$ by a new letter b'_i, we transform the sequence S into a sequence S' on $\Sigma'' = \{b'_1, b'_2, \ldots, b'_t\}$. Let $\mathcal{P}'$ be the order-t pattern with sequence Q' and ordered list of subsets $\{\{b'_1\}, \{b'_1, b'_2\}, \ldots, \{b'_1, b'_2, \ldots, b'_t\}\}$. It is easy to see that the hit probability of $\mathcal{P}$ on sequence S in Bernoulli model $\mathcal{M}(\Sigma, p_1, p_2, \ldots, p_m)$ is equal to the hit probability of $\mathcal{P}'$ on sequence S' in Bernoulli model $\mathcal{M}'(\Sigma'', p'_1, p'_2, \ldots, p'_t)$, where $p'_i = \sum_{j:b_j \in \Sigma_i - \Sigma_{i-1}} p_j$. Therefore, for simplicity, we will focus on order-t patterns with a sequence and an order list of t subsets of an alphabet with size t in the rest of paper.

3.1. *A recurrence formula for hit probability*

Let Q be an order-t pattern and S be a random sequence in Bernoulli model $\mathcal{M}(\Sigma, p_1, p_2, \ldots, p_t)$. We use E_n be the event that Q hits sequence S at position n and $\bar{E}_n$ its complement event. We use $M_Q = \{Q_1, Q_2, \ldots, Q_h\}$ to denote all $h := \prod_{i=2}^{t} i^{w_i}$ distinct sequences obtained from Q by replacing each occurrence of a_i with a letter in $\Sigma = \{b_1, b_2, \ldots, b_t\}$. Taking a transition seed $Q = 1 * \#1$ as an example, we have $M_Q = \{1111, 1211, 1311, 1121, 1221, 1321\}$.

For $1 \le i \le h$, we define $E_n^{(i)}$ to be the event that Q_i hits S at position n. Obviously, $E_n^{(i)}$ and $E_n^{(j)}$ are disjoint for $1 \le i \ne j \le h$. Define $f_n^{(i)} = P[\bar{E}_1 \bar{E}_2 \cdots \bar{E}_{n-1} E_n^{(i)}]$, the probability that Q first hits S at position n and $S[n - L_Q + 1, n] = Q_i$. Clearly, Q hits S at position n if and only if some Q_i hits S at position n and so $E_n = \bigcup_{1 \le i \le h} E_n^{(i)}$. This implies that

$$f_n = \sum_{i=1}^{h} f_n^{(i)}. \tag{2}$$

Let x be a sequence with length $|x|$. For an integer $k \le |x|$, we use $x\langle k]$ and $x[k\rangle$ to denote the length-k suffix and prefix of x respectively. For any i, j and k, $1 \le i, j \le h$, $1 \le k \le L_Q$, we define

$$p_k^{(ij)} = \begin{cases} P[\ Q_j\langle L_Q - k]\ ] & \text{if } k \le L_Q - 1 \text{ and } Q_i\langle k] = Q_j[k\rangle \\ 1 & k = L_Q \text{ and } i = j \\ 0 & \text{otherwise} \end{cases}$$

we have

$$(1 - q_n)p_j = \sum_{i=1}^{h} \sum_{k=1}^{L_Q - 1} f_{n+k}^{(i)} p_k^{(ij)} + f_{n+L_Q}^{(j)} \tag{3}$$

for $1 \le j \le h$, where p_j is the probability that Q_j occurs at the position L_Q.

Using (1) - (3), one can compute the hit probability of a pattern recursively. It was first proved for the basic spaced seeds.[10]

3.2. *An Inequality on Hit Probability*

In this section, we present an inequality that relating the first hit probability to hit probability at different positions.

Theorem 3.1. *Let Q be an order-t pattern of length L_Q. Then, for any $2L_Q - 1 \leq k \leq n$, $f_k(1 - q_{n-k+L_Q-1}) \leq f_n \leq f_k(1 - q_{n-k})$ in a Bernoulli sequence model.*

3.3. *Asymptotic Analysis of Hit Probability*

Buhler *et al.*[7] proved that for any basic spaced seed Q, there exist two constants α_Q and λ_Q such that $1 - q_n \sim \alpha_Q \lambda_Q^n$. Similar results are established by Solov'ev.[25] Such an approximation also exists for our high-order pattern.

Theorem 3.2. *For an order-t pattern Q, there exist constants α_Q and λ_Q do not depend on n, such that $q_n = 1 - \alpha_Q \lambda_Q^n (1 + o(R^n))$ with $0 < R < 1$ in a Bernoulli model $\mathcal{M}(\Sigma, p_1, p_2, \cdots, p_t)$.*

The single term $1 - \alpha_Q \lambda_Q^n$ gives a very close approximation to q_n even for relative big n. Consider a specific transition seed $1 * 1$ containing no #s. In the Bernoulli sequence model $\mathcal{M}(\{1, 2, 3\}, p = 0.6, q = 0.3, r = 0.1)$, we obtain that $\lambda_Q = 0.7291502607$ and $\alpha_Q = 1.058452825$ using Maple.

In general, it is not easy to compute α_Q and λ_Q for an order-t pattern when t and L_Q are large. However, we will establish good bounds for λ_Q using the average distance between successive non-overlapping hits.

4. The Average Distance Between Successive Non-overlapping Hits

Renewal theory studied recurrent events connected with repeated trials. A recurrent event qualifies for the theory if the number of trials between successive occurrences of the event are jointly independent random variables with identical distribution. An non-overlapping hit of a pattern Q is a recurrent event under the following assumption: If a hit at position i is selected as a non-overlapping hit, then the next non-overlapping hit is the first hit at or after position $i + L_Q$.

The average distance between successive non-overlapping hits μ_Q is a very important parameter in the renewal theory. For the purpose of studying the hit probabilities of a pattern, it is formally rewritten as

$$\mu_Q = \sum_{i=0}^{\infty} i f_i.$$

Since $\sum_{j=L_Q}^{\infty} f_j = 1$ and $1 - q_i = \sum_{j=i+1}^{\infty} f_j$ for all $i \geq L_Q$, μ_Q may also be defined as

$$\mu_Q = L_Q + \sum_{i=L_Q}^{\infty} (1 - q_i).$$

4.1. *Bounding μ_Q*

Let Q be an order-t pattern on alphabet $\Sigma' = \{a_1, a_2, \ldots, a_t\}$. For $1 \le k \le t$, we define $\mathcal{RP}(k)$ to be the ordered list of indices i such that $Q[i] = a_k$. For any $0 \le j \le L_Q - 1$, we define

$$\mathcal{RP}(k) + j = \{i + j \mid i \in \mathcal{RP}(k)\}.$$

For $1 \le k, k' \le t$ and $1 \le j \le L_Q - 1$, set

$$O_Q^j(k, k') = |\mathcal{RP}(k) \cap (\mathcal{RP}(k') + j)|,$$

which is the number of common indices in $\mathcal{RP}(k)$ and $\mathcal{RP}(k') + j$. Note that $O_Q^j(k, k') \neq O_Q^j(k', k)$ for different k' and k in general. For $1 \le k \le t$, define

$$O_Q^j(k) = O_Q^j(k, k) + \sum_{k'<k} O_Q^j(k', k) + \sum_{k'<k} O_Q^j(k, k').$$

Theorem 4.1. *With notations above,*

$$\mu_Q \le \sum_{j=0}^{L_Q-1} \frac{1}{\prod_{k=1}^{t-1} \left(\sum_{i=1}^{k} p_i\right)^{O_Q^j(k)}}. \tag{4}$$

This is a generalization of a result proved for basic spaced seeds in.[13] Applying it to transition seed, we have the following fact.

Theorem 4.2. *For any transition seed Q,*

$$\mu_Q \le \sum_{j=0}^{L_Q-1} \frac{1}{p^{O_Q^j(1)}(p+q)^{O_Q^j(2)}} \tag{5}$$

in a Bernoulli sequence model $\mathcal{M}(\{1, 2, 3\}, p, q, r)$.

The bound given above is quite tight when the generation probabilities are large. Consider transition seed $Q = 11\#\#1 * \#11$ in the Bernoulli sequence model $\mathcal{M}(\{1, 2, 3\}, p, q, r)$. Figures 1 shows both the exact μ_Q and its upper bound in Theorem 4.2. As shown in the figure, μ_Q and the bound get closer and closer when one of the generation probabilities goes to large.

4.1.1. *Bounding λ_Q in terms of μ_Q*

In this subsection, we will present lower and upper bounds for the constant λ_Q appeared in Theorem 3.2 in terms of μ_Q. A similar result was proved for basic spaced seeds in.[30]

Theorem 4.3. *For any t-order pattern Q of length L_Q,*

$$1 - \frac{1}{\mu_Q - L_Q + 1} \le \lambda_Q \le 1 - \frac{1}{\mu_Q}. \tag{6}$$

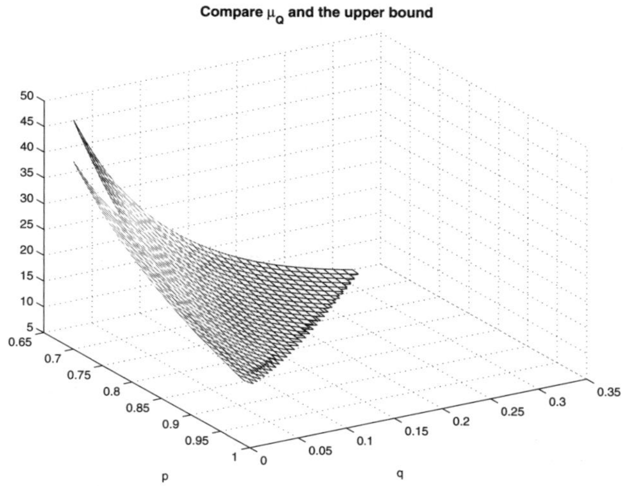

Fig. 1. μ_Q and its upper bound when p from 0.7 to 1 and q from 0 to 0.3 for $Q = 11\#\#1 * \#11$.

5. Identifying Good Transition Seeds

Transition seeds exhibit a good tradeoff between sensitivity and specificity for homology search in both coding and non-coding regions.[27,31] However, identifying good transition seeds is a hard task. This is because computing sensitivity is much harder for transition seeds than for basic spaced seeds of the same weight. The weight of a transition seed is defined as its match weight plus the half of its transition weight. By definition, an optimal seed is the seed with the highest sensitivity. In,[17] Kucherov, Noe and Roytberg gave an automata-based method for computing the sensitivity of a basic or transition seed. Such a method takes an exponential number of bit operations in the worst case. Another method for searching good spaced seeds is to use the hill-climbing strategy.[27] Here, based on our theoretical study in the previous sections, we propose an alternative method for the purpose. The efficiency of this method has been demonstrated for basic spaced seed search in[15] and.[29]

Recall that the sensitivity of a spaced seed is defined as the hit probability of the seed in a random sequence of a fixed length L (which is set to 64 traditionally). By Theorem 3.2, the sensitivity of a transition seed Q is closely related to the value of λ_Q. For two transition seeds P and Q, if $\lambda_P < \lambda_Q$, the sensitivity of P is asymptotically larger than Q. Moreover, Theorem 6 indicates that λ_Q can be approximated by a function of μ_Q. Therefore, we propose to identify good transition seeds using the tight bound of μ_Q established in Theorem 4.1. More specifically, we rank the transition seeds Q by the value of $V_Q = \sum_{j=0}^{L_Q-1} p^{-O_Q^j(1)}(p+q)^{-O_Q^j(2)}$. The smaller the value of V_Q is, the higher it is in the ranking list. Given a weight and a Bernoulli model, we identify the ten top transition seeds of the weight and the use the sensitivity in a region of length 64 to select the best one among these ten seeds.

Given a transition seed and a Bernoulli model, the value of V_Q can be simply calculated in a polynomial number of bit operations. Therefore, our heuristic method is much faster than using the sensitivity on a length-64 region to select good transition seeds. In most of cases, our selected seeds are optimal as shown in Table 1 and Table 2.

Table 1. Good transition seeds in Bernoulli model $\mathcal{M}(\{1,2,3\}, 0.7, 0.15, 0.15)$

w	Optimal seeds with w2=2	Sensitivity	Rank	Optimal seeds with w2=4	Sensitivity	Rank
9	111#*1*1**1#*11	0.73745	2	111*#*#1**1#*1#1	0.73806	5
10	111#*1*#*11**1*11	0.60424	8	111#**1#*1*#1*1#1	0.60692	5
11	111*1*1#*1**1#*111	0.47610	1	111#*1*1*#1**1##11	0.48016	1
12	111#*1*1**11*#1*111	0.36368	4	111##*11**1#*1*1#11	0.36692	1
13	111#1**11*1**1#1*111	0.27085	1	111#*1#*11*#*1*1*#111	0.27420	1
14	1111*1*1*#*11*11*#111	0.19760	6	1111#*1*1*#1*#11*#111	0.20077	1
15	1111*#1*1*11**11*1#111	0.14251	3	1111*#11*#1*1#*1*1#111	0.14494	10
16	1111*1*11#*1*11**11#111	0.10165	2	111#11*#*11*1*#11*1#111	0.10360	1
17	11111*#11**11*1*11*1#111	0.07185	3	1111#1*#11*1*1#*11#*1111	0.07333	5

Table 2. Good transition seeds in Bernoulli model $\mathcal{M}(\{1,2,3\}, 0.8, 0.1, 0.1)$

w	Optimal seeds with w2=2	Sensitivity	Rank	Optimal seeds with w2=4	Sensitivity	Rank
9	111#*1*1**1#*11	0.97266	2	111*#*#1**1#*1#1	0.97026	6
10	111*1**1*#1#*111	0.93711	8	11#1**1*1#**1#*#11	0.93405	3
11	111*#*11**1**1*1#11	0.88361	1	111#*1*#*1*#1**1#11	0.88046	1
12	111#*1*1**11*#1*1111	0.81402	4	111#*1*1*#1**1#*#111	0.81037	1
13	111#1**11*1**1#1*111	0.73263	5	111#*1#*11*#*1*1*#111	0.73019	5
14	1111*1*#*11**1*11*#111	0.64523	10	1111*#1**1#*1*1*1#*#111	0.64336	2
15	1111*#1*1*1*11**1#*1111	0.55886	9	111#1*1*1#*1*#1**11*#111	0.55729	6
16	1111*11*#1**11#*1*1*1111	0.47593	7	1111#*1#1**1*#11**1*1#111	0.47507	7
17	11111*1*1#*1*11**11*#1111	0.39955	1	1111#1*1*1#*11**1#1*#1111	0.39915	1

In these two tables, we list the ranks of the optimal transition seeds of weight nine to seventeen and transition weight two or four in model $\mathcal{M}(\{1,2,3\}, 0.7, 0.15, 0.15)$ and $\mathcal{M}(\{1,2,3\}, 0.8, 0.1, 0.1)$, respectively. In all the cases considered, the optimal transition seeds are among the top ten transition seeds selected according to V_Q.

In addition, the best transition seeds reported in Table 1 are identical to those reported in[17] for weight from nine to twelve. Here, we also list good transition seeds for weight from thirteen to seventeen.

6. Conclusion

We have studied the run probabilities of a high-order pattern in the Bernoulli sequence model. Both basic spaced and transition seeds are just order-2 and order-3 patterns respectively. We first establish a recurrence formula for computing the hit probability of a high-order pattern; then, we analyze asymptotically the hit probability. We establish a relationship between the hit probability and the average distance between two non-overlapping hits. For future work, one interesting problem is how to generalize the study to higher-order Markov sequence models.

By applying the theoretical results mentioned above, we present an efficient algorithm for identifying good transition seeds. This algorithm can also be adopted to identify multiple transition seeds.

Finally, we list good transition seeds for six different Bernoulli models. The insight gained from our theoretical study and the list of good transition seeds form a useful resource in guiding the selection of seeds in the developing practical applications.

Acknowledgments

This work was partially supported by a NUS ARF grant 146-000-068-112 and a grant from National Natural Science Foundation of China (NSF30528029).

References

1. S.F. Altschul *et al.*, Basic local alignment search tool. *J. Mol. Biology* **215** (1990), pp. 403-410.
2. N. Balakrishnan and M.V. Koutras, *Runs and Scans with Applications.* John Wiley & Sons, U.S.A. (2002).
3. S. Batzoglou, L. Pachter, J.P. Mesirov, B. Berger, E.S. Lander, Human and mouse gene structure: comparative analysis and application to exon prediction. *Genome Research* **10** (2000), pp. 950-958.
4. B. Brejovà, D. Brown, and T. Vinař, Optimal spaced seeds for homologous coding regions. *J. Bioinf. and Comp. Biol.* **1** (2004), pp. 595-610.
5. B. Brejovà, D. Brown, and T. Vinař, Vector seeds: an extension to spaced seeds allows substantial improvements in sensitivity and specificity. *Journal of Computer and System Sciences* **70(3)** (2005), pp. 364-380.
6. M. Brudno, M.A. Chapman, B. Gottgens, S. Batzoglou, B. Morgenstern, Fast and sensitive multiple alignment of large genomic sequences. *BMC Bioinformatics* (2003), pp. 4-66.
7. J. Buhler, U. Keich, and Y. Sun, Designing seeds for similarity search in genomic DNA. In *Proc. of RECOMB'03* (2003), pp. 67-75.
8. A. Califano and I. Rigoutsos, FLASH: fast look-up algorithm for string homology, in *Proc. of ISMB'93* (1993), pp. 56-64.
9. K.P. Choi, F. Zeng, and L. Zhang, Good spaced seeds for homology search. *Bioinformatics* **20** (2004), pp. 1053-1059.
10. K.P. Choi, and L. Zhang, Sensitivity analysis and efficient method for identifying optimal spaced seeds. *J. Comput. System Sci.* **68** (2004), pp. 22-40.
11. A.E. Darling *et al.*, Procrastination leads to efficient filtration for local multiple alignment. In *Proc. of WABI'06* (2006), pp. 126-137.
12. W. Feller, *An Introduction to Probability Theory and its Applications. vol. 1. 3rd edition*, John Wiley and Sons, New York (1968).
13. U. Keich, M. Li, B. Ma, and J. Tromp, On spaced seeds for similarity search. *Discrete Appl. Math.* **3** (2004), pp. 253-263.
14. W.J. Kent, BLAT-the BLAST-like alignment tool. *Genome Res.* **12(4)** (2002), pp. 656-664.

15. Y. Kong, Generalized correlation functions and their applications in selection of optimal multiple spaced seeds for homology search. *J. Comp. Biol.* **14** (2007), pp. 238-254.
16. G. Kucherov, L. Noe and M. Roytberg, Multiseed lossless filtration. *IEEE Trans. on Comput. Biol. and Bioinfor.*, **2** (2005), pp. 51-61.
17. G. Kucherov, L. Noe and M. Roytberg, A unifying frame work for seed sensitivity and its application to subset seeds. *INRIA Tech. Report: N° 5374* (2004).
18. M. Li and B. Ma, On the complexity of computing the sensitivity of spaced seeds. *J. of Comput. and Sys. Sci.***73(7)**(2007), pp. 1024-1034.
19. M. Li, B. Ma, D. Kisman, and J. Tromp, PatternHunterII: highly sensitivity and fast homogy search. *J. Bioinf. and Comp. Biol.* (2004), pp. 417-440.
20. B. Ma, J. Tromp, and M. Li, PatternHunter: faster and more sensitive homology search. *Bioinformatics* **18** (2002), pp. 440-445.
21. D. Mak, Y. Gelfand, and G. Benson, Indel seeds for homology search. *Bioinformatics* **22** (2006), pp. e341-e349.
22. L. Noe and G. Kucherov, YASS: enhancing the sensitivity of DNA similarity search. *Nucleic Acids Research* **33** (2005), pp. 540-543.
23. F.P. Preparata, L. Zhang, and K.P. Choi, Quick, practical selection of effective seeds for homology search. *Journal of Comput. Biol.* **12** (2005), pp. 1137-1152.
24. S. Schwartz, J. Kent, A. Smit, Z. Zhang, R. Baertsch, R. Hardison, D. Haussler, and W. Miller, Human-mouse alignments with BLASTZ. *Genome Research* **13** (2003), pp. 103-107.
25. A.D. Solov'ev, A combinatorial identity and its application to the problem concerning the first occurences of a rare event. *Theory of Probab. Appl.* **11** (1966), pp. 276-282.
26. Y. Sun, and J. Buhler, Designing multiple simultaneous seeds for DNA similarity search. *Journal of Computational Biology* **12** (2005), pp. 847-861.
27. Y. Sun, and J. Buhler, Choosing the best heuristic for seeded alignment of DNA sequences. *BMC Bioinformatics* **7: 133** (2006).
28. J.B. Xu, D.G. Brown, M. Li, and B. Ma, Optimizing multiple spaced seeds for homology search. In *Proc. of CPM'04* (2004), pp. 47-58.
29. I-H. Yang, S-H. Wang, Y-H. Chen, P-H. Huang, L. Ye, X.Q. Huang, K-M. Chao, Efficient methods for generating optimal single and multiple spaced seeds. In *Proc of BIBE'04* (2004), pp. 411-418.
30. L. Zhang, Superiority of spaced seeds for homology search. *IEEE Trans. Comput. Biology and Bioinformatics* **4** (2007).
31. L. Zhou, L. Florea, Designing sensitive and specific spaced seeds for cross-species mRNA-to-genome alignment. *J. Comput. Biol.* **14** (2007), pp. 113-130.

SEED OPTIMIZATION IS NO EASIER THAN OPTIMAL GOLOMB RULER DESIGN

BIN MA

Department of Computer Science
University of Western Ontario
London, ON, N6A5B7, Canada
E-mail: bma@csd.uwo.ca

HONGYI YAO

Institute for Theoretical Computer Science
Tsinghua University, Beijing, 100084, China
E-mail: thy03@mails.tsinghua.edu.cn

Spaced seed is a filter method invented to efficiently identify the regions of interest in similarity searches. It is now well known that certain spaced seeds hit (detect) a randomly sampled similarity region with higher probabilities than the others. Assume each position of the similarity region is identity with probability p independently. The seed optimization problem seeks for the optimal seed achieving the highest hit probability with given length and weight. Despite that the problem was previously shown not to be NP-hard, in practice it seems difficult to solve. The only algorithm known to compute the optimal seed is still exhaustive search in exponential time. In this article we put some insight into the hardness of the seed design problem by demonstrating the relation between the seed optimization problem and the optimal Golomb ruler design problem, which is a well known difficult problem in combinatorial design.

Keywords: spaced seeds; Golomb ruler; reduction

1. Introduction and Notations

1.1. *Seed optimization*

Similarity searches often utilize some types of filtrations to efficiently identify the similarity candidates for further examination. Normally filtration provides a tradeoff between searching sensitivity and searching speed. In DNA similarity searches, spaced seed was invented to achieve a better tradeoff.[1]

A spaced seed x is represented by a binary string such as 111*1**1*1**11*111. The positions with letter 1 are required matches, and the positions with letter * are "don't cares". The length of the string is called the *length* of the seed, denoted by $l(x)$. The number of required matches is called the *weight* of the seed, denoted by $w(x)$. A similarity is *hit* by a seed x if there is a length-$l(x)$ segment of the similarity such that all the required matches specified by x are satisfied by the segment. Figure 1 shows an example.

```
GAGTACTCAACACCAACATTAGTGGGCAATGGAAAAT
|| ||||||||| ||||| || ||||||   ||||||
GAATACTCAACAGCAACACTAATGGGCAGCAGAAAAT
           111*1**1*1**11*111
```

Fig. 1. The seed 111*1**1*1**11*111 hits the similarity region.

A spaced seed x can also be specified by the set of positions of the required matches. For example, the seed x = 111*1**1*1**11*111 can be denoted by its *set representation* $S(x) = \{0, 1, 2, 4, 7, 9, 12, 13, 15, 16, 17\}$. For a given set S and an integer, we define $S + i = \{x + i \mid x \in S\}$.

It is easy to see that spaced seeds with the same weight provide the same efficiency in filtering out random matches. However, it was observed, though not thoroughly studied, that some spaced seeds provide better filtration than the others.[2,3] Ma et al.[1] first studied the optimization of the spaced seed in their PatternHunter paper, and demonstrated that the optimized spaced seed could improve the sensitivity (hit probability) significantly over the consecutive seed (with no * in the seed) of the same weight. The term, spaced seed, was also coined in the paper.[1]

In the PatternHunter paper,[1] a length-L similarity region is modeled as a 0-1 string, where 0 means mismatch and 1 means match. Each position of the region is independently 1 with probability p. In this paper we call these regions the i.i.d. regions * and p be the *similarity level* of the region. Then[1] enumerated all the possible seeds with given weight and length, calculated their hit probabilities under certain L and p, and selected the optimal seed with the best hit probability. This is apparently an exponential time algorithm. After several years of extensive research in the seed optimization problem, many heuristic algorithms were developed to calculate the optimal seed.[5,7–12] However, the exponential-time, brute-force algorithm is still the only known algorithm that guarantees the finding of the optimal seed. We formalize the seed optimization problem under i.i.d. regions as follows:

I.I.D. Seed Optimization An instance of i.i.d. seed optimization is given by a four-tuple $\langle l, w, L, p\rangle$. The objective is to find the seed with length l and weight w that achieves the maximum hit probability in i.i.d. regions with length L and similarity level p.

Clearly, the similarity regions have another simple probabilistic model, where a length-L similarity region is uniformly drawn from all length-L 0-1 strings with exactly k matches (letter 1). In this paper we call these regions the uniform regions. Analogously, we define the following:

Uniform Seed Optimization An instance of uniform seed optimization is

* Different from this paper,[4,5] used the term "uniform" for these regions. The use of the terms i.i.d. and uniform in this paper follow.[6]

given by a four-tuple $\langle l, w, L, k\rangle$. The objective is to find the seed with length l and weight w that achieves the maximum hit probability in uniform regions with length L and exactly k matches.

Independently to the work of PatternHunter,[1] Burkhardt and Kärkkäínen[13] studied a slightly different seed optimization under uniform regions. They tried to find a seed with the maximum weight to hit *all* the uniform regions. Apparently, this problem can be reduced to the uniform seed optimization problem by trying different values of l and w.

Despite the hardness of seed optimization in practice, Li et al.[4,14] made an interesting observation that if the input parameters are given in unary forms, then the seed optimization problem can not be NP-hard. This observation is based on the theorem that a sparse language (the number of instances is bounded by a polynomial of the input size) cannot be NP-hard unless P=NP.[15] Thus, the research in seed optimization is in an awkward situation: no efficient algorithm has been designed; yet NP-hardness, the common strategy to prove the complexities of a problem, does not work here.

Much related to the seed optimization problems, researchers have studied the algorithms to calculate the sensitivity of *a given spaced seed*, under both the i.i.d. and uniform models. Under the i.i.d. model, Ma et al.[1] proposed the first exponential time algorithm and other papers[16,17] proposed algorithms with improved time complexity. Under the uniform model, Buhler et al.[12] proposed exponential time algorithm. The sensitivity calculation algorithms have been used in the brute-forth seed optimization as subroutines. Hence, sensitivity calculation appeared to be an easier problem than seed optimization. Ironically, the accurate sensitivity calculation was proved to be NP-hard under both the i.i.d. model[4,14] and the uniform model.[18]

However, the proofs of hardness of sensitivity calculation do not imply the hardness of the seed optimization. This is because the proofs required specially designed spaced seeds, which may not be the optimal seeds. In this paper we aim to provide some insight into the complexity of the seed optimization problem.

1.2. *Golomb ruler*

A w-mark Golomb ruler is a set of distinct nonnegative integers $0 = a_1, a_2, \ldots, a_w$, called "marks", such that $|a_i - a_j| \neq |a_k - a_l|$ for $\{i, j\} \neq \{k, l\}$ and $i \neq j$. The optimal Golomb ruler design problem seeks for a w-mark ruler with the least a_w.[19]

It is relatively easy to construct a w-mark Golomb ruler with polynomial a_w. In fact, because of the easy construction, Golomb ruler has been used in the reduction to prove the NP-hardness of calculating the sensitivity of a given spaced seed.[4,18]

However, the finding of the optimal Golomb ruler is much harder. Although there is no mathematical proof about the computational complexity of optimal Golomb ruler design, it is well known in combinatorial design that optimal Golomb ruler design is a very difficult problem. The largest known optimal Golomb ruler to date has $w = 24$, which was found by J. P. Robinson and A. J. Bernstein[20] in 1967 and verified to be optimal with four years of distributed computation at distributed.net (http://www.distributed.net) in 2004. Currently the finding (verifying) of the 25-mark optimal Golomb ruler is underway at distributed.net.

The optimal Golomb ruler design problem and our seed optimization problem are analogous in the simplicity of the definitions and the complexity of the algorithms in use. Indeed, in this paper we reduce the optimal Golomb ruler design problem to seed optimization, and consequently prove that seed optimization is at least as hard as optimal Golomb ruler design. Our results, together with the tremendous efforts that mathematicians have spent on optimal Golomb ruler design, justify the exponential time algorithms and heuristic algorithms for seed optimization, and suggest that the future research in this problem should still focus on these two types of algorithms.

The rest of the paper is organized as follows: Section 2 proves that in the i.i.d. regions with certain conditions, optimal seeds are Golomb rulers. A closed-form sufficient condition is given. This reduces the optimal Golomb ruler design problem to the i.i.d. seed optimization problem. Section 2 further provides a counterexample to show that without the conditions, the optimal Golomb ruler may not be the optimal seed. Section 3 studies the uniform seed optimization. Results in uniform regions are very similar to the i.i.d. regions. Section 4 discusses the results and proposes open problems.

2. I.I.D. Seed Optimization

2.1. *Reduction from optimal Golomb ruler design to i.i.d. seed optimization*

In this section we provide a polynomial time reduction from the optimal Golomb ruler design problem to the seed optimization problem. It has been believed that the sensitivity increase of the spaced seed comes from the irregularities in the positions of the letters 1 in the seed. With the irregularity, when a spaced seed hits a similarity region, an extra hit right after the first hit requires many more positions of the similarity region to be matches, as illustrated in Figure 2. This makes the concurrent

```
111*1**1*1**11*111          111*1**1*1**11*111          ...
  111*1**1*1**11*111          111*1**1*1**11*111        ...
```

Fig. 2. No matter how the seed is "slided", two overlapping 111*1**1*1**11*111 always give six or more extra required matches than one seed.

existence of more than one hits in the same similarity region a rare event; whereas for a consecutive seed, the second hit is relatively easy – only one additional required match is needed. As a result, while the total number of hits are similar, spaced seeds hit more similarity regions than a consecutive seed.

Noticing that if seed x is such that its set representation, $S(x)$ is a Golomb ruler, then $S(x) \cap (S(x) + i)$ has at most one element for any integer i. This provides the minimum level of overlap between a seed and its sliding. For the above mentioned relation between sensitivity and irregularity, a Golomb ruler is likely to the optimal spaced seed. This is not necessarily true for all conditions (Section 2.2). However, in what follows we prove that this is true under certain conditions.

We first give a very stringent condition in Theorem 2.1. Later on this condition will be relaxed in Theorem 2.2.

Theorem 2.1. *Consider the i.i.d. seed optimization problem $\langle l, w, L, p \rangle$. Let $n = L - l + 1$ be the number of positions the seed can hit the region. Suppose $p \leq \frac{1}{n^3}$ and $n \geq 2l$. Then there is a w-mark Golomb ruler with $a_w = l - 1$ if and only if the optimal spaced seed is a Golomb ruler.*

Proof. Suppose a length-l and weight-w seed is given by its set representation. When the context is clear, we also use S to refer to the seed. Define $\phi(i) = |S \cap (S + i)|$. Define $\phi = \max_i \phi(i)$.

Denote by $h(i_1, \ldots, i_k)$ the probability of that the seed hits at every position of i_1, i_2, ..., i_k. This event is equivalent to that all the positions in

$$\bigcup_{j=1}^{k} (S + i_k)$$

are matches. Therefore, it is easy to verify that Equations (1), (2) and (3) are true.

For any $0 \leq i < n$,

$$h(i) = p^w. \tag{1}$$

For any $0 \leq i < j < n$,

$$h(i, j) = p^{2w - \phi(j-i)} \leq p^{2w - \phi}. \tag{2}$$

For any $0 \leq i < j < k < n$,

$$h(i, j, k) \leq p^{2w - \phi + 1}. \tag{3}$$

We claim that

$$Pr(S \ hits) \geq \sum_{i=0}^{n-1} h(i) - \sum_{0 \leq i < j < n} h(i, j) \tag{4}$$

and

$$Pr(S \ hits) \leq \sum_{i=0}^{n-1} h(i) - \sum_{0 \leq i < j < n} h(i, j) + \sum_{0 \leq i < j < k < n} h(i, j, k) \tag{5}$$

This is because of the following two facts: (1) for any similarity region that contains $r \leq 2$ hits, the probability of the region is counted precisely once in both Eq.(4) and Eq.(5); (2) for any similarity region that contains $r > 2$ hits, the probability of the region is counted $\binom{r}{1} - \binom{r}{2} \leq 1$ time in Eq.(4) and $\binom{r}{1} - \binom{r}{2} + \binom{r}{3} \geq 1$ times in Eq.(5).

Because of Eq.(3), when $p \leq \frac{1}{n^3}$,

$$\sum_{0 \leq i<j<k<n} h(i,j,k) \leq \binom{n}{3} p^{2w-\phi+1} < p^{2w-\phi} \times \frac{1}{2} \tag{6}$$

If $\phi = 1$, Eq.(4) becomes

$$Pr(Shits) \geq np^w - p^{2w-1} \times \frac{n^2}{2} \tag{7}$$

If $\phi \geq 2$, because there is at least one pair of i and j such that $\phi(j-i) = \phi \geq 2$, as well as Eq.(6), Eq.(5) becomes

$$Pr(S\ hits) \leq np^w - p^{2w-\phi} + p^{2w-\phi} \times \frac{1}{2} = np^w - \frac{1}{2} \times p^{2w-\phi} < np^w - p^{2w-1} \times \frac{n^2}{2} \tag{8}$$

When there is a Golomb ruler of length l with w markers, the seed defined by the ruler has $\phi = 1$ and the hit probability is lower bounded by (7). Because $\phi \geq 2$ implies (8), the optimal seed must be such that $\phi = 1$. It is easy to verify that when $n \geq 2l$, $\phi = 1$ implies that the seed is a Golomb ruler. □

Corollary 2.1. *The i.i.d. seed optimization problem is at least as hard as optimal Golomb ruler design.*

Proof. Theorem 2.1 says that the finding of w-mark Golomb ruler with length l can be reduced to the seed optimization problem. Then the optimal Golomb ruler problem for a given weight w can be solved by trying different length l in polynomial steps. □

One problem of Theorem 2.1 is that the upper bound of p is $O(n^{-3})$, which is very small and not practical. We relax this upper bound in Theorem 2.2.

Theorem 2.2. *Consider the i.i.d. seed optimization problem $\langle l, w, L, p\rangle$. Let $n = L - l + 1$. Suppose $p \leq \frac{1}{4l} \cdot \frac{n-l}{n}$ and $2l \leq n \leq (2\sqrt{l})^{w-1}$. Then there is a w-mark Golomb ruler with $a_w = l - 1$ if and only if each optimal spaced seed is a Golomb ruler.*

Proof. If there is no optimal w-mark Golomb ruler with $a_w = l - 1$, then clearly the optimal spaced seed cannot be a Golomb ruler. Next we prove the "only if". Suppose there is a w-mark Golomb ruler, denoted as S^*. Denote the optimal spaced seed as S. We prove by contradiction that S is also a Golomb ruler.

Define $\phi^*(i) = |S^* \cap (S^* + i)|$. Because S^* is a Golomb ruler, $\phi^*(i) \leq 1$. Define $\phi(i) = |S \cap (S + i)|$ and $\phi = \max_i \phi(i)$. If S is not a Golomb ruler, then $\phi > 1$. Let $h^*(i_1, \ldots, i_k) = Pr(S^*\ hits\ at\ i_1, \ldots, i_k)$ and $h(i_1, \ldots, i_k) = Pr(S\ hits\ at\ i_1, \ldots, i_k)$.

Because both S and S^* have weight w, $h(i) = h^*(i) = p^w$. In addition, if $j - i \geq l$, then $h(i,j) = h^*(i,j) = p^{2w}$. Thus, replacing S by S^* and h by h^* in Eq.(4), then subtracting Eq.(5) from Eq.(4), we get the following:

$$\begin{aligned}
&Pr(S^*\ hits) - Pr(S\ hits) \\
&\geq \sum_{0\leq i<j<n} h(i,j) - \sum_{0\leq i<j<n} h^*(i,j) - \sum_{0\leq i<j<k<n} h(i,j,k) \\
&= \sum_{0\leq i<j<\min(i+l,n)} h(i,j) - \sum_{0\leq i<j<\min(i+l,n)} h^*(i,j) - \sum_{0\leq i<j<k<n} h(i,j,k) \\
&\geq (n-l)p^{2w-\phi} - nlp^{2w-1} - \sum_{0\leq i<j<k<n} h(i,j,k). \qquad (9)
\end{aligned}$$

Here the last inequality is because the following two facts: (1) There is at least one d such that $\phi(d) = \phi$. Therefore $h(i, i+d) = p^{2w-\phi}$ for at least $n-l$ different i. (2) $h^*(i,j) \leq p^{2w-1}$.

To prove the theorem, it suffices to show that when $\phi \geq 2$, Eq. (9) is greater than zero, which is a contradiction to the optimality of S. Clearly, when p is small, the absolute value of the second negative term in (9) can be bounded by a fraction of the first term in (9). We need to examine the third term $\sum_{0\leq i<j<k<n} h(i,j,k)$ more carefully.

The set of indexes $I = \{(i,j,k)|0 \leq i < j < k < n\}$ can be divided into two sets $I_1 = \{(i,j,k) \in I | j < i+l \text{ and } k < j+l\}$ and $I_2 = I \setminus I_1$. Clearly I_1 corresponds to the situations where the seed at j overlaps both the seed at i and the seed at k; and I_2 corresponds to the situations where at least one hit does not overlap the other two seeds. For any $(i,j,k) \in I_2$, $|(S+i) \cup (S+j) \cup (S+k)| \geq 3w - \phi$. Hence

$$\sum_{(i,j,k)\in I_2} h(i,j,k) \leq \frac{n^3}{6} \times p^{3w-\phi}. \qquad (10)$$

When p is small, this can also be bounded by a fraction of the first term of (9).

Again, I_1 can be divided into two sets

$$J_1 = \{(i,j,k) \in I_1 : |(S+k) \setminus ((S+i) \cup (S+j))| = 1\}$$

and $J_2 = I_1 \setminus J_1$. That is, providing that there are hits at i and j, J_1 contains the indexes where the the seed at k requires only one additional match in the similarity region, and J_2 contains the indexes where the seed at k requires at least two additional matches. Therefore, for any $(i,j,k) \in J_2$, $|(S+i) \cup (S+j) \cup (S+k)| \geq 2w - \phi + 2$. Hence

$$\sum_{(i,j,k)\in J_2} h(i,j,k) \leq nl^2 \times p^{2w-\phi+2}. \qquad (11)$$

When p is small, this can be bounded by a fraction of the first term of (9) again. The rest of the proof is to bound

$$\sum_{(i,j,k)\in J_1} h(i,j,k) \leq p^{2w-\phi+1} \times |J_1|.$$

For $(i, j, k) \in J_1$, we consider the possibilities of k for fixed i and j. As shown in Figure 3, because the last letter 1 in the seed at k already contributed an additional match, the second last letter 1 in the seed at k must coincide with the last letter 1 of either the seed at i or the seed at j. Otherwise it contributes another additional match (keep in mind that the seeds at i, j and k are the same seed), contradicting the definition of J_1.

```
i:    ...10001                 ...10001
j:       ...10001                 ...10001
k:             ...10001              ...10001

            (A)                       (B)
```

Fig. 3. The two possible choices of k for fixed i and j in J_1.

Thus, once i is fixed, there are l choices of j, and then there are at most two choices of k. As a result, $|J_1| \le 2nl$, and

$$\sum_{(i,j,k)\in J_1} h(i,j,k) \le p^{2w-\phi+1} \times 2nl \tag{12}$$

Combining Equations (9), (10), (11) and (12),

$$\begin{aligned}
&Pr(S^* \ hits) - Pr(S \ hits) \\
&> (n-l)p^{2w-\phi} - nlp^{2w-1} - \sum_{(i,j,k)\in I_2\cup J_2\cup J_1} h(i,j,k) \\
&\ge (n-l)p^{2w-\phi} - nlp^{2w-1} - \frac{n^3}{6}\cdot p^{3w-\phi} - nl^2 \cdot p^{2w-\phi+2} - 2nl \cdot p^{2w-\phi+1} \\
&= p^{2w-\phi+1}\left((n-l)\cdot p^{-1} - nl\cdot p^{\phi-2} - \frac{n^3}{6}\cdot p^{w-1} - nl^2 \cdot p - 2nl\right)
\end{aligned}$$

It is easy to verify that when $\phi \ge 2$, $p \le \frac{1}{4l}\cdot\frac{n-l}{n}$ and $n \le (2\sqrt{l})^{w-1}$, the above equation has value greater than 0. Hence the theorem is proved. □

Remark: Obviously, the main factor for the upper bound of p in Theorem 2.2 is $O(\frac{1}{l})$. With a more sophisticated analysis, it is also possible to bound p by $O(\frac{1}{w})$. The analysis is omitted here.

2.2. *Counterexample*

One natural question is to ask whether the upper bounds on p in Theorem 2.1 and Theorem 2.2 can be removed. The answer is no. With much computation, we found the following counter example. For $w = 5$, $n = 150$, $p = 0.999$, the optimal 5-mark Golomb ruler is $\{0, 2, 7, 10, 11\}$.[21] The corresponding spaced seed has sensitivity

$1 - 4.3376 \times 10^{-116}$. Whereas the spaced seed $\{0, 3, 4, 6, 11\}$, which is not a Golomb ruler, has a better sensitivity $1 - 3.3674 \times 10^{-117}$.

One may argue that this may be due to that the region is not sufficiently long, and the boundary effects cause this to happen. Figure 4 excludes this possibility. When $p = 0.999$, the curves in Figure 4 plot the trend of $\log(Pr(\textit{There is no hit}))$ for the above two seeds, with respect to the length of the region. Clearly we can see that the no-hit probability of the Golomb ruler seed $\{0, 2, 7, 10, 11\}$ goes down slower than the non-Golomb ruler seed $\{0, 3, 4, 6, 11\}$. This clearly demonstrates that the non-Golomb ruler seed $\{0, 3, 4, 6, 11\}$ is asymptotically better than the Golomb ruler seed.

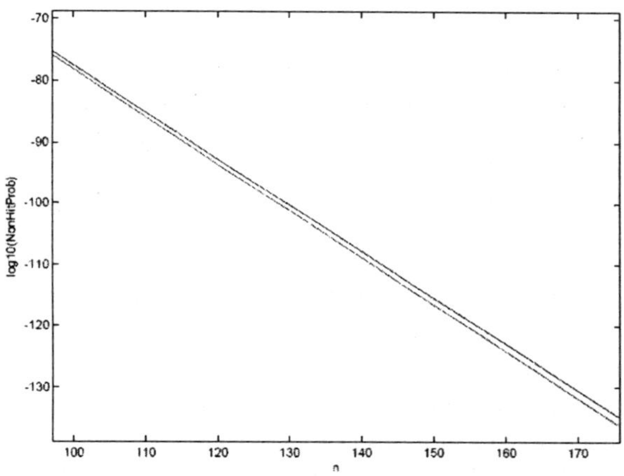

Fig. 4. The curves of $\log(Pr(\textit{There is no hit}))$ with respect to the region length. The upper curve is for the Golomb ruler seed $\{0, 2, 7, 10, 11\}$, and the lower curve is for the seed $\{0, 3, 4, 6, 11\}$.

3. Uniform Seed Optimization

For uniform regions, we can also reduce the optimal Golomb ruler design problem to seed optimization. The proof is in fact much simpler.

Theorem 3.1. *Optimal Golomb ruler design can be reduced to uniform seed optimization problem in linear time.*

Proof. Suppose the uniform regions are given with length L and k matches. Let $k = 2w - 2$ and $L \geq 3l$.

For any seed S, let $\phi(i)$ as defined in the proof of Theorem 2.1. Clearly,

$$Pr(S \textit{ hits}) \leq \sum_{i=0}^{L-l} Pr(S \textit{ hits at } i).$$

Furthermore, the equality holds if and only if $Pr(S\ hits\ more\ than\ once) = 0$. With $k = 2w - 2$, this happens if and only if $\phi(i) \leq 1$ for any i, i.e., S is a Golomb ruler seed. □

Similar to the i.i.d. regions, the conditions on the similarity level k cannot be removed. The counterexample for i.i.d. region still works here. When $L = 200$, $k = 140$, $l = 12$, and $w = 5$, the non-Golomb ruler seed $0, 3, 4, 6, 11$ has better sensitivity $1 - 1.34 \times 10^{-36}$ than the Golomb ruler seed $0, 2, 7, 10, 11$, of which the sensitivity is $1 - 9.3 \times 10^{-35}$.

4. Discussion and Open Problems

Although seed optimization was proved not to be NP-hard in the literature, in this paper we put some insight into its computational complexity by a polynomial time reduction from another well-known difficult problem, the optimal Golomb ruler design. In fact, we show that *under certain conditions*, the following statement holds:

Statement: If a Golomb ruler exists, the optimal seed is a Golomb ruler.

However, without those conditions, we also give a counterexample to show that the statement is not always true. In fact, our counterexample shows that a non-Golomb ruler seed can be asymptotically better than the optimal Golomb ruler seed. This is different from a common belief in seed design that the irregularities in the seeds increase the seed sensitivity. Our example shows that the factors that determine the seed sensitivity is more involved than just the irregularity.

For i.i.d. regions, the conditions for the statement to be true are mainly on the similarity level, p, of the similarity region. The best upper bound we give in the paper is $p = O(\frac{1}{l})$, where l is the length of the desired seed. For uniform regions, the condition is much more stringent: k must be equal to $2w - 2$. We leave it as an open question whether a significantly more relaxed condition on p or k exist for the statement to hold for i.i.d. or uniform regions.

Our seed optimization problem is given in the form of $\langle l, w, L, p \rangle$ for i.i.d. regions and $\langle l, w, L, k \rangle$ for uniform regions. In practice often the length of the seed, l, is not fixed for optimizing a seed. When l is not fixed, our reduction does not straightforwardly imply the complexity of seed optimization. This is because the optimal seed can possibly have a shorter length than the w-mark optimal Golomb ruler, as a consequence of the simple fact that shorter seeds have more positions to hit a length-L region. However, if the i.i.d. seed optimization problem is defined to maximize the hit probability at the first n positions of a region, then the length of a seed is not important anymore and our results still hold. We also point out that all our reductions work for circular regions, with or without the parameter l.

Although our results indicate that the seed optimization problem is very hard (at least as hard as optimal Golomb ruler design), whether a polynomial time algorithm exists for seed optimization is still an open problem.

Acknowledgment

The work was supported in part by China NSF 60553001, National Basic Research Program of China 2007CB807900,2007CB807901, NSERC and Canada Research Chair. BM's work was done when he visited Prof. Andrew Yao at ITCS at Tsinghua University. BM thanks Dr. Ming Li and Dr. John Tromp for commenting on an earlier version of this manuscript. The authors thank the useful discussion with the attendees of a seminar course at ITCS at Tsinghua University. In particular, Yifei Zhang pointed out the Golomb ruler design problem during the course.

References

1. B. Ma, J. Tromp and M. Li, *Bioinformatics* **18**, 440.
2. P. Pevzner and M. Waterman, *Algorithmica* **13**, 135 (1995).
3. O. Lehtinen, E. Sutinen and J. Tarhio, Experiments on block indexing, in *Proc. of the 3rd South American Workshop on String Processing*, 1996.
4. M. Li, B. Ma and L. Zhang, Superiority and complexity of the spaced seeds, in *Proc. of the 17th ACM-SIAM Symposium on Discrete Algorithms (SODA)*, 2006.
5. M. Li, B. Ma, D. Kisman and J. Tromp, *J. Bioinf. and Comp. Biol.* **2(3)**, 417 (2004).
6. M. Csürös and B. Ma, *Algorithmica* **48**, 187 (2007).
7. J. Xu, D. Brown, M. Li and B. Ma, Optimizing multiple spaced seeds for homology search, in *Proc. of the 15th Symposium on Combinatorial Pattern Matching (CPM), LNCS 3109*, 2004.
8. L. Ilie and S. Ilie, Fast computation of good multiple spaced seeds, in *Proc. of 7th Workshop on Algorithms in Bioinformatics*, 2007.
9. K. P. Choi, F. Zeng and L. Zhang, *Bioinformatics* **20**, 1053 (2004).
10. F. Preparata, L. Zhang and K. Choi, *J. Comput. Biol.* **12**, 137 (2005).
11. I.H.Yang, S.H.Wang, H. Chen, P. Huang and K.M.Chao, Efficient methods for generating optimal single and multiple spaced seeds, in *Proc. of IEEE 4th Symp. on Bioinformatics and Bioengineering*, 2004.
12. J. Buhler, U. Keich and Y. Sun, Designing seeds for similarity search in genomic DNA, in *Proc. of the 7th International Conference on Computational Biology (RECOMB)*, 2003.
13. S. Burkhardt and J. Kärkkäínen, *Fundamenta Informaticae* **23**, 1001 (2003).
14. B. Ma and M. Li, *Journal of Computer Science and System Sciences* (2007).
15. S. Mahaney, *Journal of Computer and System Sciences* **25**, 130 (1982).
16. U. Keich, M. Li, B. Ma and J. Tromp, *Discrete Appl. Math* **3**, 253 (2004).
17. K. Choi and L. Zhang, *J. Comp and Sys. Sci.* **68** (2004), 22–40.
18. F. Nicolas and E. Rivals, Hardness of optimal spaced seed design, in *Proc. 16th Annual Symposium Combinatorial Pattern Matching (CPM'05)*, 2005.
19. C. J. Colbourn and J. H. Dinitz (eds.), *CRC Handbook of Combinatorial Designs* (Boca Raton, FL: CRC Press, 1996).
20. J. Robinson and A. Bernstein, *IEEE Trans. Inform. Th.* **13**, 106 (1967).
21. A. Dollas, W. Rankin and D. McCracken, *IEEE Transaction on Information Theory* **44**, 379 (1998).

INTEGRATING HIERARCHICAL CONTROLLED VOCABULARIES WITH OWL ONTOLOGY: A CASE STUDY FROM THE DOMAIN OF MOLECULAR INTERACTIONS*

MELISSA J. DAVIS

Institute for Molecular Bioscience and ARC Center of Excellence in Bioinformatics, University of Queensland, St. Lucia Brisbane, QLD 4076, Australia

ANDREW NEWMAN

School of Information Technology and Electrical Engineering, University of Queensland, St. Lucia Brisbane, QLD 4076, Australia

IMRAN KHAN

School of Information Technology and Electrical Engineering, University of Queensland, St. Lucia Brisbane, QLD 4076, Australia

JANE HUNTER

School of Information Technology and Electrical Engineering, University of Queensland, St. Lucia Brisbane, QLD 4076, Australia

MARK A. RAGAN

Institute for Molecular Bioscience and ARC Center of Excellence in Bioinformatics, University of Queensland, St. Lucia Brisbane, QLD 4076, Australia

Many efforts at standardising terminologies within the biological domain have resulted in the construction of hierarchical controlled vocabularies that capture domain knowledge. Vocabularies, such as the PSI-MI vocabulary, capture both deep and extensive domain knowledge, in the OBO (Open Biomedical Ontologies) format. However hierarchical vocabularies, such as PSI-MI which are represented in OBO, only represent simple parent-child relationships between terms. By contrast, ontologies constructed using the Web Ontology Language (OWL), such as BioPax, define many richer types of relationships between terms. OWL provides a semantically rich structured language for describing classes and sub-classes of entities and properties, relationships between them and expressing domain-specific rules or axioms that can be applied to extract new information through semantic inference. In order to fully exploit the domain knowledge inherent in domain-specific controlled vocabularies, they need to be represented as OWL-DL ontologies, rather than in formats such as OBO. In this paper, we describe a method for converting OBO vocabularies into OWL and

* This work has been performed as part of a collaborative research and development project with Pfizer Global R&D and is partially supported by ARC grand CEO348221.

class instances represented as OWL-RDF triples. This approach preserves the hierarchical arrangement of the domain knowledge whilst also making the underlying parent-child relationships available to inference engines. This approach also has clear advantages over existing methods which incorporate terms from external controlled vocabularies as literals stripped of the context associated with their place in the hierarchy. By preserving this context, we enable machine inference over the ordered domain knowledge captured in OBO controlled vocabularies.

1 Introduction

Molecular biology as a field encompasses several dynamic sub-domains undergoing rapid expansion with resultant rapid discovery and growth in acquired data. High-throughput techniques and large scale biological research, such as genome and transcriptome sequencing projects and expression studies, generate abundant data. However, a significant gap exists between data acquisition and knowledge discovery. These massive quantities of data are frequently produced through a distributed effort, and need to be integrated for analysis and final presentation [1, 2]. Likewise, techniques such as expression profiling produce large quantities of raw data which must be recorded and described [3]. In addition to data exchange and integration issues, many projects in computational and systems biology focus on the analysis of high-level properties of biological systems. Such projects might, for example, compare the distribution of protein functional classes between genomes [4], or analyse the genetic regulatory network of an organism [5]. Such analysis requires the integration of heterogeneous information produced from multiple sources at varying levels of resolution and described using highly variable terminologies. Solutions to these exchange and integration challenges include provision of the data in delimited text files, databases, and XML documents conformant with a given XML schema. The semantic meaning of the data is not however explicit within the documents, and relies on some external definition of concepts and relationships in the data.

Analysis at the level of biological systems also requires reasoning over large and complex data sets is beyond the ability of humans. Machine reasoning has the ability to uncover implicit relationships in the data, rather than simply retrieving explicitly represented data, as is the case of querying a database. However, machine reasoning over large and complex data sets requires the use of appropriate and meaningful knowledge representations of the domain area combined with presentation of the data in machine-readable format [6]. One technique for implementing knowledge representation that has been readily adopted in biology has been the construction of bio-ontologies to establish a precisely (if not formally) defined way to model and express the knowledge of a domain in terms of defined concepts: the classes of "things", the relationships that exist between these classes, and the rules or axioms that apply to these concepts in the domain. The need for ontology development was recognized almost a decade ago in the creation of perhaps the most widely adopted biological ontology, the Gene Ontology (GO) [7].

An important consideration is why choose to create or use an ontology over traditional database schema, which are widely adopted knowledge representations in the field of molecular biology [8]. While there have been few successful demonstrations of automated inference using bio-ontologies, there are significant reasons for their adoption [9]. The key reason is that ontologies are designed to evolve over time and to facilitate integration of data, while database schemas are not [10]. Database schemas are typically considered an internal design decision for a given application and rarely, if ever, are schemas from other databases reused. A specific ontology, on the other hand, is an external, global resource that is meant to be reused, extended and integrated with other ontologies. An ontology is also more expressive than a database [11]. Finally, databases rarely allow the preservation of data. It is still common simply to add attributes to an existing schema rather than splitting it logically due of the extent of the data migration [12]. Ontologies provide a separation between the actual data and the metadata or descriptions of the datasets and their relationships. This allows the data to be migrated independently from changes within the ontology [10].

Many types of knowledge representation exist, and there are many views of what constitutes machine reasoning [13]. The majority of ontologies developed in the biological domain to date do not take advantage of this background [14]. Technologies developed by the World Wide Web Consortium (W3C; http://www.w3.org/) to support machine reasoning include the Resource Description Framework (RDF) and Web Ontology Language (OWL). Both RDF and OWL support machine inference across resources on the web.

While some bio-ontologies have been constructed using these standards, or have been converted into a form compliant with these standards [15], most do not take advantage of the W3C recommendations [14] (see OBO Foundry http://www.obofoundry.org). Many are presented as controlled vocabularies where concepts are represented taxonomically, and relationships are predominantly "is-a" or "part-of" relationships that establish the tree-like structure of the vocabulary. While these structures create well ordered catalogues of concepts relevant to a domain, they do not typically allow for the expression of rules defining other relationships between concepts. The end result is a simplified, flattened model of the domain that lacks the semantic depth or logical support to enable a reasoner to infer new relationships or new information.

A significant challenge in molecular biology is to understand the molecular interactions that occur within cells; research in cell and structural biology shows that many proteins rely on a complex network of interacting partners to achieve their correct localization and functional state in the cell. High-quality molecular interaction data are largely described in journal articles using natural language. Because of the unstructured nature of the observations, the discipline of molecular interactions is covered by several overlapping ontologies. Some of these, such as BioPax (http://www.biopax.org) and the Protein Standards Initiative Molecular Interaction vocabulary (PSI-MI; http://www.psidev.info/) provide significant coverage over concepts relevant to the domain, while others, such as GO, and the NCBI-Taxonomy (http://www.ncbi.nlm.nih.gov/) intersect with the field. The diverse formats of these overlapping ontologies make molecular interactions a useful test domain for strategies to integrate bio-ontologies and reuse domain knowledge.

2 Results

The field of bio-ontology development is active, and already populated with a number of general and domain specific ontologies that have been developed, or are under development. We reviewed ontologies listed by the OBO Foundary; and the National Center for Biomedical Ontology (NCBO; http://www.bioontology.org/). Of the ~70 ontologies listed at these sites, around three quarters are written using the OBO format [16], with the remainder using other formats, including OWL, Protégé files and plain text.

OWL is specifically designed to construct ontologies that support machine reasoning [11]. For this reason, we choose to use OWL DL (description logics) to construct a high-level ontology to integrate concepts from relevant biological ontologies and vocabularies not expressed in OWL. Given that knowledge acquisition is one of the most time consuming and necessarily manual parts of ontology construction, the knowledge captured in non-OWL ontologies constitutes a valuable resource. One approach suggested in such cases is to construct a new ontology using OWL [14], however this under estimates the value of knowledge represented in other formats. Practical strategies to rescue domain knowledge captured in non-OWL ontologies will have obvious applicability in a domain such as molecular biology where the majority of vocabularies are not expressed in OWL.

We have reviewed two ontologies used to describe molecular interactions: the OWL ontology BioPax, and the OBO ontology PSI-MI. However it is not our intention to comparatively evaluate these ontologies, as has been done recently [17, 18]. Briefly, BioPax is designed to describe pathway rather than specific molecular interaction data. However of the ~40 classes and ~70 properties that BioPax defines, many are key concepts and relationships necessary to describe molecular interactions. The PSI-MI vocabulary, on the other hand, is specifically designed to describe molecular interaction data and captures >800 concepts from the domain. However it is represented in OBO and expresses only hierarchical relationships between these classes. While BioPax lacks the descriptive power of PSI-MI, it is more suited for machine reasoning because it is represented in OWL. BioPax recognizes the value of external controlled vocabularies such as PSI-MI and GO, by providing a facility to exploit these external vocabularies through the inclusion of a class openControlledVocabulary. This class stores a term from an external vocabulary along with a cross reference holding the identifier of that term and the name of the vocabulary (as literal strings).

However, it is not sufficient merely to store the data from external controlled vocabularies as literals - the term becomes devoid of meaning if taken out of the context of the original concept tree. In order to preserve the meaning of the term its relationship to other terms throughout the hierarchy, must also be preserved. For example, most biologists would know that "mouse" is not only a label for an organism, but also that an organism which is a *mouse* is also a *mammal* and a *vertebrate*. If "mouse" is used as a text-string label, it lacks the meaning a biologist would ascribe to the concept, and all that is available is the string of text. While applications could be written that interpret the string "mouse" to have a particular meaning, there is nothing in the representation that makes the meaning explicit. However, if an OWL ontology containing a class hierarchy of taxonomic terms was used in the representation, then an organism that was an instance of the class *mouse* would inherit class membership of super-classes in the hierarchy as instances of each sub-class are also instances of super-classes. This allows an instance of *mouse* to be recognized as an instance of *mammal* and *vertebrate*. There is no need for any special encoding of this information into applications interpreting the data, since the meaning is now made explicit in the ontology itself. The meaning of the term is preserved through maintaining its relationship to other terms and its place in the hierarchy.

To illustrate this point, consider the natural language expression, "The protein Emerin is localized to the nuclear inner membrane" (Figure 1).

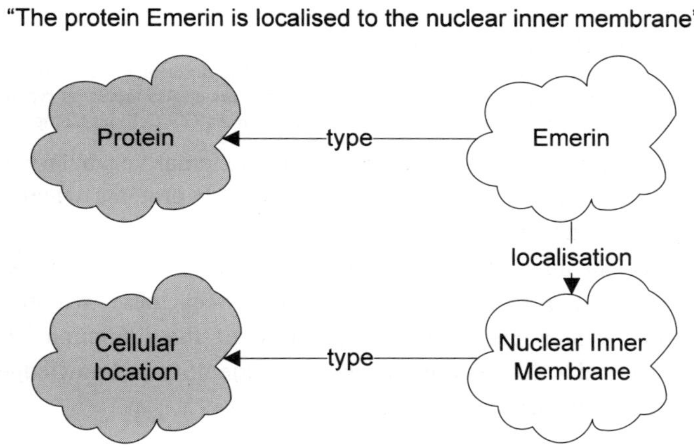

Figure 1. A natural language assertion decomposed into generic concepts (shaded) and specific instances of those concepts. The specific instance "Emerin" is of the generic type *protein*, while the specific instance "Nuclear inner membrane" is of the generic type *cellular location*. The relationship between "Emerin" and its localisation is represented by a labeled arrow, *localisation*.

This statement is composed of generic and specific concepts and relationships: two generic concepts, protein and cellular location, and two specific instances of these concepts, "Emerin", a protein, and "nuclear inner membrane", a cellular location. A relationship also exists between these two instances, namely, that "Emerin" has a property localisation, the value of which is "nuclear inner membrane". The same statement could also be expressed using the BioPax ontology, as illustrated in Figure 2, by using the openControlledVocabulary class to include a term from an external vocabulary like the GO Cellular Component hierarchy.

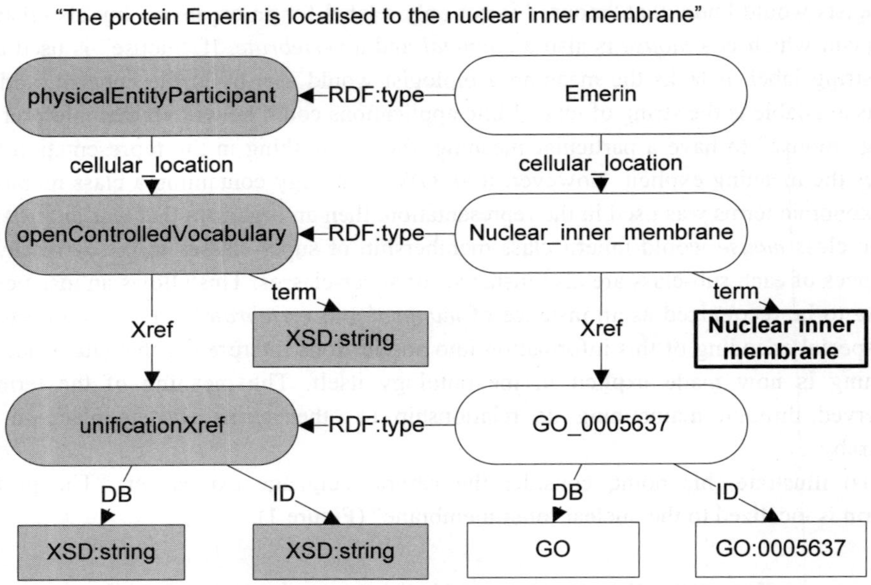

Figure 2. Expression of an assertion of subcellular location in BioPax. Classes and properties from the BioPax ontology are shaded on the left of the diagram, while the instance data describing the localisation of Emerin is on the right. Elipses represent classes, or instances of classes, while rectangles represent typed (string) literals. The string in bold, "nuclear inner membrane" has been imported from the GO Cellular Component hierarchy.

However, as the references to the term from the external vocabulary are all simple text strings, they lack context or meaning. A search for proteins where the value of cellular_location is the string "organelle membrane" would not retrieve proteins where this value was "nuclear inner membrane" unless the query application was hard-coded with additional information about the relationship between these two strings. External terms used in this fashion lack meaning. Because of the limitation of the BioPax approach, we developed a different approach to include domain knowledge captured in external controlled vocabularies (see Figure 3).

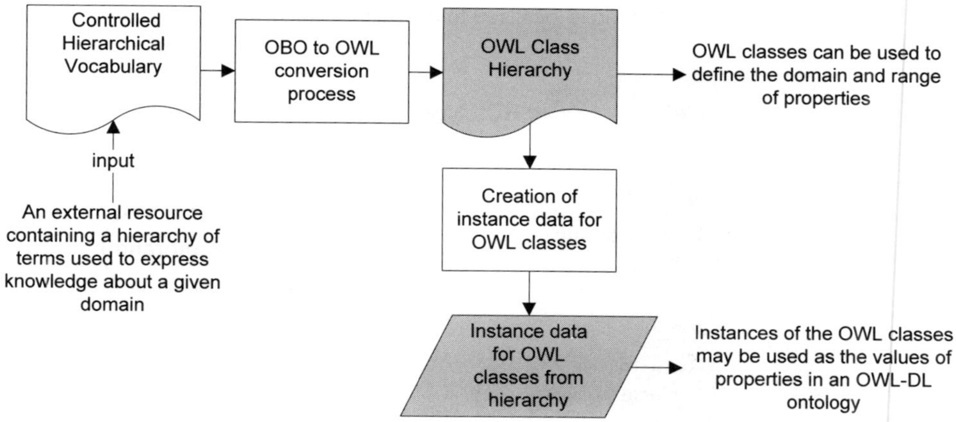

Figure 3. Integration process for an OBO Controlled Vocabulary. The OBO vocabulary is converted into OWL-DL. This results in an OWL class hierarchy where terms from the original vocabulary become OWL classes. Instance data is then created for the ontology so each class has a single instance comprising the vocabulary term that can then be used as an object or subject of triples. Outputs from this process are shaded.

An external controlled vocabulary that contains relevant descriptive terms is converted into OWL-DL. Most frequently, the external vocabulary is in OBO format, so we currently use one of several OBO to OWL conversion applications [19-21]. However this process may be more generally applied to any hierarchical controlled vocabulary such as NCBI Taxonomy, which is not in OBO format. It is important that both the class hierarchy and the instance data for this hierarchy are created. OWL DL classes are used to define restrictions for properties within the ontology, through the specification of allowable domain and range values [11]. However, a class cannot also be an instance, and only instances may be used as the values of properties. For this reason, a single instance of each class is created, taking the form of the original term from the hierarchical vocabulary. At the end of this process, both an OWL ontology representing the terms from the controlled vocabulary and a set of instance data are available (see Figure 3) to use in conjunction with an OWL ontology, as illustrated in Figure 4.

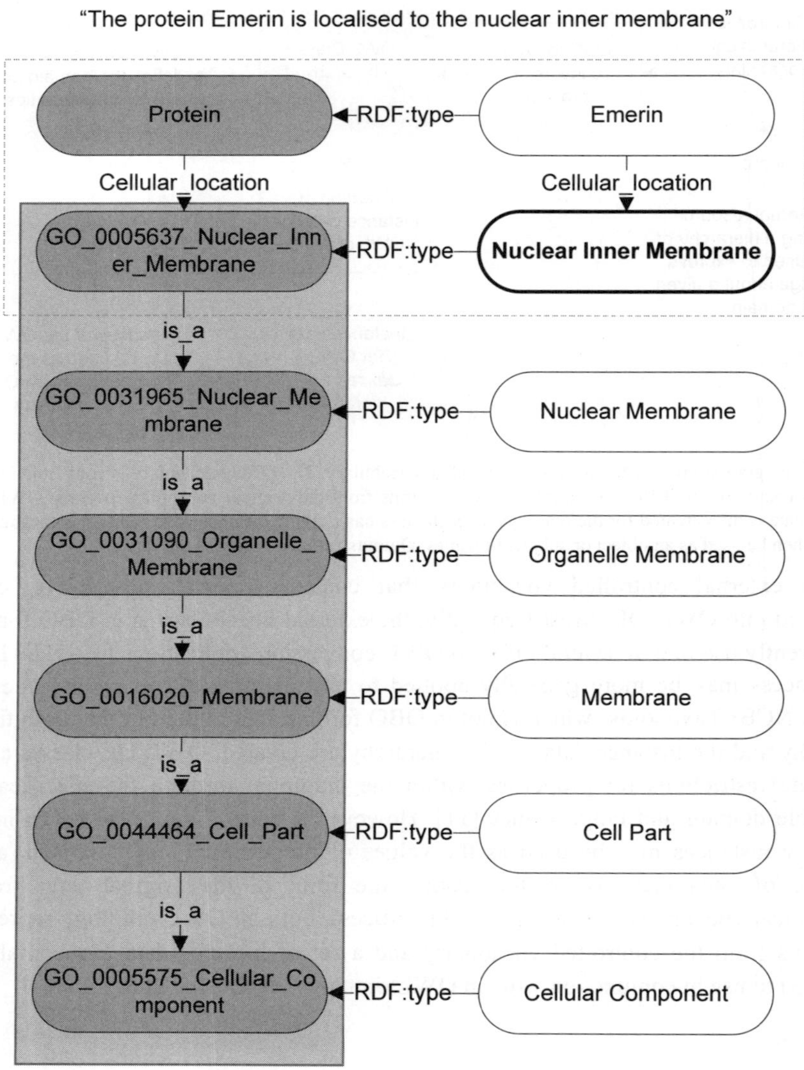

Figure 4. Application of the conversion process. Classes and instances used to express the statement are boxed in a dashed line. The term, "nuclear inner membrane" in bold is an instance of the class GO_0005637_Nuclear_Inner_Membrane, which is related to other classes in the hierarchy shaded box. Domain knowledge is explicitly captured in these hierarchical relationships, so that the relationship of the protein "Emerin" to other cellular locations can be inferred.

In this example, the instance "nuclear inner membrane" is used as the value of the cellular_location property. Not only is this value meaningful to a person who understands what is meant by the words, it is also meaningful to a machine reasoner that has access to the underlying ontology. This machine reasoner, when presented with the assertion that

Emerin is located in the "nuclear inner membrane", may correctly infer that Emerin is also located in an "organelle membrane", and located in a "membrane".

By using this process to incorporate components of external vocabularies under a high-level extensible ontology, external terms become more than text labels, and enable implicit relationships to be extracted from explicit data.

3 Discussion

Many of the existing bio-ontologies are written in the OBO format, and represent a rich source of biomedical domain knowledge. By using the approach described here, vocabularies covering specific aspects of a domain may be plugged into a high level extensible OWL ontology designed to facilitate these modular extensions. The parent-child relationships of the new vocabulary are maintained, and made available to a reasoner to infer implicit relationships from the explicitly represented data. The current strategy used by BioPax is inadequate for machine reasoning. To make bio-ontologies useful for machine reasoning, they need to be explicitly represented in languages such as OWL. Converting and importing relevant hierarchies of terms and associated instances is one solution to maintaining the meaning of terms in controlled vocabularies.

We have constructed a high-level ontology that imports classes and properties from BioPax to describe such things as entities, references and cross-references. To avoid using the BioPax openControlledVocabulary class to incorporate external information, we construct new properties with BioPax classes for domains, and ranges that specify classes in the converted OWL class hierarchy targeted by the property. For example, we create a property Cellular_location, which has a range of GO_0005575_Cellular_Component. While the property conceptually maps to the BioPax property of the same name, the range is restricted to elements from the converted GO Cellular Component hierarchy, that is, all the subclasses of the class GO_0005575_Cellular_Component. In the same way, we can exploit the rich vocabulary in PSI-MI that describes experimental methods used to determine molecular interactions by creating a property Experimental_method (which conceptually maps to the BioPax property experimental_form) which has a range restricted to values taken from subclasses of the converted PSI-MI MI_0045_experimental_interaction_detection class describing experimental methods. This strategy enables us to extend the ontology to incorporate descriptive vocabulary terms from external controlled vocabularies, without losing the context, or meaning, that those terms have in their vocabulary of origin.

One concern when creating an ontology or expanding an existing ontology is the trade off between the ability of the ontology to express concepts in the domain, and to provide tractable inference. The effects of this trade off are difficult to evaluate [23]. A strategy which we will explore to address this is to identify which branches of a given concept tree are required and include only those branches.

Since so many biological concepts are framed in terms of hierarchically inherited properties, and the majority of biological ontologies take the form of hierarchical controlled vocabularies, the process described here is a useful generic strategy for incorporating the existing wealth of ordered knowledge into a semantically rich ontology constructed using OWL. This will help to extend the utility of bio-ontologies into the arena of machine inference.

Acknowledgments

The authors thank Pfizer Inc. for support of the BioMANTA project. We also thank our colleagues in the BioMANTA project for useful suggestions and discussion.

References

1. F. Collins, M. Morgan and A. Patrinos, *Science*, **300,** 286 (2003).
2. P. Carninci et al. *Science*, **309,** 1559 (2005).
3. A. Brazma et al. *Nat. Genet.* **29,** 365 (2001).
4. M. A. Andrade et al., *J. Mol. Evol.* **49,** 551 (1999).
5. S. Li et al., *Science*, **303,** 540 (2004).
6. F. Bry and M. Marchiori, in *Proc.2nd EWIMT*, London, UK, IEEE. (2005).
7. M. Ashburner et al., *Nat. Genet.* **25,** 25 (2000).
8. M. Y. Galperin, *Nucleic Acids Res.* **35,** D3 (2007).
9. M. Keet, M. Roos and M. Marshall, in *Proc. 3rd OWLED*, Innsbruck, Austria, CEUR-WS, (2007).
10. N. Noy and M. Klein, *KAIS,* **6,** 428 (2004).
11. S. Bechhofer et al., in *W3C Recommendations,* (2004).
12. R. Elmasri, S. Navathe and C. Shanklin, *Fundamentals of Database Systems*, Adison-Wesley, Boston (2000).
13. R. Davis, H. Shrobe and P. Szolovits, *AI Magazine*, **14,** 17 (1993).
14. L. Soldatova and R. King, *Nat. Biotech.* **23,** 1095 (2005).
15. M. E. Aranguren et al., *BMC Bioinformatics,* **8,** 57 (2007).
16. R. G. Cote et al., *BMC Bioinformatics,* **7,** 97 (2006).
17. L. Stromback and P. Lambrix, *Bioinformatics,* **21,** 4401 (2005).
18. L. Stromback et al., *Brief. Bioinformatics,* **7,** 331 (2006).
19. D. A. Moreira and M. A. Musen, *Bioinformatics,* **23,** 1868 (2007).
20. C. Goldbreich and I. Horrocks, in *Proc. 3rd OWLED,* Innsbruck, Austria, CEUR-WS, (2007).
21. S. H. Tirmizi and D. P. Miranker, *Technical Report* http://www.cs.utexas.edu/~hamid/pub/tirmizi-obo2owl-tr-06-47.pdf
22. D. Martin et al., *Genome Biol.* **5,** R101 (2004).
23. H. Levesque and R. Brachman, *Comput. Intell.* **3,** 78 (1987).

SEMANTIC SIMILARITY DEFINITION OVER GENE ONTOLOGY BY FURTHER MINING OF THE INFORMATION CONTENT

YUAN-PENG LI[1] and BAO-LIANG LU[1,2*]

[1] *Department of Computer Science and Engineering, Shanghai Jiao Tong University,*
[2] *Laboratory for Computational Biology, Shanghai Center for Systems Biomedicine,*
800 Dong Chuan Road, Shanghai 200240, China
E-mail: {yuanpengli, bllu}@sjtu.edu.cn

The similarity of two gene products can be used to solve many problems in information biology. Since one gene product corresponds to several GO (Gene Ontology) terms, one way to calculate the gene product similarity is to use the similarity of their GO terms. This GO term similarity can be defined as the semantic similarity on the GO graph. There are many kinds of similarity definitions of two GO terms, but the information of the GO graph is not used efficiently. This paper presents a new way to mine more information of the GO graph by regarding edge as information content and using the information of negation on the semantic graph. A simple experiment is conducted and, as a result, the accuracy increased by 8.3 percent in average, compared with the traditional method which uses node as information source.

Keywords: Gene Ontology; Semantic Similarity; Information Content.

1. Introduction

1.1. *Gene Ontology*

Gene Ontology (GO)[1] was created to describe the attributes of genes and gene products using a controlled vocabulary. It is a powerful tool to support the research related to gene products and functions. For example, it is widely used in solving the problems including identifying functionally similar genes, and the protein subcellular or subnuclear location prediction. GO has not been completed and the number of biological concepts in it is still increasing. As GO puts its primary focus on coordinating this increasing number of concepts, at the risk of losing the characteristics of formal ontology, it has some differences from the ontology in Philosophy or Computer Science.[2,3] Gene Ontology Next Generation (GONG)[4] was established to solve this problem and discuss the maintenance of the large-scale biological ontology. Recently, as the use of similarities on GO is increasing, some convenient databases and softwares[5–8] are developed and freely available, which makes it easier to use GO semantic similarity.

*To whom correspondence should be addressed.

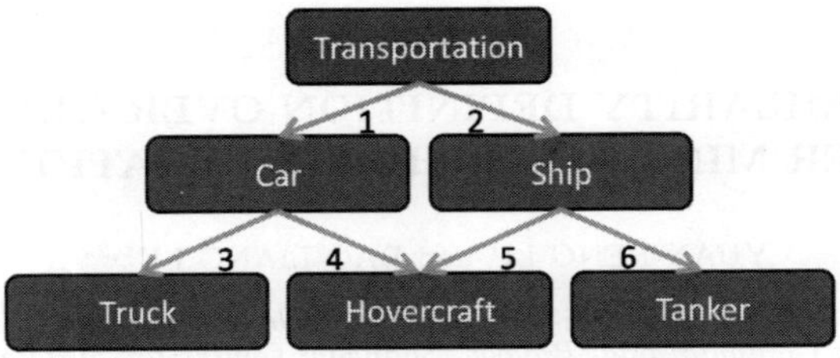

Fig. 1. Example of ontology.

The Gene Ontology[9] is made up of three ontologies: Biological Process, Molecular Function and Cellular Component. On May 2007, there are 13,552 terms for Biological Process, 7,609 for Molecular Function and 1,966 for Cellular Component.

From the graph point of view, each of these ontologies is a connected directed acyclic graph (DAG), with only one root node in that ontology. It is also true that a special node can be set to combine these three ontologies into one, i.e., the special node has the three root nodes of each ontology as its children.

Each node represents a concept, or an ontology term. If two concepts have some relationship, an edge is drawn from one to the other. Gene Ontology only has "is_a" relationship and "part_of" relationship. "is_a" relationship indicates that the concept in the in-node of the edge contains the concept in the out-node. The example in Figure 1 is not Gene Ontology, but just an ordinary ontology for explanation. In the ontology, edge 3 means that "Truck" is a kind of "Car". "is_a" relationship can also be regarded as a standard that distinguishes a concept from other concepts contained in the parent concept. Here, "Truck" is distinguished from "Hovercraft" by the standard the edge 3 provides. "part_of" relationship denotes that the in-node concept has the out-node concept as one of its parts.

If a concept is contained in another concept, then this information is considered positive information. On the other hand, when a concept is NOT contained in another concept, this information is considered negative information. In Figure 1, edge 4 is negative information for "Truck".

1.2. *GO and Similarity between Gene Products*

The final aim of this research is to define the similarities between gene products using GO information. Since each gene product has several GO terms, the similarity of gene product can be calculated from the similarities of these GO terms. There are two steps in this process.

The first step is to obtain the similarity of two GO terms from the GO graph. This is the main focus of this paper.

The second step is to get the gene product similarity from the GO term similarities. Let g_1 and g_2 be the GO term vectors of two gene products A and B, in which 1 means the gene product has the GO term, while 0 means it does not. In

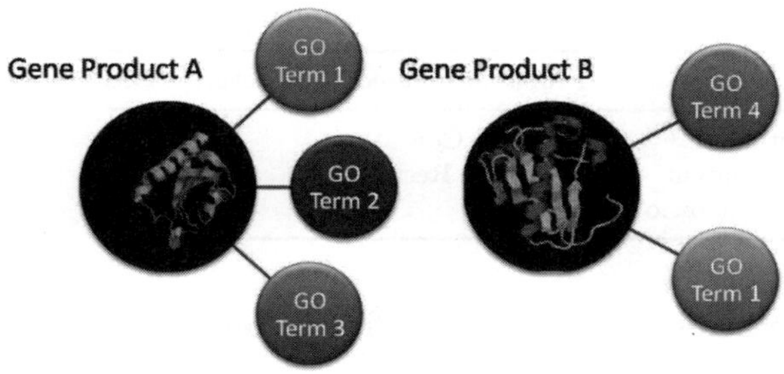

Fig. 2. Example of gene products and their corresponding GO terms.

the example of Figure 2[a], g_1 and g_2 will be as follows.

$$g_1 = (0,1,1,1,0,0)^T \qquad g_2 = (0,1,0,0,1,0)^T$$

Also, let M be a square matrix, in which the value of the ith row and the jth column represents the similarity of the ith and the jth GO terms, obtained in the first step. Then the similarity of two gene products $Sim(A,B)$,or $Sim(g_1,g_2)$, can be defined as follows.

$$Sim(g_1,g_2) = {g_1}^T M g_2 \tag{1}$$

This research is conducted to fully mine the information in GO graph and define similarities between GO terms. In other words, to get a better similarity matrix M.

1.3. *Related Work*

There are many semantic similarity definitions of GO terms. Some representative ones can be classified by two kinds of standards (Table 1).

The first standard is to divide the definitions into probability-based and structure-based ones. The probability-based methods depend on the occurrence frequency of each GO term in some database. Resinik,[10] Jiang and Conrath,[11] and Lin[12] provided their definitions from this point of view. Lord[13] introduced these definitions into Gene Ontology. Later, Couto[14] proposed a method to better apply them to DAGs rather than trees. This kind of methods is based on information theory, and seems to be reasonable. However, it relies on a particular database, SWISS-PROT. On the other hand, another idea is developed to define the similarity from the structure of ontology. The definitions proposed by Rada,[15] Wu,[16] Zhang,[6] and Gentleman[7] are examples of this idea. They made it possible to reasonably obtain the similarity of two GO terms in any database, even if the distribution of the data is highly unbalanced or the size of the database is quite small.

[a]Picture source is [http://lectures.molgen.mpg.de/ProteinStructure/Levels/index.html].

Table 1. Similarity definition methods.

	Probability-based	Structure-based
Distance	Jiang and Conrath	Rada
Info content	Resinik	Zhang, Wu
Content ratio	Lin	Gentleman

The definition measures can also be classified by another standard into three groups. The first group is to define the similarity of two nodes by the distance between them. Rada[15] proposed the original framework of this idea. Jiang and Conrath[11] investigated the weights of the edges to make it more reasonable. The second group of definitions is to calculate the shared information content of two nodes. Resinik[10] first proposed the using of information content. Zhang[6] and Gentleman[7] provided similar definitions based on the structure of ontology. The third group of definitions is to compare the shared information of the two concepts and all the information needed to describe both of these concepts. Lin[12] and Gentleman[7] did some work concerning this idea.

2. Method

2.1. *Notations*

c denotes a term, or a node, in an ontology graph. An edge e fluxes into c means that there exists a path from the root node to c which contains e. The induced graph $V(c)$ of c is the graph made up of all paths from the root node to c. $|V|_n$ and $|V|_e$ denote the number of nodes and the number of edges in V.

In Figure 1, for example, if c is "Hovercraft", the edge $e = 4$ fluxes into c, because there exists a path {{"Transportation", "Car", "Hovercraft"}, {1, 4}} from the root node "Transportation" to c, which contains e (Figure 3(a), left). The induced graph $V(c)$ is {{"Transportation", "Car", "Ship", "Hovercraft"}, {1, 2, 4, 5}}(Figure 3(b)). $|V(c)|_n = |\{$"Transportation", "Car", "Ship", "Hovercraft"$\}| = 4$ and $|V(c)|_e = |\{1, 2, 4, 5\}| = 4$.

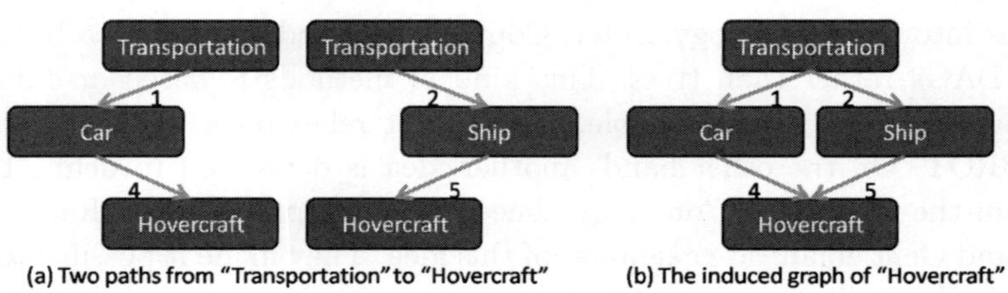

Fig. 3. The paths and the induced graph of "Hovercraft".

2.2. *Traditional Definition*

The idea of Gentleman[7] is used as a traditional definition. The similarity is defined as the number of nodes that the two induced graphs share in common, divided by the number of nodes contained in at least one of the two induced graphs.

$$SimUI(c_1, c_2) = \frac{|V(c_1) \cap V(c_2)|_n}{|V(c_1) \cup V(c_2)|_n} \tag{2}$$

In the example of Figure 1, the similarity of "Truck" and "Hovercraft" is 0.4 since they have 2 nodes in both induced graphs and 5 in at least one induced graph.

The basic idea is similar to that of Lin. Here, the information content of a node is regarded as being represented by its ancestor nodes. The shared information of two nodes is the intersection of their ancestor node sets. All information needed to describe the concepts of two nodes is the union of their ancestor node sets.

The ideas proposed in this paper can be considered as the counterparts of this method, and one of the differences is that the proposed ideas use edges, instead of nodes, to calculate information content. Therefore, SimUI should be chosen as a traditional method to be compared with the new ones.

2.3. *Proposed Similarity Definitions*

The first new method provides the positive similarity of two nodes c_1 and c_2. It is similar to SimUI, but edges are used instead of nodes.

$$SimPE(c_1, c_2) = \frac{|V(c_1) \cap V(c_2)|_e}{|V(c_1) \cup V(c_2)|_e} \tag{3}$$

Since GO is a DAG, unlike tree, edges contain more information than nodes (SEE 4.1). In Figure 1, the induced graphs of "Truck" and " Hovercraft" have one edge in common and 5 different edges altogether. Therefore the similarity is 0.2.

On the other hand, for a node c and an edge e, if e has its in-node as an ancestor of c, but e does not flux into c, it means that the node c does not meet the standard provided by the edge e. To define the negative similarity, the negative edge set should be defined first. The negative edge set of c, $NES(c)$, denotes the set of edges that have in-nodes in the induced graph of c, but not their out-nodes. This consideration of out edges of each node can also be found in the local density introduced by Jiang and Conrath.[11]

$$NES(c) = \{< c_{in}, c_{out} > \in E | c_{in} \in V(c), c_{out} \notin V(c)\} \tag{4}$$

Here, E is the set of all edges in the GO graph. Then the negative similarity can be defined as follows.

$$SimNE(c_1, c_2) = \frac{|NES(c_1) \cap NES(c_2)|}{|NES(c_1) \cup NES(c_2)|} \tag{5}$$

Here, the numerator means the size of shared negative information of both nodes, i.e., the number of the standards that c_1 and c_2 both do NOT meet. And the denominator indicates the number of standards that at least one of the nodes does NOT meet. In Figure 1, the similarity of "Truck" and "Hovercraft" is 0.

To combine these two similarities, the easiest way is to multiple them together.

$$SimEG(c_1, c_2) = SimPE(c_1, c_2) \cdot SimNE(c_1, c_2) \tag{6}$$

For an edge e that has both its in-edge and out-edge NOT in $V(c)$, whether c meets the standard provided by e is unknown, or meaningless. In Figure 1, the standard of edge 3 makes the concept "Truck" different from the concept "Car". But this standard is meaningless when applied to the concept "Tanker", since "Tanker" is not a "Car" at all. Therefore, such edge is not considered to contain either positive or negative information of c.

3. Results

To evaluate the methods UI, PE and EG, an experiment of protein subcellular location prediction was conducted. The experiment was composed of several steps. Firstly, the proteins were randomly chosen, and the corresponding GO terms were found. Secondly, the chosen proteins were divided into training and test samples. Thirdly, a classifier was used to predict the subcellular locations of test samples from the subcellular locations of the train samples, using their similarities.

3.1. *Dataset*

The Gene Ontology structural data are from the Gene Ontology.[9] As the whole ontology contains 32,297 of "is_a" relationships, but only 4,759 of "part_of" relationships, all "part_of" relationships are ignored to make the problem simple.

The training and test data were obtained by choosing from the dataset created by Park and Kanehisa.[18] The GO terms corresponding to these proteins were obtained through the InterPro. i.e., corresponding InterPros were first found from the protein, and then the GO terms of the InterPros were marked to the protein. If one protein was marked by more than one exactly the same GO terms, only one of them was left. In the experiment, several large classes (Table 2) of subcellular locations were used. To avoid the unbalance between the classes, 600 samples were randomly chosen for each of these classes. Each of these samples had at least one GO term so that the similarity of any two chosen proteins could be found via their GO term similarities.

3-fold cross validation was used to assess the performances of the definitions. Each class was divided into three sets of samples randomly. Then, two of these sets in each class were chosen and mixed as a training set and the one left over was used in a test set. Consequently, three groups of training and test sets were prepared[b].

---

[b][http://bcmi.sjtu.edu.cn/~liyuanpeng/APBC2008/{train,test}{1,2,3}.txt].

Table 2. Number of samples in each class.

Class	Subcellular location	# of Samples
1	Chloroplast	600
2	Cytoplasmic	600
3	Extracellular	600
4	Mitochondrial	600
5	Nuclear	600
Total		3000

3.2. *Classifier*

k-Nearest Neighbor (k-NN) classifier was designed to predict the subcellular locations, or classes, of the test samples. The distance of two samples was defined as the minus value of their similarity, and majority voting method was used. If two classes appeared the same number of times in the k-nearest neighbors of a test sample, one of them was selected randomly as the predicted class of that test sample.

3.3. *Tables and Graphs*

The prediction accuracies of the experiments are listed on Table 3 as percentages, followed by the corresponding k values that brought the best results. The three graphs in Figure 5 demonstrate the accuracies for each group as the change of k values. In each of these graphs, the horizontal axis represents the value of k and the vertical axis represents the accuracy percentage. The accuracies of each class, corresponding to the best k values, are listed on Table 4, for each group and the average. Their increases are plotted in Figure 6. In all tables and graphs, "Increase" means the difference between the values of the EG and UI methods.

4. Discussion

4.1. *The Use of Edges and Negative Information*

From the results, it is obvious that PE has advantage over UI, and EG has advantage over PE. The reason can be found in information gain. Consider a small ontology example in Figure 4, SimUI(B,D) will not change even if the edge from A to D is deleted. In other words, the information of the edge is ignored. SimPE(B,D) can contain this information, but the information of the edge from A to C is not

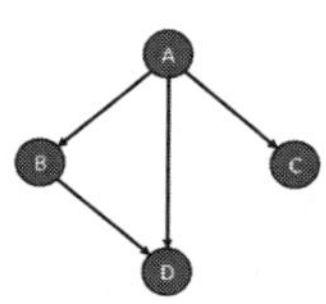

Fig. 4. Example of ontology structure.

Table 3. The accuracies of each group (%)

group	UI (k)	PE (k)	EG (k)	Increase
1	63.5 (40)	69.0 (21)	71.1 (5)	7.6
2	59.6 (8)	65.8 (6)	68.1 (7)	8.5
3	60.5 (47)	67.6 (23)	69.2 (3)	8.7
average	61.2	67.5	69.5	8.3

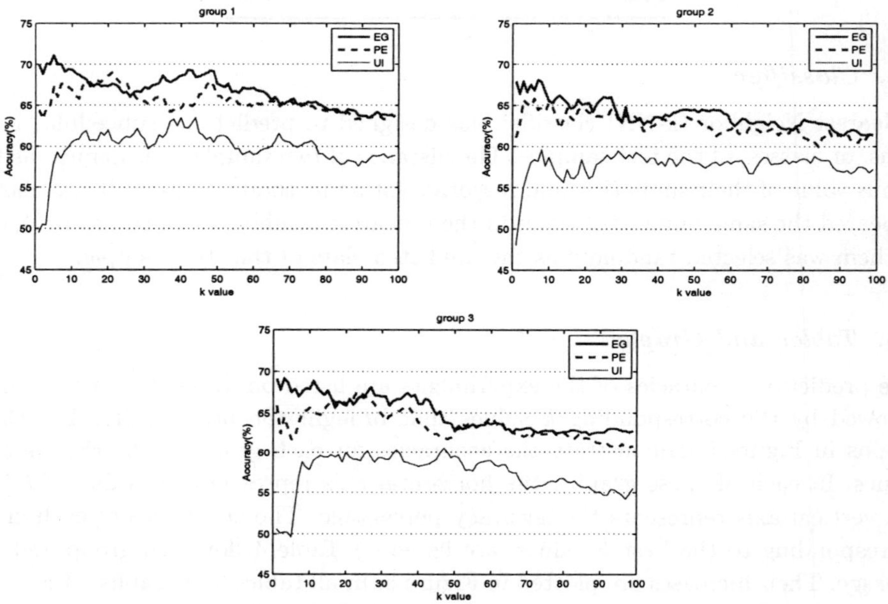

Fig. 5. The relationship of total accuracies and values of k for each group and method.

included. And when SimEG(B,D) is used, this edge information can also be included. Therefore, more information can be used in PE than in UI, and in EG than in PE.

4.2. *The Difference among Classes*

Table 4 and Figure 6 show that different classes prefer different methods of classification. For class 5, the accuracy was already close to 100% when the UI method was applied, and this could be the reason for the less change of the accuracies when the PE and EG methods were used.

4.3. *More Comparison Results*

An experiment, without cross validation, was conducted for each kind of structure-based methods. The results were 65.2% for method of Rada,[15] 61.4% for Wu,[16] 66.0% for Zhang[6] and Gentleman,[7] 64.5% for UI, 69.4% for PE ,and 70.8% for EG.

Table 4. Class accuracies corresponding to the best k values (%).

	group 1					group 2			
Class	UI	PE	EG	Increase	Class	UI	PE	EG	Increase
1	28.5	39.0	40.0	11.5	1	35.0	38.0	45.0	10.0
2	62.5	71.0	77.0	14.5	2	58.0	64.0	65.0	7.0
3	70.5	83.5	83.5	13.0	3	56.5	71.0	74.0	17.5
4	60.5	57.0	60.5	0.0	4	54.5	61.5	61.5	7.0
5	95.5	94.5	94.5	-1.0	5	94.0	94.5	95.0	1.0

	group 3					average			
Class	UI	PE	EG	Increase	Class	UI	PE	EG	Increase
1	44.5	43.0	47.5	3.0	1	36.0	40.0	44.2	8.2
2	58.5	74.5	73.5	15.0	2	59.7	69.8	71.8	12.1
3	62.0	75.5	81.5	19.5	3	63.0	76.7	79.7	16.7
4	41.0	50.0	47.5	6.5	4	52.0	56.2	56.5	4.5
5	96.5	95.0	96.0	-0.5	5	95.3	94.7	95.2	-0.1

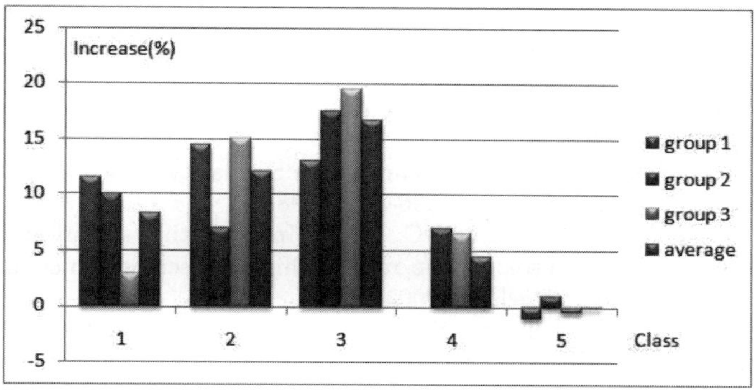

Fig. 6. Increases in each class and group.

5. Conclusions

From the experiment, it can be concluded that the use of edges as information carriers is better than the use of nodes, and that negative information, combined with positive information, provides further support for better predictability.

Acknowledgments

The authors thank Bo Yuan, Yang Yang and Wen-Yun Yang for their valuable comments and suggestions. This research is partially supported by the National Natural Science Foundation of China via the grant NSFC 60473040.

References

1. The Gene Ontology Consortium. Gene Ontology: tool for the unification of biology. *Nature Genet*, 25:25-29, 2000.
2. B. Smith, J. Williams, S. Schulze-Kremer. The Ontology of the Gene Ontology. *AMIA Symposium Proceedings*, 609-613, 2003.
3. M. E. Aranguren, S. Bechhofer, P. Lord, U. Sattler and R. Stevens. Understanding and using the meaning of statements in a bio-ontology: recasting the Gene Ontology in OWL. *BMC Bioinformatics*, 8:57, 2007.
4. Gene Ontology Next Generation [http://gong.man.sc.uk]
5. E. Camon, M. Magrane, D. Barrell, V. Lee, E. Dimmer, J. Maslen, D. Binns, N. Harte, R. Lopez, and R. Apweiler. The Gene Ontology Annotation (GOA) Database: sharing knowledge in Uniprot with Gene Ontology. *Nucleic Acids Research*, 32:D262-D266, 2004.
6. P. Zhang, J. Zhang, H. Sheng, J. J Russo, B. Osborne and K. Buetow. Gene functional similarity search tool (GFSST). *BMC Bioinformatics*, 7:135, 2006.
7. R. Gentleman. Visualizing and Distances Using GO. 2006.
8. H. Froehlich, N. Speer, A. Poustka, T. Beissbarth. GOSim - An R-package for computation of information theoretic GO similarities between terms and gene products. *BMC Bioinformatics*, 8:166, 2007.
9. the Gene Ontology [http://www.geneontology.org].
10. P. Resnik. Using Information Content to Evaluate Semantic Similarity in a Taxonomy. *Proc. the 14th International Joint Conference on Artificial Intelligence*, 448-453, 1995.
11. J. J. Jiang and D. W. Conrath. Semantic Similarity Based on Corpus Statistics and Lexical Taxonomy. *Proc. International Conference Research on Computational Linguistics, ROCLING X*, 1997.
12. D. Lin. An Information-Theoretic Definition of Similarity. *Proc. of 15th International Conference on Machine Learning*, 296-304, 1998.
13. P. Lord, R. Stevens, A. Brass and C. Goble. Investigating semantic similarity measures across the Gene Ontology: the relationship between sequence and annotation. *Bioinformatics*, 19(10):1275-1283, 2003.
14. F. Couto, M. Silva, P. Coutinho. Semantic Similarity over the Gene Ontology: Family Correlation and Selecting Disjunctive Ancestors. *Conference in Information and Knowledge Management*, 2005.
15. R. Rada, H. Mili, E. Bicknell, and M. Blettner. Development and Application of a Metric on Semantic nets. *IEEE Transaction on Systems, Man and Cybernetics*, 19(1):17-30, 1989.
16. H. Wu, Z. Su, F. Mao, V. Olman and Y. Xu. Prediction of functional modules based on comparative genome analysis and Gene Ontology application. *Nuceic Acids Research*, 33(9):2822-2837, 2005.
17. J. L. Sevilla, V. Segura, A. Podhorski, E. Guruceaga, J. M. Mato, L. A. Martinez-Cruz, F. J. Corrales, and A. Rubio. Correlation between Gene Expression and GO Semantic Similarity. *IEEE/ACM Transactions on Computional Biology and Bioinformatics*, 2(4), 2005.
18. K. J. Park and M. Kanehisa. Prediction of protein subcellular locations by support vector machines using compositions of amino acids and amino acid pairs. *Bioinformatics*, 19(13):1656-1663, 2003.
19. Z. Lei, Y. Dai. Assessing protein similarity with Gene Ontology and its use in subnuclear localization prediction. *BMC Bioinformatics*, 7:491, 2006.

FROM TEXT TO PATHWAY: CORPUS ANNOTATION FOR KNOWLEDGE ACQUISITION FROM BIOMEDICAL LITERATURE

JIN-DONG KIM[†], TOMOKO OHTA[†], KANAE ODA[†] and JUN'ICHI TSUJII[†,‡,§]

[†] *University of Tokyo, 7-3-1 Hongo, Bunkyo-ku, Tokyo, Japan*
[‡] *University of Manchester, Oxford Road, Manchester, M13 9PL, UK*
[§] *National Centre for Text Mining, 131 Princess Street, Manchester, M1 7DN, UK*
E-mail: {jdkim, okap, k.oda, tsujii}@is.s.u-tokyo.ac.jp

We present a new direction of research, which deploys Text Mining technologies to construct and maintain data bases organized in the form of pathway, by associating parts of papers with relevant portions of a pathway and vice versa. In order to materialize this scenario, we present two annotated corpora. The first, Event Annotation, identifies the spans of text in which biological events are reported, while the other, Pathway Annotation, associates portions of papers with specific parts in a pathway.

Keywords: Bioinformatics; Text Mining; Pathway; Corpus; Annotation.

1. Introduction

The importance of pathway as a means of integrating biological knowledge into a coherent system has been increasingly recognized by the community of biologists.[1] Due to the research on formal frameworks and ontologies for pathway representation such as SBML,[2] BioPAX,[3] PSI MI,[4] SBO,[5] etc., pathways have become not only graphical means of representing biological systems, but also structured data bases for storing biological knowledge, to be continuously maintained in order to keep abreast with new relevant discoveries.

However, the rapidly growing amount of literature in the field makes it extremely difficult to identify the relevant new discoveries, which should lead to revisions of the relevant portions of the pathways. Furthermore, starting from a graphical depiction of a rather small biological system, some of the current pathways, which are used as organized knowledge bases, have become a huge collection of nodes and links.[6,7] Thus, it has become increasingly difficult, if not impossible, to associate discoveries in the literature with the relevant portions of such large pathways.

On the other hand, the recent progress of text mining (TM) technologies has made it possible to perform many tasks[8–10] including: (1) identifying biological entities that appear in papers,[11] (2) extracting interactions among proteins and other biological entities,[12,13] (3) retrieving text in which specific biological entities are involved in specific types of events,[14,15] and (4) classifying literature into distinct classes, like *relevant* or *non-relevant* to a given topic.[16]

We present a new direction of research that deploys these TM technologies to construct and maintain data bases organized in the form of pathways, by associating parts of papers with relevant portions of pathways, and vice versa. In order to materialize this scenario, we have been constructing a corpus, GENIA Pathway corpus, which associates portions of papers with specific parts in a pathway. Since we have also completed another GENIA annotation, Event Annotation, the main objective of this paper is to analyze the two corpora to discuss how we can integrate events in papers with an organized whole of a pathway.

Section 2 introduces the overall construction of the GENIA corpus, while Section 3 focuses on Event Annotation, which we have recently completed. Section 4 explains the two pathway corpora that we are constructing. One is confined to the GENIA corpus. The other one, centered on a specific pathway, collected a set of all relevant sentences from full-text papers deemed to be relevant to the pathway. Section 5 reports the results of feasibility studies that links these two different streams of work, and discusses how an event recognition program can be used for pathway construction.

2. GENIA corpus

The event and pathway annotation presented here builds on our earlier work in compiling the GENIA corpus[17] and annotating it with linguistic features[18] and biological terms.[17] The documents in the corpus come from the PubMed database, which covers a broad range of domains in bio-medicine. Since we are interested in providing semantically rich annotation for text mining in molecular biology, we have focused on a much smaller, semantically homogeneous subject domain: biological reactions concerning transcription factors in human blood cells. We used the search query, *"Humans"[MeSH] AND "Blood Cells"[MeSH] AND "Transcription Factors"[MeSH]* to retrieve a set of articles, and then chose 2,000 of these articles for our annotation.

3. Event annotation

While biological entities are related with each other in various ways, we have focused on dynamic relations and have defined the GENIA event ontology: a simplified and modified version of the Gene Ontology (GO). By "dynamic", we mean that at least one of the biological entities in the relationship is affected, with respect to its properties or its location, in the reported context.

Figure 1 shows the hierarchy of the GENIA event classes. Those in dotted boxes represent the classes that we have newly created or modified to better support the text annotation. Other classes are taken from GO as they are. The number of annotation instances made to the GENIA corpus is shown in parenthesis next to the class names.

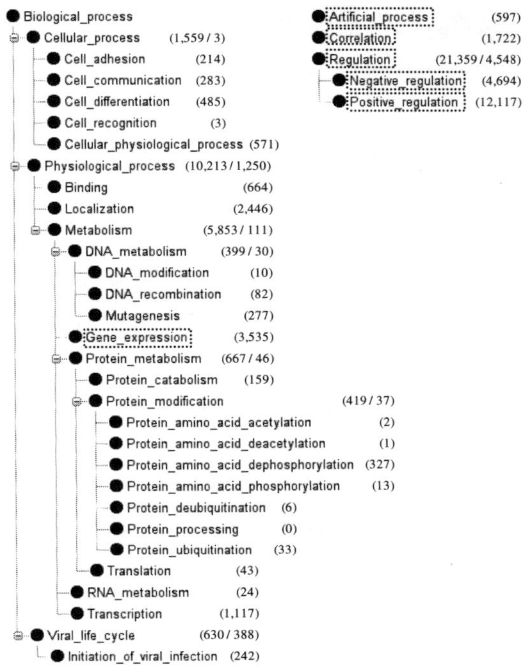

Fig. 1. GENIA event ontology

3.1. *Annotation scheme*

Figure 2A shows a screen snapshot of our annotation tool. There are four regions within the figure, each outlined by a box. The top box contains a sentence, which is undergoing annotation. Biological entities, which have been annotated during term annotation, are shown in color on the screen. Each term is assigned a term Id (T36~T40 in the example of Figure 2A). The remaining three boxes display event annotations, which are attached to the sentence.

In the GENIA framework, an individual event is identified by its type and the *theme*: an entity or entities whose properties are affected by the event. The type of an event is selected from among the classes of the GENIA event ontology, and the theme is selected from among the entities annotated to the given sentence. Each event is also assigned a unique Id, e.g. E5~E7 in Figure 2A. In the figure, The first (E5) and the second (E6) event annotation represent the binding of the two entities, T36 ("I kappa B/MAD-3") and T37 ("NF-kappa B p65"), and the localization of the protein T38 ("NF-kappa B p65"), respectively.

One of our annotation principles requires annotators to mark-up text spans that belong to the corresponding annotation. We call the text expressions or the words in such text spans *clue expressions* or *clue words*. In order to allow the mark-up for clue expressions, the original sentence without term annotation is copied inside

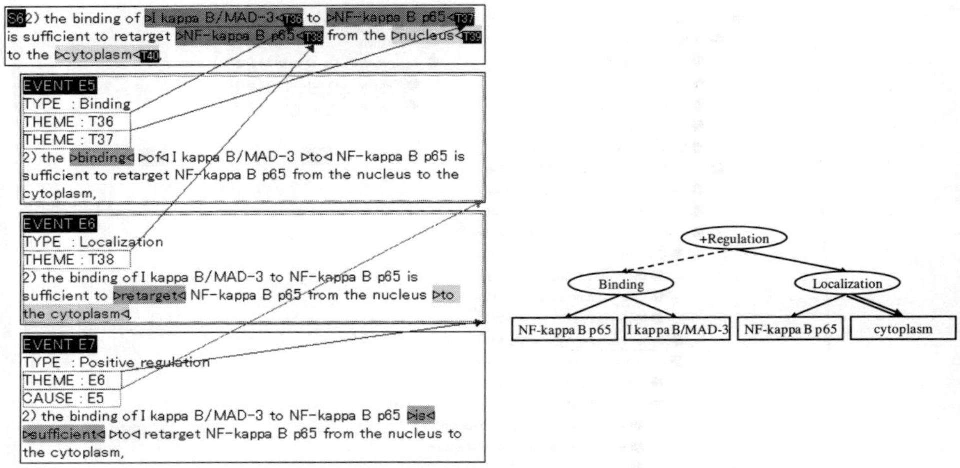


```
<sentence id="S6">2) the binding of <term id="T36" lex="I_kappa_B/MAD-3" sem="Protein_molecule">I kappa B/MAD-3</term> to <term id="T37" lex="NF-kappa_B_p65" sem="Protein_molecule">NF-kappa B p65</term> is sufficient to retarget <term id="T38" lex="NF-kappa_B_p65" sem="Protein_molecule">NF-kappa B p65</term> from the <term id="T39" lex="nucleus" sem="Cell_component">nucleus</term> to the <term id="T40" lex="cytoplasm" sem="Cell_component">cytoplasm</term>, </sentence>
<event id="E5">
<type class="Binding"/>
<theme idref="T36"/>
<theme idref="T37"/>
<clue>2) the <clueType>binding</clueType> <linkTheme>of</linkTheme> I kappa B/MAD-3 <linkTheme>to</linkTheme> NF-kappa B p65 is sufficient to retarget NF-kappa B p65 from the nucleus to the cytoplasm,</clue>
</event>
<event id="E6">
<type class="Localization"/>
<theme idref="T38"/>
<clue>2) the binding of I kappa B/MAD-3 to NF-kappa B p65 is sufficient to <clueType>retarget</clueType> NF-kappa B p65 from the nucleus <clueLoc>to the cytoplasm</clueLoc>,</clue>
</event>
<event id="E7">
<type class="Positive_regulation"/>
<theme idref="E6"/>
<cause idref="E5"/>
<clue>2) the binding of I kappa B/MAD-3 to NF-kappa B p65 <linkCause>is</linkCause> <clueType>sufficient</clueType> <linkTheme>to</linkTheme> retarget NF-kappa B p65 from the nucleus to the cytoplasm,</clue>
</event>
```


C

Fig. 2. Example of event annotation

each of the annotation boxes. In Figure 2, the words "binding" and "retarget" are marked-up as clue words for the event type *Binding* and *Localization*.

Additionaly, clue expressions for the locational, temporal, or experimental information are also marked-up. The text span "to the cytoplasm," in the event annotation E6, is an example of clue expressions indicating the location of the event.

The last event, E7, represents the causal relation between E5 and E6. That is, the binding event (E5) of the two proteins "causes" the localization event (E6) of one of the two proteins. In the GENIA event ontology, the three classes, *Positive_regulation*, *Negative_regulation*, and *Regulation*, are used to represent causal relations between events or entities; e.g. promotion, inhibition, up-/down-regulation. The events of those classes are identified by its type, its theme and its *cause*: an event or an entity that positively or negatively affects the event. Note that, although the expression "is sufficient to" is hardly a linguistic expression for causality, the annotator recognized it as such in this sentence.

To assist the reader in understanding these relationships, we present Figure 2B: a graphical depiction of the example from Figure 2A. In this representation, entities from the GENIA term ontology are shown in rectangular boxes, while entities from the GENIA event ontology are shown in circles. The solid, dotted, and double arrows indicate the link between an event, and its theme, cause, and location, respectively. Figure 2C shows the XML representation of the three event annotations. This format will be used for public distribution of the event-annotated corpus.

3.2. *Annotation results*

This new annotation was made on half of the GENIA corpus, consisting of 1,000 Medline abstracts. It contains 9,372 sentences from which 36,114 events are identified. The quality and the size of the annotated corpus make it one of the best and largest corpus, in comparison with similar attempts. The event-annotated corpus and the full specification of the annotation scheme will be publicly available in XML at http://www-tsujii.is.s.u-tokyo.ac.jp/GENIA/.

4. Pathway corpus

A pathway is a detailed graphical representation of a biological system, which encompasses a set of mutually related events.[6] It integrates pieces of information on biological events scattered in many scientific publications into a coherent system, thereby facilitating the discussion among a large group of biologists, developing consensus for what actually happens in a biological system.

As a prototype of biological knowledge, we have constructed the NF-κB pathway, modeling the lifecyle of the NF-κB protein. For the pathway representation we use *Systems Biology Mark-up Language* (SBML), which is becoming a de facto standard for biological model representation.[2] In SMBL, a pathway or biological model is a collection of chemical reactions, and a reaction is characterized by its reactants, products, modifiers, and kinetic laws - which describes how quickly it takes place. Since our focus is on the construction of event networks describing pathways, we omit the kinetic laws.

We couple a pathway with a collection of evidence sentences that support reactions in the pathway. Designed to support the development of NLP-based TM systems for pathway construction, we call the collection a *pathway corpus*. We have constructed the NF-κB pathway&corpus in two versions; the full-text version and the GENIA version, described in the following sections.

4.1. *NF-κB pathway and corpus, the full-text version*

The full-text version of the NF-κB pathway is constructed based on a set of full-text papers. The papers were collected using a traditional keyword-based search. To raise the reliability, we only considered papers cited by at least two other papers. Because the NF-κB pathway is a well-studied pathway, we could find a lot of reliable review

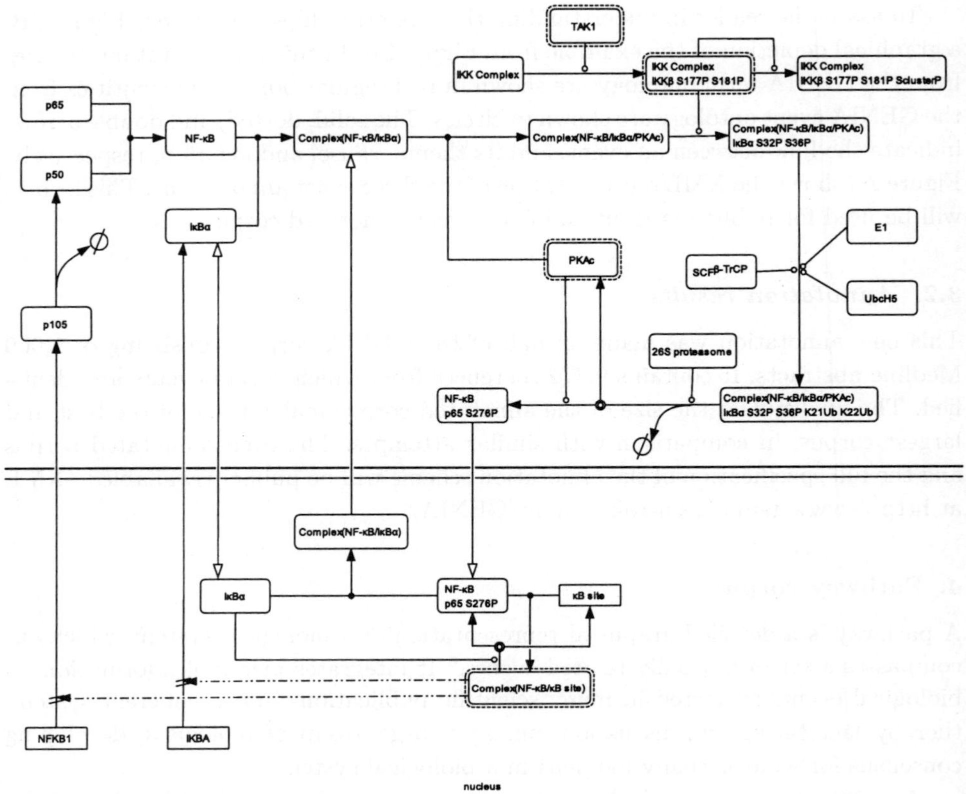

Fig. 3. The full-text version of NF-κB pathway

papers concerning its signaling.[19,20] The full-text version of the NF-κB Pathway was constructed based on the set of searched papers.

During the construction, evidence sentences supporting the pathway were collected and associated with the relavant portion of the pathway. As the result, we collected 467 sentences from the full text of 62 key papers and constructed the NF-κB pathway based on the evidence sentences. Figure 3 shows the full-text version of the NF-κB pathway.

4.2. *NF-κB pathway and corpus, the GENIA version*

As already mentioned, a pathway is a network representation of a course of focused events, which are supported by a collection of evidence texts. A pathway is thus subject to the availability of evidence texts. The GENIA version of the NF-κB pathway was constructed to explicitly address the correspondence between a pathway and a pathway corpus.

We first collected 561 abstracts from GENIA corpus that had the MeSH term

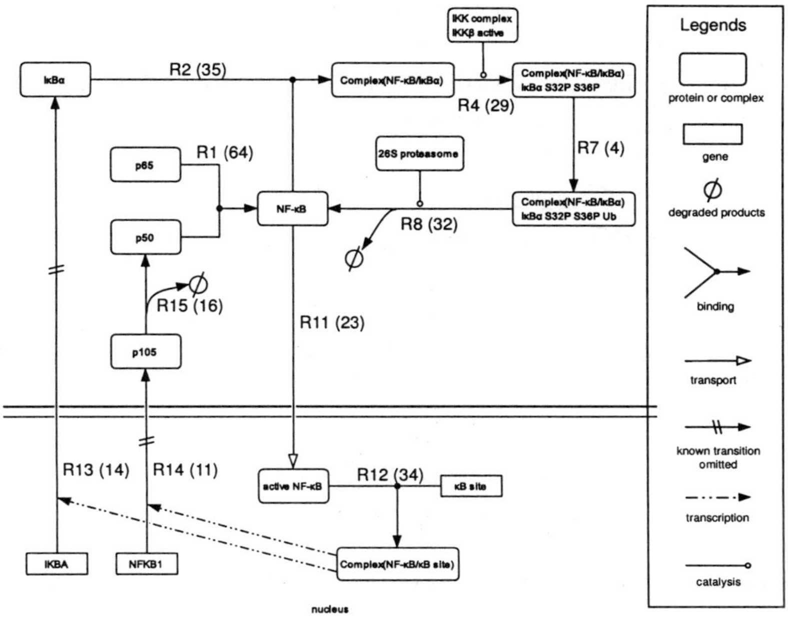

Fig. 4. The GENIA version of NF-κB pathway

"NF-kappa B" as their indexing term, and limited the source of evidence texts to this set of abstracts. We then manually examined each of the 5,223 sentence in the abstracts to collect evidence sentences for reactions in the full-text version of the NF-κB pathway. After collecting all of the evidence sentences, from the pathway, we removed the reactions without any evidence sentence from the GENIA corpus, and reorganized the pathway using only the remaining reactions to produce the GENIA version of NF-κB pathway.

Figure 4 shows the GENIA version of the NF-κB pathway. Note that the difference between the two versions of the NF-κB pathway comes from the availability of literature. Some of the evidence sentences are given in Table 1. The Id of reactions supported by the sentences are given in square brackets. Graphical depiction of the event annotation is given to some of the sentences.

Important characteristic features of the GENIA version of the NF-κB pathway-corpus include the following:

- Every sentence has been manually examined and tagged with the Id of reactions supported by the sentence.
- Every sentence that states events comes with event annotations.

Due to the first feature, the corpus can be used as a gold standard for the development and evaluation of evidence sentence retrieval systems. The second feature

Table 1. Sentences describing reactions in Figure 4.

(1) Associated with its inhibitor, I kappaB, **NF-kappaB** resides as an inactive form in the cytoplasm. [R2]

(2) Upon stimulation by various agents, I kappaB is proteolyzed and **NF-kappaB** translocates to the nucleus, where it activates its target genes. [R8, R11]

(3) *Activation* of NF-kappa B correlates with *phosphorylation* of I kappa B-alpha and requires the *proteolysis* of this inhibitor. [R4, R8]

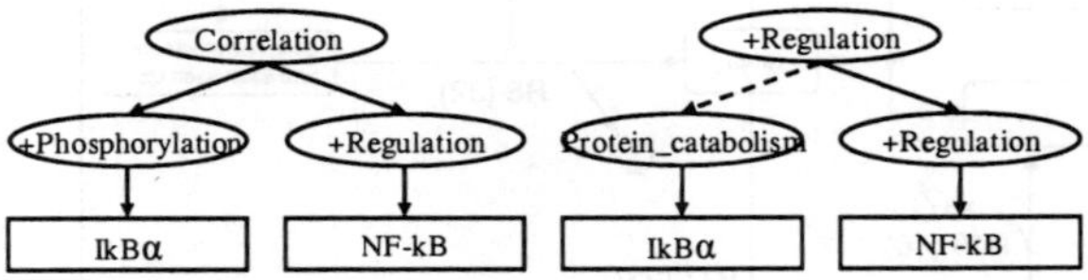

(4) The present study demonstrates that tumor necrosis factor alpha-induced *degradation* of I kappa B alpha in human T cells is preceded by its rapid *phosphorylation* in vivo. [R4, R8]

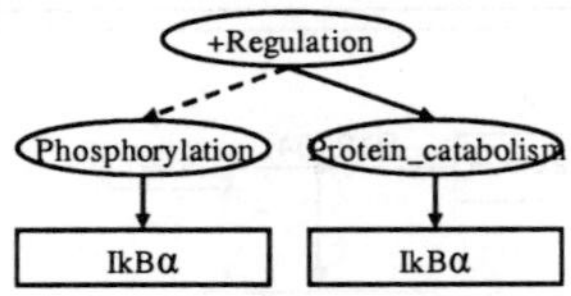

(5) NF-kappa B *activation* involves signaled *phosphorylation*, *ubiquitination*, and *proteolysis* of I kappa B. [R4, R7, R8]

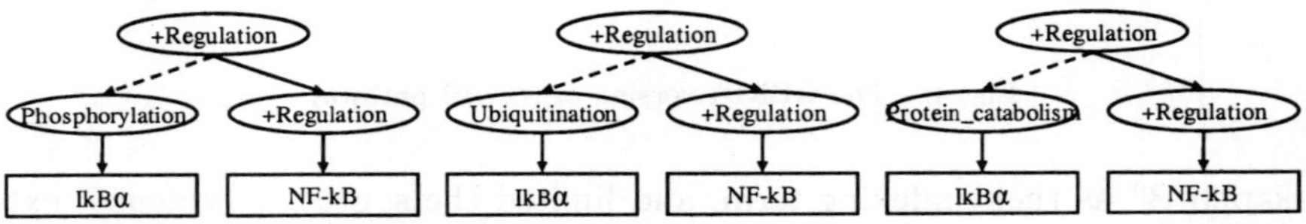

(6) IkappaB alpha *phosphorylation* on **Ser-32** and **Ser-36** is followed by its *degradation* and NF-kappaB *activation*. [R4, R8]

(7) The proteolytic degradation of the post-translationally modified I-kappa B is known to be mediated by the **26S proteasome complex**. [R8]

(8) During normal T-cell activation, IkappaBalpha is rapidly phosphorylated, ubiquitinated, and degraded by the **26S proteasome**, thus permitting the release of functional NF-kappaB. [R4, R7, R8]

enables the comparison of event representation between natural language and pathway representations.

5. Discussion: Pathway construction from Event annotation

The event annotation is intended to be used for the development of an ER (event recognition) program. While the results of ER can be used for various NLP-based TM such as intelligent text retrieval, question answering, etc., one of the major challenges is to use them to associate text fragments with the relevant part of pathways or to use them to semi-automatically construct pathways.

The sentences in Table 1 and the pathway in Figure 4 demonstrate the difference between the natural language expressions and biology-oriented representations. In this section, the difference will be characterized and the required processes to bridge the gap will be discussed.

5.1. *Finding instances from continuants*

Pathway representation is entity-centered, while language organizes information in a predicate-centered manner. To explain the difference, we apply definitions introduced in Ref. 21 that distinguishes between *continuants* and *instances*. A continuant is an entity which endures, or continues to exist throughout time, while undergoing different sorts of changes, including changes in location. We use the term *biological entity* to refer to an *instance* of a continuant at a specific time, which is also bound to a specific biological context.

The pathway representation is entity-centered since it gives an independent status to each of the biological entities or instances of the same continuant. The major players in this type of representation, e.g. nodes in a graphical representation, are biological entities that correspond to continuants in specific biological contexts. Events are expressed as directed edges between nodes, indicating the transition between biological entities.

In the pathway in the Figure 4, the reaction R11 represents translocation of NF-κB from cytoplasm to nucleus. In pathway representation, the same continuant NF-κB appears as different nodes before and after the event. These nodes denote instances of the same continuant in different biological contexts. Since these instances have different properties, it is natural that a pathway representation captures them as different nodes.

On the contrary, natural language text does not usually make explicit such distinctions among instances of the same continuant with different properties or in different contexts. In the sentence (2), the event corresponding to R11 is expressed simply as "NF-kappa B translocates to the nucleus," which indicates that NF-κB is involved in the localization event.

To construct a pathway representation that is entity-centered, distinct entities in different biological contexts must be captured at the mention of a continuant and its surrounding context in natural language expressions.

The same applies across sentences. In the sentence (1), which is followed by the sentence (2), the textual expression "NF-kappa B" refers to the continuant NF-κB, which is involved in the binding with IκB, thus suggesting the existence of two different instances of the continuant before and after the binding. Meanwhile, in the sentence (2), the preceding clause of the same textual expression "NF-kappa B," indicates a different context, in which occurs proteolysis of IκB. This suggests that a completely different set of instances from those of sentence (1) have to be introduced.

5.2. *Integration of fragmentary evidences*

While a pathway organizes a course of reactions that are carefully integrated, individual papers, and especially research papers, usually focus on a couple of reactions of the author's interest and on the causal relations between them.

In the pathway in Figure 4, the sequence of reactions R4, R7 and R8 represents how NF-κB is activated. The sentences (3), (4), (5), and (6) are evidence sentences supporting the reactions. With the exception of (5), all other sentences support the reactions only partially. For example, sentence (3) implies a causal relationship between the two events: "activation of NF-κB" and "phosphorylation of IκBα;" however, the direction of the causality is not mentioned. The sentence also states that "proteolysis of IκBα" causes "activation of NF-κB," which corresponds to the reaction R8. The order of the three events, "phosphorylation of IκBα," "proteolysis of IκBα," and "activation of NF-κB," can be determined when we consider another sentence (4) where the direction of the causal relation between "phosphorylation of IκBα," and "proteolysis of IκBα" is expressed. This exemplifies that we have to consider events in more than one sentence collectively in order to recover the integrated organization of events in a pathway representation.

The sentence (5), which mentions all three reactions: R4, R7, and R8, is from a review paper which integrates publications regarding the NF-κB pathway. Review papers, by nature, have similar properties with pathways in that they tend to pursue comprehension.

The sentence (6) is from a paper that was published later than the papers of all other sentences, and provides a novel, detailed information about the specific residue, which is phosphorylated (Ser-32 and Ser-36). The sentence (7) and (8) are the only sentences supporting the involvement of 26S proteasome complex in the proteolysis event, which means that without the two sentences, the GENIA version of the NF-κB pathway could not include the node of 26S proteasome.

6. Conclusion

In order to link the results of event recognition with pathways, we have to resolve the essential differences between the two representations. In this paper, we formulated one of the major differences as entity-centered vs. event-centered. We showed that looking at the linking problem as the problem of transforming event-centered representation to an entity-centered one helps us to formulate the technical problems in a clear manner. Another major obstacle is the underspecificity of information in text. We showed that this inevitably leads us to the problem of reconstructing actual event sequences by gathering pieces of information from more than one papers.

These foreseen problems may not be able to be automatically solved by programs, but we believe that even semi-automatic means will substantially reduce the burden of constructing and maintaining large pathway data bases.

Acknowledgments

This work was partially supported by Grant-in-Aid for Specially Promoted Research (MEXT, Japan) and Genome Network Project (MEXT, Japan).

References

1. J. Luciano and R. Stevens, e-Science and biological pathway semantics, *BMC Bioinformatics* **8**, p. S3 (2007).
2. M. Hucka, A. Finney, B. Bornstein, S. Keating, B. Shapiro, J. Matthews, B. Kovitz, M. Schilstra, A. Funahashi, J. Doyle and H. Kitano, Evolving a Lingua Franca and Associated Software Infrastructure for Computational Systems Biology: The Systems Biology Markup Language (SBML) Project, *Systems Biology* **1**, 41 (2004).
3. BioPAX, `http://www.biopax.org/`.
4. L. Martens, S. Orchard, R. Apweiler and H. Hermjakob, Human Proteome Organization Proteomics Standards Initiative: Data Standardization, a View on Developments and Policy, *Mol Cell Proteomics* **6**, 1666 (2007).
5. Systems Biology Ontology, `http://www.ebi.ac.uk/sbo/`.
6. G. D. Bader, M. P. Cary and C. Sander, Pathguide: a Pathway Resource List, *Nucl. Acids Res.* **34**, D504 (2006).
7. Kyoto Encyclopedia of Genes and Genomes, `http://www.genome.ad.jp/kegg/`.
8. S. Ananiadou and J. e. McNaught, *Text Mining for Biology and Biomedicine* (Artech House, 2006).
9. L. Hirschman, J. Park, J. Tsujii, L. Wong and C. Wu, Accomplishments and challenges in literature data mining for biology, *Bioinformatics* **18**, 1553 (2002).
10. S. Ananiadou, D. B. Kell and J. Tsujii, Text mining and its potential applications in systems biology, *Trends in Biotechnology* **24** (2006).
11. A. A. Morgan and L. Hirschman, Overview of BioCreative II Gene Normalization, in *Proceedings of Second BioCreative Challenge Evaluation Workshop*, eds. L. Hirschman, M. Krallinger and A. Valencia (2007).
12. M. Krallinger, F. Leitner and A. Valencia, Assessment of the Second BioCreative PPI task: Automatic Extraction of Protein-Protein Interactions., in *Proceedings of Second BioCreative Challenge Evaluation Workshop*, eds. L. Hirschman, M. Krallinger and A. Valencia (2007).
13. C. Nédellec, Learning Language in Logic - Genic Interaction Extraction Challenge, in *Proceedings of the 4th Learning Language in Logic Workshop (LLL05)*, eds. J. Cussens and C. Nédellec (2005).
14. C. Feng, F. Yamashita and M. Hashida, Automated Extraction of Information from the Literature on Chemical-CYP3A4 Interactions, *Journal of Chemical Information and Modeling* (2007).
15. J. G. Caporaso, J. Baumgartner, William A., D. A. Randolph, K. B. Cohen and L. Hunter, MutationFinder: a high-performance system for extracting point mutation mentions from text, *Bioinformatics* **23**, 1862 (2007).
16. W. Hersh, A. M. Cohen, P. Roberts and H. K. Rekapalli, TREC 2006 Genomics Track Overview, in *Proceeding of the TREC 2006*, (2006).
17. J. D. Kim, T. Ohta, Y. Tateisi and J. Tsujii, GENIA corpus - a semantically annotated corpus for bio-textmining, *Bioinformatics* **19**, i180 (2003), ISSN 1367-4803.
18. Y. Tateisi, A. Yakushiji, T. Ohta and J. Tsujii, Syntax Annotation for the GENIA corpus, in *Proceedings of the IJCNLP 2005, Companion volume*, (2005).
19. M. S. Hayden and S. Ghosh, Signaling to NF-kappaB, *Genes Dev.* **18**, 2195 (2004).
20. S. Ghosh and M. Karin, Missing Pieces in the NF-kappaB Puzzle., *Cell* **109**, S81 (2002).
21. B. Smith, W. Ceusters, B. Klagges, J. Köhler, A. Kumar, J. Lomax, C. Mungall, F. Neuhaus, A. L. Rector and C. Rosse, Relations in biomedical ontologies, *Genome Biology* **6**, p. R46 (2005).

CLASSIFICATION OF PROTEIN SEQUENCES BASED ON WORD SEGMENTATION METHODS

YANG YANG[1], BAO-LIANG LU[1,2*] and WEN-YUN YANG[1]

[1]*Department of Computer Science and Engineering, Shanghai Jiao Tong University,*
[2]*Laboratory for Computational Biology, Shanghai Center for Systems Biomedicine,*
800 Dong Chuan Road, Shanghai 200240, China
E-mail: {alayman, bllu, ywy}@sjtu.edu.cn

Protein sequences contain great potential revealing protein function, structure families and evolution information. Classifying protein sequences into different functional groups or families based on their sequence patterns has attracted lots of research efforts in the last decade. A key issue of these classification systems is how to interpret and represent protein sequences, which largely determines the performance of classifiers. Inspired by text classification and Chinese word segmentation techniques, we propose a segmentation-based feature extraction method. The extracted features include selected words, i.e., substrings of the sequences, and also motifs specified in public database. They are segmented out and their occurrence frequencies are recorded as the feature vector values. We conducted experiments on two protein data sets. One is a set of SCOP families, and the other is GPCR family. Experiments in classification of SCOP protein families show that the proposed method not only results in an extremely condensed feature set but also achieves higher accuracy than the methods based on whole k-spectrum feature space. And it also performs comparably to the most powerful classifiers for GPCR level I and level II subfamily recognition with 92.6 and 88.8% accuracy, respectively.

1. Introduction

The development of sequencing techniques led to an exponential growth of protein sequences in the public databases. In the last decade, sequence information has been successfully applied to unveil structure, function, evolutionary relationships, etc. To understand the functional roles or structure families of proteins, a lot of computational methods have been developed to classify protein sequences and detect remote homology based on their sequence similarity.

Machine learning approaches have been widely applied to this problem, such as hidden Markov model, neural networks and support vector machines (SVMs). In recent years, much work has focused on support vector machines for protein sequence classification and achieved better results. A key issue of these methods is how to interpret and represent protein sequences, i.e., features extraction. There are typically two trends. One trend inexplicitly presents features in kernels[1,2,3]. The other implements the feature extraction and classification separately[4,5].

*Corresponding author. This work was partially supported by the National Natural Science Foundation of China via the grant NSFC 60473040.

In this work, we analyze protein sequences similarly with natural languages, aiming to seek useful features to represent protein sequences for classification. The extracted features from amino acid sequences are usually amino acid frequencies, dimer or trimer frequencies, motifs, etc. The problem with these approaches is that we often do not know which features are important for determining the property of proteins relevant for our classification. To find the most discriminant features, we adopt some text processing techniques. In Ref. 6, we selected high-frequency k-mers and conducted a segmentation to calculate the feature vectors for predicting protein subcellular localization. Here we will examine several other criteria to select informative k-mers.

The proposed method arises from Chinese word segmentation and text classification techniques. Biological sequences and text documents are both strings of consecutive characters, written in different languages with respective words. The basic units of protein sequences are 20 kinds of amino acids, while for human languages, the basic units are usually letters or syllables. We model the protein sequences as concatenations of words without any space and punctuation, and develop an automatic segmentation technique for them. Obviously, there are many differences between text and protein sequences. Protein sequences have a much smaller character set than text, but are much longer than text sentences. Moreover, the words of protein sequences are unknown to us. Thus word selection and segmentation criteria peculiar to protein sequences are necessitated.

We applied the method to two protein family classification problems. The first data set is a well-studied collection of protein families built by Jaakkola *et al.*[7], and the second one consists of G-Protein-Coupled Receptors (GPCRs). Experiments show that the proposed method not only results in an extremely condensed feature set but also achieves higher accuracy (measured by ROC_{50} scores) than methods based on whole k-spectrum feature space in classification of SCOP protein families. The GPCR proteins have high diversity, whose sequences share little similarity and are particularly difficult to classify. Former researches on this subject, using decision tree, support vector machines and HMMs, have gained extremely high classification accuracy around 90%. We show that our method is comparably to the most powerful classifiers for GPCR level I and level II subfamily recognition with highly reduced feature space.

2. Method

In English text, spaces help to separate the words and understand the sentences well, while Chinese text contains no spaces, only punctuations indicating the pause or end of a sentence. The automatic analysis of Chinese text has been studied for tens of years. The first step of analysis is Chinese automatic segmentation, which is to separate the character strings into meaningful words or phrases. This is an important and basic step for Chinese information processing, such as information retrieval and handwriting recognition.

In addition, unlike syllabic languages, such as English or German, each single Chinese character can be treated as a word. Some researchers argue that the smallest unit in Chinese language is not a word but a single character. Similarly, considering each single amino acid is the smallest unit in protein sequence, we assume each amino acid as a single-character word, and k-mers can be regarded as so called multi-word lexemes in natural language.

The proposed method consists of three major steps as shown in Fig. 1. Firstly, a dictionary is built by collecting all the 20 amino acids and certain number of meaningful k-mers according to some criterion. Secondly, a segmentation algorithm is adopted and the corresponding matching process is conducted on the dictionary. Lastly, sequences are converted into feature vectors based on the segmentation results.

Figure 1. Flowchart of the method

2.1. *Dictionary Building*

In text, words are minimal independent and meaningful language units, and language text usually has a predefined dictionary or called lexicon, i.e., word list. However, protein sequences are written in an unknown language to us at the present state, whose words are not delineated. Any combination of letters with arbitrary length and within the given alphabet may be a word. So we first need to build a dictionary, which is the basis of segmentation.

We use statistical method to find out useful words and build dictionary based on the training data set. Firstly, a maximum word length $MaxLen$ should be set, which specifies the set of k-mers from which words are selected. Through the experiments, we find the number *four* is the best upper bound of k. Adding k-mers whose lengths are longer than four does not improve accuracy but largely increases the computational complexity. This is also mentioned in Ref. 8 that four is the typical longest distance between local interactions between amino acids.

Therefore, every k-mer with k no bigger than $MaxLen$ will be checked based on certain criterion. An intuition is that the most frequently presented strings are usually words, thus k-mers' appearance times are calculated. k-mers which are widely presented in the corpus are put into the dictionary. Considering every word will be used as a feature for classifying protein sequences, the selected words should have some discriminating ability. High-frequency words may have a balanced distribution in different classes, thus other criteria such as tf-idf and entropy value[9] could be used to choose discriminating words. These three feature ranking criteria are described in details as follows.

1) Frequency: we record the frequency for each k-mer appearing in the training sequence set and preserve a predefined proportion of the most high-frequency k-mers. The basic assumption behind this criterion is that high-frequency substrings should be segmented out and used as features, while rare strings are non-informative for classification and have little influence on global performance. Let $f_{t,s}$ be the frequency of k-mer t in

sequence s, w_t be the weight of t, and N be the size of the training set. The weight is given by the equation below:

$$w_t = \sum_{s=1}^{N} f_{t,s}. \tag{1}$$

2) *tf-idf* value: this criterion takes into account the distribution of each k-mer throughout all sequences in the training set. According to its definition in text categorization, *tf-idf* is calculated for a term in a single document. The value is in proportion to the number of occurrences of the term in the document, i.e., the *tf* (term frequency) part; and in inverse proportion to the number of documents in the training set for which the term occurs at least once, i.e., the *idf* (inverse document frequency) part. Here we refine it as the following equation. Let $w_{t,s}$ be the *tf-idf* value for a k-mer t in sequence s, and n_t be the number of sequences in which t appears.

$$w_{t,s} = f_{t,s} \times log\frac{N}{n_t}. \tag{2}$$

The weight of t, w_t, is defined as the maximum value of $w_{t,s}$:

$$w_t = \max_{s \in \mathcal{T}} w_{t,s}, \tag{3}$$

where $\mathcal{T}$ denotes the whole data set.

3) Entropy value: it is based on information theoretic ideas and is the most complex criterion on computation. We refine this criterion as the following equation, and assign the maximum value of $w_{t,s}$ to the weight of t as equation (3).

$$w_{t,s} = \log(f_{t,s} + 1.0) * (1 + \frac{1}{\log N}\sum_{s \in \mathcal{T}}[\frac{f_{t,s}}{n_t}\log\frac{f_{t,s}}{n_t}]), \tag{4}$$

Each of the three methods has some drawbacks. The k-mers selected by frequency are more likely words because they widely spread in the sequence data. However, some words irrelevant independent of the classification task are selected, like "the" or "and" in English text. They are non-informative but with extremely high frequencies. On the contrary, *tf-idf* and entropy measures tend to select particularly infrequent words. For example, suppose that a word appears many times in one or two sequences, its *tf-idf* value and entropy value could be very high. Such words also have little impact on classification due to their rare occurrences in total.

To avoid encountering unknown words, all 20 amino acids should be included in the dictionary. Domain knowledge can be also incorporated easily by adding signals or motifs into the dictionary. Here we consider motifs in protein sequences, which are amino-acid sequence patterns with certain biological significance, usually consisting of groups of conserved residues adjacent or near to each other. Motifs are rightly short strings with biological significance which are also available in the public database. We downloaded the motif sequence patterns from PROSITE database[10]. The PROSITE database has two kinds of records, patterns and profiles, to describe motifs. We only make use of the former one because it has fixed sequence patterns represented by regular expressions. Such motifs allow one or several amino-acids in a position and also a fixed or a variable number of non-fixed amino acids. That is to say, a motif pattern could match multiple sequence substrings. For example, C[DN][FY] can match CDF, CNF, CDY and CNY.

2.2. *Segmentation*

Segmentation is the process of matching sequences with words in the dictionary. There are a lot of methods for Chinese automatic segmentation[11]. Maximum Matching (MM) algorithm is the most basic method, which bases on the principle of long word preference. Scanning from the head of a character string to the end, the algorithm always matches the longest word at the current position, then skips the word and continues matching for the remaining string. A variation called reverse maximum matching (RMM) scans from the end to the head of sentences. Obviously, maximum matching obtains local-optimal solution using the greedy heuristic searching. Despite the defect, MM works very well in Chinese word segmentation system given a complete dictionary, and the algorithm is simple and fast. Therefore it is widely used in Chinese information processing.

There are thousands of characters usually used in Chinese and every sentence has tens of them at most, while protein sequences usually have hundreds of letters, which are composed of only 20 amino acids. What is more, words of protein sequences are unknown, to say nothing of a complete dictionary. Thus, there may be many more ways to segment the sequences into words. To find the best way of segmentation, we first eliminate a large portion of ways by numbers of segments generated, thus only those which have the least segments remain. That is to say, long words are preferred to be matched. This is based on the consideration that longer strings contain more sequence information and are more meaningful. In language text, short words are usually auxiliary, like “a”, “in”, “of” in English.

However, there might be multiple ways of segmentation satisfying the least segments requirement. We assign a weight for every word in the dictionary to measure its importance, and add a maximum weight product criterion to ensure unique best segmentation. After finishing the segmentation process, appearance time of each word in the sequence is recorded. The original sequence can be converted into a vector with the dimensionality of dictionary size.

Algorithm 1 describes the process for searching the optimal way of segmentation for a protein sequence. Given a dictionary $\mathcal{D}$ and a sequence S whose length is N, $segNo[1 \cdots N]$ and $wordLen[1 \cdots N]$ are two arrays recording the number of segments which have been identified from each amino acid to the end of S, and the length of word segmented from each position, respectively. They are initialized as zero arrays. $maxLen$ stands for the maximum length of words. Beginning from the start of the sequence, Search($\mathcal{D}$, S, 1) is conducted at first.

Algorithm 1 Search

Input: Dictionary: $\mathcal{D}$, Sequence: S, Position: p
Output: Number of segments:$segNum_i, 1 \leq i \leq N$
Length of the word segmented: $wordLen_i, 1 \leq i \leq N$ # N is the length of S
  if $pos = N$ # The end of S **then**
    $segNum_p \leftarrow 1, wordLen_p \leftarrow 1$
    Return $segNum_p$
  end if
  if $segNum_p \neq 0$ **then**
    Return $segNum_p$ # The position has been visited
  end if
  Initialize **len** and **num** to zero arrays with a maximum size of $maxLen$; $count \leftarrow 0$
  for $k = 1$ to $maxLen$ **do**
    if k-mer (beginning from p) $\in \mathcal{D}$ **then**
      $count \leftarrow count + 1, len_{count} \leftarrow k$
      $num_{count} \leftarrow 1+$Search$(\mathcal{D}, s, p + count)$
    end if
  end for
  if Multiple segmentation ways have the same least number of segments **then**
    Calculate weight product for each segmentation which has the least segments.
  end if
  $wordLen_p \leftarrow len_k$, $segNum_p \leftarrow num_k$, where the kth segmentation way has the maximum weight product.
  Return $segNum_p$

3. Results

To demonstrate the effectiveness of our method in extracting features for protein sequence classification, we tested it on two problems. One is SCOP families classification, and the other is G-Protein Coupled Receptor (GPCR) subfamily recognition (including level I and level II). The two data sets have different sequence diversity levels.

We selected N top ranked words according to certain measure, and converted each protein to a N-dimentional feature space. In the following experiments, N equals to 320, which is the summary of 20 amino acids and 100 words of 2-mer, 3-mer and 4-mer, respectively. In the current settings, the experimental results show little improvement by increasing the number of words per length because low-ranked informative words could be selected and deteriorate the accuracy. We also investigated the impact of maximum word length on the prediction performance. And we found that four is the most suitable length for the classification task. k-mers with lengths bigger than four have too low frequencies to contribute useful information.

Here we chose LibSVM version 2.6[12] as our classifier. RBF kernel performed the best in our experiments. The experimental results reported in the following sections were all obtained with the best kernel parameter γ and penalty parameter C from a grid search procedure. All experiments were performed on a Pentium 4 double CPU(2.8GHz) PC with 2GB RAM.

3.1. *SCOP family Classification*

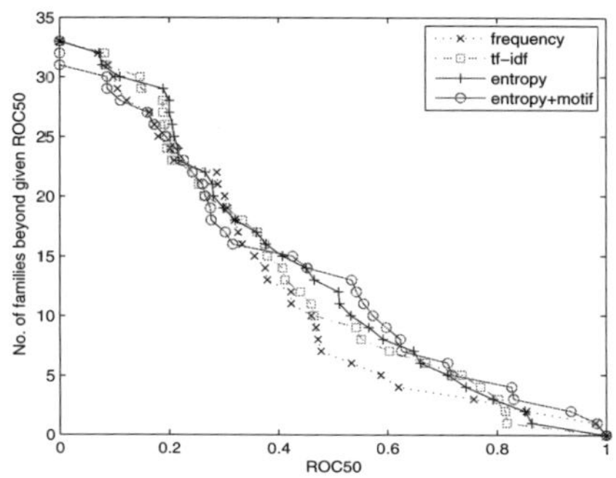

Figure 2. Results of four word selection criteria

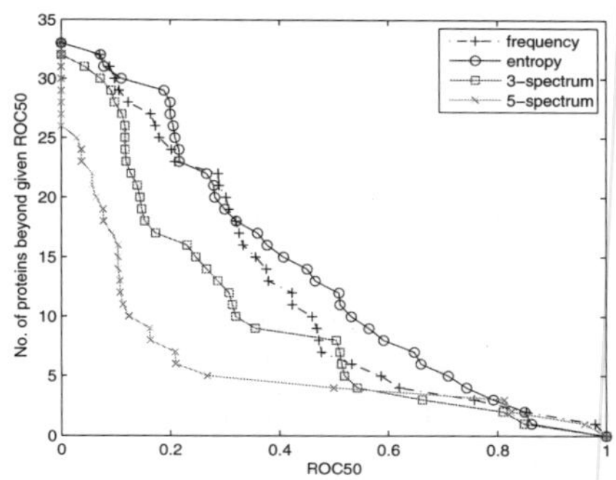

Figure 3. Comparison with spectrum-kernel

This data set consists of 33 families collected by Jaakkola *et al.* [7]. Four kinds of criteria for selecting words were compared, including a) Frequency, b) tf-idf, c) Entropy and d) Entropy plus motif. Especially, d) includes both k-mers selected by entropy value and also motifs collected from PROSITE database. Fig. 2 depicts the ROC_{50} values of 33 families in descending order of these four criteria used with segmentation.

To compare with other methods using all combinations of k-mers, we present in Fig. 3 the results of 3- and 5-spectrum kernel methods as well as our methods using frequency and entropy as the word selection criterion. A k-spectrum kernel method calculates all k-mer frequencies as features inexplicitly in the kernel function. We can see that 5-spectrum kernel does not have a satisfying result for this classification task.

In addition, we examined whether the segmentation process took effect in producing more informative features. For example, if we have a dictionary $\mathcal{D}$={A, E, P, S, SP, TPT, AAAA} and a sequence $\mathcal{S}$=TPTSPPPAAAAPAE, we can segment $\mathcal{S}$ as TPT|SP|P|P|AAAA|P|A|E. Thus the feature vector is {1,1,3,0,1,1,1}. When using the current dictionary without segmentation, i.e., the selected words are same but the feature values are calculated by counting their occurrence time overlappingly, the feature vector becomes {5,1,5,1,1,1,1}. To eliminate the influence of classifiers, we adopted linear-kernel SVMs for our methods. Fig. 4 shows results of three methods: a)3-spectrum string kernel, b) Entropy criterion with segmentation, c) Entropy criterion with overlapping counting (without segmentation).

From the experimental results, several observations can be made. Firstly, among the three measures for selecting words, frequency, tf-idf and entropy, entropy has the best ROC scores followed by tf-idf. Frequency performs slightly worse than the other two

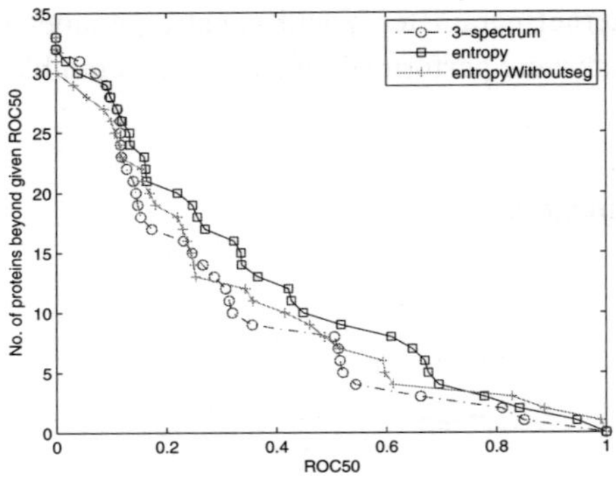

Figure 4. Comparison of methods with segmentation and without segmentation

methods, but it still achieves accuracy improvement and has minimum cost on computation. It should be noticed that adding motifs as words does not obtain an obvious better performance. Instead, it seems to widen the gap of ROC scores between the protein families which are easy to be recognized and which are not. On one hand, not all proteins have annotated motifs; on the other hand, many subfamilies share similar motifs instead of distinct motifs to discriminate them. Therefore, the motif features may even hurt the classification accuracy sometimes.

Secondly, compared with k-spectrum kernel methods which use all the k-mers, our feature extraction method gains an obvious improvement in classification accuracy, and reduces feature space dimension significantly. This is demonstrated in Fig. 3. All the four measures achieve improvement on the accuracy to a certain extent.

Thirdly, feature vectors counted through a segmentation process generally obtain better results than those counted without segmentation. Fig. 4 shows the effectiveness of segmentation. Segmentation method regards the protein sequence as a linear description of proteins, and a concatenation of words. Actually, the outcome of segmentation is to strengthen the influence of longer words in the classification by avoiding counting short words multiple times. Generally, longer words are more representative in denoting sequence features.

Finally, we could also observe from Fig. 4 that without segmentation, the feature vectors still perform better than using all k-mers, which again proves the effectiveness of our statistical measures for selecting informative words.

3.2. *GPCR protein subfamily Classification*

In Section 3.1, we present some results of protein familiy classification. Compared with family classification, the subfamilies within a certain protein family may share more similar characteristics and be more difficult to be discriminated. Therefore, to further examine the performance of the new method, we conducted another experiment to classify subfamilies of GPCRs.

GPCRs are a rich protein family of transmembrane receptors, which play a key role in a wide variety of physiological processes. They involve in many diseases, and have particular importance in drug designs. The family is usually divided into subclasses according to transmitter types, such as muscarinic receptors, catecholamine receptors, odorant receptors, etc. Classification of GPCR proteins is a very challenging task because of the large number of family members and high diversity of sequences. The GPCR family has a hierarchical organization. There are five major classes (Classes A-E), each of which can be divided into level I subfamilies, and the subfamilies can be further divided into level II subfamilies.

The data set used in our experiment consists of Class A (receptors related to rhodopsin and the adrenergic receptor) and Class C (receptors related to the metabotropic receptors). The task is to discriminate subfamilies at level I and II within the two major classes. There were totally 1418 sequences labeled as 19 and 72 classes for level I and level II subfamily classification, respectively. A same two-fold cross-validation was conducted using the training and test data split in Ref. 13.

Table 1. Comparison of methods on GPCR classification

Classifier	No. of Feature	Feature type	Accuracy(%)	
			I	II
SVM	9n[1]	Fisher score vector (FSV)[2] space	88.4	86.3
	320	Segmentation	**92.6**	**88.8**
Decision Tree	9723	n-Gram counts	77.2	66.0
	900-2800	Binary	77.3	70.2
Naive Bayes	9702	n-Gram counts	90.0	81.9
	5500-7700	Binary	**93.0**	**92.4**
BLAST			83.3	74.5
SAM-T2K HMM			69.9	70.0

[1]n is the number of match states of the given HMM; [2]The component of FSV is the gradient of the log likelihood that a protein sequence of interest is generated by the given HMM model

We performed a multi-class classification using SVMs with one-versus-rest strategy. RBF kernel and frequency criterion were adopted. The prediction accuracies of various classification methods at level I and level II are listed in table 1, denoted by I and II, respectively. The results on the first row using SVMs and last two rows using BLAST and profile HMM were reported by Karchin *et al.*[13]. Results of Decistion Tree and Naive Bayes were reported by Cheng *et al.*[8]. With the same classifiers, SVMs, our feature extraction method obtained better results both in classification of level I and II subfamilies. It can be noticed that our method used much less features than Cheng's method, and obtained nearly equal accuracy on level I classification compared with the best result using binary features with Naive Bayes. In Ref. 8, the Naive Bayes classifier with thousands of binary features achieved the highest accuracies of 92.4% on level II subfamily classification, while our method got a relatively lower accuracy. For the extremely condensed feature extraction method, it becomes more difficult to classify the data set with such a large number of class labels when the samples that may differ only slightly. However, it is still among the best classifiers for this problem.

4. Conclusion

This study focuses on seeking efficient feature extraction of protein sequences. We aim to develop a general method for mining the information encoded in enormous protein sequences. Noticing the similarity between text and protein sequences, a method combining text categorization and segmentation techniques is proposed to separate sequences of consecutive characters to words with various lengths, and represent as feature vectors by counting frequencies of the words segmented.

To demonstrate our method, we use the feature vectors to discriminate proteins of different families and subfamilies. The extremely condensed feature set not only results in a high classification efficiency, but also achieves better result than methods based on whole k-spectrum feature space. It is shown to be very competent compared with the most successful systems for detecting protein remote homology and protein sequence classification.

As a general method for feature extraction from protein sequences, this method is not limited to solve protein family classification. It could be also applied to other classification problems based on protein sequences.

References

1. C. Leslie, E. Eskin, and W.S. Noble. The spectrum kernel: A string kernel for SVM protein classification. *Proceedings of the Pacific Symposium on Biocomputing*, 7:566–575, 2002.
2. C. Leslie, E. Eskin, J. Weston, and W.S. Noble. Mismatch string kernels for SVM protein classification. *Advances in Neural Information Processing Systems*, 15:1441–1448, 2003.
3. R. Kuang, E. Ie, K. Wang, K. Wang, M. Siddiqi, Y. Freund, and C. Leslie. Profile-based string kernels for remote homology detection and motif extraction. *Proceedings. 2004 IEEE Computational Systems Bioinformatics Conference, 2004.*, pages 152–160, 2004.
4. L. Liao and W.S. Noble. Combining pairwise sequence similarity and support vector machines for remote protein homology detection. *Proceedings of the Sixth Annual International Conference on Research in Computational Molecular Biology*, pages 225–232, 2002.
5. K. Blekas, D.I. Fotiadis, and A. Likas. Motif-Based Protein Sequence Classification Using Neural Networks. *Journal of Computational Biology*, 12(1):64–82, 2005.
6. Y. Yang and B.L. Lu. Extracting Features from Protein Sequences Using Chinese Segmentation Techniques for Subcellular Localization. *Proceedings of the 2005 IEEE Symposium on Computational Intelligence in Bioinformatics and Computational Biology*, pages 1–8, 2005.
7. T. Jaakkola, M. Diekhans, and D. Haussler. A Discriminative Framework for Detecting Remote Protein Homologies. *Journal of Computational Biology*, 7(1-2):95–114, 2000.
8. B.Y.M. Cheng, J.G. Carbonell, and J. Klein-Seetharaman. Protein Classification Based on Text Document Classification Techniques. *Proteins: Structure, Function and Bioinformatics*, 58:955–970, 2004.
9. K. Aas and L. Eikvil. Text Categorization: A Survey. *Technical Report*, 941, 1999.
10. N. Hulo, A. Bairoch, V. Bulliard, L. Cerutti, E. De Castro, P.S. Langendijk-Genevaux, M. Pagni, and C.J.A. Sigrist. The PROSITE database. *Nucleic Acids Res*, 34:D227–D230, 2006.
11. M. S. Sun and B. K. Tsou. Overview of Chinese Word Segmentation. *Modern language (in Chinese)*, 3(1):22–32, 2001.
12. C. C. Chang and C. J. Lin. LIBSVM: a library for support vector machines. *Software available at http://www. csie. ntu. edu. tw/cjlin/libsvm*, 80:604–611, 2001.
13. R. Karchin, K. Karplus, and D. Haussler. Classifying G-protein coupled receptors with support vector machines. *Bioinformatics*, 18(1):147–159, 2002.

ANALYSIS OF STRUCTURAL STRAND ASYMMETRY IN NON-CODING RNAs

JIAYU WEN*† and GEORG F. WEILLER

Australian Research Council (ARC) Centre of Excellence for Integrative Legume Research and Bioinformatics Laboratory, Research School of Biological Sciences, Australian National University, Canberra, ACT 0200, Australia
† E-mail: jiayu.wen@anu.edu.au
www.anu.edu.au

BRIAN J. PARKER*

Life Sciences and Statistical Machine Learning Groups, NICTA, University of Melbourne, Melbourne, VIC 3010, Australia
E-mail: brian.bj.parker@gmail.com

Many RNA functions are determined by their specific secondary and tertiary structures. These structures are folded by the canonical G::C and A::U base pairings as well as by the non-canonical G::U complementary bases. G::U base pairings in RNA secondary structures may induce structural asymmetries between the transcribed and non-transcribed strands in their corresponding DNA sequences. This is likely so because the corresponding C::A nucleotides of the complementary strand do not pair. As a consequence, the secondary structures that form from a genomic sequence depend on the strand transcribed. We explore this idea to investigate the size and significance of both global and local secondary structure formation differentials in several non-coding RNA families and mRNAs. We show that both thermodynamic stability of global RNA structures in the transcribed strand and RNA structure strand asymmetry are statistically stronger than that in randomized versions preserving the same di-nucleotide base composition and length, and is especially pronounced in microRNA precursors. We further show that a measure of local structural strand asymmetry within a fixed window size, as could be used in detecting and characterizing transcribed regions in a full genome scan, can be used to predict the transcribed strand across ncRNA families.

Keywords: ncRNA; non-coding RNA; structural strand asymmetry; RNA secondary structure

1. Introduction

A variety of functional non-coding RNAs (ncRNAs) have been shown to play key regulatory roles in a number of cellular processes including chromatin modification, transcription initiation, mRNA and protein synthesis, as well as post-translational

*Authors contributed equally to this work

RNA modification.[1,2] MicroRNAs (miRNAs) are a class of small ncRNAs that are known to play important roles in gene regulatory networks by influencing the expression of other genes. Systematic identification of ncRNAs is important for understanding complex gene regulatory networks. However, *de novo* ncRNA prediction is a challenge for current bioinformatics due to a lack of statistically reliable characteristics in ncRNA sequences. Glusman et al.[3] has discussed a "third approach" to the problem of noncoding gene detection (in addition to the other two more usual approaches based on sequence similarity and recognition of protein-coding gene features), which involves the detection of transcribed regions by detecting evolutionary signals in the transcribed strand, including base compositional asymmetries. In a similar vein, in this paper we investigate an asymmetry in the structure forming potential between the transcribed and non-transcribed strands of a genomic sequence.

It is widely assumed that the function of a ncRNA depends on its structural features. Previous research has addressed the prospect of stability of RNA secondary structures acting as a statistical signal for RNA identification. Current opinion differs as to whether ncRNAs and/or mRNA can be recognized by their secondary structures. It has been proposed that mRNA secondary structure stability as measured by the predicted minimum free energies (MFE) is more stable than that of randomized sequences with the same base composition.[4] This hypothesis has been questioned by Workman and Krogh[5] who provide evidence that the observed stability signals disappear when sequences are shuffled so as to preserve di-nucleotide frequencies. Also, Rivas and Eddy[6] argue that ncRNA secondary structures are similar to random sequences in their stability, especially while taking local base composition effects into account, and therefore are not useful in ncRNA detection. However, Washietl et al[7] has combined thermodynamic stability and RNA structure conservation to recognize some ncRNAs. In particular, miRNA precursors have been shown to have lower MFEs than is expected by chance.[8]

In our previous work on legume ncRNA transcripts,[9] we introduced the structural strand asymmetry feature for characterizing transcribed regions. In addition to the canonical complementary bases G::C and A::U, RNA secondary structures typically include non-canonical G::U base pairs. The corresponding C::A nucleotides of the complementary strand do not pair. As a consequence, the secondary structures formed depend on the strand transcribed. Sequences that have evolved functional RNA structures should have done so predominantly on the transcribed strand. We thus hypothesize that the differential in potential secondary structures between the two complementary strands may be used as a measure of RNA structure evolution. We showed[9] that local structural strand differential is pronounced in RNAs compared to that of non-transcribed sequences. We further showed that base compositional asymmetries also contribute to distinguishing the transcribed RNA sequences from non-transcribed DNA sequences.

Here we extend our investigation of the RNA strand structural asymmetry feature by analyzing this feature in sequence sets of known ncRNAs, including miRNAs, and mRNAs. We aim to identify whether this is an independent feature, in addition to structural asymmetries induced by base compositional asymmetries alone. Also, we investigate the effectiveness of local and global measures of structural asymmetry features. Given that the stability of RNA structure depends upon di-nucleotide base stacking energies, we compared them with sequences that were shuffled so as to preserve both mono- and di-nucleotide frequencies.

2. Methods

2.1. *Datasets*

All ncRNA sequences used in this study were obtained from the Rfam database (release 8.0).[10] We retrieved sequences from Rfam.fasta.gz file which filtered Rfam members to $<$ 90% identity, including several ncRNA families: miRNA precursor, 5S rRNA, 5.8S rRNA, 7SK, Hammerhead ribozyme (type I and type III), Group I and Group II catalytic intron, IRES, RNase MRP, Nuclear RNase P, snoRNA CD-box, snoRNA HACA-box, Eukaryotic type signal recognition particle RNA (SRP), tmRNA, and tRNA. We randomly selected 100 sequences from the miRNA family and up to 50 sequences from each other RNA family; all sequences were retrieved if ncRNA family had less than 50 sequences. To select representative mRNA sequences, we selected proteins in the Swiss-Prot[11] database that were derived from a number of commonly studied organisms including *Arabidopsis thaliana, Bos Taurus, Caenorhabditis elegans, Danio rerio, Drosophila melanogaster, Escherichia coli, Homo sapiens, Mus musculus, Mycoplasma pneumoniae, Oryza sativa, Rattus norvegicus, Saccharomyces cerevisiae, Schizosaccharomyces pombe, Takifugu rubripes, Xenopus laevis,* and *Zea mays.* cDNAs for these proteins were then extracted from EMBL.[12] We randomly sampled 800 sequences and then excluded those with sequence length of more than 1000 bp (to limit computation time) to construct a set of 614 mRNA sequences. Table 1 summarizes the sequence length and GC content of the ncRNA family and mRNA sets used in this study.

2.2. *Sequence randomization*

Each sequence was permutated 100 times to generate 100 shuffled sequences that retained both mono-nucleotide and di-nucleotide base composition and length. Mono-nucleotide shuffling preserves the same nucleotide frequencies as the original sequences, whereas di-nucleotide shuffling preserves both mono- and di-nucleotide frequencies. We used the EMBOSS program shuffleseq[13] to perform mono-nucleotide shuffling. Di-nucleotide shuffling is particularly important because the stability of RNA secondary structures depends upon the stacking energies.[5] We used the program dishuffleseq.pl which implemented Altschul and Erickson shuffling algorithm[14] to perform di-nucleotide shuffling. See Altschul and Erickson for details of the careful considerations needed for di-nucleotide randomization.

2.3. *RNA secondary structures*

To compute a measure of global structure strand asymmetry, we used RNAfold in the Vienna RNA package (version 1.6.4)[15] to compute RNA global secondary structures separately for the transcribed and complementary non-transcribed strands for each RNA sequence. MFE was measured for both the transcribed and non-transcribed global structures. The values were normalized by the length of the sequences yielding the MFE densities (MFED). The MFED difference between the two strands was taken as $\Delta\text{MFED}_{tr-ntr}$. This measure was calculated on the original RNA sequences and on both mono-nucleotide and di-nucleotide shuffled sequences.

In gene-finding applications, these asymmetry features would typically be applied to local structures computed in a sliding window along the genome. For this reason, it is of interest to examine a measure of local structure strand asymmetry. To compute this, we used RNALfold (in Vienna RNA package) which can compute the MFE of locally stable secondary structures in an RNA sequence, limited to a maximum sliding window size. We used a sliding window size L = 150 (as may be typical for a genome-wide study) to predict a list of RNA local secondary structures for each sequence. The mean MFED over the predicted local structures for each strand was computed, and the difference, $\Delta\text{MFED}_{tr-ntr}$, was taken as the measure of local structure asymmetry.

2.4. *Statistical significance*

To provide evidence for whether any structural strand asymmetry ($\Delta\text{MFED}_{tr-ntr}$) is primarily due to RNA secondary structure conservation, or is primarily a side-effect of base compositional asymmetries, we performed a permutation test using the shuffled versions of the sequences from ncRNA families and mRNAs. This permutation test measures the strength of the structural asymmetry in the original sequence as compared with the structural asymmetry of shuffled versions of the sequence (which maintains base compositional asymmetries). Any overall structure is effectively removed in these shuffled sequences and any remaining structural asymmetry in these shuffled sequences would be mainly due to mono- and di-nuceotide frequency asymmetries between strands. To calculate a *p*-value for the structural asymmetry, the fraction of the shuffled sequences that achieved a $\Delta\text{MFED}_{tr-ntr}$ as great as the original version was estimated. Z-scores[6] were also calculated as a measure of the structure signal strength above random noise levels: Z-score $= (x - \mu)/\sigma$, where x is the MFED_{tr} values from the original sequences, μ is the mean value of the randomized sequences, and σ is the standard deviation.

An advantage of the structural asymmetry feature in gene-finding applications is that it also inherently provides information on the transcribed strand which is required for annotation of detected transcribed regions. Indeed, global $\Delta\text{MFED}_{tr-ntr}$ has recently been applied successfully for strand orientation detection in ncRNA multiple sequence alignments.[16] Our focus is on characterizing the regions of transcription across the genome and so the accuracy of both global and local strand asymmetry features for transcribed strand detection are computed as an additional measure of the strength of the strand asymmetry features. The accuracy in predicting the transcribed strand was computed using the sign of the differential, where accuracy is defined as the proportion of correct predictions/number of samples.

3. Results and Discussion

We used strand differences in MFEDs ($\Delta MFED_{tr-ntr}$) to compare the most stable global structures that could be formed by sequences of ncRNA families and mRNAs (See Methods). This measure should not be substantially affected by sequence regularities that equally affect both strands such as overall shifts in the absolute MFE across RNA families. Table 1 gives the MFED on the transcribed strand ($MFED_{tr}$) and the global structural strand differentials ($\Delta MFED_{tr-ntr}$) in each ncRNA family and mRNAs. On average, $\Delta MFED_{tr-ntr}$ shows large negative mean values of -0.0907 kcal/mol·bp for the miRNA set and -0.0379 kcal/mol·bp for the total ncRNA set, compared with a small (and in fact positive) mean value of 0.00365 kcal/mol·bp for the mRNA set. For miRNAs this corresponds to a very substantial 20.8% decrease in MFED from the transcribed strands; for ncRNAs this corresponds a 11.2% decrease; and for mRNAs a 1.2% increase. The preferential use of the transcribed strand for structure formation is substantially higher in miRNAs and ncRNAs than in mRNAs, which suggests a stronger signal for structure evolution in some ncRNA families.

To investigate whether the structural strand differentials are more likely to be primarily due to actual RNA secondary structural signals, or whether they could be explained by base compositional biases alone, we shuffled the original ncRNA and mRNA sequences while maintaining either mono-nucleotide or di-nucleotide base compositions. We then computed $\Delta MFED_{tr-ntr}$ for the shuffled versions (See Methods). The shuffled version would be expected to show smaller differentials between the two complementary sequence strands if the structural strand asymmetry is primarily caused by RNA secondary structures; this smaller difference in the shuffled sequences being due primarily to base composition biases. Increased G and U content in the transcribed strand can itself lead to increased probability of G::U base pairings and hence structural asymmetries. We therefore compared the original structural strand differentials with the distributions generated by the shuffled versions. Figure 1 shows the distributions of the mean $\Delta MFED_{tr-ntr}$ values of the shuffled sequences (di-nucleotide shuffling) compared with the mean $\Delta MFED_{tr-ntr}$ values of the original miRNA, ncRNA and mRNA sets separately. The mean $\Delta MFED_{tr-ntr}$ values of the original miRNA and ncRNA sets are clearly significantly different from the shuffled distribution, whereas the mean value of the mRNA set is close to the shuffled distribution. The mono-nucleotide shuffled version showed results similar to the di-nucleotide shuffled version (data are not shown).

In particular (shown in table 2), the mean $\Delta MFED_{tr-ntr}$ values for the original miRNA sequences (-0.0907 kcal/mol·bp; 20.8% decrease from the transcribed

Table 1. RNA secondary structure asymmetry in each ncRNA family and mRNA dataset

RNA type	No. of seqs	length (bp) (mean±sd)	(G+C)%	$\Delta MFED_{tr-ntr}$ (kcal/mol·bp) mean±sd	$MFED_{tr}$ (kcal/mol·bp) mean±sd	$\frac{\Delta MFED_{tr-ntr}}{MFED_{tr}}$
mRNAs	614	499±239	47.9	0.00365 ± 0.042	-0.295 ± 0.062	-1.2%
all ncRNAs	753	191±126	47.7	-0.0379 ± 0.062	-0.337 ± 0.11	11.2%
miRNA	100	87±17	46.0	-0.0907 ± 0.073	-0.435 ± 0.067	20.8%
5.8S rRNA	50	152±11	50.4	-0.0165 ± 0.041	-0.288 ± 0.067	5.7%
5S rRNA	50	116±6	52.8	-0.0237 ± 0.054	-0.336 ± 0.061	7.1%
7SK	50	316±13	51.3	0.0182 ± 0.023	-0.308 ± 0.023	-5.9%
Hammerhead_1	26	48±9	51.8	-0.00459 ± 0.064	-0.244 ± 0.08	1.9%
Hammerhead_3	17	55±2	49.4	-0.00869 ± 0.045	-0.347 ± 0.048	2.5%
Intron_gpI	50	418±89	35.1	-0.0119 ± 0.026	-0.233 ± 0.045	5.1%
Intron_gpII	50	79±14	45.8	-0.116 ± 0.067	-0.341 ± 0.091	34%
IRES	50	286±117	54.1	-0.0250 ± 0.037	-0.359 ± 0.073	6.9%
RNase MRP	21	276±35	43.8	-0.0289 ± 0.036	-0.321 ± 0.061	9%
Nuclear RNase P	39	293±55	55.1	-0.0423 ± 0.043	-0.387 ± 0.052	11%
snoRNA CD-box	50	92±34	41.8	-0.0205 ± 0.049	-0.219 ± 0.08	9.3%
snoRNA HACA-box	50	135±27	45.8	-0.0422 ± 0.047	-0.287 ± 0.055	14.7%
SRP_euk_arch	50	293±13	47.9	-0.0651 ± 0.067	-0.511 ± 0.14	12.7%
tmRNA	50	359±42	47.8	-0.0256 ± 0.026	-0.329 ± 0.08	7.8%
tRNA	50	73±5	47.7	-0.0134 ± 0.053	-0.309 ± 0.097	4.3%

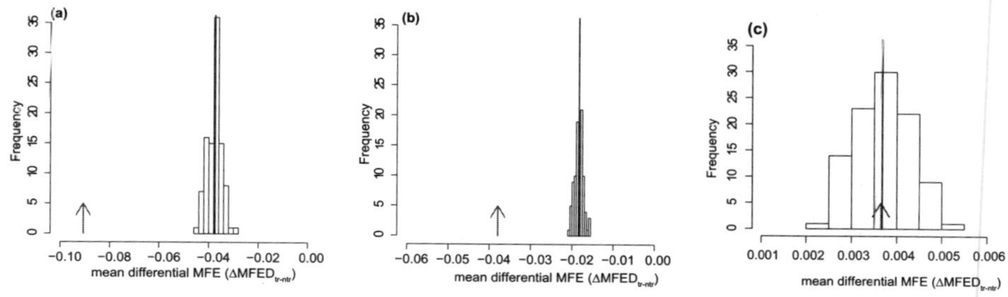

Fig. 1. The distribution of $\Delta\text{MFED}_{tr-ntr}$ values for di-nucleotide shuffled sequences compared to the mean $\Delta\text{MFED}_{tr-ntr}$ values for the original sequences for (a) miRNA set, (b) ncRNA set, and (c) mRNA set. The black arrow shows the mean $\Delta\text{MFED}_{tr-ntr}$ of the original miRNA, ncRNA, and mRNA sequences and black line shows the mean values of the distribution obtained from the di-nucleotide shuffled sequences.

strand) is substantially lower than the mean of the di-nucleotide shuffled version (-0.0375 kcal/mol·bp; 15.2% decrease from the transcribed strands). This strand asymmetry difference between the original and the di-shuffled version is statistically significant: over the 100 permutations, no sequence had a value as extreme as this, and so the differences are statistically significant at p-value < 0.01. This strand differential, to a lesser extent, is also shown by the total ncRNA set and its decrease in comparison with its shuffled version: -0.0379 to -0.0165 kcal/mol·bp (11.2% to 6.7%) on the original and di-nucleotide shuffled sequences respectively.

The stronger structural strand asymmetry signals in ncRNAs, especially pronounced in miRNA precursors, may indicate that secondary structures have predominately evolved on the transcribed strands and that the stable structure is important for the function of these ncRNAs. The MFE calculated on the transcribed strand (MFED_{tr}) of miRNA sequences showed a particularly pronounced negative shift from that of its shuffled version (Figure 2). We calculated Z-scores to indicate the thermodynamic stability of an RNA structure compared to the distribution of the shuffled versions. For each original RNA sequence, a Z-score was calculated for the transcribed sequence strand MFED_{tr} (See Methods) and the average Z-score values are shown in Table 2. Specifically, the Z-score of MFED_{tr} in the miRNA set is very substantial (mean values of -5.54) , which is consistent with the results of Bonnet et al.[8] The Z-score over all ncRNA is -2.3, which is consistent with the results of Rivas et al.[6] Figure 3 gives the Z-score distributions of MFED_{tr} in the miRNA, ncRNA, and mRNA datasets.

As discussed above, the permutation test results indicate that the potential structural strand asymmetry as measured by $\Delta\text{MFED}_{tr-ntr}$ is unlikely to be a result of global base compositional biases. We conjecture that this structural differential between strands is caused by RNA structural constraints.

To measure how consistently $\Delta\text{MFED}_{tr-ntr}$ favours the transcribed strand, as we conjecture it will, we measured the classification accuracy of this feature for predicting the transcribed strand (see Methods). Results are shown in Table 2.

Table 2. Analysis results of structural strand asymmetry features[a]

RNA type	original/ di-shuffle	$\Delta MFED_{tr-ntr}$ (mean±sd) (kcal/mol·bp)	p−value[b]	$\frac{\Delta MFED_{tr-ntr}}{MFED_{tr}}$	mean Z-score of $MFED_{tr}$	Accuracy for detecting transcribed strand: $\frac{f_G+f_U}{f_A+f_C}$ [c]	Accuracy: $\Delta MFED_{tr-ntr}$ (global)	Accuracy: $\Delta MFED_{tr-ntr}$ (local L=150)
mRNA	original	0.00365 ± 0.042	0.49	-1.2%	-0.177	0.33	0.49	0.64
	di-shuffle	0.00368 ± 0.042	—	-1.2%	—			
all ncRNAs	original	-0.0379 ± 0.062	< 0.01	11.2%	-2.3	0.64	0.73	0.70
	di-shuffle	-0.0183 ± 0.047	—	6.7%	—			
miRNA	original	-0.0907 ± 0.073	< 0.01	20.8%	-5.54	0.94	0.87	0.87
	di-shuffle	-0.0375 ± 0.048	—	15.2%	—			
5.8S rRNA	original	-0.0165 ± 0.041	< 0.01	5.7%	-0.637	0.46	0.68	0.52
	di-shuffle	-0.0062 ± 0.041	—	2.3%	—			
5S rRNA	original	-0.0237 ± 0.054	< 0.01	7.1%	-1.12	0.44	0.64	0.64
	di-shuffle	-0.0109 ± 0.043	—	3.6%	—			
7SK	original	0.0182 ± 0.023	1	-5.9%	0.733	0.70	0.20	0.72
	di-shuffle	0.00735 ± 0.023	—	-2.29%	—			
Hammerhead_1	original	-0.00459 ± 0.064	< 0.01	1.9%	-1.07	0.08	0.50	0.50
	di-shuffle	0.0284 ± 0.063	—	-15.2%	—			
Hammerhead_3	original	-0.00869 ± 0.045	0.41	2.5%	-3.19	0.53	0.65	0.65
	di-shuffle	-0.00577 ± 0.048	—	2.8%	—			
Intron_gpI	original	-0.0119 ± 0.026	< 0.01	5.1%	-1.52	0.10	0.68	0.70
	di-shuffle	-0.000384 ± 0.021	—	0.2%	—			
Intron_gpII	original	-0.116 ± 0.067	< 0.01	34%	-3.10	0.82	1.00	1.00
	di-shuffle	-0.0491 ± 0.045	—	21.6%	—			

Note: [a] global strand asymmetries unless specified otherwise; [b] p-value using permutation test; [c] nucleotide frequency

Table 2. (*Continued*)

RNA type	original/ di-shuffle	$\Delta MFED_{tr-ntr}$ (mean±sd) (kcal/mol·bp)	p−value[b]	$\frac{\Delta MFED_{tr-ntr}}{MFED_{tr}}$	mean Z-score of $MFED_{tr}$	Accuracy for detecting transcribed strand		
						$\frac{f_G+f_U}{f_A+f_C}$ [c]	$\Delta MFED_{tr-ntr}$ (global)	$\Delta MFED_{tr-ntr}$ (local L=150)
IRES	original	-0.0250 ± 0.037	< 0.01	6.9%	-1.38	0.66	0.76	0.56
	di-shuffle	-0.0148 ± 0.041	—	4.4%	—			
RNase MRP	original	-0.0289 ± 0.036	< 0.01	9.1%	-3.36	0.86	0.81	0.57
	di-shuffle	-0.0121 ± 0.031	—	4.5%	—			
Nuclear RNase P	original	-0.0423 ± 0.043	< 0.01	11%	-1.50	0.79	0.87	0.85
	di-shuffle	-0.0305 ± 0.033	—	8.5%	—			
snoRNA	original	-0.0205 ± 0.049	0.02	9.3%	-0.575	0.72	0.64	0.64
CD-box	di-shuffle	-0.0111 ± 0.047	—	5.5%	—			
snoRNA	original	-0.0422 ± 0.047	< 0.01	14.7%	-0.99	0.80	0.86	0.80
HACA-box	di-shuffle	-0.0193 ± 0.043	—	7.4%	—			
SRP_euk_arch	original	-0.0651 ± 0.067	< 0.01	12.7%	-6.78	0.80	0.84	0.80
	di-shuffle	-0.0529 ± 0.059	—	13.3%	—			
tmRNA	original	-0.0256 ± 0.026	< 0.01	7.8%	-1.97	0.45	0.84	0.49
	di-shuffle	-0.00874 ± 0.024	—	2.9%	—			
tRNA	original	-0.0134 ±0.053	0.88	4.3%	-1.84	0.66	0.60	0.60
	di-shuffle	-0.0195 ± 0.058	—	8.2%	—			

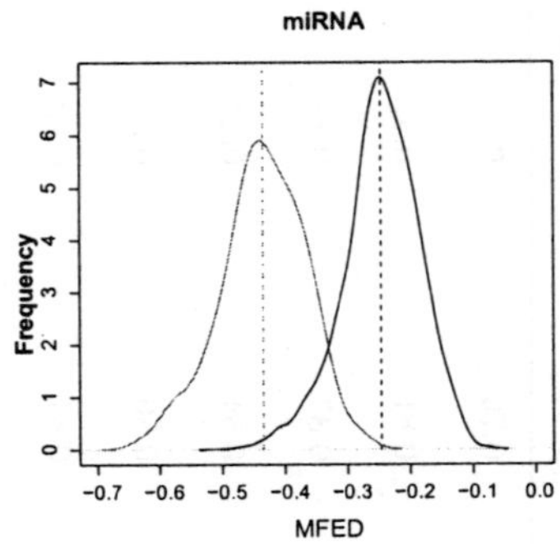

Fig. 2. The distribution comparison for the MFEDs calculated on miRNA transcribed strand (left) and that on miRNA di-nucleotide shuffled version (right). The dashed lines are the mean values of MFEDs.

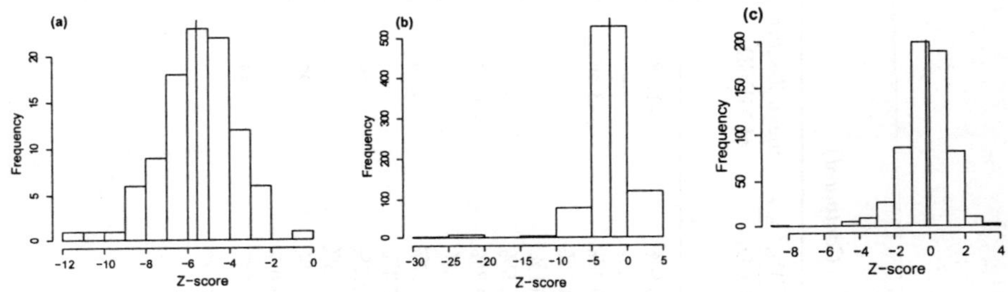

Fig. 3. The histogram of Z-score values for the di-nucleotide shuffled sequences in (a) miRNA set, (b) ncRNA set, and (c) mRNA set. The black line shows the mean z-score.

Using the global strand asymmetry feature gave a 73% overall accuracy for ncRNA. The local strand asymmetry feature for L=150 shows accuracy for ncRNAs within 3% of the global measure despite being limited to a fixed window size. Dinucleotide-shuffling the ncRNA set reduced the accuracy from 70% to 61%. For mRNAs, using global $\Delta\text{MFED}_{tr-ntr}$ gave a random level of 49%, and for local (L=150) $\Delta\text{MFED}_{tr-ntr}$, accuracy improved above random levels to 64% accuracy, which is statistically significant ($p < 0.01$). Using a di-nucleotide shuffled version of mRNA gave random accuracy results of 51%, indicating that the 64% accuracy result is unlikely to be due to global base compositional asymmetries alone.

Base composition biases can also be useful features for classification of RNA. For comparison with the structural asymmetry feature, the transcribed strand prediction results for a base composition feature are presented in Table 2. We use $\frac{f_G+f_U}{f_A+f_C}$, which as described in[3,17] shows a strand asymmetry in transcribed regions caused by differences in base mutation rates. We note that the increased G and U content in the transcribed strand is also consistent with increased structure-forming potential in this strand. This feature shows some discriminability, although inferior accuracy relative to the local structural strand asymmetry feature.

4. Conclusion

This study has not focused on the RNA optimal minimum-energy folded structures, but rather on the strand asymmetry of RNA secondary structures, extending our previous study[9] to known ncRNA datasets. We have shown that there exists a substantial asymmetry in RNA structure potential between the complementary sequence strands in ncRNAs (including miRNAs), and that this bias is in addition to that due to base compositional strand asymmetries. We conjecture that this is due to structural constraints on the transcribed strand of functional RNA sequences.

This structural strand asymmetry should be useful as an independent feature in helping to distinguish transcribed regions, including transcription orientation, for gene-finding (particularly ncRNA) purposes. This approach can be applied across an entire genome as the local structural asymmetry feature can be easily computed in this case. The non-coding gene prediction framework of Glusman et al[3] is an example which could be extended with this feature. It will be required to combine with additional statistical features to achieve higher discriminability for ncRNA prediction. The possible candidate features include base compositional biases and conserved ncRNA elements. Both our previous work[9] and other studies[18] have suggested that base compositional biases may serve as indicators of ncRNAs. Also, RNA structural conservation using comparative sequence analysis has also shown promise for ncRNA prediction.[7,19] In future work, we will investigate combining these features using machine learning approaches and apply to whole genomes, including UTR, intergenic and intronic regions.

References

1. Storz, G., An expanding universe of noncoding RNAs., *Science*, 296(5571):1260–1263, 2002.
2. Mattick, J. S., and Makunin, I. S., Small regulatory RNAs in mammals., *Hum. Mol. Genet.*, 14(Spec No 1):R121-R132, 2005.
3. Glusman, G., Qin, S., El-Gewely, M. R., Siegel, A.F., Roach, J.C., Hood, L, and Smit, A. A third approach to gene prediction suggests thousands of additional human transcribed regions., *PLOS Computational Biology*, 2(3):160–173, 2006.
4. Seffens, W., and Digby, D., mRNAs have greater negative folding free energies than shuffled or codon choice randomized sequences., *Nucleic Acids Res.*, 27(7):1578–1584, 1999.
5. Workman, C., and Krogh, A., No evidence that mRNAs have lower folding free energies than random sequences with the same di-nucleotide distribution., *Nucleic Acids Res.*, 27(24):4816–4822, 1999.
6. Rivas, E., and Eddy, S. R., Secondary structure alone is generally not statistically significant for the detection of noncoding RNAs., *Bioinformatics*, 16(2):583–605, 2000.
7. Washietl, S., Hofacker, I. L., and Stadler, P. F., Fast and reliable prediction of non-coding RNAs., *Proc. Natl. Acad. Sci. U. S. A.*, 102(7):2454–2459, 2005.
8. Bonnet, E., Wuyts, J., Rouz, P., and Van de Peer, Y., Evidence that microRNA precursors, unlike other non-coding RNAs, have lower folding free energies than random sequences., *Bioinformatics*, 16(2):2911–2917, 2004.
9. Wen, J., Parker, B. J., and Weiller, G. F., *In silico* identification and characterization

of mRNA-like noncoding transcripts in *Medicago truncatula.*, *In Silico Biology*, 7, 0034. </isb/2007/07/0034/>

10. Griffiths-Jones,S., Moxon S., Marshall,M., Khanna,A., Eddy,S.R., Bateman, A. Rfam: annotating non-coding RNAs in complete genomes. *Nucleic Acids Res.*, 33(Database issue):D121-D124, 2005.
11. Boeckmann, B., Bairoch, A., Apweiler, R., Blatter, M.-C., Estreicher, A., Gasteiger, E., Martin, M.J., Michoud, K., O'Donovan, C., Phan, I., Pilbout, S., Schneider, M. The SWISS-PROT protein knowledgebase and its supplement TrEMBL in 2003. *Nucleic Acids Res.*, 31:365–370, 2003.
12. Kulikova, T., Akhtar, R., Aldebert, et al. EMBL Nucleotide Sequence Database in 2006. *Nucleic Acids Res.*, 35:(Database issue):D16-20, 2006.
13. Rice, P., Longden, I., and Bleasby, A., EMBOSS:The European Molecular Biology Open Software Suite. *Trends Genet.*, 16(6):276-277, 2000.
14. Altschul, S. F., and Erickson, B. W., Significance of nucleotide sequence alignments: A method for random sequence permutation that preserves di-nucleotide and codon usage. *Mol. Biol. Evol.*, 2:526–538, 1985.
15. Hofacker, I. L., Fontana, W., Stadler, P.F., Bonhoeffer, S., Tacker,M., Schuster, P. Fast Folding and Comparison of RNA Secondary Structures. *Monatshefte f. Chemie*, 125:167–188, 1994.
16. Reiche, K., Stadler P. F., RNAstrand: reading direction of structured RNAs in multiple sequence alignments., *Algorithms Mol. Bio.*, 2:6, 2007.
17. Green, P., Ewing, B., Miller, W., Thomas, P. J., NISC Comparative Sequencing Program, Green, E. D., Transcription-associated mutational asymmetry in mammalian evolution. *Nat. Genet.*, 33:514–517, 2003.
18. Schattner P., Searching for RNA genes using base-composition statistics. *Nucleic Acids Res.*, 30:2076–2082, 2002.
19. Rivas, E., Eddy, S. R., Noncoding RNA gene detection using comparative sequence analysis. *BMC Bioinformatics*, 2:8, 2001.

FINDING NON-CODING RNAs THROUGH GENOME-SCALE CLUSTERING

HUEI-HUN TSENG[1], ZASHA WEINBERG[2], JEREMY GORE[2], RONALD R. BREAKER[2,3,4] and WALTER L. RUZZO[1,5]

Departments of [1]Computer Science and Engineering and [5]Genome Sciences
University of Washington, Seattle, WA 98105, USA
E-mail: {lachesis,ruzzo}@cs.washington.edu

[2]Department of Molecular, Cellular and Developmental Biology, [3]Howard Hughes Medical Institute, [4]Department of Molecular Biophysics and Biochemistry
Yale University, New Haven, CT 06520-8103, USA
E-mail: zasha.weinberg@yale.edu

Non-coding RNAs (ncRNAs) are transcripts that do not code for proteins. Recent findings have shown that RNA-mediated regulatory mechanisms influence a substantial portion of typical microbial genomes. We present an efficient method for finding potential ncRNAs in bacteria by clustering genomic sequences based on homology inferred from both primary sequence and secondary structure. We evaluate our approach using a set of Firmicutes sequences, and the results show promise for discovering new ncRNAs.

Keywords: Noncoding RNA, RNA discovery, hierarchical clustering, motif discovery

1. Introduction

1.1. *Motivation and Related Work*

Non-coding RNAs (ncRNAs) are functional transcripts that do not code for proteins. Recent findings have shown that RNA-mediated regulatory mechanisms influence a substantial portion of typical microbial genomes,[1] drawing increasing attention to their study. A major approach for computational detection of ncRNAs is through comparative genomics,[2] where conserved structures are predicted from sequences of multiple species. The key difficulty with this approach is that homologous ncRNAs are often divergent because compensatory mutation preserves structure while changing the sequence. Unfortunately, existing ncRNA-discovery algorithms that consider secondary structure are impractical for genome-scale searches since they are computationally expensive, and work best when applied to datasets in which homologous ncRNAs predominate. Together, these considerations suggest the following strategy: gather clusters of sequences so that each cluster is sufficiently small and enriched in homologous elements for successful computational motif prediction.

Recently, Yao *et al.*[3] applied this strategy to search for bacterial *cis*-regulatory RNAs. Because *cis*-regulatory RNAs are often upstream of genes, they clustered regions upstream of homologous genes (a "gene-oriented" approach). They avoided the need for accurate alignment by using a tool called CMFinder[4] that can predict RNA motifs in unaligned sequences in the face of low sequence conservation, extraneous flanking regions and unrelated sequences. The method successfully recovered most known Rfam[5] families in Firmicutes. Coupled with careful manual evaluation of top-ranking results, this paper and Weinberg *et al.*[6] identified 29 novel RNAs including several riboswitches,[7,8] some of which have been experimentally validated. However, this approach will detect ncRNAs only if they are well-represented upstream of homologous genes. For example, ncRNA genes that are independently transcribed (e.g., SRP, RNaseP, tRNAs) will tend to maintain particular neighboring genes only through a narrow phylogenetic range. This is true of some ncRNAs in the Firmicutes (and Yao *et al.* generally recovered these), but others will be missed. Another important example of the ncRNAs that might be missed by a gene-oriented approach are ones that regulate several non-homologous genes in a phylogenetically narrow range of species.

The main contribution of this paper is the development of an "IGR-oriented pipeline" that clusters intergenic regions (IGRs) based on a combination of sequence and structure similarity, independent of gene context, for purposes of ncRNA discovery. We believe it can identify ncRNAs that are difficult to find with a gene-oriented strategy. For example, an early version of our IGR-oriented approach (unpublished data) correctly predicted 7 related riboswitches regulating purine biosynthesis genes in *Mesoplasma florum*[9] with no close relatives in other sequenced species, exactly the second scenario outlined above.

1.2. *Efficient pipeline for detecting ncRNAs*

To be able to detect ncRNAs computationally, we wish to identify homologous RNA sequences. To do this without gene context, we search through entire intergenic regions (IGRs) of several species for homology. Homologous ncRNAs usually exhibit some conservation in primary sequence, but detection of this similarity is often impossible without exploiting the significant conservation of RNA structure. Traditional structure-based methods[10] perform well but are extremely slow, making them impractical for large search spaces. We design a novel lightweight approach that incorporates both secondary structure information and primary sequence homology via BLAST (referred to as the *folded*-BLAST approach). The goal is to achieve the best sensitivity possible, while maintaining feasible search time.

We wish to group sequences based on homology relationships. However, RNAs may contain multiple domains with sequence homology recognizable by BLAST, but these domains may be separated by dissimilar regions. To account for this, we design a hierarchical clustering method that, given a set of pairwise homology hits, heuristically merges and clusters overlapping sequences. Finally, as in Yao *et al.*'s pipeline,[3] the clusters can be used to predict motifs, which in turn can be used to scan genomes for more motif instances (motif scan).

Our proposed pipeline for a given input set of genomic sequences, then, consists of the following steps: (1) intergenic region extraction; (2) homology search; (3) hierarchical clustering; (4) motif discovery; and (5) motif scan.

Our pipeline shares high level goals with the work of Will *et al.*,[11] but differs in emphasis, and is somewhat complementary to it. Both cluster intergenic sequences based on homology, then attempt to predict RNA motifs in these clusters. Will *et al.*, building on Missal *et al.*,[12] need reliable sequence alignments for their motif prediction step, so they use a stringent BLAST E-value threshold for this phase. To recover broader RNA families, they apply a second clustering step to cluster the RNA motifs produced in the first step. The number of RNA predictions is much smaller than the number of IGRs, and they can afford to apply sophisticated but computationally expensive structure-based clustering methods here, and their paper develops such a method (LocARNA). In contrast, we use an RNA motif prediction tool that tolerates unaligned inputs and thus can be more aggressive in trying to gather more (and more remote) homologs, on the premise that more examples will allow inference of more accurate models. Hence, we cluster intergenic sequences based on relatively permissive BLAST searches. A novelty of our approach is incorporating secondary structure information in this clustering stage. Neither method attempts direct pairwise structure comparison among all intergenic sequences; that appears prohibitively expensive on data sets of this scale.

1.3. *Evaluation*

We clustered a set of Firmicutes genomic sequences, and evaluated them using a set of ncRNA families mainly consisting of riboswitches. Riboswitches are metabolite-sensing RNAs that regulate gene activity through binding to ligands and modifying the expression of biosynthetic and transport proteins for those ligands.[7,8] They are structurally conserved with an average family sequence identity of 55–80% and average sequence length of 60–200 nts. Primary sequence-only methods captured ~84% of the known ncRNAs, with an average cluster specificity of ~40%. Incorporating secondary structure captures about 80% of the known ncRNAs while increasing average cluster specificity to ~50%. Motifs predicted from the ncRNA-containing clusters were then used to scan a test set, and the *folded*-BLAST approach achieved median sensitivity of 76% with 99% specificity, much better than the best primary sequence-only approach (sensitivity 61%). Moreover, several motifs predicted from *folded*-BLAST clusters were more similar, and in some cases, almost identical to trusted riboswitch models. This suggests that our novel method of secondary structure-incorporation enhances clustering, which in turn increases the likelihood of inferring a strong motif.

2. Results

Full genomic sequences from a set of 212 Firmicutes species were used as input. The entire set contains 1252 known ncRNAs, 1008 of which are completely covered by our extracted intergenic regions. Primary sequence homology data were obtained using NCBI-BLAST,[13] WU-BLAST,[14] or SSEARCH.[15] To incorporate structure into our homology searches, we used WU-BLAST, since it allows convenient usage of arbitrary scoring matrices.[a]

2.1. *Clustering evaluation*

Table 1 shows evaluation for the Firmicutes clusters generated by our pipeline. Note that per-cluster specificity (p) is only a lower-bound, since unannotated members of a cluster could be undiscovered ncRNAs. NCBI-BLAST generally has the best capture count, i.e., clustering the largest number of known ncRNAs in any cluster, yet also the worst per-cluster specificity. *folded*-BLAST captures fewer, yet average cluster specificity generally tops all the rest. However, no program in Table 1 consistently surpasses the others.

Since our goal is to detect novel ncRNA families, we turn our attention to individual clusters with good specificity. For example, *folded*-BLAST produced 16 clusters that contained at least one TPP riboswitch, one of which had specificity of 35/39, another 10/10, and another 7/9. If any of these could yield a representative motif, then a motif scan would likely recover other TPP riboswitches. Thus, what is more important is our ability to produce clusters that permit RNA alignment tools like CMFinder to correctly predict structured RNAs. In the following section, we show results of predicted motifs from selected clusters.

2.2. *Motif discovery and scanning*

CMFinder predicts zero or more motifs in all ncRNA-containing clusters. For any motif, along with the covariance model (CM) produced, we do a CM scan if the number of cluster members containing this motif is at least 6, and that the average motif score (generated by CMFinder) is at least 50. These criteria are set because weak motifs/CMs will likely introduce false hits. The CMs scan our entire ncRNA dataset: ~1 Mb of ncRNAs from all available bacterial species (not just Firmicutes), plus a control set of ~5 Mb of randomly selected IGRs (from various species) not containing known ncRNAs.

In this CM scan test, a hit is considered correct only if it matches a selected ncRNA on the correct strand. Strictly speaking, we cannot be sure that our randomly selected IGRs indeed do not contain any undiscovered ncRNAs, but for the purpose of evaluation, we assume there to be none.

---

[a]Details are in Methods. Supplementary materials including additional method details and results are available at: http://bio.cs.washington.edu/supplements/lachesis/APBC2008.

Table 1. Clustering evaluation by individual ncRNA families

ncRNA (#)	RfamID or ref.	avg. seq. len. (nts)	NCBI-BLAST c	NCBI-BLAST p	WU-BLAST c	WU-BLAST p	SSEARCH c	SSEARCH p	*folded*-BLAST c	*folded*-BLAST p
t-box (452)	RF00230	223	**441**	0.42	406	0.49	423	0.51	352	**0.60**
SAMI (113)	RF00162	124	**104**	0.44	99	0.57	100	**0.59**	102	0.55
TPP (90)	RF00059	97	**77**	0.29	70	0.21	72	0.39	75	**0.61**
purine (66)	RF00167	99	**64**	0.11	52	0.23	52	**0.30**	59	0.21
ylbH (53)	RF00516	143	**37**	0.08	27	0.11	28	**0.13**	26	0.06
cobalamin (51)	RF00174	201	**51**	0.51	46	0.61	45	0.22	45	**0.67**
lysine (46)	RF00168	181	**36**	0.17	33	**0.45**	34	0.41	33	**0.45**
SRP (41)	RF00169	99	**41**	0.42	40	0.37	39	0.41	**41**	**0.45**
RNaseP (40)	RF00011	360	**37**	0.77	**37**	0.82	**37**	0.39	**37**	**0.86**
FMN (40)	RF00050	147	**40**	0.59	**40**	**0.82**	**40**	0.75	**40**	0.67
glycine (38)	RF00504	90	32	0.29	30	**0.58**	**33**	0.41	28	0.29
preQ1 (37)	RF00522	69	**33**	0.15	18	0.24	23	**0.42**	19	0.39
ydaO (29)	RF00379	171	**29**	0.85	27	**0.95**	26	0.81	24	0.86
yybP (29)	RF00080	131	**28**	0.34	23	0.36	26	0.34	23	**0.44**
6S (26)	RF00013	205	23	0.32	**24**	0.32	22	0.56	**24**	**0.66**
ykoK (25)	RF00380	178	**25**	0.29	**25**	0.46	22	0.51	**25**	**0.96**
glmS (19)	RF00234	183	18	0.46	18	**0.86**	18	0.29	**19**	0.34
ykkC (12)	RF00442	129	**10**	0.46	**10**	0.47	8	0.35	9	**0.50**
moco (10)	Weinberg et al.[6]	132	6	**0.52**	6	0.51	3	0.21	**7**	**0.52**
SMK (10)	Fuchs et al.[16]	173	**6**	0.38	2	0.30	**6**	**0.83**	2	0.26
Median**			**0.89**	0.34	0.79	0.46	0.85	0.41	0.79	**0.51**

c: captured count; p: avg. cluster specificity

Note:
For each ncRNA family F we give its name, the number of members covered by the extracted IGRs of our Firmicutes test set, the Rfam accession or other reference, the average length of members, and performance statistics for the clustering methods. For each F, we report c, the *captured count*, i.e., the number of ncRNAs in F covered by a member in some cluster, and p, the *cluster specificity*, i.e., the average over all clusters C containing members of F, of the percentage of members of F in C. We define a known ncRNA as "covered" by a IGR segment if the segment covers at least 50 nts or 50% of the ncRNA region.
**: For c, the median is taken over the ratios of capture count to family size.

Table 2. CM scan best recovery motif comparison. For each ncRNA family and each homology search program used, the motif/CM that recovered the most instances of the particular family is listed. The actual motif identifications can be cross-referenced online. *sen.* is the recovery percentage (sensitivity), and *spe.* is the specificity of the CM scan; "None" indicates that no instances were recovered.

ncRNA	NCBI-BLAST		WU-BLAST		SSEARCH		*folded*-BLAST	
family	*sen.*	*spe.*	*sen.*	*spe.*	*sen.*	*spe.*	*sen.*	*spe.*
t-box	0.69	0.98	**0.71**	**0.99**	0.41	**0.99**	0.68	**0.99**
SAMI	0.94	**0.99**	**0.96**	**0.99**	0.84	**0.99**	0.94	**0.99**
TPP	0.84	**0.99**	0.95	**0.99**	0.54	**0.99**	**0.96**	**0.99**
purine	0.36	**0.99**	0.36	**0.99**	0.32	**0.99**	**0.37**	**0.99**
ylbH	0.01	0.5	0.01	**1.00**	**0.02**	**1.00**	0.01	0.33
cobalamin	None		0.84	0.82	0.72	**1.00**	**0.86**	0.99
lysine	0.79	**1.00**	**0.84**	0.82	0.72	**1.00**	0.74	**1.00**
SRP	0.1	0.99	0.1	**1.0**	**0.84**	0.98	0.77	0.98
RNaseP	**1.0**	**0.99**	**1.0**	**0.99**	**1.0**	**0.99**	**1.0**	**0.99**
FMN	**0.96**	**0.99**	**0.96**	**0.99**	**0.96**	**0.99**	**0.96**	0.98
glycine	None		0.08	0.98	None		**0.86**	**0.99**
preQ1	**0.01**	**0.04**	None		**0.01**	0.02	None	
ydaO	**0.97**	**1.00**	0.96	**1.00**	0.96	**1.00**	0.96	0.99
yybP	**0.26**	**1.00**	0.11	**1.00**	0.11	**1.00**	0.22	0.99
6S	0.09	**1.00**	**0.42**	0.92	0.09	**1.00**	0.29	**1.00**
ykoK	**0.96**	**1.00**	**0.96**	**1.00**	0.90	**1.00**	**0.96**	**1.00**
glmS	**0.95**	**1.00**	0.93	**1.00**	0.91	**1.00**	**0.95**	**1.00**
ykkC	None		None		**0.69**	**1.00**	**0.69**	**1.00**
moco	None		None		None		None	
SMK	**0.08**	**0.67**	None		None		None	
Median	0.31	**0.99**	0.56	**0.99**	0.61	**0.99**	**0.76**	**0.99**

Table 2 summarizes the individual CM scans recovering the most instances for each particular family. For most of the more abundant families, such as FMN, SAMI, TPP, ykoK and ydaO riboswitches, all four programs had a best CM scan of ~ 0.9 sensitivity with ~ 1.0 specificity. Recovery of purine riboswitches was consistently low, and we observed that it was because non-Firmicutes purine riboswitches have much longer single-stranded terminal regions than their Firmicutes counterparts. Of particular interest are the 7 *Mesoplasma florum* purine riboswitches, a difficult case for gene-oriented pipelines. In NCBI-BLAST, 6 of the 7 were grouped in a cluster of size 44 (6/44), and although CMFinder succeeded in producing a representative motif, it had low specificity: the CM scan reported back the 6 on which it was trained, along with almost 2000 false hits. The motif discovered from the WU-BLAST cluster was more specific, but still had 96 false hits. In contrast, SSEARCH generated a 6/7 cluster and the resulting CMfinder motif scans reported back the 6 with no false hits. *folded*-BLAST produced a 7/32 cluster, but CMFinder did not predict a motif, so no CM scan was done. We examined why CMFinder failed in *folded*-BLAST's case, and determined that CMFinder's prior parameter for the expected fraction of motif-containing instances was higher than the actual percentage. If the percentage was lowered from 0.5 to 0.2, CMFinder would find a representative motif for 6 of the *M. florum* purine riboswitches. However, lowering the percentage might entail tradeoffs.

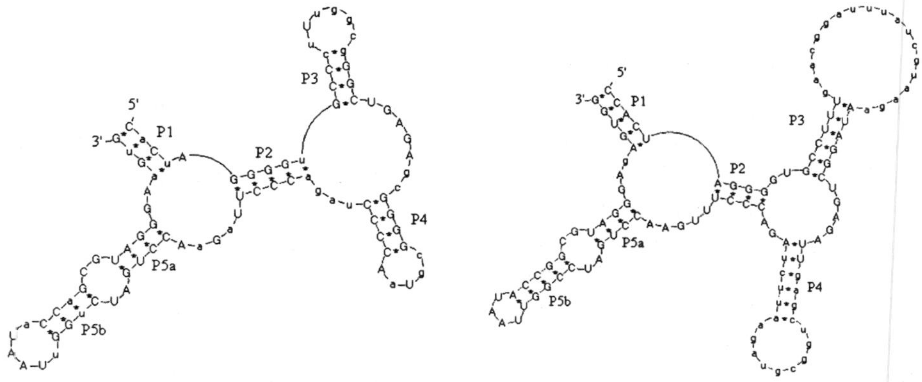

(a) *folded*-BLAST (b) Rfam

Fig. 1. A TPP riboswitch motif automatically predicted by our pipeline *vs* the (hand-curated) Rfam structure. Terminal single-stranded regions are not shown for simplicity. (a) Motif predicted using *folded*-BLAST approach that resulted in the best CM scan recovery of TPP riboswitches. (b) Consensus motif from the Rfam TPP riboswitch seed alignment (Rfam id RF00059).

Fig.1(a) depicts the motif recovering the most TPP riboswitches using the *folded*-BLAST approach, while Fig.1(b) is the (hand curated) consensus motif from the Rfam TPP seed alignment. The *folded*-BLAST motif has a longer unpaired region on the 5′ end, but shares an almost identical structure and base composition with Rfam's. It is encouraging to see how similar the two structures are, given that the cluster used to predict motif Fig.1(a) had 30 sequences, and Fig.1(b) was constructed out of 174 seed sequences. The best TPP-recovering motif produced by the other three programs (not drawn due to space limit) correctly predicted most of the structure, but varied in 5′ and 3′ ends: WU-BLAST's does not have the closing stem loop, resulting in a much longer unpaired region on both ends; NCBI-BLAST's missed both the closing stem loop and part of a multiloop; SSEARCH's predicted most of the base pairings correctly, but had a 70-nts 5′ unpaired region, which is probably why the CM scan recovery was poor. We noticed that the TPP-containing sequences in the best-TPP-recovery clusters for NCBI-BLAST and SSEARCH were 50-200 bps longer than the ones in WU-BLAST and *folded*-BLAST, and it is possible that we had tuned the parameters for the first two programs in such a way that IGR fragments are easily joined together into long sequences during clustering pre-processing.

Two small ncRNA families, the moco and preQ1 riboswitches, had poor CM scan recoveries. The moco riboswitch was discovered by manual inspection of CMfinder motifs from the gene-oriented pipeline,[6] but is not common in Firmicutes, making motif discovery difficult, even though all four homology search programs produced good clusters, grouping 5 or 6 instances in compact clusters of size 5 or 6. For preQ1, though it has more Firmicutes instances, is short (65 nts on average), which made accurate and compact clustering challenging.

3. Methods

3.1. *Extracting intergenic regions (IGR)*

Given an input genomic sequence, we remove regions annotated in RefSeq as coding regions, repeat regions, tRNAs or rRNAs. Both strands are removed when one strand contains one of the above annotations. This breaks a genomic sequence into a set of intergenic regions (IGRs). We then discard all IGRs shorter than 15 nts along with those immediately adjacent to an annotated rRNA region, for we find their $5'$ and $3'$ borders to be frequently misannotated.

Removing genomic regions encoding for genes or known RNA elements on either strand reduces search space, yet might risk missing ncRNAs. Using our ncRNA dataset, we examine how much will be missed in our Firmicutes set, and how much can be gained by extending our search space into annotated regions. Our dataset contains 1236 Firmicutes ncRNAs, and if we hypothesize that a region containing a ncRNA will have a chance of being grouped with other homologous regions if and only if there exists an extracted IGR that covers at least 50% or 50 nts of the ncRNA, then we will miss 107 ncRNAs. Even if we extend our extracted IGRs 200 nts on both ends, almost doubling the search space, by our hypothesized definition we still miss 59. We have several explanations for this: If a ncRNA overlaps an annotated coding region, the RefSeq record could have mis-annotated the location. Also, ncRNAs might overlap other functional regions either due to the evolutionary pressure of keeping genomes compact, or because their mechanism of gene regulation requires some overlap. For simplicity in this study, we do not extend IGRs.

3.2. *Homology search*

To compare performance, we used several popular search programs, including NCBI-BLAST, WU-BLAST, and SSEARCH. SSEARCH[15] implements the Smith-Waterman local alignment algorithm; it is 10 times slower than BLAST programs, but is thought to be more sensitive. NCBI-BLAST and WU-BLAST are both heuristic approximations to Smith-Waterman, and begin alignment by matching exact short words (*seeds*). In this study, we use a seed length of 11 because preliminary tests indicated that it has reasonable sensitivity and speed.

3.3. *Homology search with predicted secondary structure*

To implement *folded*-BLAST, we use RNALfold from the Vienna package[17] to compute locally stable RNA secondary structures with a maximal base span L (empirically set to 200). Given an input sequence and a defined L, RNALfold lists predicted secondary structure components. However, since it has been shown that secondary structure alone is insufficient for detecting ncRNAs,[18] we cannot entirely trust the boundaries and structures predicted. Hence, we developed a heuristic procedure to merge RNALfold's components, breaking long IGRs into small, overlapping pieces with lengths of 200-500 nt. For each piece, RNAfold predicts whether each nucleotide is paired upstream, paired downstream or unpaired. To take advantage of fast primary sequence homology search programs, we map these sequences into a 12 letter alphabet representing nucleotide plus pairing direction. The resulting sequences are treated like protein sequences, but we search using a handmade scoring matrix in which nucleotide identity (match) is favored, but when the predicted structures are the same, nucleotide mismatch penalty is mitigated. The matrix is detailed in the online supplement.

3.4. *Clustering*

Prior to clustering, we pre-process pairwise homology hits obtained from homology search programs into *nodes* and *edges*. A node represents an IGR segment, and an edge represents a homology hit. For all homology search results, hits with score less than 40 or greater than 300 are ignored. Note that, although the same score may have different statistical significance depending on the program used, we have tuned the parameters to achieve statistical similarity, and have observed that all programs generally produce the same distribution of scores. For *folded*-BLAST, an additional criterion is added: hits with percentage identity less than 0.3 or positive percentage identity (percentage of alignments contributing positive scores) less than 0.5 are ignored. The cutoff values were determined using small test sets of ncRNAs against random sequences.

When we process a homology hit, we first check whether there already is a node representing a segment overlapping the query region by 15 nts. If so, then that node is expanded to represent the union of the two regions; otherwise a new node is created. The same procedure is applied to the aligned subject region. We then create an edge to represent the hit, whose weight is the homology score. In sum, the output of pre-processing a set of homology hits is a weighted, undirected graph.

The clustering step uses WPGMA (Weighted Pair Group Method using Arithmetic averaging), also known as average-linkage clustering. Edge weights are used as scores. Missing edges are assumed to have score 0.

The output of the hierarchical clustering is a forest of trees. Some trees can be as small as only 2 leaves, which means that the homology search program did not find any other IGR segments significantly homologous to them. The largest tree can be as large as the number of nodes. Such a supersized cluster is impractical for any further evaluation, and given that most of our ncRNA families have no more than 100 instances in our species sets, we generally use a size cutoff of 50. More adaptive tree-cutting is discussed below.

3.5. *Motif prediction and scan*

Motif prediction and scan are done as in Yao *et al.*, 2007, excluding the (subjective) manual evaluation steps. Briefly, CMfinder[4] folds each sequence in its input set, and constructs an initial heuristic alignment attempting to match similar sequence and structural features between sequences. Next it builds a covariance model (CM) from the alignment, exploiting both mutual information and single-sequence structure predictions to arrive at a consensus structure prediction. Finally, it performs an EM-like iteration, alternately realigning the sequences to the model and rebuilding the model from the refined alignment. It is robust to non-motif containing sequences and extraneous regions flanking the motifs. Parts of CMfinder use the Infernal[19] software package, which was also used for the scanning step in our evaluation. On larger data sets, we would also use the RAVENNA[20] filtering package.

4. Discussion and Future Work

Refining the design of our ncRNA discovery pipeline is complicated because there is no clear winner among applicable homology and motif tools. We plan to improve our pipeline in various aspects, particularly the following: (1) Secondary structure incorporation: The heuristics used in our novel method for incorporating secondary structure were empirically determined. For example, secondary structures were predicted with a maximal base span of 200, and the scoring matrix used for *folded*-BLAST was handmade (we found scores trained from curated ncRNA alignments to perform poorly). (2) Hierarchical clustering: We merged overlapping homologous sequences based on the assumption that evolutionary divergence causes homology search programs to fail to capture full length homologous ncRNAs. We have neither deeply investigated this assumption nor determined an optimal merging strategy. (3) Adaptive tree cutting: Our fixed size-cut for partitioning large clusters may compromise motif prediction for some ncRNA families. For example, *folded*-BLAST clustered the 7 *M. florum* purine riboswitches into a compact subbranch, yet the 50 size-cut included extraneous sequences that caused CMFinder to fail in predicting a motif. To improve specificity, we could try using other evidence (e.g., homology scores) to trim a cluster, or iteratively use CMFinder to add or remove members until all cluster members are predicted as containing motif instances.

With future evaluation on other species and improvement of the existing pipeline, we hope to identify and experimentally verify novel structured RNAs.

References

1. W. C. Winkler, *Curr Opin Chem Biol* **9**, 594(Dec 2005).
2. E. Rivas and S. R. Eddy, *BMC Bioinformatics* **2**, p. 8 (2001).
3. Z. Yao, J. Barrick, Z. Weinberg, S. Neph, R. Breaker, M. Tompa and W. Ruzzo, *PLoS Comput Biol* **3**, p. e126(July 2007).
4. Z. Yao, Z. Weinberg and W. L. Ruzzo, *Bioinformatics* **22**, 445(February 2006).
5. S. Griffiths-Jones, A. Bateman, M. Marshall, A. Khanna and S. R. Eddy, *Nucleic Acids Res* **31**, 439 (2003).
6. Z. Weinberg, J. Barrick, Z. Yao, A. Roth, J. Kim, J. Gore, J. Wang, E. Lee, K. Block, N. Sudarsan, S. Neph, M. Tompa, W. Ruzzo and R. Breaker, *Nucleic Acids Res* **35**, 4809(July 2007).
7. R. L. Coppins, K. B. Hall and E. A. Groisman, *Curr Opin Microbiol* **10**, 176(Apr 2007).
8. B. J. Tucker and R. R. Breaker, *Curr Opin Struct Biol* **15**, 342(Jun 2005).

9. J. Kim, A. Roth and R. Breaker, *Proc Natl Acad Sci U S A* **104**(Oct 2 2007).
10. E. K. Freyhult, J. P. Bollback and P. P. Gardner, *Genome Res* **17**, 117(Jan 2007).
11. S. Will, K. Reiche, I. L. Hofacker, P. F. Stadler and R. Backofen, *PLoS Comput Biol* **3**, p. e65(Apr 2007).
12. K. Missal, X. Zhu, D. Rose, W. Deng, G. Skogerbo, R. Chen and P. F. Stadler, *J Exp Zoolog B Mol Dev Evol* **306**, 379(Jul 2006).
13. S. Altschul, W. Gish, W. Miller, E. Myers and D. Lipman, *J Mol Biol* **215**, 403 (1990).
14. W. Gish, (1996-2004), `http://blast.wustl.edu`.
15. W. R. Pearson, *Methods in Molecular Biology* **132**, 185 (2000).
16. R. T. Fuchs, F. J. Grundy and T. M. Henkin, *Nat Struct Mol Biol* **13**, 226(Mar 2006).
17. I. L. Hofacker, W. Fontana, P. F. Stadler, L. S. Bonhoeffer, M. Tacker and P. Schuster, *Monatshefte für Chemie* **125**, 167 (1994).
18. E. Rivas and S. R. Eddy, *Bioinformatics* **16**, 583 (2000).
19. S. R. Eddy, *BMC Bioinformatics* **3**, p. 18 (2002).
20. Z. Weinberg and W. L. Ruzzo, *Bioinformatics* **22**, 35(January 2006).

A FIXED-PARAMETER APPROACH FOR WEIGHTED CLUSTER EDITING

SEBASTIAN BÖCKER, SEBASTIAN BRIESEMEISTER, QUANG BAO ANH BUI and ANKE TRUß

Lehrstuhl für Bioinformatik, Friedrich-Schiller-Universität Jena, Ernst-Abbe-Platz 2, 07743 Jena, Germany, E-mail:{boecker, m2brse, bui, truss}@minet.uni-jena.de

Clustering objects with respect to a given similarity or distance measure is a problem often encountered in computational biology. Several well-known clustering algorithms are based on transforming the input matrix into a weighted graph although the resulting WEIGHTED CLUSTER EDITING problem is computationally hard: here, we transform the input graph into a disjoint union of cliques such that the sum of weights of all modified edges is minimized.

We present fixed-parameter algorithms for this problem which guarantee to find an optimal solution in provable worst-case running time. We introduce a new data reduction operation (merging vertices) that has no counterpart in the unweighted case and strongly cuts down running times in practice. We have applied our algorithms to both artificial and biological data. Despite the complexity of the problem, our method often allows exact computation of optimal solutions in reasonable running time.

1. Introduction

Suppose we are given an undirected graph $G = (V, E)$, and for every tuple $\{u, v\} \in \binom{V}{2}$ we are also given a *weight* that tells us the cost of deleting $\{u, v\}$ from G in case $\{u, v\} \in E$, or inserting $\{u, v\}$ into G in case $\{u, v\} \notin E$. Now the WEIGHTED CLUSTER EDITING problem is defined as follows: Transform G into a *transitive* graph, that is, a disjoint union of cliques, by applying a set of edge modifications with minimum total weight.

The problem of clustering objects according to given similarity or distance values can be found in many areas of computational biology, such as finding families of orthologous genes,[1] or analyzing gene expression data for tissue classification.[2,3] We want to partition these objects into homogeneous and well-separated subsets. In graph theoretical terms, the objects to be clustered are the vertices V of the graph, and a clustering is a disjoint union of cliques. We can transform our input matrix into a weighted graph using a simple threshold, or a stochastic model and log odds.[2,4] For perfect data, the resulting graph is a disjoint union of cliques. But for biological data the input graph is "corrupted", and we have to clean (edit) this graph under the parsimony criterion to reconstruct the best clustering.[2,5] Unlike other clustering techniques, WEIGHTED CLUSTER EDITING does not make any prior assumptions on the number of clusters or their structure.

The WEIGHTED CLUSTER EDITING problem is NP-hard because it generalizes the unweighted case.[6] In recent years, many heuristics were developed for this problem. In particular, the well-known clustering algorithms CAST, HCS, and CLICK build on its graph-theoretic intuition: CAST[4] tries to find the optimal solution with high probability, while both HCS[7] and CLICK[2] greedily use minimal cuts to find an approximate solution. In a recent article[8] we have compared our simple branching strategy (Sec. 4) with two heuristic approaches on biological and random data. To the best of our knowledge, this is the first time a fixed-parameter approach for WEIGHTED CLUSTER EDITING has been proposed.

There exist a variety of results regarding the unweighted CLUSTER EDITING problem.[5] A fixed-parameter algorithm runs in time $O(2.27^k + |V|^3)$ where k is the minimum number of edge modifications.[9] In theory, the best algorithm known for the problem has running time $O(1.92^k + |V|^3)$,[10] but this algorithm uses very complicated branching rules and has never been implemented. See Niedermeier[11] for a recent monograph on fixed-parameter algorithms.

Our contributions. In this paper, we present data reduction rules for WEIGHTED CLUSTER EDITING that are nontrivial extensions of the unweighted case. In particular, we present a new data reduction technique of "merging vertices" that has no counterpart for unweighted graphs. This technique drastically improves the running time of our algorithms in practice. We also show that our data reduction leads to a problem kernel of size $O(k^2)$.

We then adopt the $O(3^k)$ branching strategy from Gramm *et al.*[9] for WEIGHTED CLUSTER EDITING: the resulting algorithm runs in time $O(3^k + |V|^3 \log |V|)$ if every edge deletion or insertion has cost at least 1. Furthermore, we provide a strategy with running time $O(2.42^k + |V|^3 \log |V|)$, roughly following the refined branching strategy in Gramm *et al.*[9] Given an arbitrary instance of the problem with fixed k, these algorithms are guaranteed to find an optimal solution with cost at most k or return that no such solution exists. Minimum edit costs are only required to achieve a provable running time.

We have applied both branching strategies to biological and simulated graph instances. We found that without merging vertices, the $O(2.42^k)$ strategy outperforms the $O(3^k)$ strategy, as expected. But if we merge vertices, both strategies become significantly faster and, in particular, the $O(3^k)$ strategy consistently outperforms the $O(2.42^k)$ strategy. We conjecture the discrepancy between theoretical and practical running times to be an indication for the power of our merging technique. Finally, we report preliminary results of applying our method to gene expression data for tissue classification.

2. Preliminaries

For brevity, we use uv as shorthand for an unordered pair $\{u, v\} \in \binom{V}{2}$. We assume a problem instance to be given in the form of a *weight function* $s : \binom{V}{2} \to \mathbb{R}$: For $s(uv) > 0$ an edge uv is present in the graph and has deletion cost $s(uv)$, while for $s(uv) \leq 0$ the edge uv is absent from the graph and has insertion cost $-s(uv)$.

If all connected components of a graph G are cliques, this graph is called *transitive*. If we modify a graph G to obtain a graph G' which is transitive, we call G' a *clustering* of G. It is graph theory folklore that a graph G is transitive if and only if there are no three vertices which induce a 2-path in G. To this end, we call vuw a *conflict triple* in G if uv and uw are edges of G but vw is not.

When given an input graph G, we first identify its connected components. As it never makes sense to connect two components in a clustering, we calculate the best solutions for all components separately and sum up the costs to obtain the total cost for G.

Beside the input graph G, our fixed-parameter algorithms also require a cost limit k. In order to find an optimal solution, we call the algorithm repeatedly, starting with $k = 1$. If we did not find a solution with this value, we increase k by 1, call the algorithm again and so forth. Note that for real-valued edge weights, we have to traverse the complete search tree and find the *best* solution with cost at most k, if any. If we can decide in $O(c^k)$ time if there is a solution with cost at most k, the above procedure to find an optimal solution of cost at most k can also be executed in $O(c^k)$ time.

To obtain provable running times, we assume that all edges have edit costs of at least 1. There cannot be a fixed-parameter algorithm solving WEIGHTED CLUSTER EDITING problem with arbitrarily small edit costs unless P = NP.

Comment. With such an algorithm we can solve the NP-complete unweighted CLUSTER EDITING problem in polynomial time by assigning uniform edit costs of $\frac{1}{k}$ to all edges of the input graph and searching for a solution of cost 1.

3. Data reduction

In the beginning of our algorithm and in each search node, we call the following data reduction routines in order to downsize the input graph as much as possible.

Rule 1: Remove cliques. Remove already existing cliques from the input graph.

Rule 2: Check for unaffordable edge modifications. For each set of two vertices u, v from V, we calculate a lower bound for the costs induced when uv is set to "permanent" or "forbidden", e.g. when the respective edge is modified. Let $N(v) := \{u \mid s(uv) > 0\}$ denote the set of neighbors of a vertex v, and let $A \triangle B$ be the symmetric set difference of sets A and B. We define induced costs $icf(uv)$ and $icp(uv)$ for setting uv to "forbidden" or "permanent", respectively:

$$\begin{aligned} icf(uv) &= \sum_{w \in N(u) \cap N(v)} \min\{s(uw), s(vw)\} \\ icp(uv) &= \sum_{w \in N(u) \triangle N(v)} \min\{|s(uw)|, |s(vw)|\} \end{aligned} \tag{1}$$

This is how we make use of these values:

- For all $u, v \in V$ where $icf(uv) > k$: Insert uv if necessary, and set uv to "permanent" by assigning $s(uv) \leftarrow +\infty$.
- For all $u, v \in V$ where $icp(uv) > k$: Delete uv if necessary, and set uv to "forbidden" by assigning $s(uv) \leftarrow -\infty$.

If there is a pair uv such that both conditions hold simultaneously, the problem instance is not solvable.

Remark. The above conditions do not take into account edit cost of the edge uv itself. For implementation, we test whether $\max\{0, s(uv)\} + icf(uv) > k$ or $\max\{0, -s(uv)\} + icp(uv) > k$ holds. We disregard this subtlety for the sake of readability.

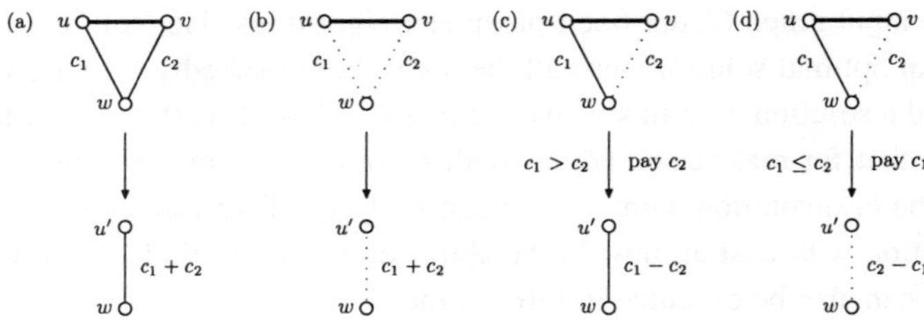

Figure 1. Merging two vertices u, v into a new vertex u': Let $c_1 = |s(uw)|$, $c_2 = |s(vw)|$ be the edit costs. Dotted edges are nonexistent.

Rule 3: Merge vertices incident to permanent edges. As soon as we set an edge uv to "permanent", we infer that u and v must be in the same clique in every solution. In this case we *merge* u and v creating a new vertex u'.

If w is a neighbor of both u and v, we create a new edge $u'w$ whose deletion costs as much as the deletion of both uw and vw. If w is neither a neighbor of u nor of v, we calculate the insertion cost of the nonexistent edge $u'w$ analogously. In case w is a neighbor of u or v but not both, uvw or vuw is a conflict triple, and we must decide whether we delete the edge connecting w with u or v, or we insert the nonexistent edge. By summing the weights (one of which is negative) to calculate $s(u'w)$ we carry out the cheaper operation decreasing k accordingly, and maintain the possibility to edit $u'w$ later.

This is how we merge u and v into a new vertex u': For each vertex $w \in V \setminus \{u, v\}$ set $s(u'w) \leftarrow s(uw) + s(vw)$. Let $k \leftarrow k - icp(uv)$, and delete u and v from the graph. See Fig. 1. Note that these reduction rules conserve the optimal solution.

To start our data reduction, we have to compute $icf(uv)$ and $icp(uv)$ for all $u, v \in V$ what takes $O(|V|^3)$ time. Setting an edge to "forbidden" or "permanent" can reduce the parameter k because we might have to delete or insert an edge. If we merge a permanent edge, this can further reduce the parameter. This, in turn, may trigger other edges to become forbidden or permanent. In addition, setting an edge to "forbidden" or "permanent" will change the induced costs of other edges. We now show how to execute our data reduction for an arbitrary input graph in time $O(|V|^3 \log |V|)$.

Let $n := |V|$. The induced costs $icf(uv)$ and $icp(uv)$ for each vertex pair $u, v \in V$ can be computed in $O(n)$ time. Therefore it initially takes $O(n^3)$ time to compute the induced costs of all $u, v \in V$. Note that during the data reduction, at most $\binom{n}{2}$ edges will be set to "forbidden", and at most $n - 1$ merge operations are executed before the graph collapses into a single vertex. For each $u \in V$ we use a binary heap to store all $icf(uv)$ and another binary heap to store all $icp(uv)$ for $v \in V$. This allows us to find $\max_v\{icf(uv)\}$ and $\max_v\{icp(uv)\}$ for each $u \in V$ in constant time.

We repeatedly do the following: Using $\max_w\{icf(uw)\}$ and $\max_w\{icp(uw)\}$ for each $u \in V$, we find the overall maximum icf and icp value in time $O(n)$. We test if there exist $u, v \in V$ with $icf(uv) > k$ or $icp(uv) > k$. If no such u, v exist, we *stop*. Otherwise, we set the corresponding edge to "forbidden" or "permanent", we update parameter $k \leftarrow k - |s(uv)|$ and also the heaps for icf and icp values as described below. The running time of this part of the algorithm is $O(n^3 \log n)$.

Setting an edge uv to "forbidden" affects the values $icf(ux)$, $icf(vx)$, $icp(ux)$, and $icp(vx)$ for all vertices $x \in V$. We concentrate on updating $icf(ux)$, the other updates can be executed similarly. Let s_0 be our weight function before the update and s_1 after the update, then these functions agree except for $s_0(uv) \neq s_1(uv) = -\infty$. Analogously, let $icf_0(ux)$ and $icf_1(ux)$ denote "induced costs forbidden" of the tuple u, x before and after the update, respectively. If $s_0(uv) \leq 0$ then no edge is deleted and we see from (1) that $icf_1(ux) = icf_0(ux)$ must hold. A similar argument resolves the case $s_0(xv) \leq 0$. If $s_0(uv) > 0$ and $s_0(xv) > 0$ then u, v as well as x, v were adjacent in the initial graph and $icf_1(ux) = icf_0(ux) - \min\{s_0(uv), s_0(xv)\}$ must hold. Clearly, computing $icf_1(ux)$ takes constant time. Updating all affected icf values and all binary heaps takes $O(n \log n)$ time: In case we have to decrease a key we can remove the corresponding entry from the heap in time $O(\log n)$ and reinsert a new entry also in time $O(\log n)$. Because every edge can be set to "forbidden" at most once, and since there are $O(n^2)$ many edges, all updates induced from setting edges to "forbidden" take total time $O(n^3 \log n)$.

When we set an edge uv to "permanent", the data reduction merges u, v into a new vertex u' and deletes u, v from the graph. We iterate over all vertices $w \in V$ and first compute $s(u'w) \leftarrow s(uw) + s(vw)$ as well as $icf(u'w)$ and $icp(u'w)$ using (1). Analogous to the previous paragraph, this affects the values $icf(wx)$ and $icp(wx)$ for all vertices $x \in V$. For every vertex w, computing $icf(u'w)$, $icp(u'w)$ and updating all heaps takes time $O(n \log n)$, and so does updating the induced costs of all icf and icp values affected by $s(u'w)$. Hence, merging an edge can be executed in total time $O(n^2 \log n)$. There can be at most $n - 1$ merge operations, so the running time of all merge operations is also bounded by $O(n^3 \log n)$. Finally, we can detect and remove all connected components which are cliques in time $O(n^2)$.

The following lemma shows that our data reduction produces a problem kernel as the size of the resulting graph is polynomial in k. We omit the proof for the sake of brevity.

Lemma 3.1. *If every edge deletion or insertion has cost at least* 1, *then our data reduction results in a problem kernel with at most* $2k^2 + k$ *vertices and* $2k^3 + k^2$ *edges.*

4. Initial branching strategy

Given a weighted graph $G = (V, E)$, we now describe a simple recursive algorithm that is guaranteed to find an optimal solution for WEIGHTED CLUSTER EDITING. Recall that an undirected graph is transitive if and only if it does not contain a conflict triple. Our algorithm takes advantage of this observation: Search for a conflict triple, and let u be the vertex of degree two and v, w be the leaves. For algorithmic reasons, we can set existent (nonexistent) edges to "permanent" ("forbidden") by assigning infinite edit costs to them. Recursively branch into three cases:

(1) Insert vw, set uv, uw, and vw to "permanent".
(2) Delete uv, set uw to "permanent" and uv and vw to "forbidden".
(3) Delete uw, set uw to "forbidden".

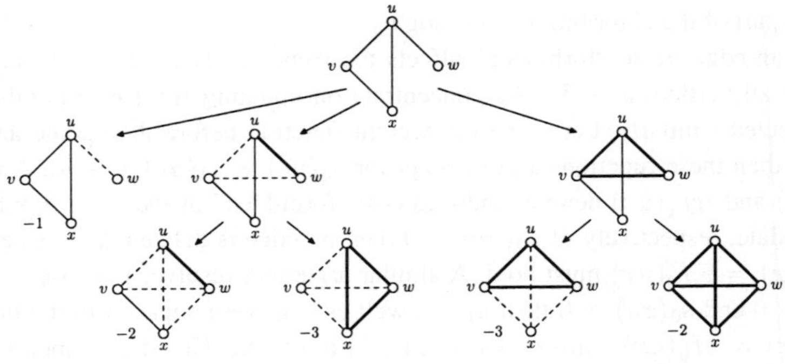

Figure 2. Case (W2a): Vertices v, w do not share a common neighbor; v has a neighbor x connected with u.

In each branch, we lower k by the insertion or deletion cost required for the executed operation. If a connected component decomposes into two components, we calculate the optimum solutions for these components separately. If k falls below zero, we discard the respective branch of the algorithm. Again, we omit the proof.

Theorem 4.1. *If every edge of the weighted graph $G = (V, E)$ has weight of at least one, the* WEIGHTED CLUSTER EDITING *problem can be solved in $O(3^k + |V|^3 \log |V|)$ time.*

5. Refined branching strategy

In the following, we will refine the simple branching strategy resulting in a search tree of size $O(2.42^k)$, considering induced subgraphs of size 4. Unfortunately, the $O(2.27^k)$ branching strategy of Gramm *et al.*[9] cannot be used in the weighted case because it is based on an observation (Lemma 5) that does not hold for weighted graphs. We now modify this branching strategy accordingly.

Note that the automated search tree generator of Gramm *et al.*[10] also found an $O(2.42^k)$ search tree for induced subgraphs of size 4, but the branching strategy is not explicitly described there. If we consider induced subgraphs of size 5, this results in an $O(2.27^k)$ search tree.[10] The latter branching strategy requires case distinction with 20 initial cases and branching vectors of size at most 16. In comparison, our branching strategy distinguishes only four initial cases and branching vectors of length five.

Let vuw be a conflict triple as above. We distinguish the following cases:

(W1) Vertices v, w have no neighbors except for u, that is, $N(v) = \{u\}$ and $N(w) = \{u\}$.

(W2) Vertices v, w do not share a common neighbor, but there exists a vertex x such that, say, $vx \in E$. We distinguish two sub-cases: (W2a) $ux \in E$ (see Fig. 2), and (W2b) $ux \notin E$ (see Fig. 3).

(W3) Vertices v, w share a common neighbor $x \neq u$, so $vx \in E$ and $wx \in E$. We distinguish two sub-cases: (W3a) $ux \in E$ (see Fig. 2 in Gramm *et al.*[9]), and (W3b) $ux \notin E$ (see Fig. 3 in Gramm *et al.*[9]).

In case (W1) holds, we *ignore* the conflict triple vuw for the moment, and continue

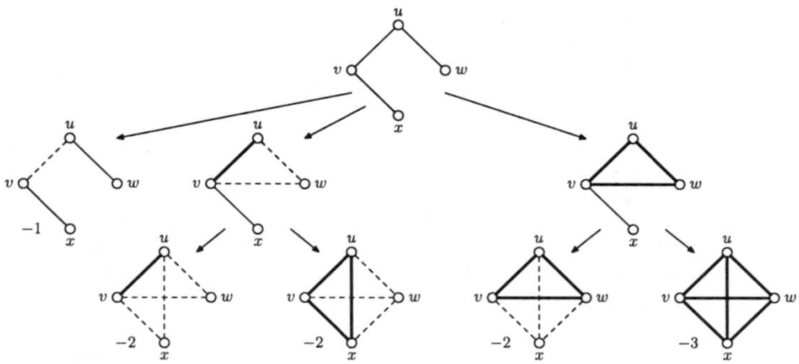

Figure 3. Case (W2b): Vertices v, w do not share a common neighbor; v has a neighbor x not connected with u.

with the next triple. In all other cases, we branch as indicated by Figs. 2, 3 in this article and Figs. 2, 3 in Gramm *et al.*[9] We describe the branching in detail for case (W2a), see Fig. 2: Here, edges uv, uw, ux, and vx are present in the induced graph. We branch into five sub-cases:

- Delete uw and set uw to "forbidden".
- Set uw to "permanent", delete uv, ux, and set uv, ux, vw, wx to "forbidden".
- Insert wx, set uw, ux, wx to "permanent", delete uv, vx, and set uv, vw, vx to "forbidden".
- Insert vw, set uv, uw, vw to "permanent", delete ux, vx, and set ux, vx, wx to "forbidden".
- Insert vw, wx and set all six edges to "permanent".

The branching strategies for case (W2b) can be easily derived from Fig. 3. Regarding branching strategies for cases (W3a) and (W3b) we refer to cases (C2) and (C3) of Gramm *et al.*[9]

One can easily check that if only conflict triples of type (W1) are present in a connected graph, this graph is a star graph, that is, a tree where all vertices but one are leaves. It is straightforward how to quickly find an optimal solution for this case.. We omit the details. The analysis of the refined branching strategy leads to Theorem 5.1.

Theorem 5.1. *If every edge of the weighted graph $G = (V, E)$ has a weight of at least one, then the total running time using our refined branching strategy is $O(2.42^k + |V|^3 \log |V|)$.*

6. Experiments and results

To explore the performance of our algorithms for WEIGHTED CLUSTER EDITING and compare the abovementioned branching strategies, we test both algorithms on the same artificial and protein similarity datasets we used in Rahmann *et al.*[8] For a given number of vertices, an artificial instance is generated by first creating a random number of clusters of random size, then setting edge weights and false edges using a Gaussian distribution. The probability to see an undesired or missing edge in the resulting graph is about 15 %. The

Table 1. Results for artificial data. Each row of the table corresponds to ten instances.

$\|V\|$	$\|E\|$	# edit	avg. cost	avg. edge	time 3^k	time 2.42^k	no merging
10	11–30	8.30	95.77	24.05	10 ms	11 ms	11 ms
20	65–165	28.10	301.9	24.41	54 ms	56 ms	69 ms
30	138–296	66.70	671.2	23.79	1.0 s	1.9 s	8.3 s
40	251–533	115.5	1238.3	24.30	29 s	52 s	31 min
50	402–821	183.2	1860.0	23.94	7.6 min	28 min	> 5 h

Note: '# edit' is the average number of edit operations, 'avg. cost' is the average total cost, 'avg. edge' is the average cost per edge, and 'no merging' is the running time of the 2.42^k branching strategy without merging permanent edges.

random instances can be seen as a worst case scenario, as we expect biological instances to be closer to a clustering. Our biological instances stem from protein similarity data, generated using more than 192 000 protein sequences from the COG dataset.[12] Edge weights are computed from log E-values of bidirectional BLAST hits using a threshold of 10^{-10}: The modification cost of each vertex pair is the difference between the logarithms of the threshold and the respective proteins' E-values. In the resulting graph 3964 connected components are not transitive, and 3788 of these have up to 100 vertices. See Rahmann *et al.*[8] for more details. Recently, Wittkop *et al.*[13] showed that the Cluster Editing model leads to valid clusterings when applied to protein similarity data and manages to outperform other methods.

We implemented the weighted data reduction and both branching strategies in C++. The results are reported in Tables 1 and 2. Running times were measured on an AMD Opteron-275 2.2 GHz with 3 GB of memory running SunOS 5.1. For protein similarity data, three instances with 84, 91, and 98 vertices did not stop after 13 days of computation with either branching strategy and are omitted from Table 2. Using the $O(2.42^k)$ strategy another five instances did not stop after 13 days of computation. For the $O(3^k)$ strategy only 30 out of 3788 instances (0.8 %) had a running time of more than ten minutes. These components are typically not near transitivity and admit many cheap edge modifications. So, despite the hardness of the WEIGHTED CLUSTER EDITING problem and the worst-case running times of our branching strategies, optimal solutions were usually computed in a matter of minutes. Note that a naïve algorithm using exhaustive enumeration cannot handle instances with more than, say, 12 vertices.

We want to evaluate whether it pays off in practice to follow a more complicated and thus time-consuming strategy to obtain a better worst-case running time. Recent experiments by Dehne *et al.*[14] show that on unweighted graphs the $O(2.27^k)$ branching is faster than the $O(3^k)$ branching.[9] As can be seen in Tables 1 and 2, in our experiments the basic strategy outperforms the refined branching strategy for all graph sizes, in contrast to the fact that its worst-case running time is inferior.

One potential cause for this unexpected result is the fact that both strategies immediately merge permanent edges. To estimate the impact of merging edges, we evaluate both branching strategies using the same implementation, but we disable merging permanent edges. Here, the refined strategy is faster than the simple branching strategy, in agreement with theoretical running times (data not shown). In addition, both algorithms are significantly slower than our branching strategies that merge edges, see Table 1.

Table 2. Results for protein similarity data.

$\|V\|$	$\|E\|$	# inst.	# edit	avg. cost	avg. edge	time 3^k	time 2.42^k	# hard
3–10	2–44	2075	2.32	6.63	19.75	2.0 ms	2.0 ms	0
11–20	14–189	725	11.78	36.49	23.23	12 ms	14 ms	0
21–30	47–434	307	30.92	95.56	23.31	103 ms	184 ms	0
31–40	62–773	178	62.02	200.1	21.52	680 ms	2.2 s	0
41–50	143–1224	181	101.6	333.9	21.26	91 s	7.1 min	3
51–60	204–1603	117	137.4	424.2	21.38	24 min	$>$ 3.2 h	8
61–70	191–2214	93	176.4	616.5	26.08	47 min	$>$ 3.6 h	1
71–80	614–3078	57	227.8	820.2	26.39	6.8 h	$>$ 16 h	10
81–90	266–3388	29	430.9	1372.7	25.87	3.7 h	$>$ 27 h	5
91–100	494–4504	23	315.0	1332.6	37.04	49 s	7.1 min	0

Note: '# inst.' is the number of instances in this group. '# hard' is the number of "hard" instances in this group where the $O(3^k)$ strategy had running time of more than ten minutes.

Finally, we report some preliminary results regarding tissue classification using gene expression data: here, one uses gene expression data to discriminate between similar cancer types. We analyzed four datasets from Monti *et al.*[3] where the correct classification is known: "Leukemia" with 38 samples, "Novartis multi-tissue" with 103 samples, "CNS tumor" with 42 samples, and "Normal tissue" with 90 samples. See Monti *et al.*[3] for details and references on these datasets. We transformed each gene expression matrix into a distance matrix between samples using normalized scalar products. For our intitial study, we did not transform these values into log-odds as suggested by Sharan *et al.*[2] but used "raw" distances. We believe that applying the probabilistic framework of Sharan *et al.*[2] will further improve the quality of our results. Thresholds for building the weighted graph were set by manual inspection, but repeated runs showed that our method was relatively insensitive to varying these thresholds. Running times of our approach range from a few seconds, to 4.6 days for the "Novartis" dataset. In three out of four cases, our results outperform that of all methods investigated in Monti *et al.*:[3] For example, using the "Novartis" dataset our clustering has an adjusted Rand index of 0.934 while the best method reported in Monti *et al.*,[3] namely Hierarchical Clustering, results in an adjusted Rand index of 0.921.[a] We will further investigate the application of our method to the problem of tissue classification in the future.

7. Conclusion

Albeit the undisputed elegance of the fixed-parameter approaches for unweighted CLUSTER EDITING, their practical use in computational biology is limited because biological data is almost always weighted in nature. Here, we have presented fixed-parameter algorithms for the WEIGHTED CLUSTER EDITING problem that guarantee to find the optimal solution with provable worst-case running times. In application, running times of our algorithms are much better than these worst-case bounds suggest: For biological data where the input graph is sufficiently close to a clustering, our algorithms finds the optimal solution

[a]The adjusted Rand index[?] is a measure of agreement between partitions with potentially different numbers of clusters. An index of 1 corresponds to perfect agreement, the expected value for two random partitions is 0.

in reasonable time even when hundreds of edge modifications are necessary. We successfully applied our algorithm to protein similarity data, and reported first results for tissue classification using gene expression data.

Different from what worst-case running times suggest, our $O(3^k)$ branching strategy almost always outperforms the $O(2.42^k)$ strategy. The reason for this unexpected outcome is certainly the power of the merge operation which outweighs advantages of complicated branching rules. We think that there may exist simple branching strategies that make even better use of edge merging, resulting in faster running times in practice and, eventually, also in improved worst-case bounds.

Acknowledgments

We thank Jan Baumbach, Sven Rahmann, and Tobias Wittkop for providing both artificial and protein similarity data.

Bibliography

1. A. Krause, J. Stoye and M. Vingron, *BMC Bioinformatics* **6**, p. 15 (2005).
2. R. Sharan, A. Maron-Katz and R. Shamir, *Bioinformatics* **19**, 1787(Sep 2003).
3. S. Monti, P. Tamayo, J. Mesirov and T. Golub, *Mach. Learn.* **52**, 91 (2003).
4. A. Ben-Dor, R. Shamir and Z. Yakhini, *J. Comput. Biol.* **6**, 281 (1999).
5. R. Shamir, R. Sharan and D. Tsur, *Discrete Appl. Math.* **144**, 173 (2004).
6. M. Křivánek and J. Morávek, *Acta Inform.* **23**, 311(June 1986).
7. E. Hartuv, A. O. Schmitt, J. Lange, S. Meier-Ewert, H. Lehrach and R. Shamir, *Genomics* **66**, 249 (2000).
8. S. Rahmann, T. Wittkop, J. Baumbach, M. Martin, A. Truß and S. Böcker, Exact and heuristic algorithms for Weighted Cluster Editing, in *Proc. of Computational Systems Bioinformics (CSB 2007)*, 2007.
9. J. Gramm, J. Guo, F. Hüffner and R. Niedermeier, *Theor. Comput. Syst.* **38**, 373 (2005).
10. J. Gramm, J. Guo, F. Hüffner and R. Niedermeier, *Algorithmica* **39**, 321 (2004).
11. R. Niedermeier, *Invitation to Fixed-Parameter Algorithms* (Oxford University Press, 2006).
12. R. L. Tatusov, N. D. Fedorova, J. D. Jackson, A. R. Jacobs, B. Kiryutin, E. V. Koonin, D. M. Krylov, R. Mazumder, S. L. Mekhedov, A. N. Nikolskaya, B. S. Rao, S. Smirnov, A. V. Sverdlov, S. Vasudevan, Y. I. Wolf, J. J. Yin and D. A. Natale, *BMC Bioinformatics* **4**, p. 41 (2003).
13. T. Wittkop, J. Baumbach, F. P. Lobo and S. Rahmann, Large-scale clustering of protein sequences with FORCE – a layout based heuristic for Weighted Cluster Editing., To appear in BMC Bioinformatics, (2007).
14. F. Dehne, M. A. Langston, X. Luo, S. Pitre, P. Shaw and Y. Zhang, The Cluster Editing problem: Implementations and experiments, in *Proc. of International Workshop on Parameterized and Exact Computation (IWPEC 2006)*, , LNCS Vol. 4169 (Springer, 2006).
15. L. Hubert and P. Arabie, *J. Classif.* **2**, 193 (1985).

IMAGE COMPRESSION-BASED APPROACH TO MEASURING THE SIMILARITY OF PROTEIN STRUCTURES

MORIHIRO HAYASHIDA and TATSUYA AKUTSU

Bioinformatics Center, Institute for Chemical Research, Kyoto University, Gokasho, Uji, Kyoto 611-0011, Japan
E-mail: {morihiro, takutsu}@kuicr.kyoto-u.ac.jp

This paper proposes series of methods for measuring the similarity of protein structures. In the proposed methods, an original protein structure is transformed into a distance matrix, which is regarded as a two-dimensional image. Then, the similarity of two protein structures is measured by a kind of compression ratio of the concatenated image. We employed several image compression algorithms: JPEG, GIF, PNG, IFS, and SPC, and audio compression algorithms: MP3 and FLAC. We applied the proposed method to clustering of protein structures. The results of computational experiments suggest that SPC has the best performance.

Keywords: Image compression; audio compression; protein structure similarity.

1. Introduction

Analysis of protein structures is an important topic in bioinformatics and computational biology. In particular, classification of protein structures is important and thus many studies have been done and several databases have been developed such as SCOP[1] and CATH.[2] Classification of protein structures is usually done based on some measure of the similarity between protein structures.

However, an agreement on which is the best similarity measure is not yet obtained and a variety of structure comparison methods have been proposed. Most of existing methods are based on *protein structure alignment.* Various methodologies have been employed for protein structure alignment, which include double dynamic programming,[3] iterative improvement,[4] combinatorial extension,[5] comparisons of distance matrices,[6] use of partial order graphs,[7] and contact map overlap.[8] In most of structure alignment methods, some scoring function is defined for measuring the quality of the obtained alignment. Then, the structure alignment problem is defined as finding a structure alignment with the optimal or near optimal score. However, score functions are defined in more or less ad-hoc manners and there is no consensus or theoretical justification. Furthermore, many of existing structure alignment methods are not very efficient.

Krasnogor and Pelta recently proposed a novel approach to measuring the similarity of protein structures.[9] Their method is similar to the contact map overlap

(CMO) approach.[8] In their method, each protein structure is transformed into a 0-1 matrix, which is further regarded as a 0-1 sequence. Then, two protein structures are compared based on the compression ratio of the sequence obtained by concatenating two 0-1 sequences. Their method is quite simple to implement and very fast. They demonstrated the usefulness of the method by means of application to clustering of protein structures. It is worthy to mention that several works have been done on measuring the similarity of biological sequences based on data compression approach.[10,11]

Though the approach by Krasnogor and Pelta is novel and useful, the distances between residues are truncated into 0 or 1. As a result, the similarity measure depends on the threshold, which should be determined by try and error. The same drawback applies to CMO,[8] In this paper, we try to overcome this drawback using a very simple idea. We employ *image compression* in place of sequence compression. Each distance matrix (not 0-1 matrix) is directly compressed by using an image compression algorithm. In this paper, we examine the following image compression algorithms: JPEG, GIF, PNG, IFS, and SPC, and audio compression algorithms: MP3 and FLAC. We apply the proposed methods to clustering of protein structures as in Ref. 9.

The organization of the paper is as follows. We begin with a brief review the method by Krasnogor and Pelta. Next, we present our proposed methods. Then, we describe details and results of computational experiments. Finally, we conclude with future work.

2. Structure Comparison Using Sequence Compression

Krasnogor and Pelta employed sequence compression to measure the similarity of two proteins. Their method is based on the universal similarity metric (USM), which was originally proposed by Li *et al.*[11] USM is based on Kolmogorov complexity. The Kolmogorov complexity $K(o)$ of an object o is defined to be the length of the shortest program P for a Universal Turing Machine U that is required to output o.[12] That is, $K(o)$ is defined by

$$K(o) = \min\{|P| \mid P \text{ is a program such that } U(P) = o\}.$$

$K(o)$ is considered to be a measure of the amount of information contained in o. Besides, the conditional Kolmogorov complexity of o_1 given o_2 is defined by

$$K(o_1|o_2) = \min\{|P| \mid P \text{ is a program such that } U(P, o_2) = o_1\},$$

where $U(P, o_2) = o_1$ means that program P outputs o_1 when o_2 is given. Based on these, information distance between two objects o_1 and o_2 can be defined as

$$InfDist(o_1, o_2) = \max(\ K(o_1|o_2),\ K(o_2|o_1)\ ).$$

Since this distance is not normalized, USM was proposed as a normalized measure:[11]

$$USM(o_1, o_2) = \frac{\max(\ K(o_1|o_2^*),\ K(o_2|o_1^*)\ )}{\max(\ K(o_1),\ K(o_2)\ )},$$

where o_i^* denotes the shortest program for o_i.

It is well-known that Kolmogorov complexity of a given object is not computable. Thus, Krasnogor and Pelta employed a sequence compression algorithm ('compress' command in UNIX). Let $C(s)$ be the size of the compressed sequence of s. They used $C(o_1)$ and $C(o_1 \cdot o_2) - C(o_2)$ in place of $K(o_1)$ and $K(o_1|o_2^*)$ respectively, where $o_1 \cdot o_2$ denotes the concatenation of two sequences o_1 and o_2. It should be noted that o_k is a 0-1 sequence obtained from a contact map M_k of protein P_k, where $M_k[i, j] = 1$ if the distance between ith residue and jth residue is less than threshold Θ, otherwise $M_k[i, j] = 0$. o_k is obtained by simple raster scanning of matrix M_k.

3. Similarity Metric Based on Image and Audio Compression

We define a contact map M_k of protein P_k as the distance matrix between residues as $M_k[i, j] = \sqrt{(r_k[i] - r_k[j])^2}$, where $r_k[i]$ denotes the three-dimensional coordinate of ith C-alpha atom of P_k.

We transform the contact map M_k to a raw image format, PPM (Portable Pixel Map), and a raw audio format, WAV. PPM can represent $(2^8)^3 = 16777216$ colors using 3 bytes memory for a pixel, where each byte is used for red, green, and blue, respectively, zero means black color, and $16777215 (= (2^8)^3 - 1)$ means white color. We transform $M_k[i, j]$ to the corresponding pixel with the color of the integer part of $cM_k[i, j]$, where c is a constant, and we set $c = 4 \cdot (2^8)^2 = 262144$ in the experiment section. If $cM_k[i, j]$ is greater than or equal to $(2^8)^3 = 16777216$, we set the color white. Fig. 1c and 1d show examples of such images for proteins 1ash and 1aa9, respectively. In order to concatenate two images horizontally, the two images must have the same height. Therefore, we fill the smaller image with black color to the height of the other (See Fig. 1c).

On the other hand, WAV can represent $(2^8)^2 = 65536$ sounds using an integer of $[-32768, 32767]$. We transform $M_k[i, j]$ to the sound with the integer part, $A_k[i, j]$, of $c'M_k[i, j] - 32768$, where c' is a constant, and we set $c' = 500$ in the experiment section. If the integer value of a sound is greater than 32767, we set it 32767. We concatenate two audios as follows: $o_1 \cdot o_2 = S(A_1, 2) \cdot S(A_2, 2) \cdot S(A_1, 3) \cdot S(A_2, 3) \cdots$, where $S(A_k, b) = A_k[1, 1+b] \cdots A_k[n_k - b, n_k]$, and n_k denotes the number of residues of protein P_k. That is, we concatenated diagonals of A_1 and A_2.

Krasnogor and Pelta approximated $K(o_1|o_2^*)$ of USM by $C(o_1 \cdot o_2) - C(o_2)$. However, $C(o_1 \cdot o_2)$ is not always equal to $C(o_2 \cdot o_1)$. Therefore, we approximate $K(o_1|o_2^*)$ by $\max(C(o_1 \cdot o_2) - C(o_2), C(o_2 \cdot o_1) - C(o_2))$. Then, the approximated

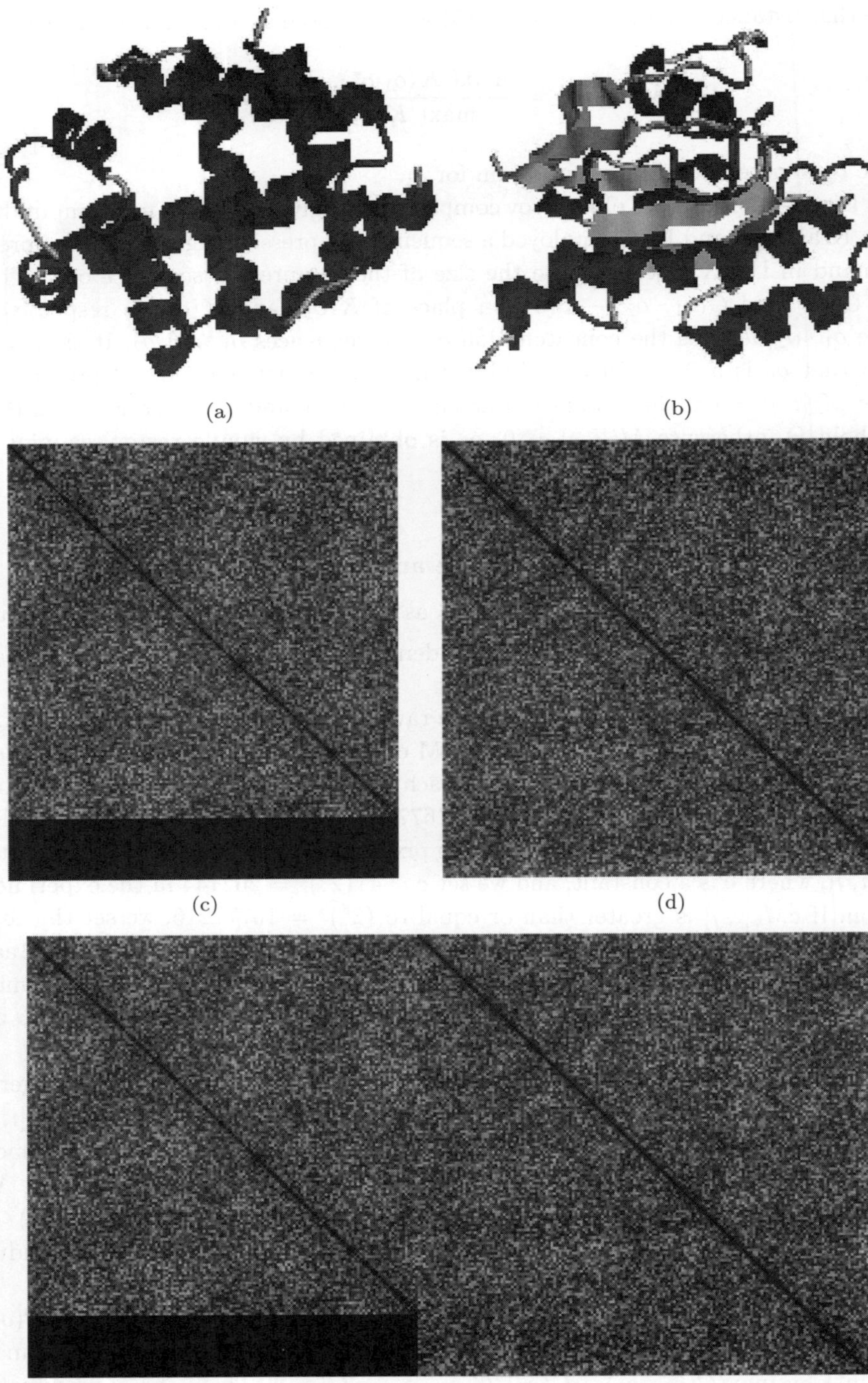

Fig. 1. An example of compressed images. (a) 1ash. (b) 1aa9. (c) The image of 1ash, which is filled with black color to the height of 1aa9. (d) The image of 1aa9. (e) The concatenated image of 1ash and 1aa9.

USM for image and audio compression, AUSM, is given as follows:

$$AUSM(o_1, o_2) = \frac{\max(\ C(o_1 \cdot o_2),\ C(o_2 \cdot o_1)\ ) - \min(\ C(o_1),\ C(o_2)\ )}{\max(\ C(o_1),\ C(o_2)\ )}.$$

4. Computational Experiments

4.1. *Image Compression Algorithms*

We employed the following image compression algorithms. In this subsection, we briefly review their algorithms.

4.1.1. *JPEG*

JPEG is usually lossy compression. An image is split into blocks of eight by eight pixels. A two-dimensional forward discrete cosine transform (DCT) is done for every blocks. After quantization, the image is compressed using Huffman coding.[13]

4.1.2. *GIF*

GIF is based on the Lempel-Ziv algorithm,[14] which is a dictionary coder. It reads an input sequence, constructs a dictionary dynamically, and replaces the sequence with words of the dictionary.

4.1.3. *PNG*

PNG is also based on the Lempel-Ziv algorithm,[14] and uses Huffman coding,[13] where PNG has been developed to replace GIF. The compression rate of PNG is often higher than that of GIF.

4.1.4. *IFS*

IFS stands for Iterated Function Systems, is a quadtree-based fractal image coder/decoder, and was implemented by Polvere.[15] The software called Mars is available from `http://inls.ucsd.edu/~fisher/Fractals/Mars-1.0.tar.gz`. Note that the software can accept only grayscale images using one byte memory for a pixel as raw image files.

4.1.5. *SPC*

SPC is a lossless image compression, and was developed by Said and Pearlman.[16] It uses a simple pyramid multiresolution scheme enhanced with predictive coding, and contains S (Sequential) transform and P (Prediction) transform.

In S transform, a sequence $c[i]$ is transformed to two sequences $l[i]$ and $h[i]$ with half the length so that the average variance of the two sequences is smaller than the

Table 1. The Chew-Kedem dataset.

Class of SCOP	Family	Proteins
All alpha	a.1.1.2[a]	1ash, 1eca, 1hlb, 1hlm, 1ithA, 1mba, 1myt, 2hbg, 2lhb, 3sdhA, 1babA, 1babB, 1flp, 1lh2, 2vhbA, 5mbn
	a.4.12.1	1jhgA
	a.39.1.2	1cnpB
All beta	b.1.1.1	1qa9B, 1cd8, 1cdb, 1ci5A1, 1hnf, 1neu, 1qfoA
Alpha and beta	c.1.11.1[b]	4enl
	c.1.11.2[b]	2mnr, 1chrA1
	c.1.15.3[b]	6xia
	c.26.2.1	1ct9A1
	c.37.1.8[c]	1aa9, 1gnp, 1qraA, 5p21, 6q21A

Note: [a] Globins, [b] TIM beta/alpha-barrel, and [c] G proteins.

variance of the original sequence if the correlation coefficient of $c[2i]$ and $c[2i+1]$ is larger than $\frac{1}{3}$. Since $h[i]$ often has small variance in image compression, we can reduce the errors of linearly predicted values for $h[i]$ using $l[i]$s and $h[i+1]$.

These transformations are done sequentially to the rows and columns of an image. Finally, these sequences are encoded using Huffman or arithmetic coding.

The software is available from `http://www.cipr.rpi.edu/research/SPIHT/EW_Code/lossless.tar.gz`. Note that the software can accept only grayscale images as raw image files.

4.2. *Audio Compression Algorithms*

In addition, we employed the following audio compression algorithms.

4.2.1. *MP3*

MP3 uses MDCT (Modified Discrete Cosine Transform), and is a lossy compression format.

4.2.2. *FLAC*

FLAC (Free Lossless Audio Codec) is a lossless compression format (`http://flac.sourceforge.net/`). FLAC is similar to SPC, and uses simple polynomial fitting and general linear predictive coding. Special Huffman coding called Rice coding can be applied for residual errors. In order to obtain the best compression, we used "–best" option.

4.3. *Data*

We used a dataset in Krasnogor and Pelta,[9] which was first used in Chew and Kedem.[17] The dataset contains proteins identified by their PDB codes. We obtained their PDB-style files (version 1.71) with coordinates from the Astral database[18] as follows (See Table 1): 16 globins (1ash, 1eca, 1hlb, 1hlm, 1ithA, 1mba, 1myt, 2hbg, 2lhb, 3sdhA, 1babA, 1babB, 1flp, 1lh2, 2vhbA, 5mbn), 2 all alpha proteins except globins (1cnpB, 1jhgA), 7 all beta proteins (1qa9B, 1cd8, 1cdb, 1ci5A, 1hnf, 1neu, 1qfoA), 4 TIM barrels (4enl, 2mnr, 1chrA1, 6xia), and 6 alpha and beta proteins except TIM barrels (1ct9A1, 1aa9, 1gnp, 1qraA, 5p21, 6q21A).

4.4. *Experiments*

For each pair of proteins included in the Chew-Kedem dataset, we generated two raw image files, o_1 and o_2, and the two concatenated image files, $o_1 \cdot o_2$ and $o_2 \cdot o_1$, from the two three-dimensional structures, and also raw audio files. We applied the above compression algorithms, JPEG, GIF, PNG, IFS, SPC, MP3, and FLAC, respectively, to the raw files, and calculated $AUSM(o_1, o_2)$. We obtained hierarchical clustering results using the nearest neighbor (single linkage) method.

5. Results

Figures 2 and 3 show the clustering results on the Chew-Kedem dataset for the compression algorithms, JPEG, GIF, PNG, IFS, SPC, MP3, and FLAC. We can see from these figures that SPC classified the dataset best. Although two all alpha proteins (1jhg and 1cnp) and three all beta proteins (1qa9, 1cdb and 1ci5) were mixed in the clustering result by Krasnogor and Pelta,[9] and an all beta protein (1hnf) was classified in globins, our SPC result classified all beta proteins correctly. The two all alpha proteins (1jhg and 1cnp) and all beta proteins were mixed in PNG similarly to the result by Krasnogor and Pelta. However, 1hnf was correctly classified in all beta proteins in PNG. Although GIF and PNG use similar compression algorithms, the result of PNG was better.

6. Conclusions

We proposed image and audio compression-based approach to measuring the similarity of protein structures, and applied them to the Chew-Kedem dataset. The clustering result by SPC image compression algorithm was the best in several image and audio compression algorithms, and was comparable to or better than that by Krasnogor and Pelta. Almost all image compression algorithms have been developed based on the property that neighbor pixels often have similar colors in images. However, similar substructures located at distant locations should also be compressed. Therefore, it is considered that the best performance was obtained with the SPC algorithm. It is expected that better similarity measure is obtained by improving the SPC algorithm. In addition, values handled by image compression algorithms are restricted to integers of a few bytes. In this paper, we transformed distances between residues of a protein to integers. In future work, we would like to develop compression algorithms for distances with real values.

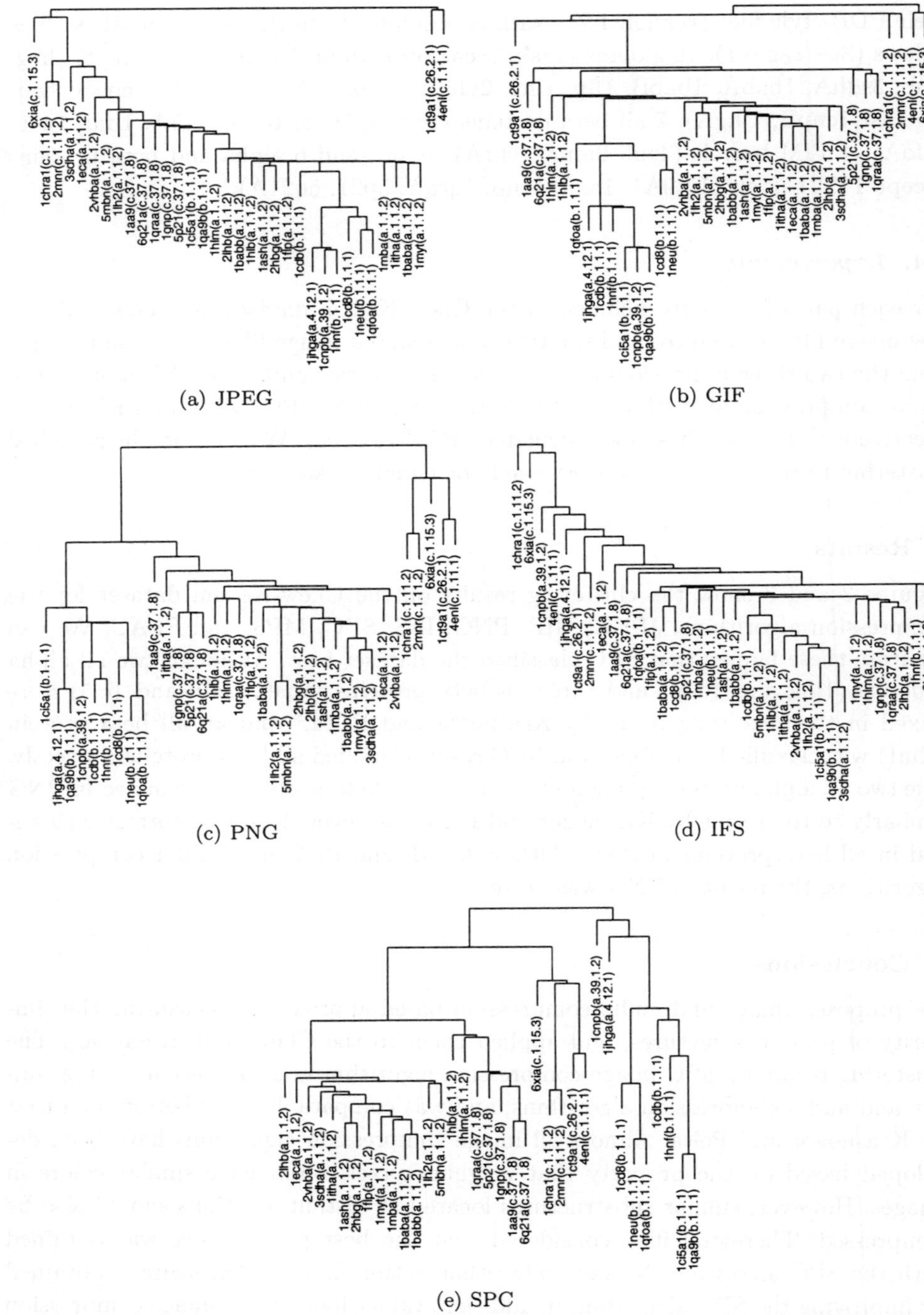

Fig. 2. The clustering results on the Chew-Kedem dataset for image compressions, (a) JPEG, (b) GIF, (c) PNG, (d) IFS, and (e) SPC.

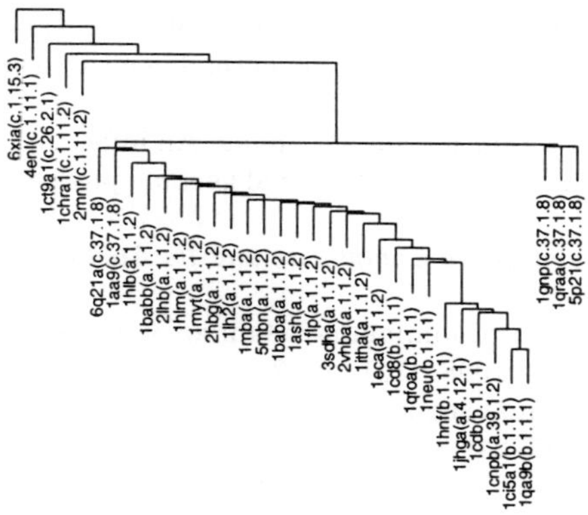

(f) MP3

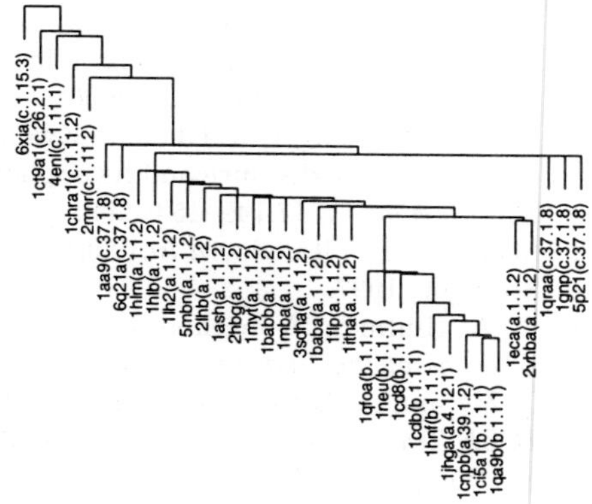

(g) FLAC

Fig. 3. The clustering results on the Chew-Kedem dataset for audio compressions, (f) MP3 and (g) FLAC.

Acknowledgments

This work was partially supported by Grants-in-Aid "Systems Genomics" and #19650053 from MEXT, Japan.

References

1. A. Andreeva, D. Howorth, S. E. Brenner, T. J. P. Hubbard, C. Chothia and A. G. Murzin. SCOP database in 2004: refinements integrate structure and sequence family data. *Nucleic Acids Research*, 32:D226–D229, 2004.
2. F. M. Pearl, C. F. Bennett, J. E. Bray, A. P. Harrison, N. Martin, A. Shepherd, I. Sillitoe, J. Thornton and C. A. Orengo. The CATH database: an extended protein family resource for structural and functional genomics. *Nucleic Acids Research*, 31:452–455, 2003.
3. W. R. Taylor and C. A. Orengo. Protein structure alignment. *Journal of Molecular Biology*, 208:1–22, 1989.
4. T. Akutsu. Protein structure alignment using dynamic programming and iterative improvement. *IEICE Trans. on Information and Systems*, E79-D:1629–1636, 1996.
5. I. N. Shindyalov and P. E. Bourne. Protein structure alignment by incremental combinatorial extension (CE) of the optimal path. *Protein Engineering*, 11:739–747, 1998.
6. L. Holm and C. Sander. Protein structure comparison by alignment of distance matrices. *Journal Molecular Biology*, 233:123–138, 1993.
7. Y. Ye and A. Godzik. Multiple flexible structure alignment using partial order graphs. *Bioinformatics*, 21:2362–2369, 2005.
8. A. Caprara, R. Carr, S. Istrail, G. Lancia and B. Walenz. 1001 Optimal PDB structure alignments: Integer programming methods for finding the maximum contact map overlap. *Journal of Computational Biology*, 11:27–52, 2004.
9. N. Krasnogor and D. A. Pelta. Measuring the similarity of protein structures by means of the universal similarity metric. *Bioinformatics*, 20:1015-1021, 2004.

10. A. Kocsor, A. Kertész-Farkas, L. Kaján and S. Pongor. Application of compression-based distance measures to protein sequence classification: a methodological study. *Bioinformatics*, 22:407–412, 2006.
11. M. Li, J. H. Badger, X. Chen, S. Kwong, P. E. Kearney and H. Zhang. An information-based sequence distance and its application to whole mitochondrial genome phylogeny. *Bioinformatics*, 17:149–154, 2001.
12. M. Li and P. Vitanyi. *An Introduction to Kolmogorov Complexity and Its Applications*, Springer, 1997.
13. D. A. Huffman. A method for the construction of minimum-redundancy codes. *Proceedings of the Institute of Radio Engineers*, 40:1098–1101, 1952.
14. J. Ziv and A. Lempel. Compression of individual sequences via variable-rate coding. *IEEE Transactions on Information Theory*, IT-24:530–536, 1978.
15. M. Polvere and M. Nappi. A feature vector technique for fast fractal image coding. *Technical report University of Salerno*, 1998.
16. A. Said and W. A. Pearlman. Reversible image compression via multiresolution representation and predictive coding. *Proc. SPIE*, 2094:664–674, 1993.
17. L. P. Chew and K. Kedem. Finding consensus shape for a protein family. *Algorithmica*, 38:115–129, 2003.
18. J. M. Chandonia, G. Hon, N. S. Walker, L. Lo Conte, P. Koehl, M. Levitt and S. E. Brenner. The ASTRAL compendium in 2004. *Nucleic Acids Research*, 32:D189–D192, 2004.

GENOME HALVING WITH DOUBLE CUT AND JOIN

ROBERT WARREN and DAVID SANKOFF

University of Ottawa

The genome halving problem, previously solved by El-Mabrouk for inversions and reciprocal translocations, is here solved in a more general context allowing transpositions and block interchange as well, for genomes including multiple linear and circular chromosomes. We apply this to several data sets and compare the results to the previous algorithm.

1. Introduction

In this paper we discuss a generalization of the genome halving process studied by El-Mabrouk.[3] Before stating and solving the problem formally in the ensuing sections, we first give some motivation for the generalization.

Models of genome rearrangement processes have permitted different repertoires of operations. Certainly, realistic models must account for inversion. Likewise, reciprocal translocations, Robertsonian translocations and other processes of chromosome fusion and fission, all of which involve transferring an entire telometric (i.e., suffix or prefix) region of at least one chromosome, are widespread across all eukaryotic domains.

Other movements of chromosomal fragments, usually not involving telomeres, are widely attested, and grouped together under the label of transpositions. They are produced by a variety of processes, such as gene duplication followed by the loss of the original copy, or retrotransposition, or recombination errors.

Of the three true movement rearrangements,[a] inversion, translocation and transposition, only the first two, separately or in combination, have proved very amenable to mathematical modeling, as exemplified by the Hannenhalli-Pevzner formula for the edit distance between two genomes, i.e., the minimum number of operations required to transform one genome into another, and the efficient algorithm for producing such a series of operations. No formula or efficient algorithm exists for transposition, either by itself or in combination with the other two operations.

Recently, Yancopoulos *et al.*[6] introduced the "double cut and join" (DCJ) operation as the basis for generating all the movement rearrangements. This allowed for the inclusion of transposition with inversion and translocation in a single model

[a]Duplications of genes or of chromosomal segments, as well as deletions and insertions are often considered as aspects of genome rearrangement, but they are not really of the same biological nature as the movements inherent in inversion, translocation and transposition, and mathematical models of rearrangement are not easily extended to encompass them.

and resulted in a simpler formula for the edit distance and a simpler algorithm for recovering a corresponding series of operations. A double cut and join operation simply cuts the chromosome in two places and joins the four ends of the cut in a new way.

The DCJ model, however, allows for the generation of a new kind of movement operation, a generalized transposition called block interchange, which is not represented in the biological genome rearrangement literature, though it has long been studied in the mathematical literature on rearrangement. Both transposition and block interchange can be thought of as the excision of a fragment, its circularization, together counting as one DCJ operation, followed by a second set of cuts, where the circle is not necessarily cut in the same place it was originally created through a join, and then reincorporated at a new site in the chromosome. Transpositions and block interchanges thus count as two DCJ operations whereas inversions and translocations each count as one.

The question arises, what is the biological significance of these chromosomal circles? On the evolutionary level, very little is known, but circular DNA structures abound in all sorts of organisms, even eukaryotes. Circular chromosomes are well-known in clinical studies[4] and the process of excision, circularization, linearization and reincorporation is exactly what happens in the configuration of the immune response in higher animals. Because the evolutionary consequences of block interchange could have come about in other ways, there has been no reason to look for evidence of this process or even to notice it. The question of its existence or importance remains open.

Yancopoulos *et al.*'s original publication[6] pointed out that the running time of their algorithm could be reduced to linear if circles were not constrained to be reincorporated into linear chromosomes as as soon as they were generated. Bergeron *et al.*[2] recently restated the DCJ model and produced a simplified (linear) algorithm ignoring the reincorporation constraint and, as in the mathematical justification of DCJ in Ref. 6, without any explicit mention of the particular operations of inversion, translocation, transposition, interchange, fusion and fission. It is thus the most general existing algorithm for movement rearrangements. As it has a form which lends itself well to constraints on the operations allowed, it can largely emulate other algorithms, e.g., the Hannenhalli-Pevzner algorithm (but without taking into account "hurdles" and "knots") or the Yancoupoulos-Attie-Friedberg algorithm (at the cost of losing its computational efficiency).

It is with this background that we ask how generalizations of the genomic distance problem, such as genome halving or rearrangement median (considered elsewhere), behave under the DCJ context.

2. Background

In this section we introduce our notation for genomes. A gene a represents an oriented sequence of DNA whose two *extremities* are its *tail* $\overset{\succ}{a}$ and its *head* $\overrightarrow{a}$.

The *adjacency* of two consecutive genes a and b is denoted by an unordered set, either $\{\overrightarrow{a}, \overset{\succ}{b}\}$, $(= \{\overset{\succ}{b}, \overrightarrow{a}\})$, $\{\overrightarrow{a}, \overrightarrow{b}\}$, $\{\overset{\succ}{a}, \overset{\succ}{b}\}$, $\{\overset{\succ}{a}, \overrightarrow{b}\}$, depending on the order and orientation of a and b. An extremity that is not adjacent to any other extremity is called a *telomere* and is represented by a singleton set $\{\overrightarrow{a}\}$ or $\{\overset{\succ}{a}\}$. A *genome* is represented by an unordered set of adjacencies and telomeres such that the head and tail of each gene appear exactly once.

A *duplicated genome* is a genome with two copies of each gene such that the head and the tail of every gene appear exactly twice. To differentiate the two genes we arbitrarily assign each gene a subscript. Thus, we say that gene a is a *unique gene* with *paralogs* a_1 and a_2 with corresponding *paralogous extremities* $\overset{\succ}{a_1}$ and $\overset{\succ}{a_2}$, and $\overrightarrow{a_1}$ and $\overrightarrow{a_2}$. To denote paralogous extremities, we have a special notation: if p is an extremity then $\overline{p}$ is its corresponding paralogous extremity. Thus, if $p = \overset{\succ}{a_1}$ then $\overline{p} = \overset{\succ}{a_2}$. Given $V \subseteq A$, where A is a duplicated genome, we can retrieve the set of all extremities using the function $\pi(V) = \bigcup_{v \in V} v$. The set of paralogous extremities in V can be retrieved using the function $\varphi(V)$ defined as follows: if p and $\overline{p}$ are in $\pi(V)$ then p is in $\varphi(V)$.

Definition 1. Let A be a duplicated genome. A is *valid* if and only if:

- If $\{u, v\} \in A$ then $\{\overline{u}, \overline{v}\} \in A$
- If $\{u\} \in A$ then $\{\overline{u}\} \in A$

A duplicated genome that is valid is a *perfectly duplicated genome*. Similarly, an invalid duplicated genome is called a *rearranged duplicated genome*.

Observant readers may notice that the above definition of validity is very general and will allow many genomes with some questionable halvings. This is intentional. One of the advantages of double cut and join is the ease with which it handles circular chromosomes. However, what is considered a valid halving in a linear multichromosonal genomes and what is considered valid for circular unichromosonal genomes are very different. For the case of an input consisting of multiple chromosomes that can be either linear of circular neither definition suffices. Our definition of validity combines both definitions of validity but, because we do not try and conserve chromosomes, it can result in some surprising results. However, typically these results are not desirable but adding additional constraints to prevent may occasionally increase the cost. For a better treatment of validity for linear multichromosonal genomes consult Ref. 3. For a better treatment of validity for circular unichromosonal genomes consult Ref. 1.

We can now define the problem:

Definition 2. The *genome halving problem* is defined as follows: given a rearranged duplicated genome A find a perfectly duplicated genome B such that the distance between A and B is minimal with respect to some distance metric.

As mentioned in the introduction, in this paper the distance metric we will use is the *double cut and join distance.* To understand the DCJ distance we must first introduce the following data structure:

Definition 3. The *adjacency graph* $AG(A, B)$ is a graph whose set of vertices are the adjacencies and telomeres of A and B. For each $u \in A$ and $v \in B$ there are $|u \cap v|$ edges between u and v.

Since every vertex in an adjacency graph has a degree of two, there are only two types of components: cycles and paths. Since the graph is bipartite, all the cycles have an even number of edges. Paths may have an odd or even number of edges. We refer to paths with an odd number of edges as *odd paths* and paths with an even number of edges as *even paths.* The difference between odd and even paths is important, thus, overall, there are three types of components to consider. Since an adjacency graph is bipartite, we can deduce the following useful lemma:

Lemma 1. *An adjacency graph $AG(A, B)$ contains a path with an odd number of edges if and only if telomere $\{u\} \in A$ and telomere $\{v\} \in B$ are endpoints of a path.*

Since double cut and join is defined for non-duplicated genomes, for the purposes of measuring distance we consider each paralog to be a different gene.

Theorem 1. ***(Ref. 2)*** *Let A and B be two genomes defined on the same set of n genes, then we have*

$$d(A, B) = n - c - \frac{i}{2}$$

where c is the number of cycles and i the number of odd paths in $AG(A, B)$.

For simplicity, throughout the rest of this paper we will use the symbol A to represent a rearranged genome and genome B to represent a perfectly duplicated genome. A and B have n unique genes, thus, they each have $2n$ paralogs, $4n$ extremities and $2n$ paralogous extremities. Thus, $d(A, B) = 2n - c - i/2$.

3. Natural Graphs

From the definition of validity, it is clear that the relationships between paralogous extremities is important. Following El-Mabrouk,[3] we define a data structure, a natural graph, to capture that relationship.

Definition 4. For each $u \in A$ we define the set V_u recursively as follows:

- Basis Step: $u \in V_u$
- Recursive Step: If $u \in V_u$ and u is an adjacency $\{p, q\}$ then $v, w \in A$ where $\overline{p} \in v$ and $\overline{q} \in w$ are also in V_u. If $u \in V_u$ and u is a telomere $\{p\}$ then $v \in A$ where $\overline{p} \in v$ is also in V_u

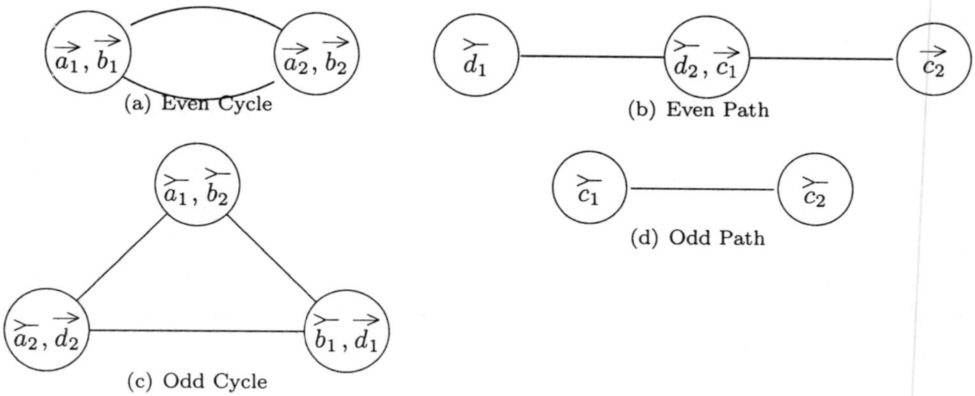

Figure 1. The natural graphs for $A = \{\{\overset{\succ}{c_1}\}, \{\overrightarrow{c_1}, \overset{\succ}{d_2}\}, \{\overrightarrow{d_2}, \overset{\succ}{a_2}\}, \{\overrightarrow{a_2}, \overrightarrow{b_2}\}, \{\overset{\succ}{b_2}, \overset{\succ}{a_1}\}, \{\overrightarrow{a_1}, \overrightarrow{b_1}\}, \{\overset{\succ}{b_1}, \overrightarrow{d_1}\}, \{\overset{\succ}{d_1}\}, \{\overset{\succ}{c_2}\}, \{\overrightarrow{c_2}\}\}$.

We define the set $E_u = \{(v, w) \in V_u | p \in v \wedge \overline{p} \in w\}$. We say that $G(V_u, E_u)$ is a *natural graph*, $G(V, E)$, generated by u. Note that if there exists an $G(V, E)$ and an $G'(V', E')$ such that $v \in V$ and $v \in V'$ then $G = G'$.

Let NG be the set of all natural graphs defined on A.

Like adjacency graphs, every vertex in a natural graph has degree of at most two. Therefore we can classify natural graphs in one of four ways:

Definition 5. Let $G(V, E) \in NG$. NG consists of four mutually exclusive subsets:

(1) If G is a cycle and $|E|$ is even then it is called an *even cycle*. We call the set of all natural graphs that are even cycles EC
(2) If G is a path and $|E|$ is odd then it is called an *odd path*. We call the set of all natural graphs that are odd paths OP
(3) If G is a cycle and $|E|$ is odd then it is called an *odd cycle*. We call the set of all natural graphs that are odd cycles OC
(4) If G is a path and $|E|$ is even then it is called an *even path*. We call the set of all natural graphs that are even paths EP

Using the properties of natural graphs we can derive some useful lemmas, the proofs of which will be included in a full version of this paper.

Lemma 2. *Let $G(V, E) \in NG$. There is no subgraph G' of G, such that G' is perfectly duplicated.*

Lemma 3. *Let $G(V, E) \in NG$ and $\{u, v\} \in V$. Let B and B' be two identical perfectly duplicated genomes except $\{u, v\}, \{\overline{u}, \overline{v}\} \in B$ and $\{u\}, \{\overline{u}\} \in B'$. $d(A, B) \leq d(A, B')$.*

Lemma 4. *Let $G(V,E) \in NG$. If $|\varphi(V)|$ is even then there exists an $H \subseteq B$ with $\pi(H) = \pi(V)$ such that $d(V,H)$ is minimal.*

From Lemma 2 and Lemma 4 we can now effectively redefine the genome halving problem to: for each natural graph, construct a subset of B such that the distance between the natural graph and its corresponding subset is minimal.

From Lemma 4 we can conclude that all the natural graphs should contain an even number of paralogous extremities. Observe that this is the case for the natural graphs in the sets EC and EP. For the remaining graphs we observe that since there are an even number of paralogous extremities in A it must be the case that $|OP| + |OC|$ is even.

4. Lower Bounds

Before we can derive lower bounds on the distance, we once again need a new data structure:

Definition 6. Let $G(V,E) \in SN$. Let $H \subseteq B$ such that $\pi(H) = \pi(V)$. Let C be a component of $AG(V,H)$. We define the *signature* S_C as follows:

(1) If $u \in \pi(C \cap V)$ then u is in S_C unless $\overline{u}$ is already in S_C;
(2) If $\{u,v\}$ is an adjacency in $C \cap V$ and u is in S_C then neither v nor $\overline{v}$ is in S_C;

A *maximal signature* is a signature which includes as many extremities as possible. Let S be a set of maximal signatures for all components in $AG(V,H)$. We define a *signature graph* $SG(S,F)$ where S is the vertices and F is the set of edges. F is defined as follows: for all $S_1, S_2 \in S$ there exists an edge $\{S_1, S_2\}$ if and only if there exists an extremity x such that $x \in S_1$ and $\overline{x} \in S_2$. We also define the function $\delta(S_C)$, where $S_C \in S$, which denotes the degree of S_C.

From the fact that $|\varphi(V)|$ is odd and from the definition of a maximal signature we get the following lemma:

Lemma 5. *Let $G(V,E) \in EP \cup OC$ and let $SG(S,F)$ be a signature graph defined on G. $\sum_{S_C \in S} |S_C| \leq |V| - 1$*

We now have enough information to establish lower bounds on the distance:

Theorem 2. *Let $G(V,E) \in SN$ with $2n$ extremities where $n \geq 1$. Let $H \subseteq B$ with $\pi(H) = \pi(V)$. The following statements are all true:*

(1) If $G \in EC$ then $d(V,H) \geq n-1$;
(2) If $G \in OP$ then $d(V,H) \geq n-2$;
(3) If $G \in OC$ then $d(V,H) \geq n-1$;
(4) If $G \in EP$ then $d(V,H) \geq n$;

Proof. From the definition of a signature graph, we know that $\sum_{S_C \in S} \delta(S_C) \leq \sum_{S_C \in S} |S_C| \leq |V|$. Therefore, $|F| = \frac{1}{2}\sum_{S_C \in S} \delta(S_C) \leq \frac{1}{2}\sum_{S_C \in S} |S_C| \leq \frac{1}{2}|V|$. It follows from Lemma 2 that all signature graphs are connected, therefore, $|S| \leq |F| + 1 \leq \frac{1}{2}|V| + 1$. Thus, $AG(V, H)$ contains $\leq \frac{1}{2}|V| + 1$ components.

When $G \in EC$, $|V| = 2n$ and thus $AG(V, H)$ contains $\leq n + 1$ components.

When $G \in OP$, $|V| = 2n$ and thus $AG(V, H)$ contains $\leq n + 1$ components.

When $G \in EP$, $|V| - 1 = 2n$ and, from Lemma 5, only $|V| - 1$ vertices contribute towards the signature graph. Thus, $AG(V, H)$ contains $\leq n + 1$ components.

When $G \in OC$, $|V| - 1 = 2n - 2$ and, from Lemma 5, only $|V| - 1$ vertices contribute towards the signature graph. Thus, $AG(V, H)$ contains $\leq n$ components.

From Theorem 1 we can observe that $d(V, S_G \cup \overline{S_G}) = |\varphi(V)| - c - i$ where c and i are the number of cycles and odd paths respectively in $AG(V, S_G \cup \overline{S_G})$. When $G \in EC \cap EP$ we know that $|\varphi(V)| = 2n$. For the remaining two cases $|\varphi(V)| = 2n - 1$. We know from the above that the maximum number of components each type of natural graph contains, to establish lower bounds on the distance we need only determine which of those components are cycles and which are odd paths.

It can be proven (proof omitted in this abstract) that some of components must be paths. In the worst case, graphs in OP contains 2 odd paths, graphs in EP contain 1 even path. For the purposes of establishing a lower bound we can safely assume that the remaining components are cycles. □

5. Upper Bounds

Using the structure of a natural graph, we can define an ordering of the vertices and extremities that will simplify ensuing developments. We define the ordering as follows:

Definition 7. Let $G(V, E) \in NG$. We relabel the extremities in V to define a *suitable order* of the vertices.

(1) $G \in EC$. $|V| = |\varphi(V)| = 2n$, for all $n \geq 1$. Let $V = \{v'_1, v_1, v'_2, v_2, \ldots, v'_n, v_n\}$ such that the following hold:
   - $v'_1 = \{\overline{x_{2n}}, \overline{x_1}\}$ and $v_1 = \{x_1, x_2\}$;
   - For each $i, 1 < i \leq n$, $v'_i = \{\overline{x_{2i-2}}, \overline{x_{2i-1}}\}$ and $v_i = \{x_{2i-1}, x_{2i}\}$;

(2) $G \in OP$. $|\varphi(V)| = 2n - 1$ and $|V| = 2n$, for all $n \geq 1$. Let $V = \{v'_1, v_1, v'_2, v_2, \ldots, v'_n, v_n\}$ such that the following hold:
   - $v'_1 = \{\overline{x_1}\}$ and $v_1 = \{x_1, x_2\}$;
   - For each $i, 1 < i < n$, $v'_i = \{\overline{x_{2i-2}}, \overline{x_{2i-1}}\}$ and $v_i = \{x_{2i-1}, x_{2i}\}$;
   - $v'_n = \{\overline{x_{2n-2}}, \overline{x_{2n-1}}\}$ and $v_n = \{x_{2n-1}\}$;

(3) $G \in OC$. $|V| = |\varphi(V)| = 2n - 1$, for all $n \geq 1$. Let $V = \{v'_1, v_1, v'_2, v_2, \ldots, v'_{n-1}, v_{n-1}, v'_n\}$ such that the following hold:
   - $v'_1 = \{x_{2n-1}, \overline{x_1}\}$;

- For each $i, 1 < i \leq n-1$, $v_{i-1} = \{x_{2i-3}, x_{2i-2}\}$ and $v'_i = \{\overline{x_{2i-2}}, \overline{x_{2i-1}}\}$;

(4) $G \in EP$. $|\varphi(V)| = 2n$ and $|V| = 2n+1$, for all $n \geq 1$. Let $V = \{v'_1, v_1, v'_2, v_2, \ldots, v'_n, v_n, v'_{n+1}\}$ such that the following hold:

- $v'_1 = \{\overline{x_1}\}$ and $v_1 = \{x_1, x_2\}$;
- For each $i, 1 < i \leq n$, $v'_i = \{\overline{x_{2i-2}}, \overline{x_{2i-1}}\}$ and $v_i = \{x_{2i-1}, x_{2i}\}$;
- $v'_{n+1} = \{\overline{x_{2n}}\}$;

From the definition of suitable order we can derive the sets S and $\overline{S}$ for each natural graph. Let $G(V, E) \in NG$ where V is suitably ordered. As noted before, $\varphi(V)$ is $2n$ when $G \in EC \cup EP$, $2n-1$ when $G \in OP \cup OC$, for all $n \geq 1$. If $G \notin OC$ then let $S_G = \{v_1, v_2, \ldots, v_n\} \in V$. We define $\overline{S_G}$ as follows: if the adjacency $\{x, y\} \in S_G$ then $\{\overline{x}, \overline{y}\} \in \overline{S_G}$, if the telomere $\{x\} \in S_G$ then $\{\overline{x}\} \in \overline{S_G}$. We call the set S_G the set of *selected vertices*. Set $\overline{S_G}$ is the set of *derived vertices*.

The case of $G \in OC$ is a special case. We $S_G = \{v_1, v_2, \ldots, v_{n-1}\} \in V$. If we define $\overline{S_G}$ as normal we end up missing the extremities x_{2n-1} and $\overline{x_{2n-1}}$. Thus, for $G \in OC$ we define $\overline{S_G}$ as $\overline{S_G}' \cup \{x_{2n-1}, \overline{x_{2n-1}}\}$ where $\overline{S_G}'$ is the same as the definition for $\overline{S_G}$ when $G \notin OC$. Note that this definition for $\overline{S_G}$ has a tendency to produce circular chromosomes which may not be desirable. Alternative definitions which avoid circles do exist but they produce a worse distance.

We can now derive a solution for B:

$$B = \bigcup_{G \in SN} S_G \cup \overline{S_G}$$

For the ensuing proofs, it is useful to keep track of the *unselected vertices* in V. Let $U_G = V \setminus S$ which is $\{v'_1, v'_2, \ldots, v'_n\}$ when $G \notin EP$ or $\{v'_1, v'_2, \ldots, v_n, v'_{n+1}\}$ when $G \in EP$.

The following useful observation describes the motivation for constructing set S in this manner:

Observation 1. Let $G(V, E) \in NG$. Each adjacency $\{x_i, x_j\} \in S_G$ corresponds to a cycle in $AG(V, S_G \cup \overline{S_G})$ and each telomere $\{x_k\} \in S_G$ corresponds to an odd path in $AG(V, S_G \cup \overline{S_G})$.

In order for B to be valid, the set of derived vertices must be constructed as above. Observe that, for $G(V, E) \in NG$, $U_G \subset V$ and that $\pi(U_G) = \pi(\overline{S_G})$. Thus, $AG(U_G, \overline{S_G}) \subseteq AG(A, B)$ which has the following properties:

Lemma 6. *The following statements are all true:*

(1) If $G \in EC$ then $AG(U_G, \overline{S_G})$ contains exactly one cycle and no paths.
(2) If $G \in OP$ then $AG(U_G, \overline{S_G})$ contains exactly one odd path and no cycles.
(3) If $G \in OC$ then $AG(U_G, \overline{S_G})$ contains exactly one cycle and no paths.
(4) If $G \in EP$ then $AG(U_G, \overline{S_G})$ contains no cycles or odd paths.

Proof. This lemma follows from the definitions of U_G and $\overline{S_G}$ as well as the definition of an adjacency graph. □

We can now define the distance between any natural graph $G(V, E)$ and $S_G \cup \overline{S_G}$:

Theorem 3. *The following statements are all true:*

(1) If $G \in EC$ then $d(V, S_G \cup \overline{S_G}) \leq n - 1$
(2) If $G \in OP$ then $d(V, S_G \cup \overline{S_G}) \leq n - 2$
(3) If $G \in OC$ then $d(V, S_G \cup \overline{S_G}) \leq n - 1$
(4) If $G \in EP$ then $d(V, S_G \cup \overline{S_G}) \leq n$

Proof. From Theorem 1 we can observe that $d(V, S_G \cup \overline{S_G}) = |\varphi(V)| - c - i$ where c and i are the number of cycles and odd paths respectively in $AG(V, S_G \cup \overline{S_G})$.

From Observation 1 and Lemma 6 we can immediately conclude that $c = n$ in all cases except when $G \in EC$ in which case $c = n + 1$ and that $i = 0$ in all cases except when $G \in OP$ in which case $i = 2$. When $G \in EC \cap EP$ we know that $|\varphi(V)| = 2n$. For the remaining two cases $|\varphi(V)| = 2n - 1$. □

Theorem 4. *Let A be a rearranged duplicated genome and B be a perfectly duplicated genome with $2n$ genes where n is the number of unique genes and $n \geq 1$ then the minimum distance between A and B is:*

$$d(A, B) = n - |EC| - \left\lfloor \frac{2 \cdot |OP| - |OC|}{2} \right\rfloor$$
$$= n - |EC| - |OP| - \left\lfloor \frac{|OC|}{2} \right\rfloor$$

Proof. This theorem follows immediatly from Theorem 2 and Theorem 3 and the fact that $|OP|$ and $|OC|$ are odd. □

6. Experiments

We have implemented the DCJ halving algorithm so that it runs in (provably minimum) linear time. We applied it to data sets on three present-day genomes that are descended from a genome doubling event: *Zea mays*,[7] with two copies of 34 markers, *Saccharomyces cerevisiae*, with two copies of 300 markers, and *Candida glabrata*, with two copies of 300 markers.[5] The number of operations from the doubling event to the present-day genome was 27, 193 and 249 respectively. The El-Mabrouk algorithm gave a result of 250 for *Candida glabrata* but otherwise the results were exactly the same. Applying the algorithm as written in the paper produced circular chromosomes in each case. However, borrowing the look-ahead routine (which prevents the formation of circular chromosomes) from the El-Mabrouk paper[3] we got the same result as El-Mabrouk with no circular chromosomes and no asymptotic increase in complexity.

7. Conclusion

We have shown that the main ideas of the El-Mabrouk algorithm carry over to the DCJ context, although the case analysis here involves both cycles and paths, instead of just cycles in the breakpoint graph. In one respect, however, the algorithm is much simpler, due to the simplifications inherent in the DCJ approach. Where El-Mabrouk had to attend to the complex components of the breakpoint graph known as hurdles and knots, the DCJ formulation avoids this completely.

Since the repertoire of movement rearrangements in the DCJ formulation is complete, the results of applying our algorithm will always be a lower bound on any result using a constrained set of operations. At the same time, constraining the DCJ operations may not yield an optimal result, since these constraints are ad hoc and may not yield the minimum number of operations. Thus the method yields both a lower bound (using unconstrained operations) and an upper bound (using the constraints) on the results of algorithms yielding optimal answers for a specific set of constraints.

Acknowledgments

We thank Julia Mixtacki, Jens Stoye, Chunfang Zheng and Nadia El-Mabrouk for helpful discussion. Research supported in part by a grant to David Sankoff from the Natural Sciences and Engineering Research Council of Canada (NSERC). David Sankoff holds the Canada Research Chair in Mathematical Genomics and is a Fellow of the Royal Society of Canada.

References

1. Max A. Alekseyev and Pavel A. Pevzner. Whole genome duplications and contracted breakpoint graphs. *SIAM Journal on Computing*, 36(6):1748–1763, 2007.
2. Anne Bergeron, Julia Mixtacki, and Jens Stoye. A unifying view of genome rearrangements. In Philipp Bücher and Bernard M.E. Moret, editors, *Algorithms in Bioinformatics: 6th International Workshop*, volume 4175 of *Lecture Notes in Computer Science*, pages 163–173. Berlin, Heidelberg: Springer-Verlag, 2006.
3. Nadia El-Mabrouk and David Sankoff. The reconstruction of doubled genomes. *SIAM Journal on Computing*, 32:754–792, 2003.
4. Fabien Kuttler and Sabine Mai. Formation of non-random extrachromosomal elements during development, differentiation and oncogenesis. *Seminars in Cancer Biology*, 17(1):56–64, 2007.
5. David Sankoff, Chunfang Zheng, and Qian Zhu. Polyploids, genome halving and phylogeny. *Bioinformatics*, 23:i431 – i439, 2007.
6. Sophia Yancopoulos, Oliver Attie, and Richard Friedberg. Efficient sorting of genomic permutations by translocation, inversion, and block interchange. *Bioinformatics*, 21(16):3340–3346, 2005.
7. Chunfang Zheng, Qian Zhu, and David Sankoff. Parts of the problem of polyploids in rearrangement phylogeny. In Glenn Tesler and Dannie Durand, editors, *Proceedings of the RECOMB 2007 Workshop on Comparative Genomics*, volume 4751 of *Lecture Notes in Computer Science*, pages 162–176. Berlin, Heidelberg: Springer-Verlag, 2007.

PHYLOGENETIC RECONSTRUCTION FROM COMPLETE GENE ORDERS OF WHOLE GENOMES

KRISTER M. SWENSON

School of Computer and Communication Sciences
Swiss Federal Institute of Technology (EPFL)
EPFL IC LCBB, INJ 232, Station 14
CH-1015 Lausanne, Switzerland
E-mail: krister.swenson@epfl.ch

WILLIAM ARNDT

Department of Computer Science & Engineering
University of South Carolina, Columbia, SC 29208, USA
E-mail: arndtw@cse.sc.edu

JIJUN TANG

Department of Computer Science & Engineering
University of South Carolina, Columbia, SC 29208, USA
E-mail: jtang@engr.sc.edu

BERNARD M. E. MORET

School of Computer and Communication Sciences
Swiss Federal Institute of Technology (EPFL)
EPFL IC LCBB, INJ 230, Station 14
CH-1015 Lausanne, Switzerland
and
Swiss Institute of Bioinformatics
E-mail: bernard.moret@epfl.ch

Rapidly increasing numbers of organisms have been completely sequenced and most of their genes identified; homologies among these genes are also getting established. It thus has become possible to represent whole genomes as ordered lists of gene identifiers and to study the evolution of these entities through computational means, in systematics as well as in comparative genomics. While dealing with rearrangements is nontrivial, the biggest stumbling block remains gene duplication and losses, leading to considerable difficulties in determining orthologs among gene families—all the more since orthology determination has a direct impact on the selection of rearrangements. None of the existing phylogenetic reconstruction methods that use gene orders is able to exploit the information present in complete gene families—most assume singleton families and equal gene content, limiting the evolutionary operations to rearrangements, while others make it so by eliminating nonshared genes and selecting one exemplar from each gene family. In this work, we leverage our past work on genomic distances, on tight bounding of parsimony scores through linear programming, and on divide-and-conquer methods for large-scale reconstruction to build the first computational approach to phylogenetic reconstruction from complete gene order data, taking into account not only rearrangements, but also duplication and loss of genes. Our approach can handle multichromosomal data and gene families of arbitrary sizes and scale up to hundreds of genomes through the use of disk-covering methods. We present experimental results on simulated unichromosomal genomes in a range of sizes consistent with prokaryotes. Our results confirm that equalizing gene content, as done in existing phylogenetic tools, discards important phylogenetic information; in particular, our approach easily outperforms the most commonly referenced tool, MGR, often returning trees with less than one quarter of the errors found in the MGR trees.

Keywords: phylogenetic reconstruction; whole-genome data; genomic distance; gene inversion; gene duplication; gene loss

1. Introduction

Phylogenetic reconstruction has for many years been based on alignments of the sequences of one or more orthologous genes and proteins. The accumulation of full genome sequences enables one to use much richer data: one can use hundreds of genes to build a more detailed picture of organismal evolution[1–3] or one can be even more ambitious and use every gene present in the genomes. In the latter category are simple content-based approaches, where the presence or absence of genes from the global inventory are the informational characters;[4,5] as these approaches represent the data in the form of bit strings where each position is a character, they can make use of existing software packages for analysis. Obviously, however, a complete genome sequence contains much information besides the individual sequences of constituent genes or the presence or absence of these genes: the genome sequence identifies an ordering of these genes along the chromosomes, as well as a direction of transcription. Moreover, disruption of this ordering is a relatively rare occurrence—a "rare genomic event".[6] Thus changes in the ordering are valuable study tools in phylogenetics as well as comparative genomics.

Phylogenetic methods based on gene orders are still in their infancy—see the survey of Moret and Warnow:[7] the problems faced are mathematically and computationally much more challenging than in sequence-based reconstruction and the models not as well understood. These methods have been applied to simple data, such as organellar genomes across sets of taxa where the gene content is highly conserved (and where, of course, the number of genes is quite small, on the order of 40 genes for animal mitochondria and 120 genes for plant chloroplast).[8–12] As one attempts to scale such analyses to cellular organisms, several problems arise. One is simply a problem with the data: annotations of complete cellular genomes are still in various stages of completion, so that identifying homologous gene families with high accuracy is a challenge. Another is the highly variable gene content (just in bacteria, obligate endosymbionts may have under 1,000 genes, while free-living bacteria may have over 5,000). A third is the widespread occurrence of gene duplications and losses: gene families, while mostly containing a single gene, may contain up to 100 homologs in bacteria and over 1,000 homologs in eukaryotes. Finally, the difference in scale is a huge challenge given that most algorithms proposed for the task exhibit exponential growth in running time as a function of the size of the problem.

Only a few attempts to use gene orders for reconstructing the phylogeny of a group of cellular organisms have been made to date. The first few reduced the gene-order data (which forms a single phylogenetic character with an immense number of possible states)

to a collection of much simpler characters, such as the presence or absence of adjacent gene pairs,[13] an approach later broadened into formal encodings of gene orders used in parsimony analyses (see the survey of Wang et al.[14]); other approaches used phylogenetically informative clusters of genes.[15–17] More recently Belda et al.[18] used a variant of these approaches on a set of 30 γ-proteobacteria: they chose 247 specific orthologs present in all 30 bacteria, thereby both reducing the size of the problems and sidestepping the issue of gene families. Many papers have appeared on phylogenetic reconstruction from gene-order data when each gene order is a (signed) permutation of a reference set—for a recent survey, see Moret and Warnow[7]—, the two most notable ones being our own GRAPPA[19] and the multichromosomal tool MGR.[20] Finally, Blin *et al.*[21] went one step further on a subset of 13 of the aforementioned γ-proteobacteria, by using a local, pairwise restriction on gene content rather than a global one. None of these attempts to date has explicitly taken duplications and losses into account nor attempted to model them as evolutionary events. Bayesian MCMC methods, such as BADGER,[22] suffer from similar issues.

We earlier developed a measure of genomic distance that, given a pair of genomes, returns an estimate of the total number of evolutionary events under the iDLR (insertions, duplications, losses, and rearrangements) separating these two genomes.[23,24] (An alternative based on the closely related notion of common intervals recently appeared.[25]) Simulations results show very high accuracy up to a high threshold of saturation (where the estimated distance starts lagging behind the true number of events). Pairwise distances alone can be used as a basis for distance-based reconstruction, as was done for 13 γ-proteobacteria (the same that would later be used by Blin *et al.*[21]) in the MS thesis of Earnest-DeYoung,[26] who found that the reconstructed phylogeny differed from the reference one of Lerat *et al.*[2] by a single SPR event—that is, a single subtree was misplaced, as would also later be the case in the reconstruction of Blin *et al.*[21] This work served as a proof of concept, but used a number of *ad hoc* measures to keep the computational work down, such as identifying groups of genes that always formed a contiguous group and taking advantage of the reference phylogeny to establish an asymmetric cost for gene gains and losses.

In this paper, we combine our genomic distance[24] with our tight bounding based on linear programming (LP)[27] to produce the first phylogenetic reconstruction method that attempts to return a most parsimonious tree in terms of a palette of evolutionary events that include insertion, duplication, and loss of genes (or gene segments) as well as inversions, using the complete gene orders with full gene families and no prior known orthologies (as the orthologies will obviously depend on the returned tree). We provide experimental results comparing our new approach to reconstruction based on the genomic distances alone (using neighbor-joining), to reconstruction by our same tool, but from genomes reduced to equal gene contents, and to reconstruction, again on the basis of equalized gene contents, by the MGR server[20] and by neighbor-joining (NJ).

Our results indicate that computing under the iDLR model (i.e., using the full genomic gene ordering) regularly improves results over using equalized gene contents, often significantly so—errors are commonly reduced by a factor of 4 or more. They also indicate that the parsimonious trees returned by our LP-based procedure are as good as or better than those returned by neighbor-joining. Under parameter settings with relatively modest numbers of events, the two exhibit similar accuracy, indicating that the iDLR distance estimates are both close to additivity and quite distinct from each other. These findings echo practice with sequence data, and, as with sequence data, we find that increased deviations from ultrametricity (in the form of widely different total amount of evolution on different paths from the root to the leaves) create situations where NJ does increasingly worse than our LP-based procedure—until the pairwise distances grow large enough to prevent accurate reconstruction by any means. We kept the number of taxa low (13 or fewer) in order to run large series of experiments with the LP-based method and with MGR, but we know from our past work[28] that the LP-based method can be scaled up to much larger numbers of genomes with very little loss of accuracy by using a disk-covering method.

2. Methods and Models

Our phylogenetic reconstruction algorithm is based on GRAPPA,[19,29] which we developed for analyzing chloroplast gene orders. GRAPPA examines every tree topology, computes a bound for each, and, for each tree that passes the bound, scores it by computing ancestral gene orders that minimize the total length of the tree, as measured in terms of inversions. The original GRAPPA is limited to singleton gene families and equal gene content, just like the various inference programs developed since, such as MGR, BADGER, etc. Its exhaustive examination of all trees also limits the maximum number of genomes it can handle, to about 15 taxa for single runs, 12-14 taxa when running benchmarks, while its method for scoring a tree requires the repeated computation of an inversion median at each internal node, an NP-hard problem that limits the lengths of tree edges it can handle. To extend it to larger numbers of taxa, Tang and Moret used a disk-covering method (in effect, a specialized divide-and-conquer approach) and showed that the resulting DCM-GRAPPA scaled gracefully to at least 1,000 genomes.[28] To date, the best way to extend the approach to larger genomes has been to avoid scoring trees. The original bounding computes a shortest cycle on the leaves of the tree and was found to eliminate well over 99% of the candidates.[29] Tang and Moret[27] later proposed a linear programming (LP) formulation where variables are the lengths of the tree edges and the constraints are simple metric inequalities; this approach eliminated well over 99.99% of the candidates in their experiments. Their LP formulation was later improved into a pure covering LP,[30] which offers efficient solutions (running in $O(n^{2.5})$ time, where n is the number of genes) and even more efficient approximations.

The LP score was close enough to the actual score that Tang and Moret proposed using the LP score in lieu of scoring the tree, avoiding any median computation. The resulting reconstruction lacks ancestral orderings, but gives a topology, an estimated score, and estimated edge lengths (the values of the LP variables), much as a maximum-likelihood reconstruction does for sequence data. We still lack a good approach to the inference of ancestral gene orders under the iDLR model, both from the point of view of computational effort (medians again) and from that of accuracy. Indeed, Earnest-DeYoung *et al.*,[31] in a study of the 13 γ-proteobacteria, found that internal gene orders were seriously underconstrained and so could not be reliably inferred—we need a more detailed and sensitive model of the evolutionary operations on a gene ordering.

The triangle inequalities that form the LP rely on a direct computation of the distances between selected pairs of leaves. Thus we can generalize the LP formulation directly to the iDLR model by using an estimate of the distance between two arbitrary genomes with varied gene families. We had proposed and tested just such a measure,[23,24] which estimates the total number of insertions (including duplications), losses, and inversions needed to transform one unichromosomal genome into another. The measure is readily extended to multichromosomal genomes by replacing inversions with double-cut-and-join operations,[32] since the latter cover fusion, fission, and translocation among chromosomes, yet can be handled just like inversions.

Our final algorithm thus combines DCMs for scaling to large numbers of genomes, a specific LP formulation to estimate branch lengths and total score of the trees, and the intergenomic distance of Swenson *et al.*[24] to provide input values to the LP. More specifically, we first compute the pairwise intergenomic distances; we then enumerate all possible trees, following the strategy of GRAPPA, attempting to eliminate as many trees as possible. The bounding is done first using the circular lower bound as described in Moret *et al.*;[33] if the tree passes that test, we then proceed to set up a linear program for it. In the linear program, the variables are the edge lengths; the constraints are derived using the triangle inequality—basically, a leaf-to-leaf path in the tree, corresponding to a particular sum of variables, should have length no less than the pairwise intergenomic distance between the two leaves. It should be noted that, whereas the constraints in the original use of the LP approach[27] were mathematically correct because all measures used were edit distances, the constraints used here have no such guarantee, since we are now using estimates of the true evolutionary distance. On the basis of the results of Swenson *et al.*,[24] we expect most of them to be correct, with a few possibly off by small deviations. Then again, we also expect the LP score to be even closer to optimal than in its original use, as the distances used in the constraints are much closer to the true evolutionary distances than was the case in the study of Tang and Moret. Finally, the score returned by the LP, rounded up to the nearest integer, is assigned as the score of that tree and the algorithm returns the trees with the lowest score.

3. Experimental Design

Our objective is to verify that computing under the full iDLR mode, i.e., handling both rearrangements and changes in gene content, allows for better reconstruction than handling only rearrangements on genomes reduced to signed permutations. Relative accuracy is thus our main evaluation criterion. However, absolute accuracy is needed in order to put the comparison in perspective. Since, in phylogenetic reconstruction, error rates larger than 10% are considered unacceptable, there is obviously little use in improving the error rate by a factor of two if the result is just bringing it from 60% down to 30%. We also need to test a wide range of parameters in the iDLR model, as well as to test the sensitivity of the methods to the rate of evolution. These considerations argue for testing on simulated data, where we can conduct both absolute and relative evaluations of accuracy, before we move to applying the tools to biological data, where only relative assessments of scores can be made. The range of dataset sizes need not be large, however, as we know that applying DCM methods scales up results from datasets of fewer than 15 taxa to datasets of over one thousand taxa with little loss in accuracy and very little distortion over the range of parameters. As we can run many more tests on small datasets and as our primary interest is the effect of model parameters on accuracy, we generated datasets in the range of 10 to 13 taxa.

Simulated trees are often generated under the Yule-Harding model—they are birth-death trees. Many researchers observed that these trees are better balanced than most published ones. Other simulations have used trees chosen uniformly at random from the set of all tree topologies, so-called "random" trees; these, in contrast, are more imbalanced than most published trees. Aldous[34] proposed the β-split model to generate trees with a tailored level of balance; depending on the choice of β, this model can produce random trees ($\beta = -1.5$), birth-death trees ($\beta = 0$), and even perfectly balanced trees. We use all three types of trees in our experiments; for β-split trees, Aldous recommended using $\beta = -1$ to match the balance of most published trees; instead, we chose the parameter to match the computational effort on the datasets from which those trees were computed, which led us to using $\beta = -0.8$. On random and β-split trees, expected edge lengths are set after the tree generation by sampling from a uniform distribution on values in the set $\{1, 2, \ldots, r\}$, where r is a parameter that determines the overall rate of evolution. In the case of birth-death trees, we used both the same process and the edge lengths naturally generated by the birth-death process, deviated from ultrametricity and then scaled to fit the desired diameter.

We generate the true tree by turning each edge length into a corresponding number of iDLR evolutionary events on that edge. The events we consider under the iDLR model are insertions, duplications, losses, and inversions of genes or contiguous segments made of several genes—in particular, inserting, duplicating, or deleting a block of k consecutive genes has the same cost regardless of the value of k. We forced the expected number of inserted and duplicated elements to equal the expected number of deleted elements, in order to keep genome sizes within a general range. We varied the percentage of inversions as a function of the total number of operations from 20% to 90%. The remaining percentages were split evenly between insertions/duplications and losses, with the balance of insertions and duplications tested at one quarter, one half, and three quarters. The expected Gaussian-distributed length of each operation filled a range of combinations from 5 to 30 genes. These are conditions similar to, but broader in scope than, those used in the experiments reported in Swenson *et al.*[24]

In all our simulations, we used initial (root) genomes of 1'000 genes. The resulting leaf genomes are large enough to retain phylogenetic information while exhibiting large-scale changes in structure. These sizes correspond to the smaller bacterial genomes and allow us to conclude that our results will extend naturally to all unichromosomal bacterial genomes.

The collections of gene orders produced by these simulations are then fed to our various competing algorithms. These are of two types: (i) algorithms running on the full gene orders, namely NJ and our new LP-based algorithm; and (ii) algorithms running on equalized gene contents, which include NJ again (running on the inversion distance matrix produced by GRAPPA), GRAPPA, and MGR. Gene contents are equalized by removing gene families with more than one gene, then keeping only singleton genes common to all genomes. On some of these datasets, the equalized gene content is minuscule—with high rates of evolution, the number of genes shared by all 12 taxa is occasionally in the single digits, obviously leading to serious inaccuracies on the part of reconstruction algorithms. We collect the data (including running times, the actual trees, internal inferred gene orders, inferred edge lengths, etc.) and compute basic measures, particularly the Robinson-Foulds[35] distance from the true tree—the most common error measure in phylogenetic reconstruction.

4. Results and Discussion

We ran collections of 100 datasets of 10 to 13 genomes, each of 1'000 genes, under various models of tree generation and various parameters of the iDLR model. We used birth-death, random, and β-split (with $\beta = -0.8$) models, with evolutionary diameters (the length of the longest path, as measured in terms of evolutionary operations, in the true tree) of 200, 400, 500, and 800 operations. (We ran tests with diameters of 800, but noted that most resulting instances exhibited strong saturation—that is, that many of the true edge lengths were significantly larger than the edit distances between the genomes at the ends of the edge; since no reconstruction method can do well in the presence of strong saturation, we did not pursue diameters larger than 800.) For each tree returned, we measured its RF error rate (the percent of edges in error with respect to the true tree) and then averaged the ratios over the set of test instances for each fixed parameter. We computed the ratio of the RF rate for our approach with that for NJ on full genomic distances and with those for the three approaches with equalized gene contents, binning the results into one "losing" bin (the other method did better), one bin of ties, and 5 bins of winners, according to the amount of improvement. Not all 100 instances are included in these averages, because some instances had equalized gene contents of just 2 or 3 genes and could not be run with GRAPPA.

We present below a few snapshots of our results. Table 1 shows the results of using full genomic distances for β-split trees on datasets of diameters 200, 400, and 500, using 80% inversions. In this case, no difference was found between the results returned by our LP-based method and those returned by NJ using full genomic distances. The average RF error rate for MGR was 23% for diameter 200, 32% for diameter 400, and 42% for diameter 500. As simple a method as NJ handily beats existing methods that must rely on equalized gene contents, often by large factors (e.g., factors of 4 or more in 26% of the cases with diameter 200 with respect to MGR). The reduction in error rate was sufficient in many cases to turn unacceptable results (with error rates well in excess of 10%) into acceptable ones.

Table 1. *Accuracy results for NJ on full genomic distances and for three evolutionary diameters compared to three methods on equalized gene contents. Column triples show wins, ties, and losses, in percent. Quintiles in the winning columns denote error reductions by factors larger than 4, 3, 2, 1.5, and 1.*

Dataset	NJ			GRAPPA			MGR		
200	16-4-25-1-0	50	4	14-0-11-4-0	1	3	26-6-21-4-1	36	6
400	4-0-5-4-0	23	0	3-0-6-1-0	0	0	5-1-7-6-12	1	4
500	5-5-5-8-0	69	8	11-2-14-17-15	18	23	17-7-14-17-14	24	7
	w	t	l	w	t	l	w	t	l

Experience with sequence data leads us to expect that an MP method, should do better than NJ when the diameter and deviation from ultrametricity get large. Our LP-based approach is a hybrid: unlike an MP method, it does not reconstruct ancestral labels, but like an MP method, it attempts to minimize the total length of the tree; thus it should at least occasionally outperform NJ. We tested this hypothesis on random trees and birth-death trees where, in both cases, we generated edge lengths by uniform sampling from the set $\{1, 2, \ldots, r\}$, for values of r ranging from 20 to 100, still using 80% inversions. Tables 2 and 3 present the results, this time limited to the reference MGR and to the two methods using full genomic data. Both tables show gains for the LP-based method over simple NJ

Table 2. *Error rates, in percent, on random trees for the two approaches using full genomic data and for MGR on equalized gene contents.*

	20	40	60	80	100
LP	0.9	8.0	7.8	6.0	26.0
NJ	0.5	8.5	8.7	9.5	25.5
MGR	11.3	31.8	34.0	35.0	49.0

Table 3. *Error rates, in percent, on birth-death trees for the two approaches using full genomic data and for MGR on equalized gene contents.*

	20	40	60	80	100
LP	0.2	8.5	7.6	5.7	19.4
NJ	1.4	9.0	8.5	8.0	18.0
MGR	9.7	31.7	31.8	33.7	51.4

as evolutionary rates increase, until both methods start failing at $r = 100$. Note that the accuracy gains over MGR are consistently very high.

Keeping the proportions of inversions to 80%, however, is neither very realistic, as gene duplications and losses are presumably more frequent in nature than rearrangements, nor very challenging, as, given a bounded set of possible gene choices, duplications and losses will saturate sooner than inversions. The experiments of Swenson *et al.*[24] did not test low percentages of inversions, so we ran sets of tests with 20% inversions only, keeping all other relative percentages of events identical. Table 4 shows these results. We were pleased, and somewhat surprised, to observe actual improvements in the quality of trees for rates up to $r = 40$; the threshold effect to $r = 60$ corresponds to a type of saturation caused by too many insertions and deletions. (Approaches with equalized gene contents are not reported, since they failed completely, as expected.)

Table 4. *Error rates, in percent, on birth-death trees with only 20% inversions.*

	20	40	60	80
LP	3.8	3.0	21.0	37.8
NJ	3.1	4.9	18.9	33.7

Finally, we reproduced the results of Earnest-DeYoung[26] on the dataset of 13 bacteria, with genome sizes ranging from 1'000 to over 5'000 genes and gene families of up to 70 members, this time without any special preprocessing, and using our LP-based approach rather than NJ. Once again the resulting phylogeny is one SPR (subtree) move away from that of Lerat et al. The large disparity in gene content between species in this dataset was handled automatically, for the first time for this dataset (or, indeed, for any other set of cellular genomes).

5. Conclusion

Our algorithm offers, for the first time, the possibility to evaluate the phylogenetic information present in the gene families and in the change in gene content among genomes while at the same time taking into account the complete gene orders; and they can do so on scales compatible with the smaller cellular genomes, such as bacterial genomes. Most importantly, our experiments indicate clearly the benefit to be derived from considering the full gene orderings of the genomes rather than some simplified subset—in almost all of our test cases, even the simple NJ procedure outperformed, often by large margins, the best reconstruction algorithms running on data with equalized gene contents. Much work remains to be done, of course: we need to generalize the distance computation of Swenson *et al.* to multichromosomal genomes (not particularly difficult using the DCJ model, but the introduction of additional parameters means further modelling questions) and to start using the algorithm on biological data, which should enable us to refine the model. And, while being able to estimate the true edge lengths of the tree is a help, we are still very far from being able to reconstruct ancestral genomes, because we have no viable algorithm to solve the vexing problem of the median of three genomes and because the iDLR model remains underconstrained.

Acknowledgments

JT and WA were supported by US National Institutes of Health (NIH) grant R01 GM078991-01 and by the University of South Carolina; WA also acknowledges support from the Rothberg Foundation.

References

1. J.R. Brown, C.J. Douady, M.J. Italia, W.E. Marshall and M.J. Stanhope, *Nature Genetics* **28**, 281 (2001).
2. E. Lerat, V. Daubin and N. Moran, *PLoS Biology* **1** (2003).
3. A. Rokas, B.L. Williams, N. King and S.B. Carroll, *Nature* **425**, 798 (2003).
4. S.T. Fitz-Gibbon and C.H. House, *Nucleic Acids Res.* **27**, 4218 (1999).
5. X. Gu and H. Zhang, *Mol. Biol. Evol.* **21**, 1401 (2004).
6. A. Rokas and P. Holland, *Trends in Ecol. and Evol.* **15**, 454 (2000).
7. B.M.E. Moret and T. Warnow, Advances in phylogeny reconstruction from gene order and content data, in *Molecular Evolution: Producing the Biochemical Data, Part B*, eds. E. Zimmer and E. Roalson, Methods in Enzymology **395**, 673 (Elsevier, 2005).
8. J.L. Boore and W.M. Brown, *Curr. Opinion Genet. Dev.* **8**, 668 (1998).
9. M.E. Cosner, R.K. Jansen, B.M.E. Moret, L.A. Raubeson, L.-S. Wang, T. Warnow and S.K. Wyman, An empirical comparison of phylogenetic methods on chloroplast gene order data

in Campanulaceae, in *Comparative Genomics*, eds. D. Sankoff and J.H. Nadeau, 99 (Kluwer Academic Publishers, 2000).
10. S.R. Downie and J.D. Palmer, Use of chloroplast DNA rearrangements in reconstructing plant phylogeny, in *Molecular Systematics of Plants*, eds. D. Soltis, P. Soltis and J.J. Doyle, 14 (Chapman and Hall, New York, 1992).
11. D. Sankoff, G. Leduc, N. Antoine, B. Paquin, B.F. Lang and R. Cedergren, *Proc. Nat'l Acad. Sci., USA* **89**, 6575 (1992).
12. D.B. Stein, D.S. Conant, M.E. Ahearn, E.T. Jordan, S.A. Kirch, M. Hasebe, K. Iwatsuki, M.K. Tan and J.A. Thomson, *Proc. Nat'l Acad. Sci., USA* **89**, 1856 (1992).
13. Y.I. Wolf, I.B. Rogozin, A.S. Kondrashov and E.V. Koonin, *BMC Evol. Biol.* **1** (2001).
14. L.-S. Wang, R.K. Jansen, B.M.E. Moret, L.A. Raubeson and T. Warnow, *J. Mol. Evol.* **63**, 473 (2006).
15. T. Kunisawa, *J. Theor. Biol.* **213**, 9 (2001).
16. T. Kunisawa, *J. Theor. Biol.* **222**, 495 (2003).
17. T. Kunisawa, *J. Theor. Biol.* **239**, 367 (2006).
18. E. Belda, A. Moya and F.J. Silva, *Mol. Biol. Evol.* **22**, 1456 (2005).
19. B.M.E. Moret, S.K. Wyman, D.A. Bader, T. Warnow and M. Yan, A new implementation and detailed study of breakpoint analysis, in *Proc. 6th Pacific Symp. on Biocomputing (PSB'01)*, 583 (World Scientific Pub., 2001).
20. G. Tesler, *J. Comp. Syst. Sci.* **65**, 587 (2002).
21. G. Blin, C. Chauve and G. Fertin, Genes order and phylogenetic reconstruction: Application to gamma-proteobacteria, in *Proc. 3rd RECOMB Workshop on Comparative Genomics (RECOMBCG'05)*, LNCS **3678**, 11 (Springer Verlag, 2005).
22. B. Larget, D.L. Simon, J.B. Kadane and D. Sweet, *Mol. Biol. Evol.* **22**, 486 (2005).
23. M. Marron, K.M. Swenson and B.M.E. Moret, *Theor. Comp. Sci.* **325**, 347 (2004).
24. K.M. Swenson, M. Marron, J.V. Earnest-DeYoung and B.M.E. Moret, Approximating the true evolutionary distance between two genomes, in *Proc. 7th SIAM Workshop on Algorithm Engineering & Experiments (ALENEX'05)*, 121 (SIAM Press, Philadelphia, 2005).
25. S. Angibaud, G. Fertin, I. Rusu and S. Vialette, *J. Comp. Biol.* **14**, 379 (2007).
26. J. Earnest-DeYoung, Reversing gene erosion: Reconstructing ancestral bacterial gene orders, Master's thesis, University of New Mexico (2004).
27. J. Tang and B.M.E. Moret, Linear programming for phylogenetic reconstruction based on gene rearrangements, in *Proc. 16th Ann. Symp. Combin. Pattern Matching (CPM'05)*, LNCS **3537**, 406 (Springer Verlag, 2005).
28. J. Tang and B.M.E. Moret, Scaling up accurate phylogenetic reconstruction from gene-order data, in *Proc. 11th Int'l Conf. on Intelligent Systems for Mol. Biol. (ISMB'03)*, Bioinformatics **19**, i305 (2003).
29. B.M.E. Moret, J. Tang, L.-S. Wang and T. Warnow, *J. Comp. Syst. Sci.* **65**, 508 (2002).
30. A. Bachrach, K. Chen, C. Harrelson, R. Mihaescu, S. Rao and A. Shah, Lower bounds for maximum parsimony with gene order data, in *Proc. 3rd RECOMB Workshop on Comparative Genomics (RECOMBCG'05)*, LNCS **3678**, 1 (Springer Verlag, 2005).
31. J.V. Earnest-DeYoung, E. Lerat and B.M.E. Moret, Reversing gene erosion: reconstructing ancestral bacterial genomes from gene-content and gene-order data, in *Proc. 4th Int'l Workshop Algs. in Bioinformatics (WABI'04)*, LNCS **3240**, 1 (Springer Verlag, 2004).
32. S. Yancopoulos, O. Attie and R. Friedberg, *Bioinformatics* **21**, 3340 (2005).
33. B.M.E. Moret, L.-S. Wang, T. Warnow and S. Wyman, New approaches for reconstructing phylogenies from gene-order data, in *Proc. 9th Int'l Conf. on Intelligent Systems for Mol. Biol. (ISMB'01)*, Bioinformatics **17**, S165 (2001).
34. D. Aldous, *Stat. Sci.* **16**, 23 (2001).
35. D. Robinson and L. Foulds, *Math. Biosci.* **53**, 131 (1981).

SPR-BASED TREE RECONCILIATION: NON-BINARY TREES AND MULTIPLE SOLUTIONS*

CUONG THAN and LUAY NAKHLEH

Department of Computer Science
Rice University
6100 Main Street, MS 132
Houston, TX 77005, USA
Email: {cvthan,nakhleh}@cs.rice.edu

The SPR (subtree prune and regraft) operation is used as the basis for reconciling incongruent phylogenetic trees, particularly for detecting and analyzing non-treelike evolutionary histories such as horizontal gene transfer, hybrid speciation, and recombination. The SPR-based tree reconciliation problem has been shown to be NP-hard, and several efficient heuristics have been designed to solve it. A major drawback of these heuristics is that for the most part they do not handle non-binary trees appropriately. Further, their computational efficiency suffers significantly when computing multiple optimal reconciliations. In this paper, we present algorithmic techniques for efficient SPR-based reconciliation of trees that are not necessarily binary. Further, we present divide-and-conquer approaches that enable efficient computing of multiple optimal reconciliations. We have implemented our techniques in the PhyloNet software package, which is publicly available at `http://bioinfo.cs.rice.edu`. The resulting method outperforms all existing methods in terms of speed, and performs at least as well as those methods in terms of accuracy.

Keywords: Subtree prune and regraft; phylogenetic tree reconciliation; horizontal gene transfer.

1. Introduction

Comparing phylogenetic trees and quantifying the similarities and differences among their topologies play important roles in studying the quality of phylogeny reconstruction methods and understanding gene evolution within species trees. As such, several tree transformation operations have been introduced, and their induced distance measures have been studied extensively.[1] One such operation is the *subtree prune and regraft* (SPR) operation. An SPR operation, or move, transforms a phylogenetic tree by cutting (pruning) a subtree and attaching (regrafting) it from its root to a different branch in the tree. Studies of this operation and the distance measure it induces on pairs of trees have increased significantly in recent years, mainly due to the central role it plays in detecting *reticulate*, i.e., non-treelike, evolutionary histories, such as horizontal gene transfer, hybrid speciation, and recombination.[2–6] In a nutshell, the occurrence of reticulate evolutionary events results in different genomic regions having incongruent, or disagreeing, trees. One way of identifying these

*This work is supported in part by a Department of Energy grant DE-FG02-06ER25734 and a National Science Foundation grant CCF-0622037.

events is based on the comparison of such trees and determining the minimal set of SPR moves that reconcile the incongruities among these trees, as well as their disagreements with the species tree, if such a tree is known. Therefore, the computational problem that has been addressed in this context is: given two trees T_1 and T_2, find a minimal set of SPR moves that transform T_1 into T_2.

While recently developed methods have made significant progress in terms of accuracy (number and location of the SPR moves) and efficiency (time), there remain two central issues that have not been addressed appropriately by these methods:

Non-binary trees. Reconstructed phylogenetic trees often contain multi-furcating nodes; i.e., nodes with more than two children (in the rooted tree setting); see Figure 1. The way these nodes are handled by methods for estimating SPR moves affects the number and location of those moves, and currently most algorithms and tools do not handle non-binary trees.

Multiple minimal sets of SPR moves. It is often the case that a minimal set of SPR moves that reconciles two trees is not unique, and the number of such sets may be exponential in the size of the set;[7] see Figure 1. Current tools that compute multiple solutions take several days on moderate-sized trees, and run out of memory on larger ones.

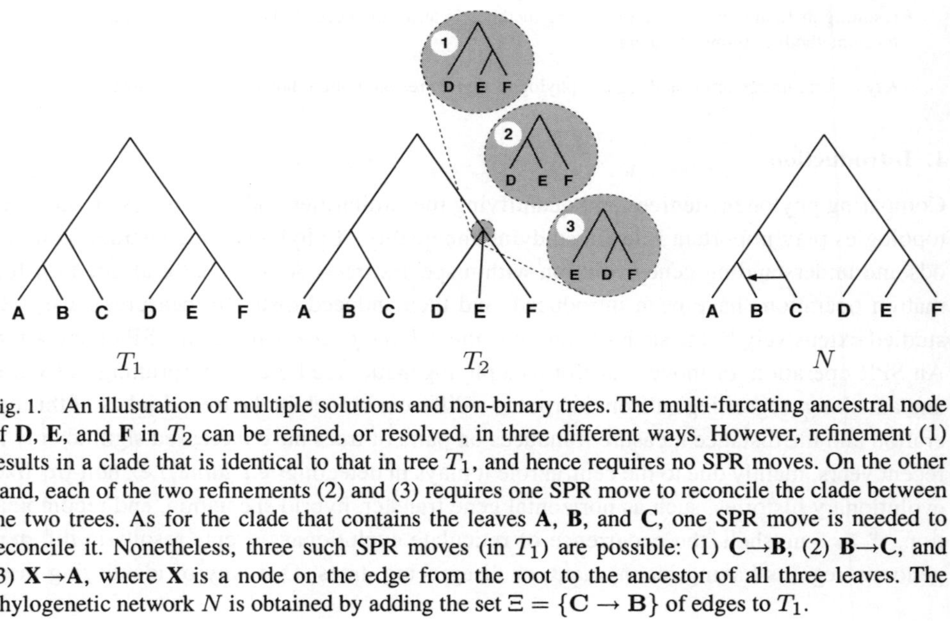

Fig. 1. An illustration of multiple solutions and non-binary trees. The multi-furcating ancestral node of **D**, **E**, and **F** in T_2 can be refined, or resolved, in three different ways. However, refinement (1) results in a clade that is identical to that in tree T_1, and hence requires no SPR moves. On the other hand, each of the two refinements (2) and (3) requires one SPR move to reconcile the clade between the two trees. As for the clade that contains the leaves **A**, **B**, and **C**, one SPR move is needed to reconcile it. Nonetheless, three such SPR moves (in T_1) are possible: (1) **C**→**B**, (2) **B**→**C**, and (3) **X**→**A**, where **X** is a node on the edge from the root to the ancestor of all three leaves. The phylogenetic network N is obtained by adding the set $\Xi = \{\mathbf{C} \rightarrow \mathbf{B}\}$ of edges to T_1.

In this paper we address these two issues by introducing algorithms for efficient refinement of trees to yield minimal sets of SPR moves, as well as algorithms for collapsing identical components of the trees to enable efficient handling of large trees in terms of time and space requirements, while not affecting the accuracy of the computed set of SPR moves. For computing multiple minimal sets of SPR moves, we utilize the sharing among solutions and present algorithmic techniques for efficient computing and displaying of multiple solutions. Besides their value in taming the computational complexity of the problem, the outcome of these techniques has biological significance since it summarizes the "essentiality" of the SPR moves (i.e., which moves *must* be considered in order to account for incongruence, and which ones have *alternatives* that can be considered and have the same effect), and hence the support for the corresponding reticulate evolutionary event.

We have extended the RIATA-HGT method[6] for reconciling trees by incorporating our new algorithms and techniques. The resulting method outperforms all existing methods in terms of computing time, and performs at least as well in terms of accuracy (number of SPR moves in a minimal set, and number of such minimal sets) on binary trees. For non-binary trees, most existing methods are not able to handle them (the tools simply quit, giving an error message stating that the input trees are not binary). The extended method has been implemented in the PhyloNet software package, which is available at `http://bioinfo.cs.rice.edu`. In this paper we use HGT as the guiding biological example of a reticulate evolutionary event, but the method can be applied to trees even when other reticulate evolutionary events have occurred.

2. Preliminaries

Trees and networks. Let $T = (V, E)$ be a tree, where V and E are the *tree nodes* and *tree edges*, respectively, and let $\mathscr{L}(T)$ denote its leaf set. Further, let $\mathscr{X}$ be a set of taxa (species). Then, T is a phylogenetic tree over $\mathscr{X}$ if there is a bijection between $\mathscr{X}$ and $\mathscr{L}(T)$. A tree T is said to be *rooted* if the edges in E are directed and there is a single internal node r with in-degree 0. Let $T = (V, E)$ be a rooted tree, and u be a node in V. We denote by T_u the subtree of T whose root is node u, and $L(u)$ the set of leaves in T_u. A phylogenetic tree t is a *clade* of a phylogenetic tree $T = (V, E)$ if there exists a node $v \in V$ such that $t = T_v$. Given two phylogenetic trees $T = (V, E)$ and $T' = (V', E')$, with $\mathscr{L}(T) = \mathscr{L}(T')$, a *maximal pair of matching clades* is a pair $\langle t, t' \rangle$ such that $t = T_u$ and $t' = T'_{u'}$ for $u \in V$ and $u' \in V'$, $t = t'$, and (1) either u and u' are the roots of the two trees, or (2) $(x, u) \in E$, $(x', u') \in E'$, and $T_x \neq T'_{x'}$. Given a set $X \subseteq \mathscr{L}(T)$, we denote by $lca_T(X)$ the *least common ancestor* of X in T.

A phylogenetic network $N = N(T) = (V', E')$ over the taxa set $\mathscr{X}$ is derived from $T = (V, E)$ by adding a set Ξ of edges to T, where each edge $h \in \Xi$ is added in three steps: (1) split an edge $e = (u, v) \in E$ by adding a new node, v_e, and replacing e by two edges (u, v_e) and (v_e, v); (2) split an edge $e' = (u', v') \in E$ by adding a new node, $v_{e'}$, and replacing e' by two edges $(u', v_{e'})$ and $(v_{e'}, v')$; and (3) add a directed *HGT edge* from v_e to $v_{e'}$. In this case, we write $N = T + \Xi$. Figure 1 shows a phylogenetic network obtained by adding a single HGT edge to the tree T_1. Finally, we denote by $\mathscr{T}(N)$ the set of all trees contained inside network N. Each such tree is obtained by the following two steps: (1) for each node of in-degree 2, remove one of the incoming edges; and (2) For every node x of in-degree and out-degree 1, whose parent is u and child is v, remove node x and its two incident edges, and add a new edge from u to v. This operation is called a *forced contraction*. For example, in Figure 1, the tree T_1 and tree T_2 (with clade refinement (1)) are the only members of $\mathscr{T}(N)$.

Reticulate evolution and the SPR operation. Let $T = (V, E)$ be a rooted tree. An SPR move involving edges $e = (u, v)$ and $e' = (u', v')$ in E (u' is not reachable from the root of the tree T through node v) deletes edge e, splits edge e' into two edges by adding a new node $v_{e'}$, as described above, and adds a new edge from $v_{e'}$ to v. Equivalently, the SPR move may involve a node x instead of edge e', in which case, the move deletes edge e and adds a new edge from x to v. As mentioned above, when HGT occurs, the evolutionary history of the species may not be represented by phylogenetic trees; rather, *phylogenetic networks* are the appropriate model.[8] In the phylogeny-based HGT detection problem, a pair of trees T_1 and T_2 (usually, a species/gene tree pair) is given, and a minimal set Ξ of edges is sought so that $T_2 \in \mathscr{T}(N)$, where $N = T_1 + \Xi$. The minimization requirement simply reflects a parsimony criterion: in the absence of any additional biological knowledge, the simplest solution is sought. This problem has been shown to be related to finding the minimal set of SPR moves that transform T_1 into T_2[a] and several heuristics for solving the problem using SPR moves have been recently introduced.[4–6,9–12]

Non-binary trees and tree compatibility. An edge $e = (u, v)$ in a rooted tree T is contracted by deleting it and merging the two nodes u and v into a single node x (the edges incident from x are the union of the edges incident from u and v). We say a tree T' is a *contraction* of tree T, if T' is obtained by contracting a set of edges in T. Equivalently, we say that T is a *refinement* of tree T'. An edge $(u, v) \in E$ induces a *split* $A|B$, where $A = L(v)$, and $B = \mathscr{L}(T) - A$. A split $A|B$ is non-trivial if $|A| > 1$ and $|B| > 1$. We say that two splits $A|B$ and $C|D$ are *compatible* if at least one of the four intersections $A \cap C$, $A \cap D$, $B \cap C$ and $B \cap D$ is empty. We denote by $\pi(T)$ the set of all splits induced by the edges of tree T. We say that two trees T_1 and T_2 are compatible if $\pi(T_1)$ and $\pi(T_2)$ are pairwise compatible. When one or both of the trees T_1 and T_2 are not necessarily binary, the phylogeny-based HGT detection problem is slightly modified, since we seek a minimal set of SPR moves that makes T_1 compatible with, and not necessarily identical to, tree T_2. In other words, a minimal set Ξ of HGT edges is sought so that (1) $N = T_1' + \Xi$, (2) $T_2' \in \mathscr{T}(N)$, and (3) T_1', T_2' are refinements of T_1 and T_2, respectively, that result in the minimum size of such a set Ξ. The network N in Figure 1, with a single HGT edge, is an example of a solution to the problem for the pair of trees T_1 and T_2.

3. Algorithmic Techniques

As mentioned above, existing methods for solving the phylogeny-based HGT detection problem do not handle non-binary tree appropriately (in fact, most tools do not run on non-binary trees), nor do they handle multiple minimal solutions efficiently. In this section we present algorithmic techniques for efficient handling of these two cases.

---

[a] An HGT edge involving two edges e and e', or an edge e and a node x is obtained by computing the SPR move as defined, with the only difference that edge e is not deleted.

3.1. *Handling Non-binary Trees: Refine and Collapse*

As illustrated in Figure 1, when multi-furcating nodes are present, different refinements may lead to different estimates of the minimum number of SPR moves needed to reconcile two trees. Since we seek the minimum number of SPR moves to reconcile two trees T_1 and T_2, which are not necessarily binary, our proposed solution is to *maximally refine* both trees to obtain two trees T_1' and T_2', respectively, such that the number of SPR moves required to reconcile T_1' and T_2' is minimum among all possible refinements of T_1 and T_2. For example, under this approach, refinement (1) of tree T_2 in Figure 1 is preferred over the other two possible refinements. We now present an efficient algorithm for solving this problem.

(1) Generate all nontrivial splits of T_1 and T_2.
(2) For each split $A|B$ of T_1 that is not a split of T_2 but compatible with every split of T_2:
    (a) Let $u = lca_{T_2}(A)$, and let $x_1, x_2, \ldots x_k$ be the children of u such that $A = \cup_{i=1}^{k} L(x_i)$. If no such set of children exists, redo this step for $u = lca_{T_2}(B)$.
    (b) Delete all edges (u, x_i), for $1 \leq i \leq k$, add a new node x' with new edge (u, x'), and add k new edges (x', x_i) for $1 \leq i \leq k$.
(3) Repeat Step 2 for all splits of T_2 with respect to T_1.

The algorithm takes $O(|\mathscr{L}(T_1)|^2)$ time and maximally refines the two trees T_1 and T_2 without affecting the number of SPR moves required to reconcile them; details are omitted due to space constraints.

Once the two trees are refined, we *collapse* them to achieve reduction in the size of the trees, without affecting the set of SPR moves. The idea is that clades that are identical in both trees do not require any SPR moves to reconcile them, and hence we can preprocess the trees by collapsing them into single leaf nodes. Formally, for every maximal pair of matching clades $\langle t, t' \rangle$ in the two trees, replace both t and t' by a single node that is labeled with the same label ℓ_t, where ℓ_t is unique (per tree). If k is the minimum number of SPR moves required to transform tree T_1 into tree T_2, then the same SPR moves are required to transform T_1' into T_2', where T_1' and T_2' are obtained from T_1 and T_2, respectively, through the application of any number of collapse operations.[5]

A special case that requires special handling is that of *identical chains* in the two trees. Allen and Steel[2] handled chains in binary trees. We now generalize that to include trees that are not necessarily binary (assuming that the collapse operation has been applied maximally to the trees, as described above). Two sequences $\mathscr{P} = \langle u_1, u_2, \ldots, u_k \rangle$ where $u_i \in V(T_1)$ and $\mathscr{P}' = \langle u_1', u_2', \ldots, u_k' \rangle$ where $u_i' \in V(T_2)$, $k \geq 2$, are said to be identical if (1) u_{i+1} is parent of u_i and u_{i+1}' is parent of u_i', $1 \leq i \leq k-1$; (2) all clades whose roots are children of u_i, u_i' and not in $\mathscr{P}, \mathscr{P}'$ are identical, $2 \leq i \leq k$; and (3) all clades whose roots are children of u_1 and u_1', except for exactly one, are identical. (Note that the requirement in (3) is to distinguish identical chains from identical clades.) The value $k-1$ is the chain length. Bordewich and Semple[5] showed that an identical chain can be replaced in both binary trees by an identical chain of only three leaves $\langle a, b, c \rangle$, without affecting the SPR moves. With the definition of identical chains above, this rule can be applied to non-binary trees. The reason is that clades whose roots are children of u_i (u_i') and not in $\mathscr{P}$ ($\mathscr{P}'$) can be thought of as being "contained" in *one* big clade, and therefore the rule for binary trees can be used. Figure 2 shows an example of identical chains in non-binary trees and how they can be replaced. Applying this operation can further reduce the size of trees, as the collapse operation does not apply in this case.

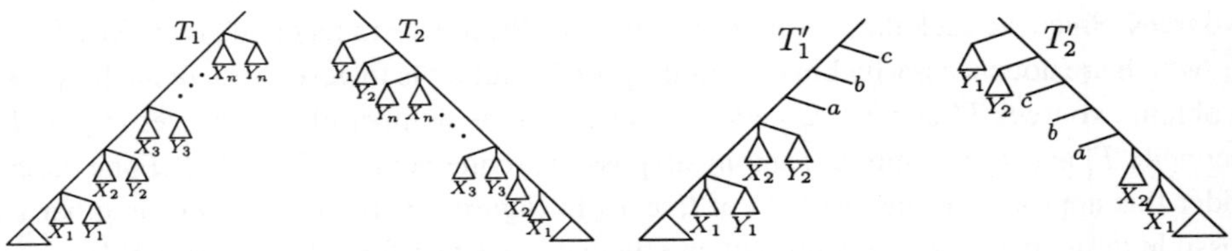

Fig. 2. Replacement of identical chains in non-binary trees. The identical chain $\langle (X_3, Y_3), \ldots, (X_n, Y_n) \rangle$ in the trees T_1 and T_2 is replaced by the chain $\langle a, b, c \rangle$, which results in a significant decrease in the size of trees, without affecting the number of SPR required to reconcile the two trees.

While Bordewich and Semple stated this operation can be done in time that is polynomial in the number of leaves in a binary tree, we now present an $O(|\mathscr{L}(T_1)|^2)$ time algorithm for applying the collapse operation maximally to a pair of trees (not necessarily binary) T_1 and T_2 to replace all identical chains by 3-leaf chains. In order to collapse identical maximal chains in the two trees, a bottom-up scan of the clades of T_1, comparing them for compatibility with clades of T_2, and replacing such pairs of identical clades by leaf nodes, can be achieved in $O(|\mathscr{L}(T_1)|^2)$ as well.

(1) Compute the set of leaves $L(v)$ for every internal node v in T_1 and T_2.
(2) Starting from a deepest node v in T_1 , let u be the parent of v, and do the following:
   (a) Compute $L(u) - L(v)$ and find an edge $(u', v') \in E(T_2)$ such that $L(u') - L(v') = L(u) - L(v)$. If no such edge (u', v') exists, let v be the next deepest node that has not been examined, and go to step (2).
   (b) Restrict tree T_1 and T_2 to the leaves $L(u) - L(v)$. If the two restricted trees are identical, replace u, u' by their parents in T_1 and T_2, and repeat this substep.
   (c) Identical chains obtained in (2)(b) are maximal. If their length is at least 3, replace them by 3 new leaves that preserve orientation relative to the roots of T_1 and T_2.
(3) Repeat step (2) with the next deepest node that has not been examined.

The algorithm replaces all maximal identical chains in trees by 3-leaf chains, and takes $O(|\mathscr{L}(T_1)|^2)$ time; details are omitted due to space constraints.

3.2. *Algorithmic Techniques for Efficient Enumeration of Minimal Solutions*

Than *et al.* have recently shown that the number of solutions to the phylogeny-based detection problem is $O(3^k)$, where k is the minimum number of SPR moves required to reconcile the two trees.[7,13] In this section, we exploit the fact that there are only $O(n^2)$ possible distinct SPR moves that can be applied to a tree on n taxa, to design strategies for efficient enumeration of all minimal solutions.

We denote by $T - t$ the tree obtained from T by removing the clade t and applying forced contractions. Let T_1 and T_2 be two phylogenetic trees on the same set of taxa $\mathscr{X}$, with sets $S(T_1)$ and $S(T_2)$ of clades. In this section we assume the trees have been maximally refined and collapsed. We denote by $\mathbf{Sol}(T_1, T_2)$ the set of all minimal sets of SPR moves that reconcile the two trees (i.e., the set of all solutions to the HGT detection problem). We define a mapping $f : S(T_1) \to S(T_2)$, such that $f(t_1) = t_2$ when $L(t_1) = L(t_2)$, and $f(t_1) = nil$ when there does not exist $t_2 \in S(T_2)$ such that $L(t_1) = L(t_2)$. Given two trees T_1 and T_2, and the mapping f, we process the trees as follows. Suppose there are m clades $t_1^1, \ldots, t_1^m$ in T_1 such that $f(t_1^i) \neq nil$. Then, we generate $m+1$ pairs of trees $\langle \alpha^i, \beta^i \rangle$, $1 \leq i \leq m+1$,where α^i and β^i, $1 \leq i \leq m$, are obtained from t_1^i and $f(t_1^i)$ by replacing in each of them every clade t' and $f(t')$ ($f(t') \neq nil$), respectively, by a single leaf with the same name in both clades. The last pair $\langle \alpha^{m+1}, \beta^{m+1} \rangle$ is obtained from T_1 and T_2 by removing from them all clades t_1^i and $f(t_1^i)$, $1 \leq i \leq m$, respectively. We call these $m+1$ pairs, the *decomposition* of the pair of trees T_1 and T_2, denoted $\mathscr{D}(T1, T2)$.

Lemma 3.1. *Let $T1$ and $T2$ be two phylogenetic trees whose decomposition is $\mathscr{D} = \{\langle \alpha^i, \beta^i \rangle : 1 \leq i \leq p\}$. Then,* $\mathbf{Sol}(T_1, T_2) = \mathbf{Sol}(\alpha^1, \beta^1) \times \cdots \times \mathbf{Sol}(\alpha^p, \beta^p)$.

This lemma states that a minimal solution for a pair of trees can be obtained by taking the union of minimal solutions from each of the pairs in the decomposition $\mathscr{D}$, and gives the basis for our divide-and-conquer strategy. In this strategy, a decomposition of the two trees is first performed, the HGT detection problem is solved on each pair in the decomposition separately, and then the cartesian product of the sets of minimal solutions of these pairs is taken as the set of all minimal solutions of the trees.

Notwithstanding the gains achieved by the divide-and-conquer approach, it may be the case that a few pairs have large clades in them. However, empirical performance shows that large clades may have fewer solutions, given the lack of "locality" in the HGT events involved.[7] To enable efficient handling of these clades, we consider HGT event equivalence, and describe how this concept may lead to further reductions in time for computing minimal solutions.

Equivalence of minimal sets of SPR moves. Given a tree T and a set Ξ of SPR moves defined on T, we denote by $T \uparrow \Xi$ the tree obtained from T by applying the SPR moves, followed by forced contractions, that correspond to the HGT edges in Ξ.

Definition 3.1. Given two sets Ξ_1 and Ξ_2 of HGT edges defined on a tree T, we say that Ξ_1 is equivalent to Ξ_2 (with respect to tree T), denoted $\Xi_1 \equiv \Xi_2$, if $T \uparrow \Xi_1$ is compatible with $T \uparrow \Xi_2$.

The $\equiv$ relation on sets of SPR moves is an equivalence relation. Further, equivalent sets of SPR moves from two minimal solutions have the same cardinality, as we now show.

Lemma 3.2. *Let $T1$ and $T2$ be two trees, and Ξ_1 and Ξ_2 be two sets of SPR moves in* $\mathbf{Sol}(T_1, T_2)$. *If $X' \equiv Y'$ for $X' \subseteq \Xi_1$ and $Y' \subseteq \Xi_2$, then $|X'| = |Y'|$.*

Based on the above observations and the defined equivalence relation, we have the following strategy for efficient enumeration of multiple equivalent solutions to the HGT detection problem:

(1) Find a solution Ξ to the problem.

(2) Partition Ξ into $\Xi_1, \ldots \Xi_m$ such that for any other solution Y, Y can be partitioned into $Y_1, \ldots, Y_m$ where

   (a) $\forall \Xi_i$, $\exists Y_j$ such that $\Xi_i \equiv Y_j$, and

   (b) m is the maximum cardinality of such a partition of Ξ.

(3) For each Ξ_i, compute its equivalence class $[\Xi_i]$.

(4) The set of solutions is $Z_1, Z_2, \ldots, Z_m$, where $Z_i \in [\Xi_i]$.

As described above, when the HGT events (SPR moves) are "local", i.e., do not span a large portion of the tree, the decomposition process yields small components, and hence the number of equivalence classes is large and their sizes are small. However, when the HGT events are more global, we expect the number of solutions to be smaller, and hence the number of equivalence HGT edge-sets (SPR moves) to be small as well.[7]

4. Empirical Performance

We used the `r8s` tool[14] to generate four random trees, T_i, $i \in \{10, 25, 50, 100\}$, where i denotes the number of taxa in the tree. The `r8s` tool generates molecular clock trees; we deviated the trees from this hypothesis by multiplying each edge in the tree by a number randomly drawn from an exponential distribution. The expected evolutionary diameter (longest path between any two leaves in the tree) is 0.2. Then, from each model "species" tree T_i, we generated five different "gene" trees, $T_{i,j}$, $j \in \{1, 2, 3, 4, 5\}$, where j denotes the number of simulated HGT events (SPR moves) applied to T_i to obtain $T_{i,j}$.For each T_i and $T_{i,j}$, $i \in \{10, 25, 50, 100\}$ and $j \in \{1, 2, 3, 4, 5\}$, and for each sequence length $\ell \in \{250, 500, 1000, 2000, 4000, 8000\}$, we generated 30 DNA sequence alignments $S_i^\ell[k]$ and $S_{i,j}^\ell[k]$, $1 \leq k \leq 30$, whose evolution was simulated down their corresponding trees under the GTR+Γ+I (gamma distributed rates, with invariable sites) model of evolution, using the Seq-gen tool.[15] We used the parameter settings of.[16] Then, from each sequence alignment, we reconstructed a tree TNJ using the Neighbor Joining (NJ) method.[17] At the end of this process we had 4 trees T_i, 20 trees $T_{i,j}$, 720 NJ trees $TNJ_i^\ell[k]$, and 3600 NJ trees $TNJ_{i,j}^\ell[k]$ ($i \in \{10, 25, 50, 100\}$, $j \in \{1, 2, 3, 4, 5\}$, $1 \leq k \leq 30$, and $\ell \in \{250, 500, 1000, 2000, 4000, 8000\}$). To compute solutions to the HGT detection problem, as well as the number of such solutions, we applied two methods to pairs of species and gene trees: LatTrans[10,18] and the extended RIATA-HGT,[6] which implements the strategies for handling non-binary trees and computing multiple minimal scenarios, as described in the previous section. Both tools were applied to pairs $(TNJ_i^\ell[k], TNJ_{i,j}^\ell[k])$ of binary trees; since LatTrans cannot handle non-binary trees, we do not report any comparisons for that. Due to space limitations, we only show results for 50-taxon data sets, shown in Figure 3. In each run of a tool on a pair of trees, we computed two values: the minimum number of inferred HGT events (SPR moves), and the number of such minimal solutions found by the method. We report the average of all 30 runs and actual running times for each combination of i, j, and ℓ. In Figure 3 we observe a similar relative performance between the two methods in terms of the number of HGT events estimated and the number of minimal solutions computed. Notice that both methods almost identically overestimate

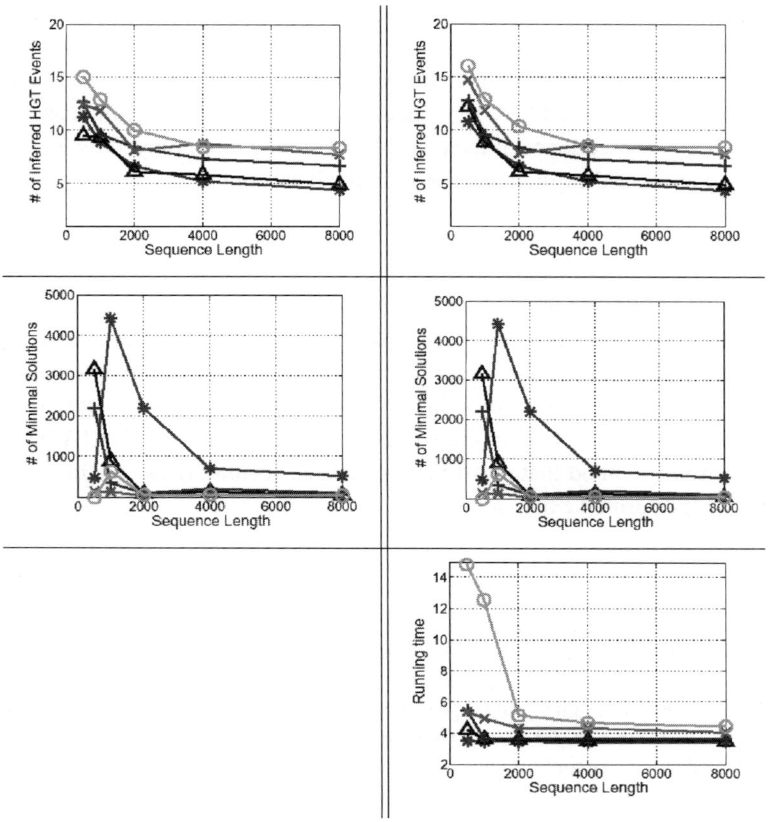

Fig. 3. The performance of LatTrans (left column) and RIATA-HGT (right column) in terms of the minimum number of HGT edges (top row), number of minimal solutions (middle row), and actual running times in seconds (bottom row), as functions of the sequence length. All results are obtained from 50-taxon NJ trees. LatTrans took several days on each pair of 50-taxon trees, and for sequence length 250 it crashed after 4 days without returning results (hence we omit its running time graph). Each curve corresponds to one of the five actual numbers of HGT events: $\star$: 1 HGT; $\triangle$: 2 HGTs; $+$: 3 HGTs; $\times$: 4 HGTs; and $\circ$: 5 HGTs.

the minimum number of HGT events, or SPR moves needed to reconcile the two trees, and this overestimation decreases as the sequence length increases. This is a result of the large amount of wrong edges in the trees inferred by NJ, and the fact that these errors made by NJ decrease as the sequence length increases, since NJ is *statistically consistent*. Further, notice that as the sequence length increases and the estimates of the number of HGT events decreases, the number of minimal solutions decreases drastically, which is in agreement with the results showing that the number of solutions is proportional to their size, and can be exponential in these sizes.[7] However, where the big difference is pronounced between the two methods is in terms of running times. RIATA-HGT and LatTrans found the same number of minimal solutions; yet, RIATA-HGT found these solutions in a few seconds, whereas LatTrans ran for several days on each of these data sets, and crashed on all data sets for sequence length 250—which is the case where a large number of HGT events are identified. Notice that even though the number of solutions for the case of 1 HGT is much larger than that of 5 HGTs, RIATA-HGT finds the solutions in the former case much more quickly, which is a consequence of the algorithmic strategies employed by RIATA-HGT to exploit sharing and avoid explicit enumeration of all solutions.

5. Conclusions

In this paper, we considered the problem of reconciling a pair of phylogenetic trees, mainly to estimate the amount of non-treelike evolutionary events in the evolution of a set of organisms. We addressed the two issues of appropriate handling of non-binary trees and efficient enumeration of equally optimal solutions. We developed a set of algorithmic techniques for handling both issues, and incorporated these techniques into the RIATA-HGT method. The outcome was a method that performed at least as accurately as existing methods, and significantly outperformed existing methods.

References

1. C. Semple and M. Steel, *Phylogenetics*
2. B. Allen and M. Steel, *Annals of Combinatorics* **5**, 1 (2001).
3. M. Baroni, C. Semple and M. Steel, *Annals of Combinatorics* **8**, 391 (2004).
4. L. Nakhleh, T. Warnow and C. Linder, Reconstructing reticulate evolution in species–theory and practice, in *Proc. 8th Ann. Int'l Conf. Comput. Mol. Biol. (RECOMB04)*, 2004.
5. M. Bordewich and C. Semple, *Annals of Combinatorics* , 1 (2005).
6. L. Nakhleh, D. Ruths and L. Wang, RIATA-HGT: A fast and accurate heuristic for reconstrucing horizontal gene transfer, in *Proceedings of the Eleventh International Computing and Combinatorics Conference (COCOON 05)*, ed. L. Wang2005. LNCS #3595.
7. C. Than, D. Ruths, H. Innan and L. Nakhleh, *Journal of Computational Biology* **14**, 517 (2007).
8. B. Moret, L. Nakhleh, T. Warnow, C. Linder, A. Tholse, A. Padolina, J. Sun and R. Timme, *IEEE/ACM Transactions on Computational Biology and Bioinformatics* **1**, 13 (2004).
9. V. Makarenkov, *Bioinformatics* **17**, 664 (2001).
10. M. Hallett and J. Lagergren, Efficient algorithms for lateral gene transfer problems, in *Proc. 5th Ann. Int'l Conf. Comput. Mol. Biol.*, (ACM Press, New York, 2001).
11. R. Beiko and N. Hamilton, *BMC Evolutionary Biology* **6** (2006).
12. D. MacLeod, R. Charlebois, F. Doolittle and E. Bapteste, *BMC Evolutionary Biology* **5** (2005).
13. C. Than, D. Ruths, H. Innan and L. Nakhleh, Identifiability issues in phylogeny-based detection of horizontal gene transfer, in *Proceedings of the Fourth RECOMB Comparative Genomics Satellite Workshop*, eds. N. El-Mabrouk and G. Bourque, Lecture Notes in Bioinformatics (LNBI), Vol. 42052006.
14. M. Sanderson, Analysis of rates (r8s) of evolution (2006), Available from http://loco.biosci.arizona.edu/r8s/.
15. A. Rambaut and N. C. Grassly, *Comp. Appl. Biosci.* **13**, 235 (1997).
16. D. Zwickl and D. Hillis, *Systematic Biology* **51**, 588 (2002).
17. N. Saitou and M. Nei, *Mol. Biol. Evol.* **4**, 406 (1987).
18. L. Addario-Berry, M. Hallett and J. Lagergren, Towards identifying lateral gene transfer events, in *Proc. 8th Pacific Symp. on Biocomputing (PSB03)*, 2003.

ALIGNMENT OF MINISATELLITE MAPS: A MINIMUM SPANNING TREE-BASED APPROACH

MOHAMED I. ABOUELHODA

Cairo University, Giza, Egypt
Email: mohamed.ibrahim@uni-ulm.de

ROBERT GIEGERICH

University of Bielefeld, 33501 Bielefeld, Germany

BEHSHAD BEHZADI

Google Switzerland GmbH, Switzerland and Ecole Polytechnique, France

JEAN-MARC STEYAERT

LIX, Ecole Polytechnique, Palaiseau cedex 91128, France

In addition to the well-known edit operations, the alignment of minisatellite maps includes duplication events. We model these duplications using a special kind of spanning trees and deduce an optimal duplication scenario by computing the respective minimum spanning tree. Based on best duplication scenarios for all substrings of the given sequences, we compute an optimal alignment of two minisatellite maps. Our algorithm improves upon the previously developed algorithms in the generality of the model, in alignment quality and in space-time efficiency. Using this algorithm, we derive evidence that there is a directional bias in the growth of minisatellites of the MSY1 dataset.

Keywords: Minisatellite maps; Sequence Analysis; Run length encoding.

1. Introduction

1.1. *Alignment of minisatellite maps*

A genomic region is classified as a minisatellite locus if it spans more than 500 bp and is composed of tandemly repeated DNA stretches. Each stretch, called *unit*, is a sequence of nucleotides whose length ranges between 7-100 bp. A potential mechanism responsible for the evolution of minisatellites is the *unequal cross-over*, where the paired homologous chromosomes exchange unequal segments during the cell division. This gives rise to a repeated segment in one chromosome and to a deletion in the other; see Figure 1 (b).

A minisatellite map represents a minisatellite region, where each unit is encoded by a character and handled as one entity; see Figure 1 (a). For one minisatellite locus, both the type and the number of units vary between individuals in a population. Therefore, minisatellite maps provide a means for studying the evolution of

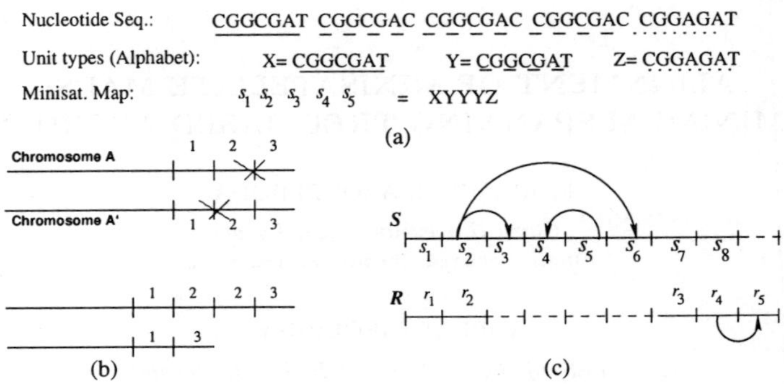

Fig. 1. (a): A minisatellite locus: five units, their nucleotide sequences, and the respective map are shown. (b): The unequal cross over producing duplication of the unit 2. (c): Alignment of two sequences. The matched copies are put above each other. The arcs represent duplication events.

populations. The key algorithm for this study is to align maps of individuals from different populations.

The traditional model of sequence alignment is based on the edit operations of replacement (including matches) and deletions/insertions (indels). In aligning minisatellite maps, one has to also consider that regions of the map have arisen as a result of duplication events from the neighboring units. The single copy duplication model, where only one unit can duplicate at a time, is the most popular, and its biological validation was asserted for the MSY1 minisatellites; see [1, 2]. The scoring of minisatellite map alignment accounts for common aligned units as well as for individual duplication histories. For example, the score of the alignment in Figure 1 (c) is composed of (1) replacement scores for the unit pairs (s_1, r_1), (s_2, r_2), (s_7, r_3) and (s_8, r_4), (2) costs of duplication of the units s_3, s_6 originated from the unit s_2, duplication of s_4 from s_5, and duplication of r_5 from r_4, and (3) insertion of the unit s_5. That is, the comparison delivers a three-stages scenario: The aligned units refer to common ancestors, the duplications refer to differences in the individual duplication histories, and the indels refer to units (possibly emerged by a transposition [3]) not homologous to the map units.

1.2. *Previous and new results*

The problem of comparing two minisatellite maps under the single copy duplication model was investigated for the first time by Bérard and Rivals [1] who presented an algorithm that takes $O(n^4)$ time and $O(n^3)$ space, where n is the average map length. Subsequently, Behzadi and Steyaert [4] followed a different approach and presented a transformation-distance based algorithm: one sequence is considered as a source and the other as a target. The optimal transformation distance is the minimum cost set of operations required to transform the source to the target. Their algorithm takes $O(n^3|\Sigma|)$ time and $O(n^2|\Sigma|)$ space, where $|\Sigma|$ is the alphabet size. Based on a run length encoding scheme, Behzadi and Steyaert [5] improved the running time of their algorithm to $O(n^2 + nn'^2 + n'^3|\Sigma|)$, where n' is the length of the run-length compressed sequence. Very recently, Bérard et al. [6] argued that the alignment distance for two minisatellite sequences S and R is symmetric, while the transformation distance is not. They correspondingly refined the algorithm of [1] by incorporating ideas of [5] and presented an algorithm that takes $O(n^3 + n'^3|\Sigma|)$.

In this paper, we present an *alignment* algorithm that improves upon the previous ones in many respects: The alignment model is more general and our algorithm relaxes the constraint that the mutation distance $M(a, b)$ between two units is symmetric. The time complexity of our algorithm is alphabet-independent and we show that the run length encoding scheme can be incorporated, which yields an $O(n^2 + nn'^2 + n'^3)$ time and $O(n^2)$ space algorithm. This makes our algorithm the fastest map alignment algorithm in theory, and in practice as well, as we demonstrate by experiments.

From the biological point of view, our algorithm is so flexible that investigators can put constraints on the direction of duplications to study the structural variation and duplication dynamics. Based on this feature, we could quantitatively verify the assumption in [2] that the units duplicate in a biased fashion at the $5'$ end.

Our algorithm computes an optimal alignment by first computing and storing the cost of an optimal duplication history for each interval in each sequence separately. Then it computes the optimal alignment based on the precomputed costs.

In fact, reconstructing duplication histories of a minisatellite map which occurs here as a subproblem of map alignment is related, but in a somewhat modified form, to the problem of inferring the duplication history of tandemly repeated genes; see [7–10], and in particular [11], which is closely related to our work here. What distinguishes the general construction of duplication histories from their use in the map alignment problem is that here, it is required to compute, for each interval of units, a duplication history originated either from the leftmost or the rightmost unit. Furthermore, some units may be inserted (not duplicated from other units) and may then undergo further duplications.

2. The Duplication History

Let $S = s_1, s_2, \ldots, s_n$ denote a minisatellite map of n units. We write $S[i..j]$ to denote the $j - i + 1$ contiguous units $s_i, s_{i+1}, \ldots, s_j$. A sequence T is a subsequence of S, if there is a set of indices $i_1 < i_2 < \cdots < i_m$, $m \leq n$, such that $T = s_{i_1}, s_{i_2}, \ldots, s_{i_m}$. The cost of mutating a unit s' to s'' is denoted by $M(s', s'') \geq 0$.

We denote the cost of a duplication event affecting a unit s' by $DUP(s')$. We call the total cost of duplicating and mutating the unit s' to produce $s's''$ or $s''s'$, where $s', s'' \in S$, the *duplication cost* $d(s', s'') = DUP(s') + M(s', s'')$. Note that if $M(s', s'') = M(s'', s')$, and $DUP(s')$ is constant for every unit $s' \in S$, then $d(s', s'') = d(s'', s')$, i.e., we speak of a symmetric distance function. Our algorithms work for both symmetric and non-symmetric distance functions. We assume that $d(s', s'') \leq d(s', s''') + d(s''', s'')$, for all $\{s', s'', s'''\} \subset S$.

dup. (Edge)	subsequence
$s_3 \to s_1$	$s_1\ s_3$
$s_3 \to s_2$	$s_1\ s_2\ s_3$
$s_3 \to s_4$	$s_1\ s_2\ s_3\ s_4$
$s_4 \to s_6$	$s_1\ s_2\ s_3\ s_4\ s_6$
$s_6 \to s_7$	$s_1\ s_2\ s_3\ s_4\ s_6\ s_7$
$s_6 \to s_5$	$s_1\ s_2\ s_3\ s_4\ s_5\ s_6\ s_7$

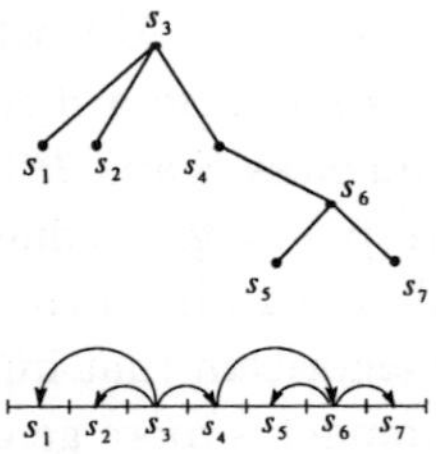

(a)

dup. (Edge)	subsequence
$s_3 \to s_1$	$s_1\ s_3$
$s_3 \to s_2$	$s_1\ s_2\ s_3$
$s_3 \to s_4$	$s_1\ s_2\ s_3\ s_4$
$s_2 \to s_6$	$s_1\ s_2\ s_6\ s_3\ s_4$

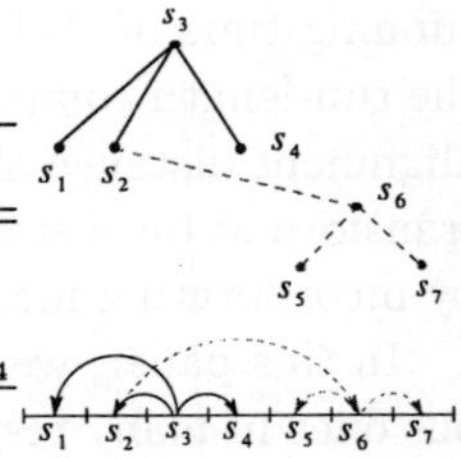

(b)

Fig. 2. (a) The table on the left shows the duplication events producing the sequence $s_1, \ldots, s_7$, we show also the resulting intermediate subsequences. We show the corresponding ORDST and its arched-arrow representation (shown under the tree). (b) A spanning tree that cannot be an ORDST, along with the duplication events and the resulting subsequences.

The duplication history of a sequence of units describes a series of duplication events that created the observed sequence of units. Here, we represent these histories by *ordered directed spanning trees* (ORDSTs). The nodes of an ORDST are labeled with units of S, and an edge from s' to s'' corresponds to a duplication event where the unit s' is duplicated to $s's''$ or $s''s'$. (For brevity, we will write "node s" to mean the "node labeled with s".)

Figure 2 (a) shows an example of an ORDST, its arched-arrow representation, and the corresponding events for generating it. Note that the resulting sequences after each duplication event are subsequences of S; we call this property *the ordering constraint.* However, not every spanning tree on S is an ORDST describing the duplication history. For example, in the same figure part (b) we show a subtree that cannot lead to an ORDST. This is because the duplication of the unit s_2 produces s_6. Note that the resulting sequence s_1, s_2, s_6, s_3, s_4 is not a subsequence of S. In terms of the single copy duplication model, it is impossible to duplicate the unit s_2 to produce s_6 after the emergence of s_3 or s_4. That is, the ordering constraint preserves the properties of the evolutionary mechanism. The following lemma specifies properties of the ORDST used in our algorithm.

Lemma 2.1. *For an ORDST of n units, the following statements are equivalent:*

a. *For each node s_k in an ORDST, we can write the nodes of its full subtree as $S[i..j]$, $1 \le i \le k \le j \le n$. Moreover, this subtree can be divided into two subtrees sharing the root s_k and partitioning the interval $[i..j]$ into $S[i..k]$ and $S[k..j]$.*
b. *Let the interval $[i..j]$ include the nodes in the subtree of node s_k, and let the intervals $[l_1..r_1]$, .., $[l_t..r_t]$ include the subtrees of the child nodes of s_k. Then $k \notin [l_i..r_i]$, and $[l_i..r_i] \cap [l_j..r_j] = \phi$ for all $1 \le i, j \le t, i \ne j$.*
c. *Let the interval $[i..j]$ include the nodes in the subtree of node s_k, and let s_l be a descendant node of s_k. If every element in $S[l+1..j]$ is neither an ancestor nor a right brother of s_l, we can divide the tree over $[i..j]$ at node s_l into two subtrees intersecting at s_l: The first includes the nodes $S[i..l]$ with s_k as a root, and the second includes $S[l..j]$ with s_l as a root.*

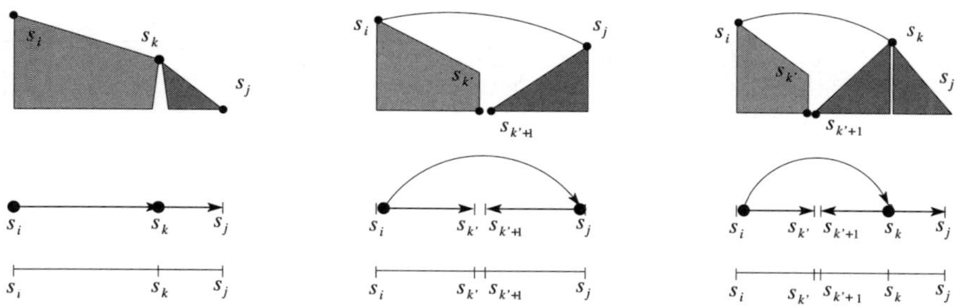

Fig. 3. The interval partitioning of the recurrence (3.1). The horizontal arrows correspond to *left*-ORDMST or *right*-ORDMST extending from the arrow beginning to its end. The arched arrows correspond to a duplication event between two units. The black circles correspond to roots of trees. (a) The recurrence for $C_l^A(i,j)$ partitioning the tree at node s_k. (b) The recurrence for $C_l^B(i,j)$. (c): A general topology for *left*-ORDMST.

Proof. This follows from the ordering constraint on the units of S. □

Definition 2.1. Each edge $s' \to s''$ in an ORDST for a sequence S is assigned a duplication cost $d(s', s'')$, where $s', s'' \in S$. The *ORDST cost* is the total sum of all duplication costs. An optimal ORDST for S is one of minimum cost.

3. Left- and Right Ordered Directed Spanning Trees

The alignment of minisatellites includes duplication events, where the duplicated units originate from the left- or rightmost unit of the interval containing the duplications. Therefore, we compute for each interval two optimal histories: one originated from the leftmost unit and the other from the rightmost unit. For ease of presentation, we will first present an algorithm that computes optimal histories from duplications only, without insertions. Then we will extend this algorithm to incorporate such insertions.

3.1. *Computation of optimal trees without insertion*

Based on Lemma 2.1, any ORDST can be expressed in terms of contiguous intervals. This suggests that a dynamic programming algorithm can be used to construct an optimal tree by recursively searching for an optimal partitioning of the intervals enclosing the units.

Given an interval $S[i..j] \subseteq S[1..n]$, we call an ORDST over this interval *left*-ORDST if the root of the respective tree is the leftmost unit s_i, and we call it *right*-ORDST if the root is the rightmost unit s_j. Let $C_l(i,j)$ denote the cost of a minimum *left*-ORDST defined over the units $S[i..j]$. Similarly, let $C_r(i,j)$ denote the cost of a minimum *right*-ORDST over $S[i..j]$. The values of $C_l(i,j)$ and $C_r(i,j)$ can be computed iteratively by examining subintervals of $[i..j]$ as follows.

Figure 3 (c) shows the general decomposition of a *left*-ORDST. s_k denotes the rightmost son of s_i, and the overall *left*-ORDST decomposes into two *left*-ORDSTs

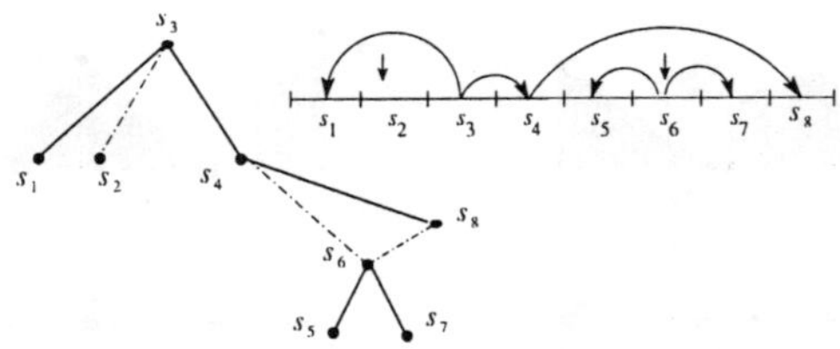

Event	subsequence
dup. $s_3 \rightarrow s_1$	$s_1\ s_3$
Ins. s_2	$s_1\ s_2\ s_3$
dup. $s_3 \rightarrow s_4$	$s_1\ s_2\ s_3\ s_4$
dup. $s_4 \rightarrow s_8$	$s_1\ s_2\ s_3\ s_4\ s_8$
Ins. s_6	$s_1\ s_2\ s_3\ s_4\ s_6\ s_8$
dup. $s_6 \rightarrow s_7$	$s_1\ s_2\ s_3\ s_4\ s_6\ s_7\ s_8$
dup. $s_6 \rightarrow s_5$	$s_1\ s_2\ s_3\ s_4\ s_5\ s_6\ s_7\ s_8$

Fig. 4. An ORDST for $S = (s_1, .., s_8)$ with the inserted units: $\{s_2, s_5, s_6, s_7\}$. The unit s_2 was inserted but did not undergo further duplications. s_6 was inserted then duplicated to s_5 and s_7. An ORDST with insertions should be regarded as a forest; the dash-dotted lines are virtual links between the forest roots and their potential parent nodes. On the left, the respective events. The resulting sequences are subsequences of S, i.e., the ordering constraint property is conserved

over intervals $S[i..k']$ and $S[k..j]$, and a *right*-ORDST over the interval $S[k'+1..k]$. However, this most natural decomposition leads to a recurrence of time complexity $O(n^4)$, by iterating over k and k'. Therefore, we use different decompositions to reach $O(n^3)$ time complexity.

Consider the two cases $k < j$ and $k = j$ in the interval partitioning of Figure 3 (c). Figure 3 (a) deals with the case $k < j$. We see a partitioning at s_k into two *left*-ORDSTs over $S[i..k]$ and $S[k..j]$, where the first one accounts internally for the cost of generating s_k from s_i. Figure 3 (b) deals with the case $k = j$. We see a *left*-ORDST for $S[i..k']$ and a *right*-ORDST for $S[k'+1..j]$. Neither of the two accounts for the generation of s_j from s_i, so this must be added in the decomposition. (In the recurrence given below for this case, k' will be renamed to k.) In other words, the decomposition in Figure 3 (c) can be regarded as the concatenation of the two cases in Figure 3 (b) and (a). These two cases cover all tree topologies, and we reach the following recurrences.

If $i = j$, then $C_l(i,j) = C_r(i,j) = 0$. If $i = j-1$, then $C_l(i,j) = d(s_i, s_j)$ and $C_r(i,j) = d(s_j, s_i)$. (Note that in general $d(s_i, s_j)$ can be different from $d(s_j, s_i)$.)

For $j - i > 1$, we have, according to the above case analysis,

$$C_l(i,j) = \min\{C_l^A(i,j), C_l^B(i,j)\} \tag{3.1}$$

where

$$\begin{aligned} C_l^A(i,j) &= \min_k\{C_l(i,k) + C_l(k,j)\} & i < k < j \\ C_l^B(i,j) &= \min_k\{C_l(i,k) + C_r(k+1,j) + d(s_i,s_j)\} & i \le k < j \end{aligned}$$

By symmetry, $C_r(i,j)$, can be computed as follows:

$$C_r(i,j) = min \begin{cases} \min_k\{C_r(i,k) + C_r(k,j)\} & i < k < j \\ \min_k\{C_l(i,k) + C_r(k+1,j) + d(s_j,s_i)\} & i \le k < j \end{cases}$$

Computing this set of recurrences requires, for every i and j, to iterate over all k, which has a total time complexity of $O(n^3)$. We store for every interval the optimal $C_l(i,j)$ and $C_r(i,j)$, which takes $O(n^2)$ space.

3.2. *Incorporating insertions in optimal trees*

In the case of an insertion, a unit is not derived from a neighboring unit, but imported as an unrelated DNA fragment. This inserted unit may undergo duplication events. Figure 4 shows an example of an ORDST with insertions over a sequence S. It is not difficult to see that Lemma 2.1 still holds for ORDST with insertions provided that the inserted units are taken into account. It is also clear that the tree topology itself is not affected by whether $s_x \in S$ is inserted or copied from some s_k. In other words, we may consider s_x as derived from some (arbitrary) s_k. The emergence of s_x has an insertion cost $I(s_x)$ instead of $d(s_k, s_x)$. To accommodate insertions in Recurrence 3.1, $d(s_i, s_j)$ must be replaced by $\min\{d(s_i, s_j), I(s_j)\}$. For $C_r(i, j)$, $\min\{d(s_j, s_i), I(s_i)\}$ replaces $d(s_j, s_i)$.

4. The Alignment Algorithm

4.1. *Formal model of map alignment*

Let two sequences $S = s_1, s_2, \ldots, s_n$ and $R = r_1, r_2, \ldots, r_m$ be given. In the alignment of two minisatellite maps, we have the following operations: (1) Match of two units s_i and r_j, (2) Indels in S/R, (3) left duplications in S/R, and (4) right duplications in S/R. Figure 5 (left) shows generation rules formalizing the alignment of of minisatellite maps.

Relating models for map alignment is difficult. For comparison, we also show in Figure 5 (right) the model of Bérard et al. It is not difficult to see that it lacks a rule for simultaneous right duplications in S and R. In the example shown, there is no rule for generating $R[5..6]$ from r_7, and at the same time generating $S[7..9]$ from r_7.We clarify this point with an additional example. Consider $S = ab$ and $R = dc$, where $d(a, b) = d(b, c) = d(c, d) < d(a, c) = d(b, d) < d(a, d)$. The optimal alignment is to match b with c. Then b produces a by right duplication, and c produces d by right duplication. In Bérard et al. model there is no way for generating d from c, while b producing a. Our model overcomes this minor omission. Note also that although the last two rules of Bérard et al. seem to have no counterpart in our model, their effect is achieved by a combination of match, left and right duplication rules.

4.2. *An algorithm for computing an optimal alignment*

Let two sequences $S' = s_1, s_2, \ldots, s_n$ and $R' = r_1, r_2, \ldots, r_m$, $m \leq n$, be given. For ease of presentation, we prepend the character \$ to both S' and R', i.e. the alignment algorithm runs for $S = \$S' = \$, s_1, s_2, \ldots, s_n$ and $R = \$R' = \$, r_1, r_2, \ldots, r_m$. From left-to-right the units of S appear at positions 0, 1, 2, ..., n in S, and similarly for R. The mutation and duplication costs between the unit \$ and any other unit in S' and R' is much larger than the costs between the units of S' and R', i.e., $d(\$, v) \gg d(v, v')$, where $v, v' \in \{S' \cup R'\}$. (The rationale for introducing this unit is to allow that prefixes of S' and/or R' appear as insertions.) Let $\mathcal{C}_s(l, i, x)$ ($\mathcal{C}_r(l', j, y)$) denote the cost of an optimal duplication history of the units $S[l..i]$ ($R[l'..j]$) originating from the unit s_x (r_y). Note that $\mathcal{C}_s(l, i, l)$ ($\mathcal{C}_r(l', j, l')$) corresponds to an optimal *left*-ORDST over the interval $S[l..i]$ ($R[l'..j]$), because the duplications are originated from the leftmost unit s_l ($r_{l'}$). Note also that $\mathcal{C}_s(l, i, i)$ ($\mathcal{C}_r(l', j, j)$) corresponds to

Our Model

Operation	Generation rule
Match:	$\genfrac{}{}{0pt}{}{A}{B} \rightarrow \genfrac{}{}{0pt}{}{s_i\ A}{r_j\ B}$
Insertion in S:	$\genfrac{}{}{0pt}{}{A}{B} \rightarrow \genfrac{}{}{0pt}{}{s_i\ A}{\ \ B}$
Insertion in R:	$\genfrac{}{}{0pt}{}{A}{B} \rightarrow \genfrac{}{}{0pt}{}{\ \ A}{r_j\ B}$
Left dup. in S and/or R:	$\genfrac{}{}{0pt}{}{A}{B} \rightarrow \genfrac{}{}{0pt}{}{\overrightarrow{S[l..i]}}{\overrightarrow{R[l'..j]}}\genfrac{}{}{0pt}{}{A}{B}$
Right dup. in S and/or R:	$\genfrac{}{}{0pt}{}{A}{B} \rightarrow \genfrac{}{}{0pt}{}{\overleftarrow{S[l..i]}}{\overleftarrow{R[l'..j]}}\genfrac{}{}{0pt}{}{A}{B}$
Termination:	$\genfrac{}{}{0pt}{}{A}{B} \rightarrow \genfrac{}{}{0pt}{}{\epsilon}{\epsilon}$

Example :

$s_1\ \overrightarrow{s_2\ s_3\ s_4\ s_5}\ s_6\ \overleftarrow{s_7\ s_8\ s_9}$

$r_1\ \_\ \sim\ \sim\ \sim\ r_2\ \overleftarrow{r_3\ r_4\ r_5}$

Bérard et al. Model

Operation	Generation rule
Insertion in S:	$\genfrac{}{}{0pt}{}{A}{B} \rightarrow \genfrac{}{}{0pt}{}{s_i\ A}{\ \ B}$
Insertion in R:	$\genfrac{}{}{0pt}{}{A}{B} \rightarrow \genfrac{}{}{0pt}{}{\ \ A}{r_j\ B}$
Left dup. in S and/or R:	$\genfrac{}{}{0pt}{}{A}{B} \rightarrow \genfrac{}{}{0pt}{}{\overrightarrow{S[l..i]}}{\overrightarrow{R[l'..j]}}\genfrac{}{}{0pt}{}{A}{B}$
Dup. in S from a unit in R:	$\genfrac{}{}{0pt}{}{A}{B} \rightarrow \genfrac{}{}{0pt}{}{\overleftrightarrow{S[l..i]}}{r_j}\genfrac{}{}{0pt}{}{A}{B}$
Dup. in R from a unit in S:	$\genfrac{}{}{0pt}{}{A}{B} \rightarrow \genfrac{}{}{0pt}{}{s_i}{\overleftrightarrow{R[l'..j]}}\genfrac{}{}{0pt}{}{A}{B}$
Termination:	$\genfrac{}{}{0pt}{}{A}{B} \rightarrow \genfrac{}{}{0pt}{}{\epsilon}{\epsilon}$

Example :

$s_1\ \overrightarrow{s_2\ s_3\ s_4\ s_5}\ s_6\ \overleftrightarrow{s_7\ s_8\ s_9}$

$r_1\ \sim\ \sim\ \sim\ \sim\ \overrightarrow{r_2\ r_3\ r_4}\ r_5$

Fig. 5. The generation rules produces the alignment from left to right, where A and B are the alignment strings for S and R. (Think of them as a Turing machine writing on two tapes A and B.) The units of the interval $S[l..i]$ ($R[l'..j]$) are produced by duplication events, and the arrows $\rightarrow$, $\leftarrow$, and $\leftrightarrow$ above them specify if the duplications originate from the leftmost, rightmost, or any unit in the interval, respectively. In the example under our model, the alignment is generated through the following rules: Matching s_1 to r_1, insertion of s_2, left duplication of $S[2..4]$ originated from s_2, matching s_6 to r_2, right duplication of $S[7..9]$ from s_9 and right duplication of $R[3..5]$ from r_5, and finally matching s_9 to r_5. On the right, we show the Bérard et al. model. In this model there is neither explicit match nor right duplication. Instead, the duplication in S (R) from a unit in R (S) includes these events. In the respective example, r_5 produces $S[7..9]$. If r_5 produces, say s_9, then we can tell that r_5 matches s_9 and there is a right duplication of $S[7..9]$ from s_9.

an optimal *right*-ORDST over the interval $S[l..i]$ ($R[l'..j]$), because the duplications are originated from the rightmost unit s_i (r_j). We have $\mathcal{C}_s(t,t,t) = \mathcal{C}_r(t',t',t') = 0$, where $t \in [0..n], t' \in [0..m]$. Moreover, let $M'(s',s'') = \min\{M(s',s''), M(s'',s')\}$.

Let $\mathcal{A}(i,j)$ be the cost of aligning $S[0..i]$ to $R[0..j]$, where $0 \le i \le n$ and $0 \le j \le m$. We have the boundary values $\mathcal{A}(0,0) = M(\$,\$) = 0$. $\mathcal{A}(i,j)$ is computed by the following recurrence (noting the limits of i, j):

$$\mathcal{A}(i,j) = \min \begin{cases} M'(s_i,r_j) + \mathcal{A}(i-1,j-1) & \forall\, i>0 \textbf{ and } \forall\, j>0 \\ \mathcal{C}_s(l,i,l) \quad + \mathcal{A}(l,j) & \forall l \in [0,i-1],\ i>0,\ j \ge 0 \\ \mathcal{C}_r(k,j,k) + \mathcal{A}(i,k) & \forall k \in [0,j-1],\ i \ge 0,\ j>0 \\ \mathcal{C}_s(t_s,i,i) + \mathcal{C}_r(t_r,j,j) + & \\ M'(s_i,r_j) + \mathcal{A}(t_s-1,t_r-1) & \forall t_s \in [1..i], \forall t_r \in [1..j], \forall i>0, \forall j>0 \end{cases}$$

The optimal duplication costs for each interval originated from the unit either on the left or right boundary are computed in a pre-processing step, using the algorithm of Section 3. This takes totally $O(n^3)$ time and $O(n^2)$ space. Note that indels are not explicitly incorporated in this recurrence, because the duplication histories already take them into account. For computing the recurrences including $\mathcal{C}_s(l,i,l)$ and $\mathcal{C}_r(k,j,k)$, one iterates for all i and j over all l (concurrently k), which takes $O(n^3)$ time. For computing the recurrence involving $\mathcal{C}_s(t_s,i,i)$ and $\mathcal{C}_r(t_r,j,j)$, one iterates for all i and j over all t_s and t_r. This naively takes $O(n^4)$, but the time complexity can be reduced to $O(n^3)$, as follows. The righthand side $\mathcal{C}_s(t_s,i,i) + \mathcal{C}_r(t_r,j,j) + M(s_i,r_j) + \mathcal{A}(t_s-1,t_r-1)$ can be rewritten as $\mathcal{C}_s(t_s,i,i) +$

$M(s_i, r_j) + \mathcal{A}'(t_s - 1, j)$, where $\mathcal{A}'(t_s - 1, j) = \mathcal{A}(t_s - 1, t_r - 1) + \mathcal{C}_r(t_r, j, j)$. If the cost $\mathcal{C}_s(t_s, i, i) + \mathcal{C}_r(t_r, j, j) + M(s_i, r_j) + \mathcal{A}(t_s - 1, t_r - 1)$ is to be minimal, then $\mathcal{A}'(t_s - 1, j)$ must also be minimal. If the value $\mathcal{A}'(t_s - 1, j)$ is already precomputed and stored, then it takes only $O(n^3)$ time to compute the respective recurrence. To this end, an optimal $\mathcal{A}'(t_s - 1, j)$ is computed earlier when computing the alignment $\mathcal{A}(t_s - 1, j)$. For given t_s and j, this computation minimizes over t_r. Altogether, for computing $\mathcal{A}(i, j)$, one has to pre-compute and store an optimal $\mathcal{A}'(t_s - 1, j)$; which takes totally $O(n^3)$ and requires one extra table. Adding the complexity of the preprocessing step, the total complexity is $O(n^3)$ time and $O(n^2)$ space.

The run length encoding (RLE) scheme can be readily incorporated to further reduce the time complexity to $O(n^2 + nn'^2 + n'^3)$, where n' is the length of the run-length compressed sequence. We layout this improvement in Appendix I.

5. Experimental Results

5.1. *Performance evaluation*

For constructing the duplication history, we compare our algorithm to the alphabet-dependent algorithm using different alphabet sizes. We use random sequences between 80 and 60 units long, where each unit can duplicate at most 50 times. Figure 6 (left and middle) shows the results of the experiment over 1000 such random sequences. It is clear that our algorithm is invariant against the alphabet size, while the other is linearly dependent. We show also the results of applying the run-length encoding (RLE) scheme. The usage of RLE scheme has actually attenuated the running time of both algorithms, but our algorithm is still invariant against the alphabet size. Naturally, the effect of RLE decreases when increasing the alphabet *without* also increasing sequence length.

Regarding the alignment phase, we compared our algorithms `MSATcompare` (without RLE) and `MSATcompareRLE` (with RLE) to the only existing program for this task, `MS_ALIGN` [6], that runs without RLE. The table on the right of Figure 6 summarizes the comparison results for simulated and the largest real datasets. Our algorithms are faster than `MS_ALIGN`. Clearly, `MSATcompareRLE` is superior to other algorithms: It reduced the time of analyzing large datasets to a few minutes, including output of results. It is worth mentioning that for the MSY1 dataset our algorithms and `MS_ALIGN` yielded identical pairwise scores; that is, the limitation in the alignment model of Bérard et al. discussed in Section 4.1, did not matter. However, for other datasets, this may not be the case.

5.2. *Detecting directional duplication bias in minisatellites*

We analyzed the updated (and the largest available) dataset MSY1 [2, 12] that contains 465 maps of individuals from different populations. We excluded repeated maps (to have a non-redundant dataset) and also excluded those maps including ambiguous unit types. In the remaining 345 maps, the number of distinct unit types, including *null* repeats, is eight, i.e., the alphabet size is eight. The types are assigned

Without RLE		
$\|\Sigma\|$	Dep.	Indep.
5	147	65
10	262	65
20	472	61
30	703	65
50	1165	65
60	1428	67

With RLE		
$\|\Sigma\|$	Dep.	Indep.
5	0.46	0.46
10	0.59	0.55
20	0.95	0.59
30	1.11	0.56
50	1.5	0.48
60	1.7	0.6

Data	Algn. No.	MS_ALIGN	MSATcompare	MSATcompareRLE
rand 50	1225	5.58	2.3	0.23
rand 100	4950	24.2	10.2	0.98
rand 150	11175	49.8	21.4	2.1
rand 250	3112	161.5	70	5.9
rand 350	61075	317	140	12
MSY1 345	59340	87	25	4.8

Fig. 6. Left and middle tables: The running times, in *seconds*, of constructing duplication history for different alphabet sizes $|\Sigma|$ without and with RLE scheme, respectively; this is measured on a PC with 1.5GHz CPU and 512M RAM. The columns "Dep." and "Indep." are for Alphabet-dependent and independent algorithms, respectively. Right table: Running times, in *minutes*, of our algorithms compared to MS_ALIGN. The first four rows contain results for random data, and the last row is for the MSY1 real dataset. ("rand 50" means 50 random sequences). The second column contains the number of pairwise alignments for each set of sequences.

the codes $\{null, 1, 1a, 2, 3, 3a, 4, 4a\}$. The pairwise hamming distances d_H between the units (except for *null*) range between 1 and 3. The *null* repeats are unidentified types due to further base substitutions [2]; hence, we assumed it has $d_H = 4$ to other units. We consider three versions of this dataset. The first includes all 345 unique maps. The second (329 maps) excludes maps with more than 3 adjacent *null* repeats, as suggested in [6]. The third (249 maps) contains no *null* repeats.

The MSY1 dataset was previously analysed for studying population evolution [6]. Here, we ask whether the units duplicate at the $5'$ end (to the left) more than at the $3'$ end (to the right), or not; i.e., to verify the assumption given in [2]. In terms of our alignment model, Subsection 4.1, we want to examine if *left and right duplications* contribute equally to the duplication history. To answer this question, we measured the effect of ignoring left/right duplications in all pairwise alignments by comparing the respective costs to the optimal costs considering both kinds. The rationale is that if both kinds contribute equally, we obtain nearly the same number of alignments with increased cost. Otherwise, the numbers will be different. To this end, we used five cost schemes given in Figure 7 (left). These schemes sample the range from 1 to ∞ of the ratio M/DUP. The fourth scheme is the one recommended by [6], and the final scheme reduces the comparison of two maps to the comparison of their modular structures [2], e.g., the map "*aaabbc*" reduces to "*abc*".

We then ran three experiments under these cost schemes: In experiment E_{full}, we performed all-against-all comparisons over the above-mentioned three dataset versions. In Experiment E_{left}, we allowed only left-duplications (achieved by switching off the recurrence corresponding to right duplications). In experiment E_{right}, we allowed only right duplications (achieved by switching off the recurrences corresponding to left duplications). In the latter two experiments, we counted the number of alignments whose costs are higher than in E_{full}. All results of running E_{left} and E_{right} compared to E_{full} using various distance functions and versions of the dataset are shown in Figure 7 (right).

Interestingly, the number of alignments with optimal cost increased relative to E_{full} is clearly smaller in E_{right} than in E_{left}. Since our model is symmetric, this

	(M,DUP,I)
1	$(1 \times d_H, 1, 40)$
2	$(2 \times d_H, 1, 40)$
3	$(5 \times d_H, 1, 40)$
4	$(10 \times d_H, 1, 40)$
5	$(1 \times d_H, 0, 40)$

Dataset with	Total Algn.	$r=1 \times d_H$		$r=2\times d_H$		$r=5\times d_H$		$r=10\times d_H$		$r=\infty$	
		L	R	L	R	L	R	L	R	L	R
nulls	59340	186	0	616	16	3005	57	1977	127	3219	107
max. 3 *nulls*	53956	148	0	398	0	2403	8	1487	10	2604	44
no *nulls*	30876	0	0	0	0	869	0	876	0	1040	0

Fig. 7. Left: The cost schemes used in our experiments. The triple (M,DUP,I) denotes the costs for mutation $M(x,y)$ as a function of $d_H(x,y)$, duplication $DUP(x)$, and insertion $I(x)$, $x,y \in \{null, 1, .., 4a\}$. For all schemes, the insertion cost $I = 40$. Right: The results for three versions of the dataset with the inclusion of all, at most three, and no *null* types. The 2nd column contains the total number of pairwise alignments for each dataset version. The other table entries are the number of alignments with costs higher than the optimal under different cost schemes, where $r = M/DUP$ is the ratio between the mutation and duplication costs. The columns titled with "L" and "R" correspond to E_{left} and E_{right}, respectively.

directly suggests a bias *in vivo* in the contribution of left- and right-duplications. To check against artefacts, we generated sequences representing minisatellites of random structural variations. The result was as expected: The number of optimal alignments with increased cost in E_{left} and E_{right}, compared to E_{full}, over this random dataset is nearly the same.

Back to the MSY1 dataset of [2], we can state that left and right duplications do not contribute equally to the duplication history, which (1) supports the idea that the types appear non-randomly at that locus and they are generated at the 5′ ends with a limitation regarding type mutations [2], or (2) assumes the existence of further unkown dynamic constraints limiting the duplication of the MSY1 units. This observation calls for closer investigation.

References

[1] S. Bérard and E. Rivals, *Computational Biology* **10**, 357 (2003).

[2] M. Jobling, N. Bouzekri and P. Taylor, *Human Molecular Genetics* **7**, 643 (1998).

[3] C. Alkan, J. Bailey, E. Eichler and et al., *Genome Informatics* **13**, 93 (2002).

[4] B. Behzadi and J. M. Steyaert, An improved algorithm for generalized comparison of minisatellites, in *Proc. of the 14th CPM, LNCS 2676*, (Morelia, Mexico, 2003).

[5] B. Behzadi and J. M. Steyaert, The minisatellite transformation problem revisited: A run length encoded approach, in *Proc. of the 4th WABI, LNBI Vol. 3240*, (Bergen, Norway, 2004).

[6] S. Bérard, F. Nicolas, J. Buard and et al., *Evolutionary Bioinformatics* **2**, 327 (2006).

[7] D. Jaitly, P. Kearney, G. Lin and et al., *Computer and Systems Sciences* **65**, 494 (2002).

[8] O. Gascuel, M. D. Hendy, A. Jean-Marie and et al., *Systematic Biology* **52**, 110 (2003).

[9] M. Tang, M. Waterman and S. Yooseph, *Computational Biology* **9**, 429 (2002).

[10] L. Zhang, B. Ma, L. Wang and et al., *Bioinformatics* **19**, 1497 (2003).

[11] G. Benson and L. Dong, Reconstructing the duplication history of a tandem repeat, in *Proc. of the 7th ISMB*, (Heidelberg, Germany, 1999).

[12] N. Bouzekri, P. Taylor, M. Hammer and et al., *Human Molecular Genetics* **7**, 655 (1998).

Appendix I: The Inclusion of Run-length Encoding Scheme

In Run-Length Encoding (RLE) of sequences, i consecutive occurrences of symbol x is shown by x^i, which is called a *run*. For example, $S = aaaabbbbcccabbbcc$ is encoded as $a^4b^4c^3a^1b^3c^2$. For brevity, we call the sequence of symbols in the encoded sequence the *compressed sequence* and denote it with S', in the previous example $S' = abcabc$. Minisatellites are ideal patterns for using run-length encoding technique, because they consists of a large number of tandem repeats. As a result, the length of run-length encoded representation of minisatellite sequences is generally much shorter than the initial length.

The RLE technique helps us in both computing the duplication history and the alignment algorithm. If we compute the duplication histories for the RLE version of strings, then $C_l(i,j)$ and $C_r(i,j)$ for the non-compressed sequences, can be computed in constant time for any i and j. More precisely, let $e(i)$ denote the position of $S[i] = x_i$ in the compressed sequence S', and let $\hat{C}_l(i,j)$ ($\hat{C}_r(i,j)$) denote the cost for constructing the history in S'. If $DUP(x_i)$ is constant for all x_i, then $C_l(i,j) = \hat{C}_r(e(i),e(j)) + (j-i) - (e(j) - e(i))$. Otherwise, $C_l(i,j) = \hat{C}_r(e(i),e(j)) + F(j) - F(i) - \sum_{y=e(i)}^{y=e(j)} DUP(S'[y])$, where $F(i) = \sum_{r=1}^{r=i} DUP(s_r)$. Note that $e(i)$ and $F(i)$ can be pre-computed in linear time.

The score $\mathcal{A}(i,j)$ in the alignment algorithm can be correctly determined without iterating over all possible values of k, l, t_s and t_r. More precisely, it is enough to iterate over the block boundaries in S and R. Moreover, these iterations are launched only if i or j are block boundaries. We show how this works for the second clause that involves left duplications in S in the alignment recurrence of Subsection 4.2. Let $\hat{E}_l(i)$ and $\hat{E}_r(i)$ be the set of positions in S of all leftmost and rightmost boundaries of each block before $s_i \in S$. Let $\hat{E}(i) = \hat{E}_l(i) \cup \hat{E}_r(i)$ be an ordered list, w.r.t. positions in S, and let E_x denote the xth entry in this list. During the iteration, if $s_i = s_{i-1}$, i.e., s_i is not the leftmost unit of a block. Then it is enough to consider only one value of l, namely $l = i-1$. The idea is that $s_{i-1} = s_i$, and the iteration over other values of $l \in [0..i]$ yields no better score. If $s_i \neq s_{i-1}$, it is enough that l iterates over all values of $\hat{E}(i)$. The idea is as follows: Let $l_x < i$ denote the position of an element within a block E_x (E_x is the position of the leftmost unit of block $x/2+1$), then $\mathcal{C}(l_x, i, l_x) = \mathcal{C}(E_{x+1}, i, E_{x+1}) + Dup(l_x) \times (E_{x+1} - l_x)$, where E_{x+1} is the rightmost unit. The optimal value $A(l_x, j) + \mathcal{C}(l_x, i, l_x) = A(l_x, j) + \mathcal{C}(E_{x+1}, i, E_{x+1}) + Dup(l_x) \times (E_{x+1} - l_x) = A(E_{x+1}, j) + \mathcal{C}(E_{x+1}, i, E_{x+1})$. The time complexity is derived as follows: for each s_j, we examine each s_i and s_{i-1}. We iterate over all the block boundaries in S only if i is a block boundary. This yields $O(n^2 + nn'^2)$ time. With similar arguments, we can prove how the RLE scheme works for the other clauses in the recurrence.

Computing the duplication histories for the RLE version of strings using our algorithm takes $O(n'^3)$. The aligment phase will take $O(n^2 + nn'^2)$. That is, our algorithm in this paper with the RLE techniques takes $O(n^2 + nn'^2 + n'^3)$ time and $O(n^2)$ space, which is also independent of the alphabet size.

METABOLIC PATHWAY ALIGNMENT (M-PAL) REVEALS DIVERSITY AND ALTERNATIVES IN CONSERVED NETWORKS

YUNLEI LI,
DICK DE RIDDER , MARCO J. L. DE GROOT and MARCEL J. T. REINDERS

Information and Communication Theory Group
Faculty of Electrical Engineering, Mathematics and Computer Science
Delft University of Technology, Mekelweg 4, 2628 CD Delft, The Netherlands

We introduce a comparative analysis of metabolic reaction networks between different species. Our method systematically investigates full metabolic networks of multiple species at the same time, with the goal of identifying highly similar yet non-identical pathways which execute the same metabolic function, i.e. the transformation of a specific substrate into a certain end product via similar reactions. We present a clear framework for matching metabolic pathways, and propose a scoring scheme which combines enzyme functional similarity with protein sequence similarity. This analysis helps to gain insight in the biological differences between species and provides comprehensive information on diversity in pathways between species and alternative pathways within species, which is useful for pharmaceutical and industrial bioengineering targets. The results also generate hypotheses for improving current metabolic networks or constructing such networks for currently unannotated species.

1 Introduction

The metabolic network of a species represents all known chemical reactions of metabolism within a cell. A single, relatively isolated cascade of such reactions is normally called a metabolic pathway. Most metabolic reactions are catalyzed by specific groups of enzymes. These enzymes are annotated by EC numbers[1], hierarchically organized numbers indicating the type(s) of reaction they catalyze. Studying the metabolic network is a powerful tool to elucidate the cellular machinery. Therefore, it has been an active research field for the last decade[2-13].

Comparing pathways between multiple species provides valuable information to understand evolutionary conservation and variation. Kelly *et al.*[14] align protein interaction networks and predict protein function and interaction using conserved pathways. We extend their alignment concept to the metabolic level, to discover conserved metabolic pathways. Such a pathway transforms a specific substrate into a specific end product via very similar reactions in multiple species. These reactions are similar since they have common substrates and common products. However, they may have different co-substrates or co-products, be catalyzed by different enzymes, need different numbers of reactions to complete the transformation, or reactions may occur in a different order.

Although many comparative analyses at the metabolic level have been performed, little work focuses explicitly on the discrete differences between conserved pathways, and to our knowledge no global search has been carried out yet. For example, Forst *et al.*[5] perform a phylogenetic analysis on four pre-chosen pathways by combining the sequence information of a set of enzymes and gene-coded metabolites in a pathway. Dandekar *et al.*[6] also limit their study, to the glycolysis pathway. As for the similarity measure for matching pathways, Tohsato *et al.*[7] align pathways based on enzyme EC number similarity, discarding information on the involved metabolites. In Clemente *et al.*[8, 9], sets of reactions in multiple pathways are compared, omitting connectivity between the reactions.

Inspired by the PathBLAST algorithm of Kelly *et al.*[14], we propose a novel approach to align metabolic pathways. Our method, Metabolic Pathway ALignment (M-Pal), aligns entire metabolic networks of different species in order to explore highly conserved pathways. In the resulting aligned pathways, most reactions are identical; the remaining reactions are not identical, yet similar. These conserved pathways are very likely to be essential or efficient pathways. More importantly, our method sheds light on differences between species in the use of non-identical but similar reactions, revealing between-species diversity and within-species alternatives. We introduce *diversity* in a pathway as a term indicating that each species has its own unique mechanism to allow a certain biochemical transformation to take place. If both species share a common reaction, but one of the species has a second, unique reaction to perform the same transformation, then this last transformation forms part of a unique *alternative* pathway.

Diversity and alternatives across species give insight into biological differences between species, provide potential candidate enzymes for bioengineering, and generate hypotheses on missing enzymes or incorrect annotations in current metabolic networks. Moreover, the resulting pathways give more options in pathway engineering and constructing metabolic networks for unannotated species. Finally, this method unites reactions in isolated metabolisms into a large network, relating reactions with upstream substrates and downstream products which might be elusive if we only look at a subset of the network.

We apply M-Pal to *Saccharomyces cerevisiae* and *Escherichia coli*, and find 2518 short conserved pathways. In each conserved pathway, 4–5 reactions from one species are aligned with similar reactions from another species. Among the results, ~1500 pathways are diverse or contain unique alternative enzyme activities. We categorize the differences between pathways and refine the search result by scoring each pathway according to functional and sequence similarity of the enzymes involved. This scoring scheme enables us to focus on highly conserved pathways with similar enzymes. We show that a number of metabolic annotations can be attached to each of the resulting pathways, demonstrating the strength of our systematic search in unearthing novel cross-links in metabolic networks.

We describe M-Pal in detail in Section 2. The results are presented and discussed in Section 3. Section 4 ends with some conclusions and an outlook to further work.

2 Methods

Since we seek to investigate diversity and alternatives in highly conserved metabolic pathways, we align the pathways from two species into a conserved pathway in a rather strict way. That is, we align two pathways only if most of the involved reactions in this two species use similar enzymes to catalyze common substrates into common products, introducing only a limited amount of freedom into the alignment. More specifically, let P_1 and P_2 denote two metabolic pathways in two species containing reactions $[R_{11}, R_{12}\ldots R_{1L}]$ and $[R_{21}, R_{22}\ldots R_{2L}]$, respectively. P_1 and P_2 can be aligned into a conserved pathway only if the individual reactions are aligned in the right order. That is, R_{11} is aligned with R_{21}, R_{12} is aligned with R_{22} *etc.* until R_{1L} is aligned with R_{2L}. We call each pair of matching reactions, e.g. R_{11} and R_{21}, a *building block.*

Given the restrictions mentioned above, we propose an efficient matching mechanism which constructs all building blocks first, and then assembles them into pathways of a desired length, taking reaction directions into account. After the aligned pathways are obtained, we compute an enzyme similarity score for each aligned pathway. In this way, we eventually get a list of conserved pathways, ordered by this score.

This sequential procedure of matching and scoring (see Figure 1) ensures the search for all matching pathways is complete and allows for a flexible scoring function. The exhaustive search results can be pre-computed and, as scoring is performed separately, no potential match will be missed because of prematurely discarding a pathway in the search. Our method is explained in detail in the remainder of this section.

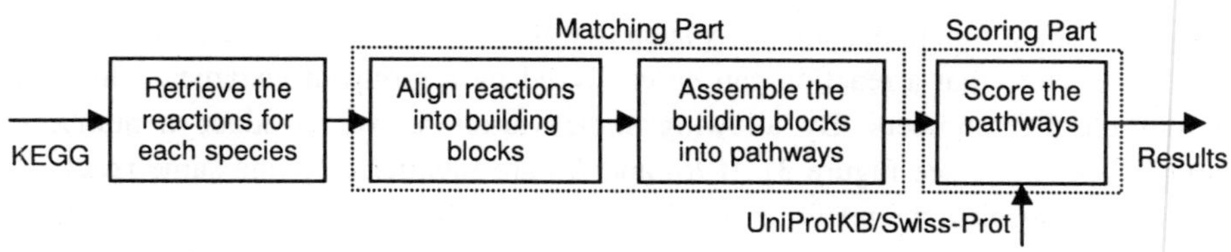

Figure 1. M-Pal flow chart.

2.1 *Reaction Retrieval*

We obtained the general reaction definitions from Release 42.0 of the KEGG LIGAND composite database[15], updated on May 14, 2007. For each species, we acquired the subset of reactions present in that species, together with the EC numbers and ORF names of the enzymes which catalyze each reaction, from the KEGG/XML and KEGG/PATHWAY databases.

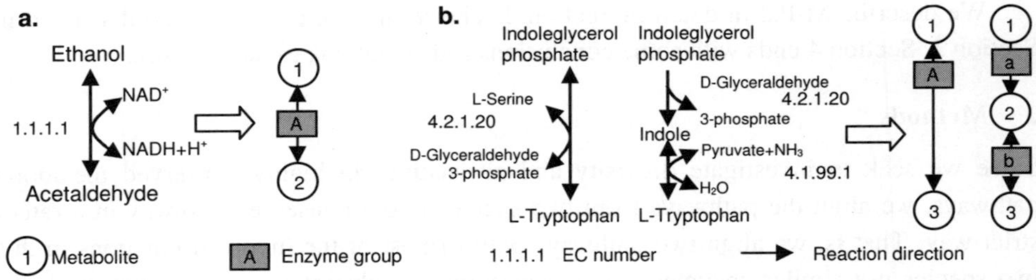

Figure 2. Reaction representation. **a.** Illustration of two representations of reactions in our method. **b.** One reaction from *S. cerevisiae* (on the left) and two reactions from *E. coli* (on the right) share a common substrate (Indoleglycerol phosphate) and product (L-Tryptophan). This situation forms one "gap", i.e. the difference in the number of reactions to transform Indoleglycerol phosphate into L-Tryptophan is one.

In M-Pal, reactions are represented as a combination of the classic "enzyme-centric" and "compound-centric" representations. Thus, a reaction is represented by all elements involved: metabolites, (a group of) enzyme(s), and its direction. Figure 2a gives an example. To allow us to compare reactions from different species, we plot them next to each other, with the matching substrate or product in the same row. Sometimes, a single reaction and a series of reactions connected in tandem may share common substrates and products. This introduces "*gaps*", indicating that the number of reactions to transform the specific substrates into the specific products differs between species. Figure 2b illustrates this: one reaction from *S. cerevisiae* and two reactions from *E. coli* form a "gap".

2.2 Building Block Alignment

Two reactions R_{1l} and R_{2l} can be aligned to form a building block when they have a common substrate and a common product, and at least one pair of enzymes (one from each species) share functional similarity such that the first two digits of their EC numbers are the same. Note that a reaction can be catalyzed by a group of enzymes, which may have multiple EC numbers. By allowing some variation, we introduce a number of *building block types* (see Figure 3). If R_{1l} and R_{2l} are identical, i.e. the same reaction is present in both species, the resulting building block is called "*identical*" (*i*). If R_{1l} and R_{2l} are different reactions, because of different co-substrates or co-products according to the definition in Section 2.1, they form a "*direct*" building block (*d*). To incorporate alternative pathways, evolutional diversity and annotation errors, we also allow one "mismatch" or one "gap" in a building block. Thus, in an "*enzyme mismatch*" building block (*em*), the first two digits of the EC numbers of the enzymes involved are not the same. The building blocks containing one "gap" are "*direct-gap*" (*dg*) and "*enzyme mismatch-gap*" (*eg*). Furthermore, we include "*enzyme crossover match*" building blocks (*ec*) to accommodate possible variation in the order of the catalyses: there are two reactions in each species sharing common substrates and end products with the EC numbers of the first and second reaction in one species being similar to those of the second and first reaction in the other species, respectively.

To summarize, the reaction alignment method described above results in six types of building blocks, each containing one or two reactions from each species. Note that 26 "current metabolites"[10,13], listed below[a], were excluded from consideration as common substrate or product to avoid finding large numbers of trivial conserved pathways.

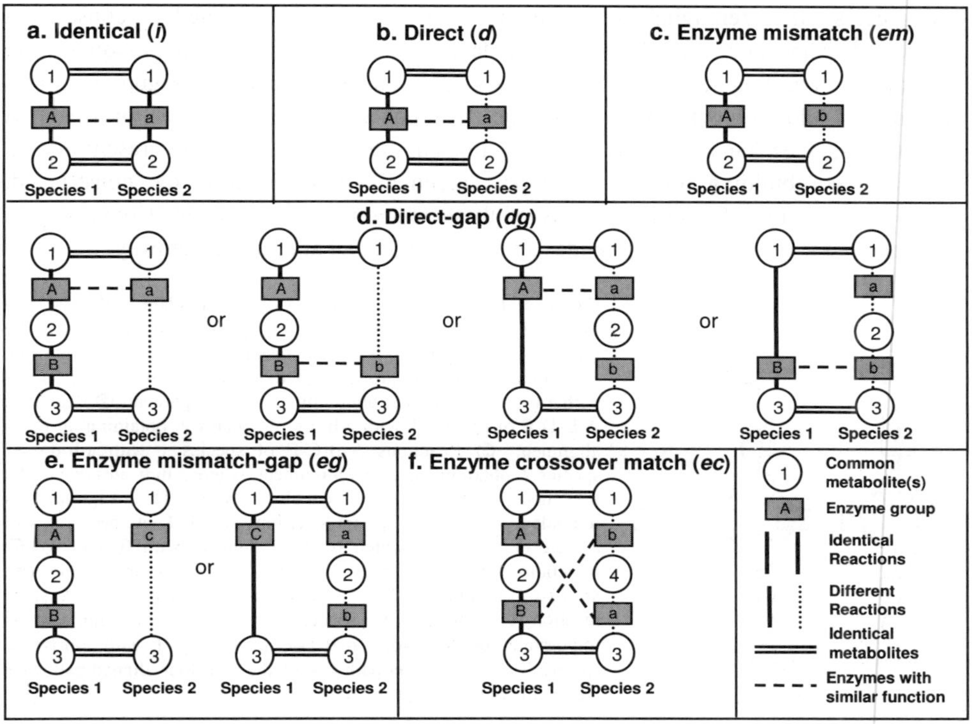

Figure 3. Illustration of the six types of building blocks. The reaction directions are omitted in the figure for simplicity. A dashed link is drawn between two groups of enzymes if they share the same first two digits of their EC numbers.

2.3 *Pathway Assembly*

Next, we focus on finding conserved short acyclic pathways. We only assemble four building blocks into a pathway, ensuring that one reaction does not appear more than once in a pathway. Moreover, we demand that out of these four building blocks, at least three must be of the type "identical" or "direct", representing the conserved part of the pathway. Only a single building block of type "enzyme mismatch", "direct-gap", "enzyme mismatch-gap" or "enzyme crossover match" is allowed in a pathway. Abbreviations are used to denote the pathway composition of building blocks regardless of the order,

---

[a] ATP, ADP, UTP, UDP, GTP, GDP, AMP, UMP, GMP, NAD, NADH,NADP, NADPH, Acetyl-CoA, CoA, Propanoyl-CoA, L-Glutamine, L-Glutamate, 2-Oxoglutarate, CTP, CDP, CMP, H_2O, CO_2, NH_2, Phosphate.

e.g. "*i-i-i-d*" indicates a pathway with three reactions of type "identical" and one of type "direct", in any order. In total, there are 21 such compositions possible for pathway alignment. These are used as 21 *pathway categories* in the discussion of our results.

To enhance the informativeness of our resulting set of pathways, we remove some redundant pathways. First, building blocks whose substrate and product are identical in one species (after removing current metabolites) will not be selected to construct a pathway. Furthermore, we reduce the redundancy in the result by enforcing uniqueness in choosing the building blocks of the five types other than "identical", see Figure 4. A non-"identical" building block can be chosen only if it contains at least one reaction absent in one of the species. This is because if all reactions in the building block are present in both species, two building blocks of type "identical" will already be constructed. Consequently, any other combinations of these reactions are redundant. Conversely, a reaction unique to one species provides an interesting novel alternative pathway.

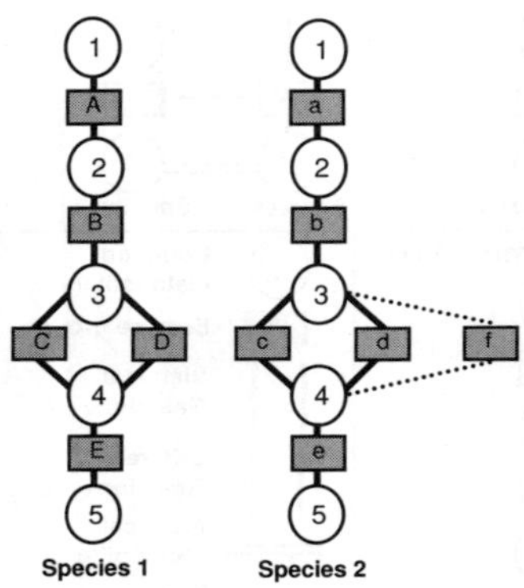

Figure 4. Illustration of the removal of redundant pathways. See Figure 3 for legends. Six possible pathway alignments can be induced in this example (each reaction is represented by the corresponding enzyme groups): ➊ Reactions A-B-C-E of species 1 with a-b-c-e of species 2, obtaining an *i-i-i-i* alignment. ➋ Reactions A-B-D-E of species 1 with a-b-d-e of species 2, obtaining an *i-i-i-i* alignment. ➌ Reactions A-B-C-E of species 1 with a-b-d-e of species 2, obtaining an *i-i-x-i* alignment, where *x* indicates one of the five non- "identical" building block types. This alignment is redundant with ① and ②. ➍ Reactions A-B-D-E of species 1 with a-b-c-e of species 2, which is also redundant with ① and ②. ➎ Reactions A-B-C-E of species 1 with a-b-f-e of species 2, obtaining an *i-i-x-i* alignment.. This is a novel alternative pathway, since reaction f is unique in species 2, hence *i-i-i-i* alignment is impossible. ➏ Reactions A-B-D-E of species 1 with a-b-f-e of species 2 also is a novel pathway. In the end, four aligned pathways are obtained: ①, ②, ⑤ and ⑥.

2.4 Scoring Function

Two factors indicate the extent to which an aligned pathway is conserved. One is the pathway category, i.e. the building block composition. For instance, we consider an "*i-i-i-d*" pathway to be more conserved than an "*i-i-i-dg*" pathway. The other factor is enzyme similarity, which we evaluate here based on functional similarity (EC numbers) and sequence similarity. Since they are not fully correlated, we integrate them to introduce a more informative measure of true orthology. In the following, we explain how to calculate functional similarity and sequence similarity of a building block, followed by their integration.

Given a building block containing one reaction from each species, enzyme functional similarity E_f is taken to be the maximum number of digits of EC numbers that the two groups of enzymes share. This is a simple and straightforward manner to measure enzyme functional similarity[12, 17, 18], since EC numbers form a functional hierarchy. Although more complex methods exist[7-9], their validity is still under research. Let the EC numbers in the reaction for species 1 be EC_{11}, EC_{12}, …, EC_{1m}, and for species 2 EC_{21}, EC_{22}, …,

EC_{2n}, we count the number of shared digits for each possible pair of EC numbers, and use the maximum as the functional similarity E_f for this building block. For "direct-gap" and "enzyme mismatch-gap" building blocks, for which one group of enzymes should be compared to two groups of enzymes, we compute E_f for both pairs of groups, and choose the larger E_f. For "enzyme crossover match" building blocks, E_f is taken to be the averaged value of the crossover enzyme group comparisons.

For the sequence similarity E_s between two reactions, we take the minimum BLAST E-value between all possible enzyme pairs. For "direct-gap" and "enzyme mismatch-gap" building blocks, E_s is computed between the two groups of enzymes which have the larger E_f. For "enzyme crossover match" building blocks, E_s is averaged. BLAST (version 2.2.15) is performed with e = 100 on the protein sequences in UniProtKB / Swiss-Prot Release 51.6.

After computing the E_f and E_s scores for all building blocks in a pathway, we sum all E_fs in a pathway and transform the result into a score $S_f \in [0, 1]$; likewise for all E_ss in the pathway to obtain $S_s \in [0, 1]$. Tables 1 and 2 detail these transformations. Since the original values of E_f and E_s have very different ranges, this transformation step actually scales these two measures into the same range in a sensible way, so that they are comparable and easy to combine. The intervals in the transformation tables are chosen to reflect our objective in finding conserved pathways with similar enzymes: high functional similarity values are examined in more detail in the score. For sequence similarity, we focus on the traditional cutoff value 10^{-2} for weak sequence similarity[14], thus the intervals around 10^{-2} are smaller than those for high sequence similarities. We do not restrict ourselves to highly similar sequences because our main interest is to reveal the alternatives and diversities in the pathways. Since the maximum value for E_s is 100 (due to the parameter setting used for BLAST), the intervals for $S_s \geq 0.8$ indicate the number of building blocks with very dissimilar enzyme sequences.

Table 1. Transformation of the total functional similarity $\sum_{b\in P} E_f(b)$ into the score S_f.

$\sum_{b\in P} E_f(b)$	16	15.5	15	14.5	14	13.5	13	12	11	10	8
S_f	0	0.1	0.2	0.3	0.4	0.5	0.6	0.7	0.8	0.9	1

Table 2. Transformation of the total sequence similarity $\sum_{b\in P} E_s(b)$ into the score S_s.

$\sum_{b\in P} E_s(b)$	$(0, 10^{-80})$	$[10^{-80}, 10^{-60})$	$[10^{-60}, 10^{-40})$	$[10^{-40}, 10^{-20})$	$[10^{-20}, 10^{-10})$	$[10^{-10}, 10^{-6})$
S_s	0	0.1	0.2	0.3	0.4	0.5
$\sum_{b\in P} E_s(b)$	$[10^{-6}, 10^{-2})$	$[10^{-2}, 100)$	$[100, 200)$	$[200, 300)$	$[300, \infty)$	
S_s	0.6	0.7	0.8	0.9	1.0	

Finally, the two scores are summed so as to combine the functional and sequence similarity:

$$S(P) = S_f\left(\sum_{b\in P} E_f(b)\right) + S_s\left(\sum_{b\in P} E_s(b)\right) \tag{1}$$

in which b denotes a building block and P denotes an aligned pathway. The lower this score, the more similar the enzymes in P are.

3 Results and Discussion

Of 881 enzymatic reactions in *S. cerevisiae* and 1106 in *E. coli*, 588 reactions are present in both species (Figure 5a). Based on the total of 1399 unique reactions, six types of building blocks are assembled into 2518 unique pathways of length 4. Figure 5b shows the number of reaction involved in the resulting pathways. Table 3 summarizes the number of building blocks of each type found. There results indicate that the reactions and building blocks in the resulting pathways reasonably cover all available reactions and building blocks, demonstrating the strength of our systematic search.

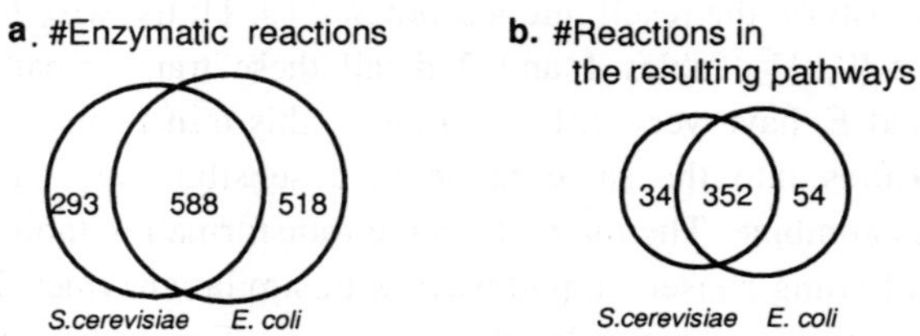

Figure 5. Venn diagrams showing **a.** the total number of enzymatic reactions in the two species and **b.** the number of reactions involved in the results.

Table 3. The number of each of the six types of building blocks.

Type	Identical (*i*)	Direct (*d*)	Direct-gap (*dg*)	Enzyme mismatch (*em*)	Enzyme crossover match (*ec*)	Enzyme mismatch-gap (*eg*)
#Building blocks	516	116	108	27	40	52
#Building blocks in the resulting pathways	352	67	64	11	12	29

For each pathway category containing a specific composition of building blocks, the total number of resulting pathways is shown in Figure 6a and their average functional similarity score S_f and sequence similarity score S_s are shown in Figure 6b. As shown in Figure 6a, ~1000 completely conserved pathways of type "*i-i-i-i*" are found. Not surprisingly, their enzyme sequences are highly similar, with BLAST *E*-values ranging from 10^{-10} to 10^{-6} on average. The pathway with the best score, 0, is depicted in Figure 7a. However, the variance of the sequence similarity score is also large, indicating that some reactions in these pathways do not have enzymes with similar sequences. This might arise because of different specificity, horizontal gene transfer, gene fusions, or the fact that only subunits of the enzymes are the same.

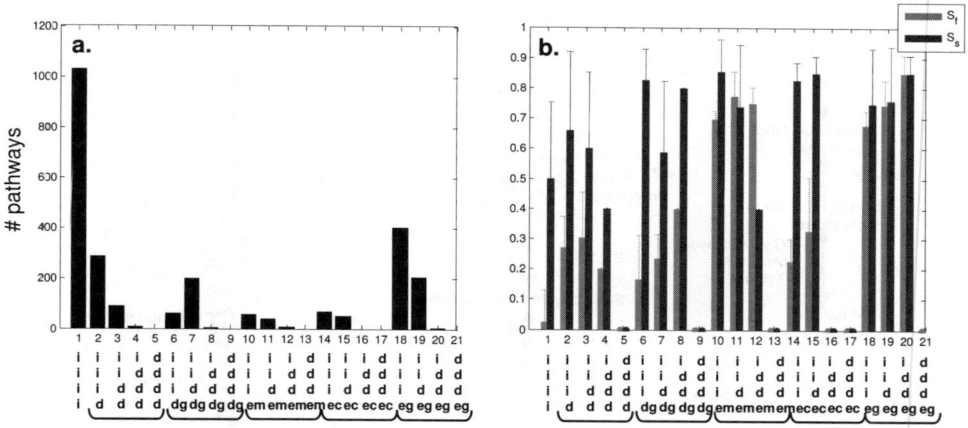

Figure 6. **a.** Total number of pathways in 21 pathway categories. Note that long conserved pathways may result in multiple short overlapping pathways. **b.** The average enzyme functional similarity score and sequence similarity score of each pathway category. Whiskers indicate standard deviations.

We also found ~1500 highly conserved pathways which contain some diversity between both species or unique alternatives within one species. Each of these pathways has a building block of type "direct", "direct-gap", "enzyme mismatch", "enzyme mismatch-gap", or "enzyme crossover match". Examples are given in Figure 7b-7f. These pathways are of great interest in bioengineering as they manifest the hidden information about pathway diversity and alternatives, which will not be found if we only look at a subset of the metabolic network in one species.

The results are useful in many applications. First, some resulting pathways suggest a more exact EC number annotation of their enzymes is possible and call for detailed comparison of the enzymes. For example, the enzymes in the pathways of type "*i-d-d-em*" in Figure 6b have dissimilar EC numbers, but their sequences are actually very similar (low S_s and high S_f). They might be incorrectly annotated, since they both transform a common substrate into a common product. Another example is given in Figure 7c, in which the enzymes with EC number 4.2.1.20 in *E. coli* (trpA and trpB) could also be annotated as 4.1.2.8, which is the α-subunit of 4.2.1.20. Comparing the enzymes in alternative pathways in different species can also be beneficial to understand their structural difference and relationship. In Figure 7c for instance, the two enzymes in *E. coli*, 4.2.1.20 and 4.1.99.1, might be different subunits of the enzyme 4.2.1.20 in *S. cerevisiae*. The same can be observed in Figure 6b, where the sequence similarity in the pathways with "*dg*" is generally worse than in those with "*d*" only, implying that the enzymes in "*dg*" are only subunits of the corresponding enzymes in "*d*".

Second, the results can help to understand diversity in metabolism and evolution. Reactions which are unique to one species are highlighted in Figure 7. Investigation of the biological difference between the two species is expected to explain their uniqueness

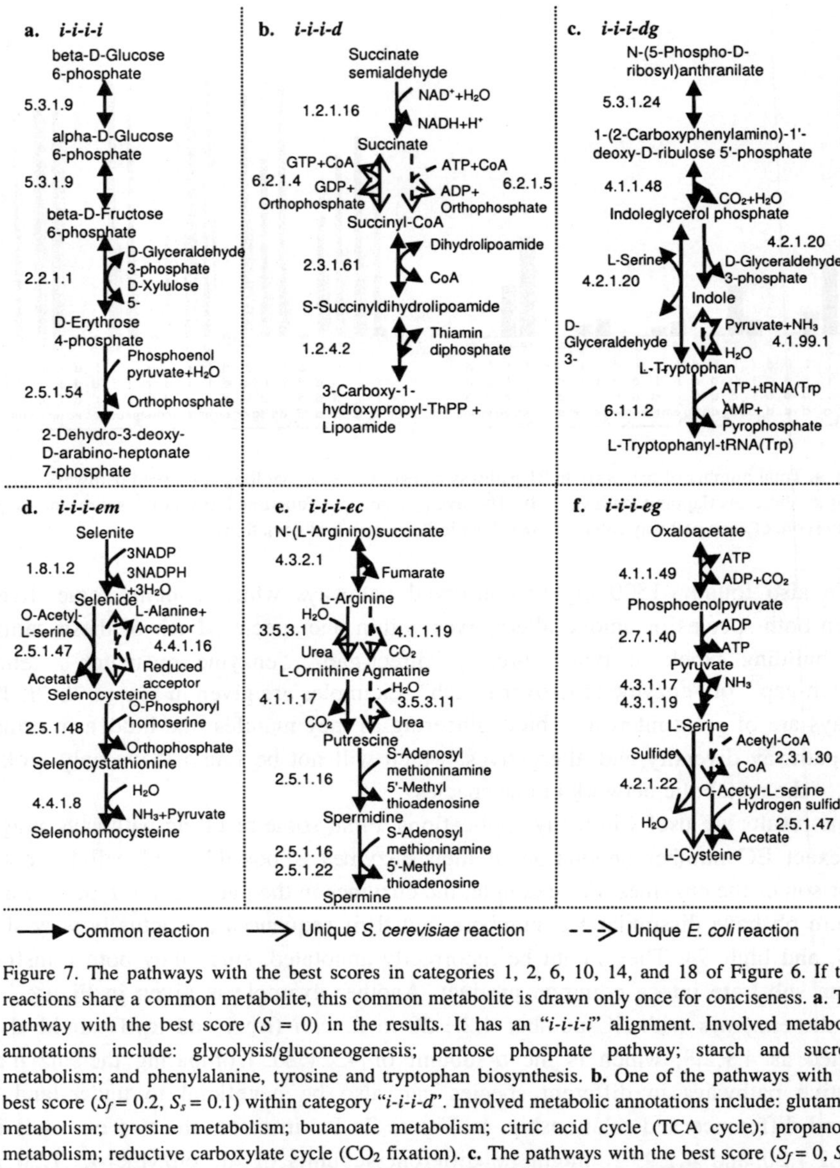

Figure 7. The pathways with the best scores in categories 1, 2, 6, 10, 14, and 18 of Figure 6. If two reactions share a common metabolite, this common metabolite is drawn only once for conciseness. **a.** The pathway with the best score (S = 0) in the results. It has an "*i-i-i-i*" alignment. Involved metabolic annotations include: glycolysis/gluconeogenesis; pentose phosphate pathway; starch and sucrose metabolism; and phenylalanine, tyrosine and tryptophan biosynthesis. **b.** One of the pathways with the best score (S_f = 0.2, S_s = 0.1) within category "*i-i-i-d*". Involved metabolic annotations include: glutamate metabolism; tyrosine metabolism; butanoate metabolism; citric acid cycle (TCA cycle); propanoate metabolism; reductive carboxylate cycle (CO_2 fixation). **c.** The pathways with the best score (S_f = 0, S_s = 0.5) within category "*i-i-i-dg*". Involved metabolic annotations include: phenylalanine, tyrosine and tryptophan biosynthesis; tryptophan metabolism; aminoacyl-tRNA biosynthesis; benzoxazinone biosynthesis; and nitrogen metabolism. **d.** One of the pathways with the best score (S_f = 0.7, S_s = 0.6) within category "*i-i-i-em*". Involved metabolic annotations include: selenoamino acid metabolism. **e.** The pathways with the best score (S_f = 0.2, S_s = 0.7) within category "*i-i-i-ec*". Involved metabolic annotations include: urea cycle and metabolism of amino groups; alanine and aspartate metabolism; arginine and proline metabolism; and beta-alanine metabolism. **f.** One of the pathways with the best score (S_f = 0.7, S_s = 0.3) within category "*i-i-i-eg*". Involved metabolic annotations include: TCA cycle; pyruvate metabolism; carbon fixation; glycolysis/gluconeogenesis; purine metabolism; glycine, serine and threonine metabolism; cysteine metabolism; and sulfur metabolism.

Further, we can project the knowledge to a new species. For instance, if the new species has the enzymes which catalyze a unique reaction of *S. cerevisiae*, then probably they are very closely related in the phylogenetic tree, and therefore share more common properties. Nevertheless, the revealed diversity might be an artifact of current metabolic network databases. Therefore it is recommended to examine whether the other species also has this unique enzyme, or whether some enzymes (and reactions) are missing in the pathways with "gaps". Another interesting result which might be worthy of further research is shown in Figure 6b, for the group containing enzyme crossover match building blocks (*ec*). Although the crossover enzymes have similar functions, their sequences are very dissimilar. Possible reasons could be that the enzymes have different substrate specificities, or the intermediate substrates are very different. They could also have been isoenzymes in parallel pathways, having become specialized to one species in evolution.

Third, the unique alternative pathways revealed by M-Pal provide potential candidate enzymes for bioengineering. Certain natural enzymes can be removed or changed so that we can choose between different alternative pathways, or enforce the reaction direction to produce the product of our interest. In the pathway shown in Figure 7c, *E. coli* has two alternative pathways to transform Indoleglycerol phosphate into L-tryptophan, one being reversible (catalyzed by 4.2.1.20) and the other one reported to be irreversible (catalyzed by 4.2.1.20 and 4.1.99.1). If the enzymes of 4.2.1.20 in the irreversible pathway are indeed also possibly annotated as 4.1.2.8, we can remove the 4.2.1.20 enzyme activity to enforce the direction towards producing tryptophan, which is an essential amino acid in human nutrition[16].

Finally, our results provide additional opportunities to construct the metabolic networks for currently unannotated species. As discussed above, our method points out possible missing enzymes and suggests related enzymes in well-studied species. The alternative pathways also provide more possibilities for optimizing the network to fit the found enzymes and reactions better.

4 Conclusions

The systematic search of M-Pal associates different parts of metabolic networks with each other and combines information from multiple species to discover diversity and alternatives in highly conserved pathways. The results shed light on the small differences found in the conserved pathways and provide useful information for many applications. Gene knock-out experiments can be performed to test our hypotheses, and the essentiality of the resulting pathways should be examined.

Our research is still at an early stage, and can be refined in a number of ways. Possible extensions include increasing the freedom in the alignment, e.g. allowing for more gaps or mismatches, further separated crossover matches, and longer pathways. This implies the search algorithm will have to become more sophisticated, as exhaustive enumeration will become infeasible. Next, the scoring function can be modified to prefer certain types of alignment. Non-identical metabolites could be included in the matching, implying a need for a compound similarity measure to be added to the scoring function.

The enzyme sequence similarity measure could also be refined using protein domain information. The current scoring mechanism assumes functional and sequence similarity is equally important. Weights could be added to model a trade-off between the two[8]. The scoring function itself could be enhanced by using a probabilistic framework such as in Kelly *et al.*[14], allowing us to look for relatively rather than absolutely conserved pathways and to attach a *p*-value to the pathways found. Other possible enhancements to the score are to take reversibility of reactions and the presence of isoenzymes into account.

Currently, this method is performed on two species only and is expected to give more informative results if applied on species not closely related. An extension could be to apply M-Pal on multiple species, at different evolutionary distances. We expect that larger differences will be found as evolutionary distance increases. The results will give insight to understand evolution and specialization, and will provide new building blocks and alternatives for pathway engineering. Applying this method for prediction of unannotated genes will be of great value. Finally, by relating different sets of enzymes in different species to a common metabolic function, this work provides an infrastructure based on which the regulatory factors can be associated, and functional hypothesis can be generated.

Acknowledgments

The authors would like to thank Rogier J. P. van Berlo, Domenico Bellomo, Wouter van Winden and Peter van Nes for their help and the constructive discussions. This work was part of the BioRange programme of the Netherlands Bioinformatics Centre (NBIC), which is supported by a BSIK grant through the Netherlands Genomics Initiative (NGI).

References

1. Nomenclature Committee of the International Union of Biochemistry and Molecular Biology (NC-IUBMB). http://www.chem.qmul.ac.uk/iubmb/enzyme/.
2. R. Overbeek, N. Larsen, W. Smith, N. Maltsev and E. Selkov, *Gene* **191**, GC1–GC9 (1997).
3. S. Schuster, D.A. Fell and T. Dandekar, *Nat. Biotechnol.* **18**, 326-332 (2000).
4. C. Francke, R.J. Siezen and B. Teusink, *Trends Microbiol.* **13**(11), 550-558 (2005).
5. C.V. Forst and K. Schulten, *J. Mol. Evol.* **52**, 471-489 (2001).
6. T. Dandekar, S. Schuster, B. Snel, M. Huynen and P. Bork, *Biochemical J.* **343**(1), 115-124 (1999).
7. Y. Tohsato, H. Matsuda, and A. Hashimoto, *Proc. of the 8th Inter. Conf. on Intel. Sys. for Mol. Biol.*, 376-383 (2000).
8. J.C. Clemente, K. Satou and G. Valiente, *Bioinformatics* **23**(2): e110-e115 (2006).
9. J.C. Clemente, K. Satou and G. Valiente, *Genome Informatics* **17**(2), 46-56 (2006).
10. D. Zhu and Z.S. Qin, *BMC Bioinformatics* 6-8 (2005).
11. A.G. Malygin, *Biochemistry Moscow* **69**(12), 1379-1385 (2004).
12. M. Heymans and A.K. Singh, *Bioinformatics* **19**, i138-i146 (2003).

13. H.W. Ma and A.P. Zeng, *Bioinformatics* **19**, 270-277 (2003).
14. B.P. Kelley, R. Sharan, R.M. Karp, T. Sittler, D.E. Root, B.R. Stockwell and T. Ideker, *Proc. Natl. Acad. Sci. U S A* **100**, 11394–11399 (2003).
15. S. Goto, T. Nishioka, and M. Kanehisa, *Bioinformatics* **14**, 591-599 (1998).
16. R.J. Wurtman, W.J. Shoemaker and F. Larin, *Proc. Natl. Acad. Sci. U S A* **59**(3), 800-807 (1968).
17. K. Pawlowski, L. Jaroszewski, L. Rychlewski and A. Godzik, *Pac. Symp. Biocomput.* 42-53 (2000).
18. Z. Li, S. Zhang, Y. Wang, X. Zhang and L. Chen, *Bioinformatics* **23**(13), 1631-1639 (2007).

AUTOMATIC MODELING OF SIGNAL PATHWAYS FROM PROTEIN-PROTEIN INTERACTION NETWORKS

XING-MING ZHAO[1,2,3], RUI-SHENG WANG[4], LUONAN CHEN[1,3,4] and KAZUYUKI AIHARA[1,3]

[1] *ERATO Aihara Complexity Modeling Project, JST, Tokyo 151-0064, Japan*
E-mail: xmzhao@aihara.jst.go.jp
[2] *Intelligent Computing Lab, Hefei Institute of Intelligent Machines, Chinese Academy of Sciences, Hefei, Anhui, 230031, China*
[3] *Institute of Industrial Science, The University of Tokyo, Tokyo 153-8505, Japan*
[4] *Department of Electrical Engineering and Electronics, Osaka Sangyo University, Osaka 574-8530, Japan*

This paper presents a novel method for recovering signaling pathways from protein-protein interaction networks automatically. Given an undirected weighted protein interaction network, finding signaling pathways is treated as searching for the optimal subnetworks from the network according to some cost function. To approach this optimum problem, an integer linear programming model is proposed in this work to model the signal pathways from the protein interaction network. The numerical results on three known yeast MAPK signal pathways demonstrate the efficiency and effectiveness of the proposed method.

1. Introduction

Signal transduction is the primary means that cells response to the external stimulus of the environment such as growth factors, nutrients, and so on. Furthermore, signal transduction plays an important role in coordinating metabolism, cell proliferation and differentiation. Generally, external signal or stimulus is transduced into a cell through an ordered sequence of biochemical reactions inside the cell. In many signal transduction processes, the number of proteins and other molecules participating in these events increases as the process proceeds from the initial stimulus, which results in a "signal cascade". Despite the success of traditional methods in detecting components involved in signaling networks, they can only generate specific linear signal pathways. The knowledge of complex signaling networks and their internal interactions is still unclear now. Therefore, it is necessary to develop new computational methods to capture the details of signaling pathways by exploiting high-throughout genomic and proteomic data.

Recently, with the advance in high-throughput bio-technology, the large-scale genomic and proteomic data provide insights into the components involved in signal transduction. For example, protein interactions and microarray data have been utilized to reconstruct signaling networks.[1–4] Since signal transduction is a process of biochemical reactions achieved by a cascade of protein interactions, protein interaction data can provide an alternative approach to understanding signaling networks. Ideker *et al.*[4] have proposed a variant of the color coding algorithm to reconstruct signaling networks from yeast protein interaction networks. In the color coding method, a number of candidate pathways are found, with a score assigned to each candidate. The highest scoring candidate is assumed to be the putative pathway and the top scoring pathways are then assembled into a signaling network. Steffen *et al.*[2] have developed an algorithm, namely Netsearch, to reconstruct signaling networks by utilizing both gene expression data and protein interaction data. In the Netsearch method, they also rank the candidate pathways and aggregate top scoring pathways into a signaling network. Zhao *et al.*[1] have also proposed a method for ranking signal transduction pathways by utilizing both protein interaction and microarray data. In the methods described above, signaling network is not detected as a whole, on the other hand, the separate linear pathways are detected and used to assemble the signaling network.

In this work, we present a new simple and efficient method for detecting signaling pathways from protein interaction data by an integer linear programming technique. In our method, we treat the finding of signal pathways as an optimization problem and wish to find out an optimal subnetwork starting from membrane proteins and ending at transcription factors with respect to some cost functions. The objective of our method is similar to the color coding method. The difference lies in that our method treats a signaling network as a whole entity and detect it by running the model once instead of ranking individual linear pathways and assembling them into a network. The numerical experiments on yeast protein interaction data demonstrate the effectiveness of the proposed method.

The rest of the paper is organized as follows: Section 2 describes the proposed integer linear programming model; Section 3 presents the experimental results; Section 4 draws conclusions.

2. Methods

In this section, we present a method for detecting a signaling network given the possible end points (e.g. membrane proteins and transcription factors (TFs)) of signal pathways and a protein interaction network. Given a protein interaction network, it can be represented as a weighted undirected graph $G(V, E)$, where the vertices are proteins and the edge $E(i, j)$ denotes the experimentally observed interaction between proteins i and j. In this study, the weight of each edge represents the interaction reliability between the corresponding proteins. In literature, there are many methods proposed for estimating the reliability of protein interactions.[5–7] In this work, we utilize the method proposed by Sharan *et al.*[8] to estimate the reliability of protein interactions. With the assumption that proteins in the same signal pathway will interact with one another with high probability, the weighted protein interaction network can be utilized to find putative signaling pathways.

In the weighted network, given a starting node, the linear path of a specific length of m from the starting node to another node can be assigned a score which equals to the sum or the product of the weights for the edges in the path. With a series of paths of length m starting from specific proteins generated this way, the top ranking paths are possible candidates for true linear signal transduction pathways. In this case, the specific starting proteins are membrane proteins because the signal transduction process starts from receptor proteins.

In this work, the weight of each edge $E(i, j)$ is defined as $a_{i,j} = -p(i, j)$, where $p(i, j)$ is the interaction reliability between proteins i and j. The score for each linear path is the sum of the weights for the edges in the path, and the length of the path is the number of proteins involved in the path. Similarly, the score of a subnetwork is the sum of the weights for the edges it contains, and the network size is the number of proteins it contains. Given an undirected weighted network $G(V, E, w)$ and the possible end points of signal pathways, i.e. membrane proteins and TFs, we wish to find out the minimum-weight subnetwork of specific size from the network G.

To accomplish the above mission, we proposed a novel integer linear programming (ILP) model to find out signal pathways, given membrane proteins, TFs and a weighted protein interaction network. The model is described as follows:

$$
\begin{aligned}
\text{Min} \quad & \sum_{i=1}^{|V|}\sum_{j=1}^{|V|} a_{ij}e_{ij} + \lambda \sum_{i=1}^{|V|}\sum_{j=1}^{|V|} e_{ij} \\
\textbf{s.t.} \quad & e_{ij} \leq x_i \\
& e_{ij} \leq x_j \\
& \sum_j e_{ij} \geq 1, \quad \text{if } i \text{ is a membrane protein or TF} \\
& \sum_j e_{ij} \geq 2x_i, \text{if } i \text{ is not a membrane protein or TF} \\
& x_i = 1 \text{ , if } i \text{ is a membrane protein or TF} \\
& x_i \in \{0, 1\}, \ i = 1, 2, \cdots, |V| \\
& e_{ij} \in \{0, 1\}, \ i, j = 1, 2, \cdots, |V|
\end{aligned}
$$

where a_{ij} is the weight for edge $E(i, j)$ of the undirected weighted network, x_i is a binary variable for protein i to denote whether protein i is selected as a component of the signaling network or not, e_{ij} is also a binary variable to denote whether the biochemical reaction represented by protein-protein interaction $E(i, j)$ is a part of the signaling network or not. λ is the punishment parameter to control the subnetwork size. The constraint $\sum_j e_{ij} \geq 2x_i$ is to ensure that x_i has at least two linking edges once it is selected in the signaling network so that the selected subnetwork is as connected as possible, whereas the constraint $\sum_j e_{ij} \geq 1$ makes sure that each membrane protein or TF has at least one link to other proteins. On the other hand, the constraints $e_{ij} \leq x_i$ and $e_{ij} \leq x_j$ ensure that only when protein i and protein j are selected as components of the signaling network, the biochemical reaction denoted by the edge e_{ij} would be considered.

The first term of the above cost function implies that we want to find out the minimum-weight subnetwork, while the second term is used to control the subnetwork size and the number of biochemical reactions involved in the subnetwork because each protein interaction is actually a biochemical reaction. The idea behind the model is that we want to find out a minimum-weight subnetwork of specific size which accomplishes the signal transduction process with as few biochemical reactions as possible, where biochemical reactions are represented by protein interactions, i.e. e_{ij} in the cost function. The assumption is reasonable because cells usually accomplish their missions with as less energy as possibl. This criterion is also consistent with the parsimony principle widely adopted in other areas of biology such as phylogeny tree construction and haplotype inference.[11,12]

The model described above is a standard integer linear programming which can be solved efficiently in polynomial time. To make the model suit for large-scale interaction networks, we can relax the constraints $x_i \in \{0,1\}, e_{ij} \in \{0,1\}$ to $0 \leq x_i \leq 1, 0 \leq e_{ij} \leq 1$ which make the ILP model become a linear programming (LP) model. Experiment results show such a relaxation does not reduce the performance, and at the same time highly improve the computation efficiency. Although the model has a parameter λ, it can be tuned in a relatively easy manner.

3. Experimental Results

Our proposed ILP model was applied to find the signaling networks in the yeast protein-protein interaction network. In this work, the protein interaction data were obtained from the DIP database,[9] which includes 4839 proteins and 14319 interactions. This data set has also been used by Ideker *et al.*[4] To evaluate the performance of the proposed methods, we applied it to find the three known yeast MAPK signaling pathways. The three yeast signal pathways are pheromone response, filamentous growth invasion and cell wall integrity, respectively. To reduce the computation complexity, the ILP model was applied to a smaller protein interaction network generated by depth first search (DFS) algorithm starting from membrane proteins and ending at TFs. This smaller network consists of the paths of length 6-8, and the interactions among proteins in this network were borrowed from the original protein interaction network. Therefore, three smaller protein interaction networks were generated by DFS for the three MAPK signal pathways, respectively. The sequential experiments were conducted on these three smaller protein interaction networks.

For the pheromone response pathway, the ILP model was applied to look for the signaling network starting from membrane protein STE3 and ending at transcription factor STE12. By varying the λ in the ILP model, we can get signaling networks of different size, e.g. linear pathway or signaling network. Fig.1 (a) shows the main chain of pheromone response pathway deposited in KEGG, Fig.1 (b) shows the linear signaling pathway found by color coding, and Fig.1 (c) shows the linear path found by ILP model, where the gray points are end points. Comparing (b) against (c), we can see that in the linear path we found, AKR1 links directly to STE5 instead of through STE4, CDC24 and BEM1 like that detected by color coding because there is a direct interaction between AKR1 and STE5. Although we failed to detect STE4, CDC24 and BEM1 in the main chain compared with color coding, we can successfully detect the linear signaling pathway with fewer components involved in the main chain. Fewer proteins imply fewer biochemical reactions which is biologically reasonable because signals may be transduced in a parsimonious way that consume less energy.

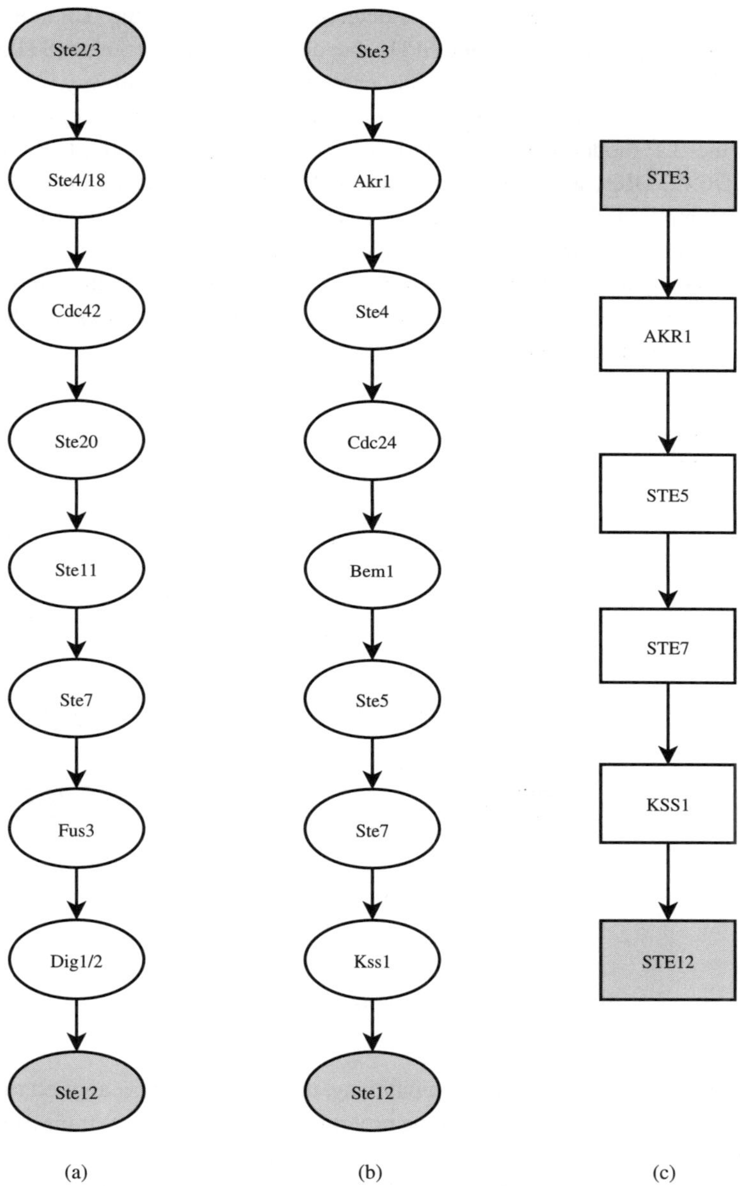

Fig. 1. The linear signal pathways for pheromone response: (a) the pathway from KEGG; (b) the pathway detected by color coding; (c) the pathway detected by ILP model.

Figure 2 shows the signaling network detected by our method, where the gray points are end points. This signaling network consists of 19 genes. By comparing the detected signaling network with those found by Netsearch[2] and color coding,[4] we can learn that most of the components of the three signaling networks are the same. Compared with the signaling network of the same size as ours detected by Netsearch, the ILP model failed to detect proteins SST2, DIG1, DIG2 and SPH1, but detect four new proteins (STE50, BEM3, BEM4 and CDC28) which are related to the pheromone response pathway.[10] Furthermore, protein STE50 has also been detected by color coding method,[4] which confirms the effectiveness of the ILP model. Compared with the color coding model, the ILP model failed to detect CDC42, DIG1 and DIG2, but detected MPT5 which has also been detected by the Netsearch method. Such a result demonstrates that our method can be a helpful complement to existing algorithms. The ILP model failed to detect DIG1 and DIG2 due to our assumption that signal transduction is assumed to be accomplished with as few biochemical reactions as possible, whereas DIG1 and DIG2 introduce many links to other proteins that have already been detected by the ILP model.

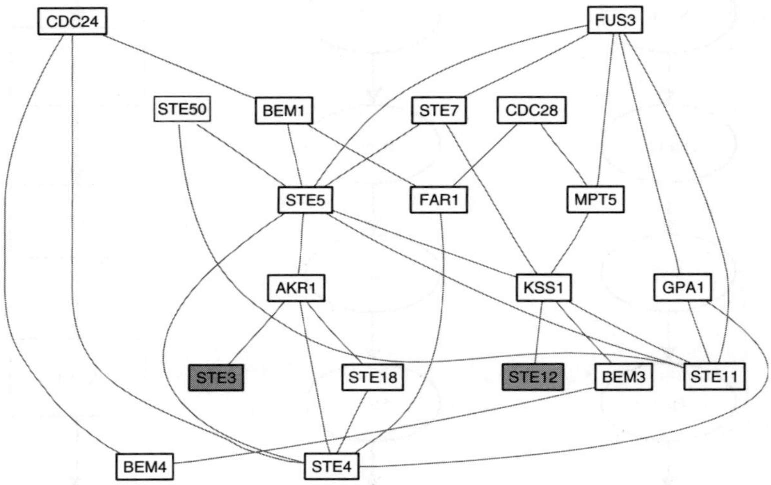

Fig. 2. The signaling network for pheromone response.

For the filamentous growth invasion pathway, the ILP model was applied to detect the signaling network starting from membrane protein RAS2 and ending at transcription factor STE12. Figure 3 respectively shows the signal pathway of the same size that are deposited in KEGG, detected by color coding and ILP model, where the gray points are end points. It can be seen from Fig.3 (a) and (c) that the signaling pathway recovered by the ILP model matches the known signal pathway to a large extent. The CDC25 and HSP82 were detected due to the missing links between RAS2 and CDC42 in the protein interaction network. Comparing Fig.3 (b) with Fig.3 (c), we can see that the ILP model can find the identical signaling pathway of the same size as that detected by color coding. Furthermore, the ILP model found out the additional links compared with the color coding method, where the additional links may imply alternative signal pathways.

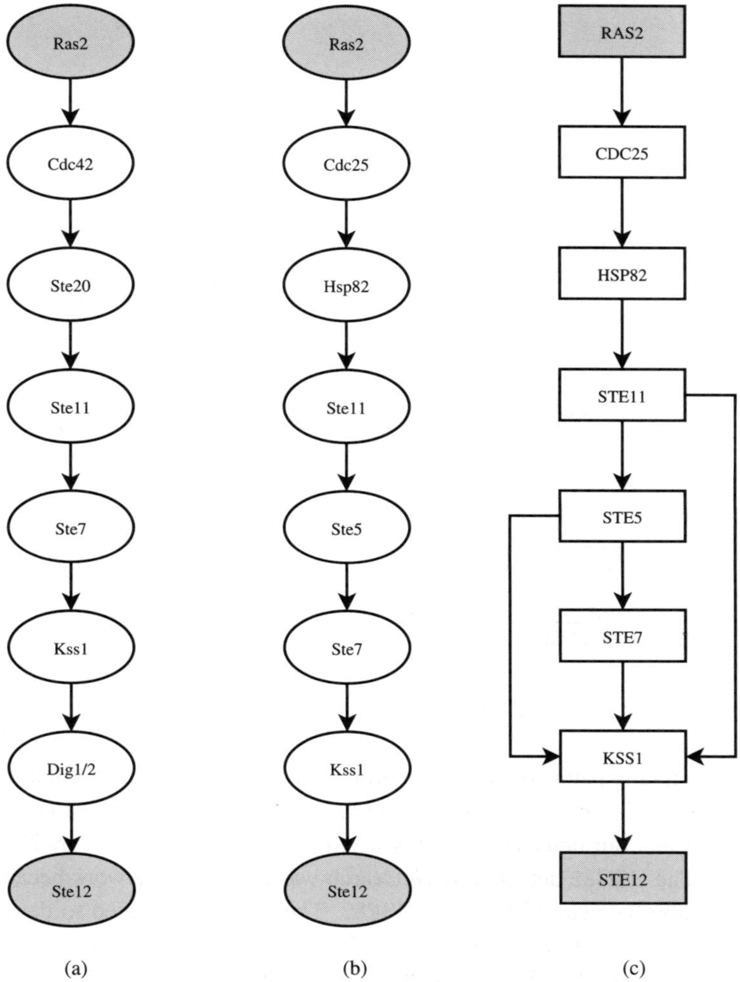

Fig. 3. The signal pathways for filamentous growth invasion: (a) the pathway from KEGG; (b) the pathway by color coding; (c) the pathway by ILP model.

Furthermore, Fig.4 shows the signaling network of larger size detected by the ILP model, where the gray points are end points. The left figure in Fig.4 shows a signaling network of size 13. Compared to the network generated by Netsearch,[2] all of the proteins involved in the detected signaling network by ILP have also been found by Netsearch except GIN4, NAP1 and RIM11. The GIN4, NAP1 and RIM11 were detected because they appear in the same complex together with CDC25,[10] and GIN4 and NAP1 have the function of cell polarity and filament formation.[10] Therefore, they are related to the filamentous signaling pathway. The right figure in Fig. 4 shows another signaling network of size 19, where we assume that the proteins SPA2, CYR1, FUS3 and BEM1 are known to be involved in the signaling pathway. Although it is difficult to know exactly all the proteins

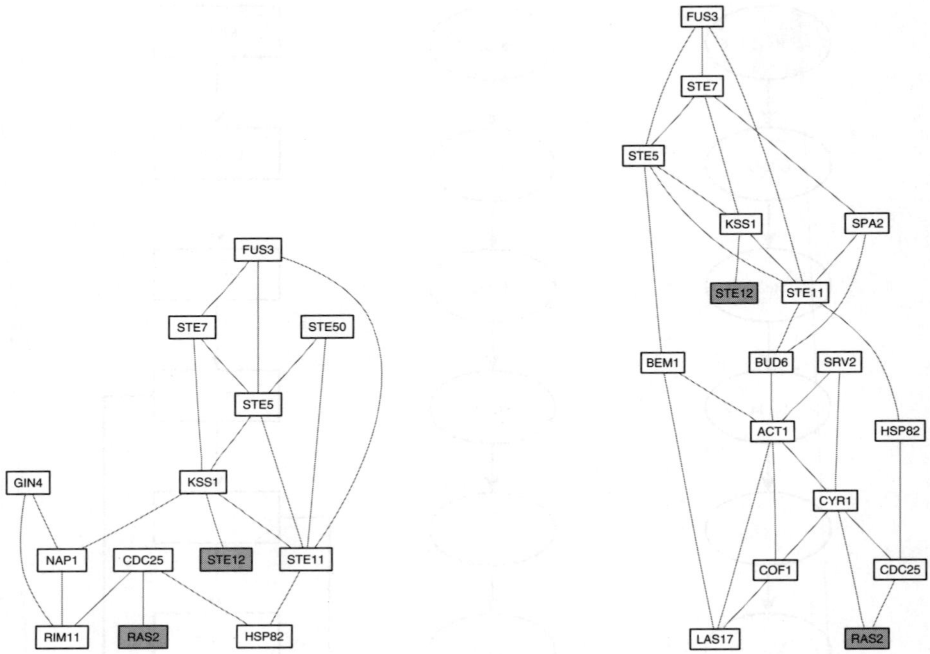

Fig. 4. The signaling network for filamentous growth invasion.

involved in a signaling pathway, our assumption is reasonable because we can know some proteins in the signaling pathway from the published results by other researchers. It can be seen from Fig.4 that our detected signaling network matches that found by Netsearch[2] to a large extent. The HSC82 detected by Netsearch was not in our network because there is a direct interaction between STE11 and HSP82. The ILP model failed to detect proteins ABP1, DIG1, DIG2 and BNI1, while included two other proteins COF1 and LAS17 because COF1, LAS17, BEM1, BUD6 and SRV2 occur in the same complex[10] and therefore may have similar functions.

For the cell wall integrity pathway, the ILP model was applied to detect the signalling network starting from MID2 and ending at RLM1. Fig.5 shows the linear signal pathways detected by the ILP model and color coding, and the one deposited in KEGG, where the gray points are end points. It can be seen from Fig.5 that the ILP model can detect the identical signaling pathway as that by color coding. It is not surprising to see the same results because we use the same interaction data set as the one used by color coding. The detected signal pathway matches most of the known pathway except ROM2 due to the missing links between MID2 and RHO1.

From the results described above, we can see that the proposed ILP model is indeed effective for finding signaling networks from protein interaction networks. Furthermore, the ILP model is very simple and can detect the signalling network directly instead of working

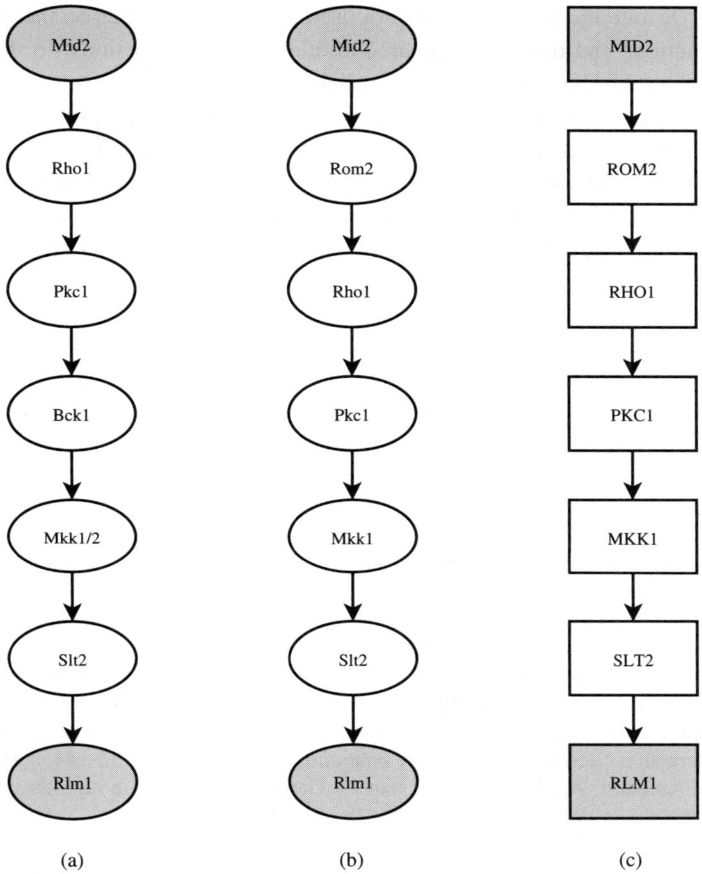

Fig. 5. The linear signal pathways for cell wall integrity: (a) the pathway from KEGG; (b) the pathway by color coding; (c) the pathway by ILP model.

in multiple-stage like Netsearch and color coding: find the candidate signal pathways, rank the candidate pathway, and assemble the top scoring pathways.

4. Conclusions

In this paper, we presented a new method for recovering signaling networks from protein interaction networks. The proposed method utilizes integer linear programming to find out the subnetwork with minimum weight of specific size. The results on three known MAPK signal pathways using yeast protein interaction network show that the ILP model can recover most of the signaling pathway and the reconstructed signaling networks match most of those published results, which confirm the effectiveness and efficiency of the proposed method. Compared with existing methods, our method is much simpler because it can detect the signaling networks from protein interaction network directly instead of ranking the candidate signal pathways and assembling the top scoring signal pathways into a signaling network. Despite the success of the proposed method, it depends on the quality of the protein interactions and the estimated probabilities of the interactions. In this work, the probability of protein interactions are estimated precisely. However, most of the protein interactions are not assigned reliable scores to represent exactly the probability of protein interactions. One alternative approach to this problem is to utilize the microarray data information because there are large amount of microarray data available nowadays, and the combination of protein interactions and microarray data may provide insights into signal transduction discovery. In the future, we will explore this point in reconstructing signaling networks.

Acknowledgment

This work was partly supported by the National High Technology Research and Development Program of China (2006AA02Z309)

References

1. Liu Y, Zhao H. A computational approach for ordering signal transduction pathway components from genomics and proteomics Data. *BMC Bioinformatics*, 5: 158, 2004.
2. Steffen M, Petti A, Aach J, D'haeseleer P, Church G. Automated modelling of signal transduction networks. *BMC Bioinformatics*, 3:34, 2002.
3. Zien A, Kuffner R, Zimmer R, Lengauer T. Analysis of gene expression data with pathway scores. *Proc Int Conf Intell Syst Mol Biol*, 8:407-417, 2000.
4. Scott J, Ideker T, Karp RM, Sharan R. Efficient Algorithms for Detecting signaling pathways in Protein Interaction Networks. *Journal of Computational Biology*, 13: 133-144, 2006.
5. Bader, J., Chaudhuri, A., Rothberg, J., Chant, J. Gaining confidence in high-throughput protein interaction networks. *Nature Biotechnol*, 22: 78 - 85, 2003.
6. Deng M, Sun F, Chen T. Assessment of the reliability of protein-protein in-teractions and protein function prediction. *In: Proceedings of the Eighth Pacific Symposium on Biocomputing*, 140-51, 2003
7. von Mering C, Krause R, Snel B, Cornell M, Oliver SG, Fields S, Bork P. Comparative assessment of large-scale data sets of protein-protein interactions. *Nature*, 417:399-403, 2002.
8. Sharan R, Suthram S, Kelley RM, Kuhn T, McCuine S, Uetz P, Sittler T, Karp RM, Ideker T. Conserved patterns of protein interaction in multiple species. *PNAS*, 102: 1974-1979, 2005.
9. Xenarios I, *et al.* DIP, the database of interacting proteins: a research tool for studying cellular networks of protein interactions. *Nucleic Acids Res.*, 30: 303-305, 2002
10. Mewes HW, Amid C, *et al.* MIPS: analysis and annotation of proteins from whole genomes. *Nucleic Acids Research*, 32: Database issue:D41-4, 2004.
11. Wang L, Xu Y. Haplotype inference by maximum parsimony. *Bioinformatics*, 19,1773-1780, 2003.
12. Tobias H, Andor L, Robert F, Helgi B Genetic algorithm for large-scale maximum parsimony phylogenetic analysis of proteins. *Biochim. Biophys. Acta*, 1725, 19-29, 2005.

SIMULTANEOUSLY SEGMENTING MULTIPLE GENE EXPRESSION TIME COURSES BY ANALYZING CLUSTER DYNAMICS

SATISH TADEPALLI[1,*], NAREN RAMAKRISHNAN [1], LAYNE T. WATSON[1,2], BHUBANESHWAR MISHRA[3] and RICHARD F. HELM[4]

[1] *Department of Computer Science,* [2] *Department of Mathematics,* [4] *Department of Biochemistry*
Virginia Polytechnic Institute and State University , Blacksburg, VA 24061
* *E-mail: stadepal@vt.edu*
www.vt.edu
[3] *Courant Institute of Mathematical Sciences, New York University, NY 10012*
www.nyu.edu

We present a new approach to segmenting multiple time series by analyzing the dynamics of cluster rearrangement around putative segment boundaries. By directly minimizing information-theoretic measures of segmentation quality derived from Kullback-Leibler (KL) divergences, our formulation reveals clusters of genes along with a segmentation such that clusters show concerted behavior within segments but exhibit significant regrouping across segmentation boundaries. This approach finds application in distilling large numbers of gene expression profiles into temporal relationships underlying biological processes. The results of the segmentation algorithm can be summarized as Gantt charts revealing temporal dependencies in the ordering of key biological processes. Applications to the yeast metabolic cycle and the yeast cell cycle are described.

Keywords: Time series segmentation, clustering, KL-divergence, temporal regulation.

1. Introduction

Time course analysis has become an important tool for the study of developmental, disease progression, and cyclical biological processes, e.g., the cell cycle,[1] metabolic cycle,[2] and even entire life cycles. Recent research efforts have considered using static measurements to "fill in the gaps" in the time series data,[3] quantifying timing differences in gene expression,[4] and reconstructing regulatory relationships.[5]

One of the attractions of time series analysis is its promise to reveal temporal relationships underlying biological processes:[6] which process occurs before which other, and what are the "checkpoints" that must be satisfied (and when). Although similar analysis can also be conducted by tracking individual genes whose function is known, we desire to automatically mine, in an unsupervised manner, temporal relationships involving *groups* of genes, which are not yet characterized *a priori*. In particular, given multiple gene expression profiles over a time course, we desire to identify both segments of the time course where groups show concerted behavior and boundaries between segments where there is significant functional "regrouping" of genes. We cast this problem as a form of time series segmentation where the segmentation criterion is driven by measures over cluster dynamics.

It is important to contrast our goals with prior work. Typical published results on time series segmentation[7] are focused on segmenting a single time series with homogeneity assumptions on successive time points. We are focused on simultaneously segmenting multiple time series by modeling each segment as a heterogeneous mix of multiple clusters which can themselves be redefined across segments. Our work is hence directly targeted to mining datasets involving thousands of genes where there are complex inter-relationships and reorganizations underlying the dataset. As an example, consider the yeast metabolic cycle (YMC), using the dataset of

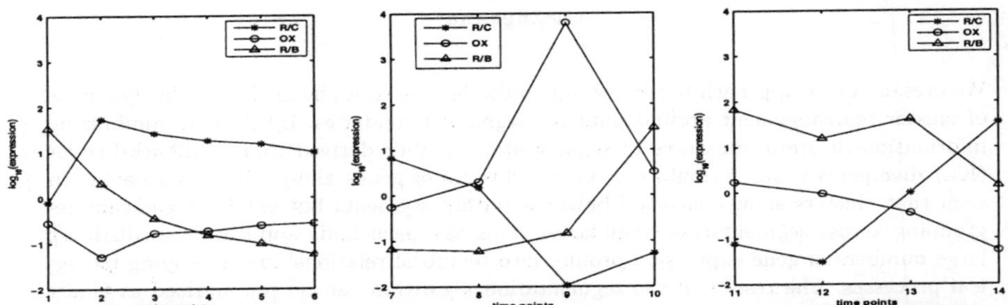

	R/C	Ox	R/B	others
W1-C1	920	102	72	100
W1-C2	80	427	377	309
W1-C3	102	374	403	336

	R/C	Ox	R/B	others
W1-C1	421	60	370	328
W1-C2	200	795	150	106
W1-C3	481	48	332	311

	R/C	Ox	R/B	others
W1-C1	509	337	34	285
W1-C2	422	458	20	280
W1-C3	191	108	778	180

	W2-C1	W2-C2	W2-C3
W1-C1	421	401	372
W1-C2	380	448	365
W1-C3	378	402	435

	W3-C1	W3-C2	W3-C3
W2-C1	380	355	428
W2-C2	415	435	410
W2-C3	368	390	421

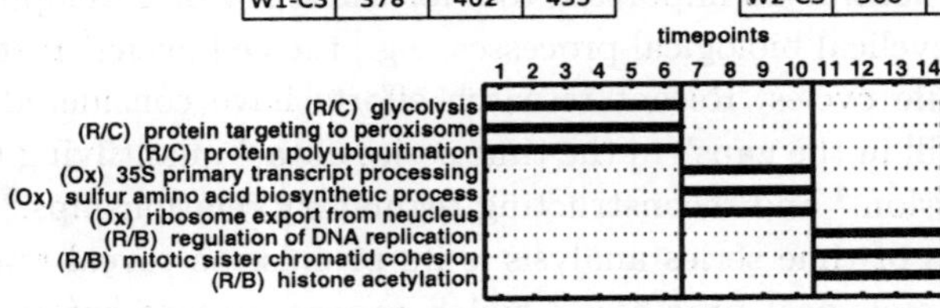

Fig. 1. Preview of results from the segments in the Yeast metabolic cycle.

Tu *et al.*[2] The YMC is a carefully coordinated mechanism between a reductive charging (**R/C**) phase involving non-respiratory metabolism (glycolysis, fatty acid oxidation) and protein degradation, followed by oxidative metabolism (**Ox**), where respiratory processes are used to generate adenosine triphosphate (ATP), culminating in reductive metabolism (**R/B**), characterized by a decrease in oxygen uptake and emphasis on DNA replication, mitochondrial biogenesis, and cell division. Different genes are central to each of these phases. Tu *et al.* analyzed this 36-pt time course—spanning approximately three cycles (**R/B** phase is not sampled in the last cycle)—by tracking 'sentinel' genes showing periodic behavior across the time course. We analyzed this dataset of 3602 gene expression profiles over a 15 hour period using our segmentation algorithm and arrived at the same segmentation corresponding to three cycles, one of which is shown in Fig. 1. The top row depicts the prototypes of the clusters in the **R/C**, **Ox**, and **R/B** phases respectively. The second row shows how the genes corresponding to each phase are coming together in a particular cluster *within a segment* (intersection of the highlighted row and column) while the genes of other phases are spread apart across other clusters. This regrouping of the clusters of genes *across the segments* is captured by the contingency tables in the third row. Observe that the contingency tables in the second row involve significant enrichments whereas the tables in the third row approximate a uniform distribution. These time-bounded enrichments for the clusters in each segment identify the biological processes with modulated activities as shown in the Gantt chart in the bottom row of Fig. 1. We reiterate that the time point boundaries, the groups of genes important in each segment, and the functions enriched in them, are inferred automatically. No explicit modeling of periodicity or other prior biological knowledge has been imparted to the segmentation algorithm.

2. Problem Formulation

We are given multiple l-vectors of measurements $\mathcal{G} = \{\mathbf{g}_1, \mathbf{g}_2, \ldots, \mathbf{g}_\nu\}$, where each $\mathbf{g}_i$ is a time series over $\mathcal{T} = (t_1, t_2, \ldots, t_l)$. The problem of segmentation is to express $\mathcal{T}$ as a sequence of segments or windows: $(w_{t_1}^{t_a}, w_{t_{a+1}}^{t_b}, \ldots, w_{t_k}^{t_l})$ where each window $w_{t_s}^{t_e}$, $t_s \leq t_e$, is a sequence of consecutive time points beginning at (and inclusive of) time point t_s and ending at (and inclusive of) time point t_e.

We first describe a way to evaluate a given segmentation before presenting an algorithm for identifying segmentations. We begin by studying the case of just two adjacent windows: $w_{t_a}^{t_b}$ and $w_{t_{b+1}}^{t_c}$. Given two clusterings of genes, one for each of the windows, our evaluation criterion requires that these two sets of clusters are highly dissimilar, i.e., genes clustered together in some cluster of $w_{t_a}^{t_b}$ move out of their clusters and are clustered together with different genes in $w_{t_{b+1}}^{t_c}$. For instance, given a dataset with 18 genes and 3 clusters in either window, the evaluation criterion prefers contingency table (a) below over tables (b) and (c).

(a)

2	2	2
2	2	2
2	2	2

(b)

6	0	0
0	6	0
0	0	6

(c)

0	6	0
6	0	0
0	0	6

Here the rows refer to clusters of $w_{t_a}^{t_b}$ and the columns refer to clusters of $w_{t_{b+1}}^{t_c}$. We achieve this by enforcing that the (projected) row-wise and column-wise distributions from the contingency table resemble a uniform distribution. Formally, given two windows $w_{t_a}^{t_b}$ and $w_{t_{b+1}}^{t_c}$, which have been clustered into r and c clusters (respectively), we define the $r \times c$ contingency table over the clusterings. Entry n_{ij} in the $(i, j)^{th}$ cell of the table represents the overlap between the genes clustered together in cluster i of $w_{t_a}^{t_b}$ and in cluster j of $w_{t_{b+1}}^{t_c}$. The sizes of the clusters in $w_{t_a}^{t_b}$ are given by the column-wise sums across each row: $n_{i\cdot} = \sum_j n_{ij}$, while the sizes of clusters in $w_{t_{b+1}}^{t_c}$ are given by row-wise sums down each column: $n_{\cdot j} = \sum_i n_{ij}$. Using these, we define $(r + c)$ probability distributions, one for each row and one for each column. The distribution corresponding to row i, $\mathbf{R}_i$, takes values from the column indices, i.e., $1, \ldots, c$, with value j $(1 \leq j \leq c)$ occurring with probability $\frac{n_{ij}}{n_{i\cdot}}$. Similarly, the column distribution for column j, $\mathbf{C}_j$, takes values from the row indices, i.e., $1, \ldots, r$, with value i $(1 \leq i \leq r)$ occurring with probability $\frac{n_{ij}}{n_{\cdot j}}$. We capture the deviation of these row-wise and column-wise distributions w.r.t. the uniform distribution as

$$\mathcal{F} = \frac{1}{r}\sum_{i=1}^{r} D_{KL}(\mathbf{R}_i||U(\frac{1}{c})) + \frac{1}{c}\sum_{j=1}^{c} D_{KL}(\mathbf{C}_j||U(\frac{1}{r})), \qquad (1)$$

where, $D_{KL}(p||q) = \sum_x p(x) \log_2 \frac{p(x)}{q(x)}$ is the Kullback-Leibler (KL) divergence between two probability distributions $p(x)$ and $q(x)$, and $U(\cdot)$ denotes the uniform distribution whose argument is the probability of any outcome. The optimization problem is then to minimize $\mathcal{F}$. This function can also be interpreted in terms of entropies of the row-wise and column-wise distributions, and also in terms of conditional entropies of the clusters in windows $w_{t_a}^{t_b}$ and $w_{t_{b+1}}^{t_c}$. Also, $\mathcal{F}$ has connections to the principle of minimum discrimination information (MDI).[8] The MDI principle states that if q is the assumed or true distribution, the estimated distribution p must be chosen such that $D_{KL}(p||q)$ is minimized. In our case, q is the uniform distribution desired and p is the distribution estimated from observed data. Observe that the combinations of the r row-wise KL-divergences and c column-wise KL-divergences are averaged to form $\mathcal{F}$. This is to mitigate the effect of lopsided contingency tables ($r \gg c$ or $c \gg r$) wherein it is possible to optimize $\mathcal{F}$ by focusing on the "longer" dimension without really ensuring that the other dimension's projections are close to uniform. Finally, note that Eq. (1) can be readily extended to the case where we have more than two segments.

Minimizing $\mathcal{F}$ will yield row-wise and column-wise distribution estimates that are close to the respective uniform distributions and, hence, result in independent clusterings across the neighboring windows. Maximizing $\mathcal{F}$ leads to highly dependent clusters across the windows which is the same as associative clustering described by Kaski *et al.*[9] However, for our current problem of time series segmentation, we are concerned with only minimizing $\mathcal{F}$ to obtain independent clusters.

3. Clustering across windows

We now turn our attention to the clustering algorithm that must balance two conflicting criteria: namely, the clusters across neighboring windows must be independent and, yet the clusters must exhibit concerted behavior within a window. In typical clustering algorithms, each cluster has a prototype and the data vectors are assigned to the nearest cluster based on some distance measure from these prototypes. The prototypes are iteratively improved to find the best possible clusters.

Again, we develop our notation for two adjacent windows and the extension to greater numbers of windows is straightforward. Given a gene vector $\mathbf{g}_k$, let its projection onto the 'left' window $w_{t_a}^{t_b}$ be referred to as $\mathbf{x}_k$, and its projection onto the 'right' window $w_{t_{b+1}}^{t_c}$ be referred to as $\mathbf{y}_k$. Recall that sets of such projections are clustered separately such that the clusters are maximally dissimilar. Let r and c be the number of clusters for $\mathbf{x}$ and $\mathbf{y}$ vectors, which results in a $r \times c$ contingency table. Let $\mathbf{m}_i^{(x)}$ be the prototype vector for the ith cluster of the $\mathbf{x}$ vectors. The assignment of a data vector to the clusters is given by the probability distribution $\mathbf{V}(\mathbf{x}_k) = \{V_i(\mathbf{x}_k)\}$, where $\sum_{i=1}^{r} V_i(\mathbf{x}_k) = 1$. The probabilities $V_i(\mathbf{x}_k)$ are the cluster membership indicator variables, i.e., the probability that data vector k is assigned to cluster i. Similar cluster prototypes $\mathbf{m}_j^{(y)}$, distributions $\mathbf{V}(\mathbf{y}_k)$, and cluster indicator variables $V_j(\mathbf{y}_k)$ are defined for $\mathbf{y}$ vectors as well. Then the contingency table counts can be calculated as $n_{ij} = \sum_k V_i(\mathbf{x}_k)V_j(\mathbf{y}_k)$. Assigning a data vector to the nearest cluster with a probability of one and calculating n_{ij} renders the objective function $\mathcal{F}$ in Eq. (1) nondifferentiable at certain points, as a result of which we cannot leverage classical numerical optimization algorithms to minimize $\mathcal{F}$. To avoid this problem, cluster indicator variables are typically parametrized as a continuously differentiable function that assigns each data vector to its nearest cluster with a probability close to one and to the other clusters with a probability close to zero, i.e. $V_i(\mathbf{x}_k), V_j(\mathbf{y}_k) \in (0,1)$. For this purpose, we define

$$\gamma_{(i,i')}(\mathbf{x}_k) = \frac{||\mathbf{x}_k - \mathbf{m}_{i'}^{(x)}||^2 - ||\mathbf{x}_k - \mathbf{m}_i^{(x)}||^2}{D}, \qquad 1 \le i, i' \le r, \tag{2}$$

where, $D = \max_{k,k'} ||\mathbf{x}_k - \mathbf{x}_{k'}||^2, 1 \le k, k' \le \nu$ is the pointset diameter. A well known approximation to $\min_{i'} \gamma_{(i,i')}(\mathbf{x}_k)$ is the Kreisselmeier-Steinhauser (KS) envelope function[10] given by

$$KS_i(\mathbf{x}_k) = \frac{-1}{\rho} \ln \Big[\sum_{i'=1}^{r} \exp(-\rho\, \gamma_{(i,i')}(\mathbf{x}_k)) \Big],$$

where $\rho \gg 0$. The KS function is a smooth function that is infinitely differentiable. Using this function the cluster membership indicators are redefined as

$$V_i(\mathbf{x}_k) = Z(\mathbf{x})^{-1} \exp \Big[\rho\, KS_i(\mathbf{x}_k) \Big],$$

where $Z(\mathbf{x})$ is a normalizing function such that $\sum_{i=1}^{r} V_i(\mathbf{x}_k) = 1$. The cluster membership indicators for the "right" window, $V_j(\mathbf{y}_k)$, are also smoothed similarly. The membership probability of a vector to a cluster is assigned relative to its distance from all the other clusters as captured by the function γ in Eq. (2). This approach tracks the membership probabilities better than using individual Gaussians for each cluster as suggested by Kaski et al[9].

Minimizing the function $\mathcal{F}$ in Eq. (1) should ideally yield clusters that are independent across windows and local within each window. However, using smooth cluster prototypes gives rise to an alternative minimum solution where each data vector is assigned with uniform probability to every cluster. Recall the contingency table example from Sec. 2; each of the 18 genes can be assigned to the three row clusters (and three column clusters) with probability $[1/3, 1/3, 1/3]$ and the estimate of the count matrix from these soft counts would still be uniform in each cell ($\sum_k V_i(\mathbf{x}_k)V_j(\mathbf{y}_k) = 2$). To avoid degenerate solutions such as these, we require maximum deviation of each data vector's cluster membership probabilities ($V_i(\mathbf{x}_k)$ and $V_j(\mathbf{y}_k)$) from the uniform distribution over the number of clusters. This leads to the regularized objective function

$$\begin{aligned}\mathcal{F} =& \frac{\lambda}{r}\sum_{i=1}^{r} D_{KL}(\mathbf{R}_i||U(\frac{1}{c})) + \frac{\lambda}{c}\sum_{j=1}^{c} D_{KL}(\mathbf{C}_j||U(\frac{1}{r})) \\ & - \frac{1}{\nu}\sum_{k=1}^{\nu} D_{KL}(\mathbf{V}(\mathbf{x}_k)||U(\frac{1}{r})) - \frac{1}{\nu}\sum_{k=1}^{\nu} D_{KL}(\mathbf{V}(\mathbf{y}_k)||U(\frac{1}{c})),\end{aligned} \quad (3)$$

where λ is the weight, set to a value greater than 1, gives more emphasis to minimizing the row and column wise distributions. This also enforces equal cluster sizes. Optimization of $\mathcal{F}$ is performed using the augmented Lagrangian algorithm with simple bound constraints on the cluster prototypes using the FORTRAN package LANCELOT.[11] The initial cluster prototypes are set using individual k-means clusters in each window and are iteratively improved till a local minimum of the objective function is attained.

4. Segmentation Algorithm

Let $\mathcal{T} = (t_1, t_2, \ldots, t_l)$ be the given time series data sequence, and l_{min} and l_{max} be the minimum and maximum window lengths, respectively. For each time point t_a, we define the set of windows starting from t_a as $S_{t_a} = \{w_{t_a}^{t_b} | l_{min} \leq t_b - t_a + 1 \leq l_{max}\}$. Given a window $w_{t_a}^{t_b}$, the choices for the next window are given by $S_{t_{b+1}}$, the set of windows starting form t_{b+1}. These windows can be organized as nodes of a directed acyclic graph, where directed edges exist between $w_{t_a}^{t_b} \in S_{t_a}$ and every $w_{t_{b+1}}^{t_c} \in S_{t_{b+1}}$. The edge weights are set to be the objective function from Eq. (3) realized by simultaneously clustering the windows $w_{t_a}^{t_b}$ and $w_{t_{b+1}}^{t_c}$, as discussed in the previous section. Since local optimization procedures are sensitive to initialization, we perform 100 random restarts of the optimization procedure (each time with different k-means prototypes found in individual windows) and choose the best (minimum) of the local optimum solutions as the weight for the edge between the two windows. Given this weighted directed acyclic graph, the problem of segmenting the time series is equivalent to finding the minimum path (if one exists) between a node representing a window beginning at t_1 and a node corresponding to a window that ends in t_l (recall that there can be several choices for nodes beginning at t_1 as well as for those ending at t_l, depending on l_{min} and l_{max}). We find the shortest path using dynamic programming (Dijkstra's algorithm) where the path length is defined as D_{avg}, given by Eq. (4), described later.

5. Experiments

Datasets: Our experimental datasets constitute gene expression measurements spanning the yeast metabolic cycle (YMC) and the yeast cell cycle (YCC). As stated earlier, the YMC dataset[2] consists of 36 time points collected over three continuous cycles. The YCC was taken from the well known α-factor experiment of Spellman et al.[1] The original YMC dataset consists of 6555 unique genes from the *S. cerevisiae* genome. We first eliminated those genes that do not have an annotation in any GO biological process category (revision 4.205 of GO released on 14 March 2007), resulting in a universal set of 3602 genes. The gene expression values were log transformed (base 10) and normalized such that the mean expression of each gene across all time points is zero. To segment this dataset we experimented with the number of clusters in each segment ranging from three to 15, $l_{min} = 4$, $l_{max} = 7$, $\rho = 20$, and $\lambda = 1.4$. The λ and ρ values were adjusted to give approximately equal sized clusters with good intracluster similarities. For the YCC dataset which originally had 6076 genes, we considered the genes with no missing values and mean centered each gene's expression across all time points to zero. From this data, we removed the genes that do not have any annotation in the GO biological process category resulting in a final set of 2196 genes. To segment this dataset, again we ranged from three to 15 clusters in each window, $l_{min} = 3$, $l_{max} = 5$, $\rho = 20$, and $\lambda = 1.4$ (ρ and λ adjusted as before).
Evaluation metrics: We evaluate our clusterings and segmentations in five ways: cluster stability, cluster reproducibility, functional enrichment, segmentation quality, and segmentation sensitivity. We assess **cluster stability** using a bootstrap procedure to determine significance of genes brought together.

Recall that each window except the first and last windows has two sets of clusters, one set independent with respect to the previous window and the other independent with respect to the next window. We are interested in the genes that are significantly clustered together in these two sets of clusters, as they represent the genes that are specific to the window under consideration. We calculate a contingency table between these two clusterings for each window (excluding the first and the last window). Each cell in the contingency table represents the number of genes that are together across the two independent sets of clusters. We randomly sample 1000 pairs of clusterings within each window (with cluster sizes same as the two independent clusterings) and compute their contingency tables. By the central limit theorem, the distribution of counts in each cell of the table is approximately normal (also verified using a Shapiro-Wilk normality test with $p = 0.05$). We now evaluate each cell of the actual contingency table with respect to the corresponding random distribution and retain only those cells that have more genes than that observed at random with $p < 0.05$ (Bonferroni corrected with the number of cross clusters to account for multiple hypothesis testing). To ensure **reproducibility** of clusters, we retain only those genes in each significant cell of the contingency table that are together in more than 150 of the 200 clusterings (conducted with different initializations).

For the first and last windows, which have only 100 randomly initialized clusterings, we retain those genes that are clustered together in more than 75 of the 100 clusterings. After the above two steps, we perform **functional enrichment** using the GO biological process ontology (since we are tracking biological processes) over the selected clusters of genes. A hypergeometric p-value is calculated for each GO biological process term, and an appropriate cutoff is chosen using a false discovery rate (FDR) q-level of 0.01.

The **segmentation quality** is calculated as a partition distance[12] between the "true" segmentation (from the literature of the YMC and YCC) to the segmentations computed by our algorithm. We view each window as a set of time points so that a segmentation is a partition of time points. Given two segmentations S_1 and S_2, whose windows are indexed by the variables $w_{t_a}^{t_b}$ and $z_{t_c}^{t_d}$ respectively, the partition distance is given by:

$$P_D = -\sum_{w_{t_a}^{t_b} \in S_1} \sum_{z_{t_c}^{t_d} \in S_2} |w_{t_a}^{t_b} \cap z_{t_c}^{t_d}| \log_2 \frac{|w_{t_a}^{t_b} \cap z_{t_c}^{t_d}|}{|w_{t_a}^{t_b}|} - \sum_{z_{t_c}^{t_d} \in S_2} \sum_{w_{t_a}^{t_b} \in S_1} |w_{t_a}^{t_b} \cap z_{t_c}^{t_d}| \log_2 \frac{|w_{t_a}^{t_b} \cap z_{t_c}^{t_d}|}{|z_{t_c}^{t_d}|}.$$

The **segmentation sensitivity** to variations in the number of clusters is calculated as the average of the ratios of KL-divergences between the segments to the maximum possible KL divergence between those segments. This latter figure is easy to compute as a function of the number of clusters, which is considered uniform throughout the segmentation. Suppose we have $|S|$ windows in a given segmentation $S = \{w_{t_1}^{t_a}, w_{t_{a+1}}^{t_b}, \ldots, w_{t_{j+1}}^{t_k}, w_{t_{k+1}}^{t_l}\}$ with c clusters in each window. Let $\mathcal{F}_{max}$ be the objective function value for the maximally similar clustering (the $c \times c$ diagonal contingency table (b) in the example in Sec. 2). Then the measure we compute is

$$D_{avg} = \frac{1}{|S|-1}\left[\frac{\mathcal{F}_{\{w_{t_1}^{t_a}, w_{t_{a+1}}^{t_b}\}}}{\mathcal{F}_{max}} + \frac{\mathcal{F}_{\{w_{t_{b+1}}^{t_c}, w_{t_{c+1}}^{t_d}\}}}{\mathcal{F}_{max}} + \ldots + \frac{\mathcal{F}_{\{w_{t_j}^{t_k}, w_{t_{k+1}}^{t_l}\}}}{\mathcal{F}_{max}}\right], \quad (4)$$

where $\mathcal{F}_{\{w_{t_a}^{t_b}, w_{t_{b+1}}^{t_c}\}}$ is the optimal objective function value obtained by clustering the pair of adjacent windows $w_{t_a}^{t_b}, w_{t_{b+1}}^{t_c}$. Lower values of this ratio indicate that the segmentation captures maximal independence between adjacent segments while higher values indicate the clusters obtained are more similar in adjacent segments.
Results: The YMC segmentation generated for the minimum number (3) of clusters is: 1-6, 7-10, 11-14, 15-18, 19-22, 23-26, 27-31, 32-36, which correspond to alternating R/C, Ox, and R/B phases. The GO categories enriched ($p < 10^{-7}$) in one cycle for this dataset have already been depicted in Fig. 1. This segmentation is stable up to eight clusters after which it begins to deviate from the "true" segmentation (discussed further below). The segmentation (Fig. 2) generated for YCC—1-3, 4-6, 7-9, 10-12, 13-15, 16-18—is also periodic with the stages approximately corresponding to alternating M/G1, {G1,S}, {G2,M} phases. Note that each phase is of very

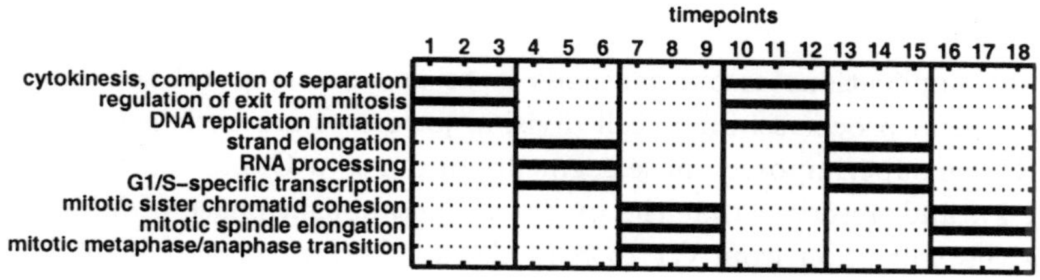

Fig. 2. Gantt chart from segmentation of Spellman et al. dataset. To preserve space, only some of the enriched GO biological process terms are shown.

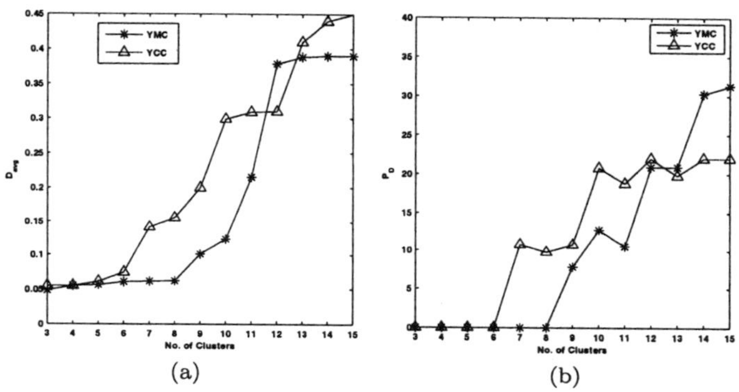

Fig. 3. (a)Segmentation sensitivity and (b)Segmentation quality

short length in this experiment as compared to YMC: the phases M/G1, G1, S each last for approximately two time points, while the G2 phase lasts only for one time point. Because our minimum window length is three (set so that we recover significant clusterings and regroupings), we cannot resolve these short-lived phases. A possible approach is to use continuous representations such as spline fits to gain greater resolution of data sampling. Nevertheless, the key events occurring in these segments are retrieved with high specificity ($p < 10^{-7}$) as shown in Fig. 2.

The effect of the number of clusters on segmentation characteristics is studied in Fig. 3. In Fig. 3 (a), we see that as the number of clusters increases, it is increasingly difficult to obtain independent clusterings and hence, for higher values of the number of clusters, the segmentation problem actually resembles associative clustering (observe that this curve tends toward a D_{avg} value of 0.5). Figure 3 (b) tracks the segmentation quality, and shows that the correct segmentation is recovered for many settings in the lower range for number of clusters, but as the number of clusters increases, the best segmentations considerably deviate from the true segmentation. Nevertheless, comparing the two plots, we see that D_{avg} tracks the segmentation quality P_D well and hence can be a useful surrogate for determining the "right" number of clusters.

6. Discussion

One of the applications of our methods is to decode temporal relationships between biological processes. Since cell division processes are enriched in both YCC and YMC, we superimposed those segments of our two Gantt charts (from Fig. 1 and Fig. 2), and observed that the oxidative metabolism phase of YMC typically precedes the transition from G1 to S in the YCC. This is significant because it permits the DNA replication process to occur in a reductive environment. These and other connections between the YMC and the YCC are presently under intense investigation.[13–15]

Temporal modeling of biological process activity is a burgeoning area of research. For instance, Shi *et al.*[16] present an approach to detect the activity levels of biological processes in a time series dataset. Such ideas can be combined with our segmentation algorithm to get a temporal activity level model of biological processes. In particular, we can develop richer models of cluster reorganization, e.g., dynamic revisions in the number of clusters, split-and-merge behaviors of clusters, and a HMM for cluster re-organization, leading to inference of complete temporal logic models.

References

1. P. Spellman, G. Sherlock, M. Zhang, V. Iyer, K. Anders, M. Eisen, P. Brown, D. Botstein and B. Futcher, *Molecular Biology of the Cell* **9**, 3273 (1998).
2. B. Tu, A. Kudlicki, M. Rowicka and S. McKnight, *Science* **310**, 1152 (2005).
3. I. Simon, Z. Siegfried, J. Ernst and Z. Bar-Joseph, *Nature Biotechnology* **23**, 1503 (2005).
4. T. Yoneya and H. Mamitsuka, *Bioinformatics* **23**, 842 (2007).
5. Y. Shi, T. Mitchell and Z. Bar-Joseph, *Bioinformatics* **23**, 755 (2007).
6. N. Ramakrishnan, M. Antoniotti and B. Mishra, Reconstructing formal temporal logic models of cellular events using the go process ontology, in *Proceedings of the Eighth Annual Bio-Ontologies Meeting (ISMB'05 Satellite Workshop)*,
7. E. Keogh, S. Chu, D. Hart and M. Pazzani, Segmenting time series: A survey and novel approach, in *Data Mining in Time Series Databases*, (World Scientific Publishing Company, 2003)
8. S. Kullback and D. Gokhale, *The Information in Contingency Tables* (Marcel Dekker Inc., 1978).
9. S. Kaski, J. Nikkilä, J. Sinkkonen, L. Lahti, J. Knuuttila and C. Roos, *IEEE/ACM TCBB* **2**, 203 (2005).
10. G. Wrenn, An Indirect Method for Numerical Optimization using the Kreisselmeier-Steinhauser Function NASA Contractor Report 4220(March, 1989).
11. A. Conn, N. Gould and P. Toint, *LANCELOT: A Fortran Package for Large-scale Nonlinear Optimization (Release A)* (Springer Verlag, 1992).
12. R. D. Mántaras, *Machine Learning* **6**, 81 (1991).
13. B. Futcher, *Genome Biology* **7**, 107 (2006).
14. Z. Chen, E. Odstrcil, B. Tu and S. McKnight, *Science* **316**, 1916 (2007).
15. D. Murray, M.Beckmann and H. Kitano, *PNAS* **104**, 2241 (2007).
16. Y. Shi, M. Klustein, I. Simon, T. Mitchell and Z. Bar-Joseph, *Bioinformatics (Proceedings of ISMB 2006)* **23**, i459 (2007).

SYMBOLIC APPROACHES FOR FINDING CONTROL STRATEGIES IN BOOLEAN NETWORKS

CHRISTOPHER JAMES LANGMEAD* and SUMIT KUMAR JHA

Carnegie Mellon University
5000 Forbes Ave.,
Pittsburgh, PA 15213, USA
E-mail: {cjl,sumit.jha}@cs.cmu.edu

We present algorithms for finding control strategies in Boolean Networks (BN). Our approach uses symbolic techniques from the field of *model checking*. We show that despite recent hardness-results for finding control policies, a model checking-based approach is often capable of scaling to extremely large and complex models. We demonstrate the effectiveness of our approach by applying it to a BN model of embryogenesis in *D. melanogaster* with 15,360 Boolean variables.

Keywords: Systems Biology, Model Checking, Control, Boolean Networks.

1. Introduction

Computational cellular and systems modeling is playing an increasingly important role in biology, bioengineering, and medicine. The *promise* of computer modeling is that it becomes a conduit through which reductionist data can be translated into scientific discoveries, clinical practice, and the design of new technologies. The *reality* of modeling is that there are still a number of unmet technical challenges which hinder progress. In this paper, we focus on the specific problem of automatically devising control policies for Boolean Networks (BN). That is, given a BN model with external controls, we seek a sequence of control signals that will drive the network to a pre-specified state at (or by) a pre-specified time.

Recently, it has been shown that finding control strategies for arbitrary BNs is NP-hard,[1] but that polynomial-time algorithms exist for deterministic BNs if the network topology forms a tree. In this paper, we consider a more general family of BNs with arbitrary network topologies. Our algorithm uses techniques from the field of *model checking*.[14] Model checking refers to a family of algorithms and data structures for verifying systems of concurrent reactive processes. Historically, model checking has been used to verify the correctness and safety of circuit designs, communications protocols, device drivers, and C or Java code. Abstractions of these systems can be encoded as finite-state models that are equivalent to Boolean net-

*Corresponding author

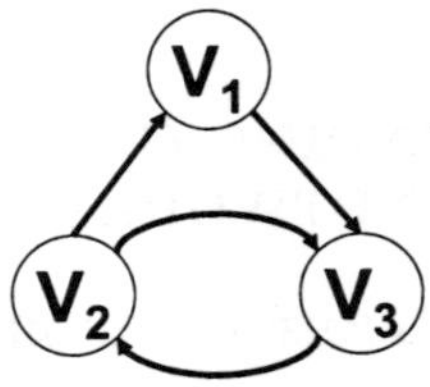

$$\psi_1 \equiv V_1(t+1) = V_1(t) \wedge V_2(t)$$
$$\psi_2 \equiv V_2(t+1) = \neg V_3(t)$$
$$\psi_3 \equiv V_3(t+1) = V_1(t) \wedge V_2(t)$$

time t			time t + 1		
V_1	V_2	V_3	V_1	V_2	V_3
0	0	0	0	1	0
0	0	1	0	0	0
0	1	0	0	1	0
0	1	1	0	0	0
1	0	0	0	1	0
1	0	1	0	0	0
1	1	0	1	1	1
1	1	1	1	0	1

Fig. 1. (**Left**) A Boolean Network (BN). A BN consists of a graph and a set of Boolean functions. The vertices of the graph correspond to Boolean variables and the edges describe functional dependencies. The Boolean functions describe the evolution of the model from time t to $t+1$. The functions can contain any combination of Boolean connectives. (**Right**) A transition relation encoding the same dynamics as the BN. Notice that the BN is a compact encoding of the transition relation.

works. We show that existing model checking algorithms can be used to find control strategies for BNs.

Two important features of model checking algorithms are that they are exact *and* scale to real-world problem instances. For example, model checking algorithms for finite-state systems have been able to reason about systems having more than 10^{20} states since 1990,[8] and have been applied to systems with as many as 10^{120} states.[7] More recently, model checking techniques have been created for stochastic systems.[5] These algorithms can be either exact or approximate, and have also been shown to scale to systems with as many as 10^{30} states.[16] In this paper, we will show that model checking can be used to devise control strategies for very large Boolean networks (up to 15,360 nodes) within seconds or minutes. These techniques are useful in their own right, but will also lay the groundwork for future techniques for finding control strategies in models with asynchronous and stochastic dynamics.

2. Boolean Networks

A BN is a pair, $B = (\mathcal{G}, \Psi)$, where $\mathcal{G} = \{V, E\}$ is a directed graph, and $\Psi = \{\psi_1, \psi_2, ..., \psi_{|V|}\}$ is a set of Boolean functions. Each vertex, $v_i \in V$, represents a Boolean random variable. The state of variable v_i at discrete time t is denoted by $v_i(t)$. The state of all vertices at time t is denoted by $\mathbf{v}(t)$. The directed edges in the graph specify causal relationships between variables. Let $Pa(v_i) \subset V$ be the parents of v_i in the directed graph and let $k_i = |Par(v_i) \cup \{v_i\}|$. A node can be its own parent if we add a self-edge. Each Boolean function $\psi_i : \{0,1\}^{k_i} \mapsto \{0,1\}$ defines the dynamics of v_i from time t to $t+1$ based on the state of its parents at time t. Thus, the set Ψ defines the dynamics of the entire BN. An example BN is shown in Figure 1-left. Notice that a BN is simply a compact encoding of a transition relation over V (Fig. 1-right).

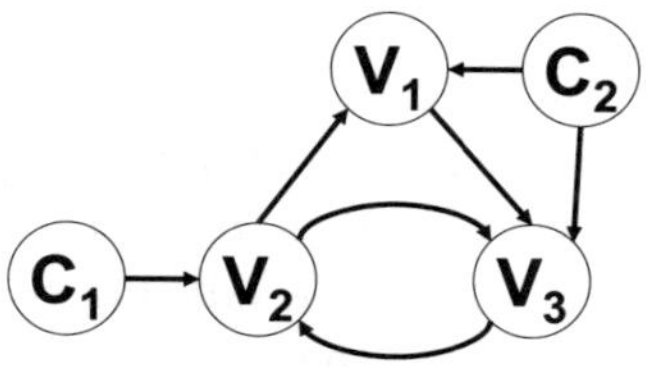

$$\psi_1 \equiv V_1(t+1) = V_1(t) \wedge V_2(t) \vee \neg C_2(t)$$
$$\psi_2 \equiv V_2(t+1) = \neg V_3(t) \wedge C_1(t)$$
$$\psi_3 \equiv V_3(t+1) = V_1(t) \wedge V_2(t) \wedge C_2(t)$$

	V_1	V_2	V_3
Start (t=0)	0	0	0
Goal(t=3)	1	0	0

t	V_1	V_2	V_3	C_1	C_2
0	0	0	0	1	0
1	1	1	0	0	1
2	1	0	1	0	0
3	1	0	0		

Fig. 2. **(Left)** A BN with two control nodes (C_1 and C_2). **(Right top)** An initial state and time-sensitive goal. **(Right bottom)** A control policy (last two columns) that achieves the goal at the specified time.

This basic model can be extended to define a BN with external controls by augmenting our graph with special control nodes, $\mathcal{G} = \{V, C, E\}$. Each control node, c_i, is connected to one or more nodes in V by a directed edge going from c_i to v_j (Fig. 2). The control nodes themselves are externally manipulated. That is, there is no ψ_i that defines the dynamics of c_i.

Consider a set of initial states, I, for the nodes in V specified in terms of a Boolean expression. For example, the expression $I = (v_1 \wedge \neg v_2 \wedge v_3)$ defines the set $\{(1,0,1)\}$, and $I = (v_1 \wedge v_3)$ defines the set $\{(1,0,1),(1,1,1)\}$. We define a set of goal states, F, in a similar fashion. A *control policy*, $\Gamma = \langle \mathbf{c}(0), \mathbf{c}(1), ..., \mathbf{c}(t)\rangle$, is a set of Boolean vectors that defines a sequence of signals to be applied to the control nodes. The BN control problem is to find a control policy that drives the BN such that $\mathbf{v}(0) = I$ and $\mathbf{v}(t) = F$. Our goal in this paper is to algorithmically generate Γ for a given, B, I, F, and t, or to indicate that no such policy exists.

3. Model Checking

The term *model checking*[14] refers to a family of techniques from the formal methods community for verifying systems of concurrent reactive processes. The field of model checking was born from a need to formally verify the correctness of hardware designs. Since its inception in 1981, it has expanded to encompass a wide range of techniques for formally verifying finite-state transition systems, including those with non-deterministic (i.e., asynchronous) or stochastic dynamics. Model checking algorithms are simultaneously theoretically very interesting and very useful in practice. Significantly, they have become the preferred method for formal verification in industrial settings over traditional verification methods like theorem proving, which often need guidance from an expert human user. A complete discussion of model checking theory and practice is beyond the scope of this paper. The interested reader is directed to [14] for a detailed treatment of the subject.

3.1. *Modeling Concurrent Systems as Kripke Structures*

An *atomic proposition*, a, is a Boolean predicate referring to some property of a given system. Let AP be a set of atomic propositions. A *Kripke structure*, M, over AP is a tuple, $M = (S, R, L)$. Here, S is a finite set of states, $R \subseteq S \times S$ is a total transition relation between states, and $L : S \mapsto 2^{AP}$ is a labeling function that labels each state with the set of atomic propositions that are true in that state. Variations on the basic Kripke structure exist. For example, if the system is stochastic, then we replace the transition relation, R, with a stochastic transition matrix, T where element $T(i, j)$ contains either a transition rates (for continuous-time Markov models) or a transition probability (for discrete-time Markov models).

It is easy to see that, in principle, BNs can be encoded as Kripke structures. The state space, S, corresponds to the $2^{|V \cup C|}$ possible states of the BN . We will use the atomic propositions to reveal the state of each variable in the model. That is, $|AP| = |V \cup C|$ and the propositions will be of the form: "*is the state of* v_i 1?" The labeling function, L, can thus be used to define the set of initial states, I, and goal states, F (see Sec. 2). The transition relation, R, corresponds to the table in Figure 1-right. Alternatively, a stochastic transition matrix, T, can be used to encode the stochastic dynamics of the PBN. Naturally, it is generally not possible to explicitly instantiate the Kripke structure for an arbitrary BN because the state space is exponential in the number of nodes. In the next section, we discuss how Kripke structures can be efficiently encoded symbolically.

3.2. *Symbolic Encodings of Kripke Structures*

The basis for symbolic encodings of Kripke structures, which ultimately facilitated industrial applications of model checking, is the reduced ordered Binary Decision Diagrams (BDDs) introduced by Bryant[6] (Fig. 3). BDDs are directed acyclic graphs that symbolically and compactly represent binary functions, $f : \{0,1\}^n \mapsto \{0,1\}$. While the idea of using decision trees to represent boolean formulae arose directly from Shannon's expansion for Boolean functions, two key extensions made by Bryant were i) the use of a fixed variable ordering, and ii) the sharing of sub-graphs. The first extension made the data structure canonical, while the second one allowed for compression in its storage. A third extension, also introduced in [6], is the development of an algorithm for applying Boolean operators to pairs of BDDs, as well as an algorithm for composing the BDD representations of pairs of functions. Briefly, if f and g are Boolean functions, the algorithms implementing operators APPLY(f,g,op) and COMPOSE(f,g) compute directly on the BDD representations of the functions in time proportional to $O(|f||g|)$, where $|f|$ is the size of the BDD encoding f. In this paper, BNs and the desired behaviors are encoded symbolically using BDDs. Model checking algorithms, which call APPLY and COMPOSE as subroutines, are then used to find a valid control strategy, or prove that none exists.

In practice, the construction of the BDDs is done automatically from a high-level language describing the finite-state system and its behavior. In this paper, we use the specification language used in the symbolic model checking tool NuSMV.[12]

x_1	x_2	x_3	f
0	0	0	1
0	0	1	0
0	1	0	0
0	1	1	1
1	0	0	0
1	0	1	0
1	1	0	1
1	1	1	1

(A)

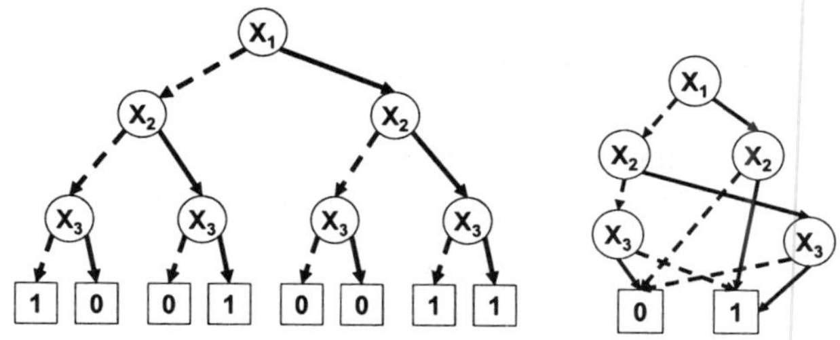

(B) (C)

Fig. 3. (**A**) A truth table for the Boolean function $f(x_1, x_2, x_3) = (\neg x_1 \wedge \neg x_2 \wedge \neg x_3) \vee (x_1 \wedge x_2) \vee (x_2 \wedge x_3)$ (**B**) A Binary Decision Tree of the truth table in (A). A dashed edge emanating from variable/node x_i indicates that x_i is false. A solid edge indicates that x_i is true. (**C**) A Binary Decision Diagram of the truth table in (A). Notice that it is a more compact representation that the Binary Decision Tree.

We note that BDDs can be generalized to Multi-terminal BDDs[13] (MTBDD), which encode discrete, real-valued functions of the form $f : \{0,1\}^n \mapsto \mathbb{R}$. Significantly, MTBDDs can be used to encode real-valued vectors and matrices, and algorithms exist for performing matrix addition and multiplication over MTBDDs.[13] These algorithms play an important role in several model checking algorithms for stochastic systems[5] which, in turn, we have used to develop algorithms for finding control strategies in BNs with stochastic behaviors. Due to space limitations, we will focus on algorithms for deterministic BNs in this paper and report the algorithms for stochastic BNs elsewhere.

3.3. *Temporal Logics*

Temporal logic is a formalism for describing behaviors in finite-state systems. It has been used since 1977 to reason about the properties of concurrent programs.[23] There are a number of different temporal logics from which to chose, and different logics have different expressive powers. In this paper, we use a small subset of the Computation Tree Logic (CTL). CTL formulae can express properties of *computation trees.* The root of a computation tree corresponds to the set of initial states (i.e., I) and the rest of the (infinite) tree corresponds to all possible paths from the root. A complete discussion of CTL and temporal logics is beyond the scope of this paper. The interested reader is directed to [14] for more information.

The syntax of CTL is given by the following minimal grammar:

$$\phi ::= a \mid true \mid (\neg\phi) \mid (\phi_1 \wedge \phi_2) \mid \mathbf{EX}\phi \mid \mathbf{E}[\phi_1 \mathbf{U} \phi_2]$$

Here, $a \in AP$, is an atomic proposition; "true" is a Boolean constant; $\neg$ and $\vee$ are the normal logical operators; **E** is the existential path quantifier (i.e., "there exists some path from the root in the computation tree"); and **X** and **U** are temporal operators corresponding to the notions of "in the next state", and "until", respectively. Given these, additional operators can be derived. For example, "false" can be derived from "$\neg true$" and the universal quantifier, $\mathbf{AX}\phi$, can be defined as $\neg\mathbf{EX}\neg\phi$.

Given some path through the computation tree, $\pi = \langle \pi[0], \pi[1], \ldots \rangle$, the semantics of a CTL formula are defined recursively:

$$\pi \models a \text{ iff a} \in \mathrm{L}(\pi[0])$$
$$\pi \models true, \forall \pi$$
$$\pi \models \neg\phi \text{ iff } \pi \not\models \phi$$
$$\pi \models \phi_1 \wedge \phi_2 \text{ iff } \pi \models \phi_1 \text{ and } \pi \models \phi_2$$
$$\pi \models \mathbf{EX}\phi \text{ iff } \pi[1] \models \phi$$
$$\pi \models \mathbf{E}[\phi_1 \mathbf{U} \phi_2] \text{ iff } \exists i \geq 0, \pi[i] \models \phi_2 \wedge \forall j < i, \pi[j] \models \phi_1$$

Here, the notation"$\pi \models \alpha$" means that π *satisfies* α.

3.4. *Model Checking Algorithms*

A model checking algorithm takes a Kripke structure, $M = (S, R, L)$, and a temporal logic formula, ϕ, and finds the set of states in S that satisfy ϕ: $\{s \in S \mid M, s \models \phi\}$. The complexity of model checking algorithms varies with the temporal logic and the operators used. For the types of formulas used in this paper (see Sec. 4), an explicit state model checking algorithm requires time $O(|\phi|(|S| + |R|))$, where $|\phi|$ is the number of sub-formulas in ϕ ([14] p. 38).

Of course, for very large state spaces, even linear time is unacceptable. *Symbolic model checking* algorithms operate on BDD encodings of the Kripke structure and CTL formula. Briefly, the temporal operators of CTL can be characterized in terms of fixpoints. Let $\mathcal{P}(S)$ be the powerset of S. A set $S' \subseteq S$ is a fixpoint of a function $\tau : \mathcal{P}(S) \mapsto \mathcal{P}(S)$ if $\tau(S') = S'$. Symbolic model checking algorithms define an appropriate function, based on the formula, and then iteratively find the fixpoint of the function. This is done using set operations that operate directly on BDDs. The fixpoint of the function corresponds exactly to $\{s \in S \mid M, s \models \phi\}$. The interested reader is encouraged to read [14], ch. 6 for more details.

The symbolic model checking algorithms used in this paper are exact. We note that there are also approximation algorithms for model checking (e.g., [27]), which employ sampling techniques and hypothesis testing. Such algorithms provide guarantees, in terms of the probability of the property being true, and can scale to much larger state spaces. These do not use BDDs, but rather operate on the high-level language description of the finite-state model.

4. A Symbolic Model Checking Approach to Finding Control Policies

The use of model checking algorithms for finding control strategies requires three steps:

First, the BN must be encoded using a high level language for describing finite-state models. Different model checking software use different modeling languages. In Figure 4, we show pseudo-code for encoding the BN in figure 2. This pseudo-code is based on the language used in the model-checking tool NuSMV. The code contains a block of variable definitions. In the example, we declare Boolean variables for v_1, v_2, v_3, c_1,and c_2. The set of initial states, I, is encoded using "init" statements. The update rules, Ψ, are encoded using "next" statements. A single variable COUNTER is declared that marks the passage of time. A "next" statement for COUNTER updates the counter.

Second, a CTL formula must be written. In this paper, we are concerned with CTL formulae that ask whether it is possible to end up in the goal state(s), F, at time t. Let ϕ_F be a formula describing the goal state. This formula can describe any subset of the variables in the BN. For example, $\phi_F := v_1 \wedge \neg v_2 \wedge v_3$ or $\phi_F := v_1 \wedge v_3$ are both valid formulas. The former chooses to specify the state of each variable, the latter does not. Let ϕ_t :=COUNTER= t be a Boolean formula that evaluates to true if the variable COUNTER is t. The formula $\phi := \mathbf{E}[\neg\phi_F \mathbf{U}(\phi_F \wedge \phi_t)]$ can be used to find a control policy. In English, this formula says: "*There exists a path that enters state F for the first time at time t*". Alternatively, if we wish to relax the restriction that the BN cannot enter state F before time t, we would use the formula $\phi' := \mathbf{E}[true\mathbf{U}(\phi_F \wedge \psi_t)]$, which translates as "*In the future, the model will be in F at time t.*" Temporal logics are very expressive and can encode a number of complex behaviors. For example, it is possible to specify particular milestones through which the model should pass en route to the final goal. That is, one can construct formula that say that the BN should enter state X_1 before X_2, must enter X_2 by time t_1, and must reach the goal state at exactly time t_2. This expressive power is one of the key advantages of a model checking based approach to the design of control policies.

```
MODULE BN
VAR
  V1: boolean;          // variable node 1
  V2: boolean;          // variable node 2
  V3: boolean;          // variable node 3
  C1: boolean;          // control node 1
  C2: boolean;          // control node 2
  COUNTER: 0 .. T+1;// counter
ASSIGN
  init(V1) := 1;
  init(V3) := 1;
  next(V1) := (V1 & V2) | !C2;
  next(V2) := ! V3 & C1;
  next(V3) := V1 & V2 & C2;
  next(COUNTER) := COUNTER+1 ;
```

Fig. 4. Pseudocode based on the language used in the symbolic model checking program NuSMV. This code implements the BN in Figure 2. The code consists of a module with variable declaration statements, "init" statements that initialize the variables, and "next" statements that implement each ϕ_i and increment a counter.

Finally, we apply an appropriate symbolic model checking algorithm to find a control policy. If a control policy exists (i.e., if ϕ is true), then we ask the model checking algorithm for a *witness*, π_w, to the formula. The control policy, Γ, is then simply extracted from π_w by reading off the values of $\langle \mathbf{c}(0), \mathbf{c}(1), ..., \mathbf{c}(t) \rangle$[a].

5. Related Work

Boolean Networks have been used extensively to model complex biological systems (e.g., [2,3,17,18]). The design of control strategies for Boolean networks and related models has been considered by a number of different authors (e.g.,[1,11,15,24]). Akutsu and co-workers[1] were the first to show that the design of control policies is NP-hard. They also provide a polynomial-time algorithm that works on the special case where the topology of the BN forms a tree. The primary difference between our work and these is that our method is based on symbolic model checking and we place no restriction on the topology of the network. We will show in the next section that despite the fact that the problem is NP-hard, in practice model checking based approaches to control policy design can scale to very large models. Of course, the hardness result implies that our approach will not apply to every BN.

Recently, there has been growing interest in the application of formal methods, including model checking to biology. Most applications of model checking in biology have been directed to modeling biochemical and regulatory networks, (e.g.,[4,9,10, 19,22]), although not for the design of control policies. In our own work, we have applied model checking,[20] and a related technology based on decision procedures[21] to the protein folding problem.

6. Results

We present results from two kinds of experiment. The first experiment is designed to highlight the scalability of a model checking based approach to control policy design. The second experiment applies our approach to an existing BN model of embryo development in drosophila.

6.1. *Scalability*

We have performed a large-scale study on randomly generated BNs in order to characterize the scalability of our approach. In total, we considered 13,400 separate BNs. We considered several different network topologies, which are shown in Figure 5. These topologies are meant to reflect different kinds of networks ranging from simple feedback loops (*chains*), feedback loops with complex topologies (*random chains*), loosely coupled modules (*modular*), to a dense network (*small diameter*). Within each network category, we performed separate experiments randomly generating graphs by varying: a) the number of non-control variables over

[a]Equivalently, as we performed in our experiments, we can request a counterexample to $\neg\phi$.

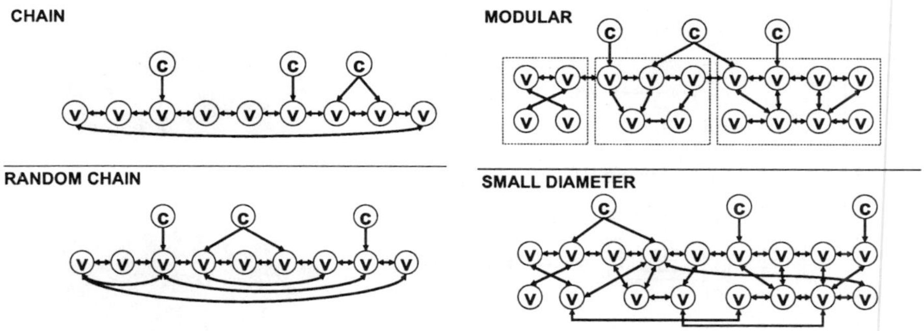

Fig. 5. Network topologies used in our experiments on scalability. *Chain* describes a model where the variables form a circular chain. *Random Chain* describes a model where the variables form a circular chain, but a random number of "long-range" edges are added. *Modular* describes a model with coupled modules. Each module is outlined. *Small Diameter* describes a model where a graph has a small diameter. In each case, the placement of the control nodes is random.

the interval [10,640]; b) the average number of parents for each node over the interval [2, 8]; c) the number of control nodes over the interval [2,64]; d) the number of variables specified in the goal state, F, over the interval [4,80]; and e) the target time, t, over the interval [1,32]. For each combination of parameters, we generated 100 BNs randomly, constructed a CTL formula, and identified a control strategy using NuSMV. Due to space limitations, we will simply report that each experiment took less than 12 minutes on a single Pentium 3 processor with 2 GB of memory. The mean and median runtimes were 2 and 0.6 seconds, respectively. The longest runtime (693 seconds) was on a random chain topology model with 80 nodes, an average in-degree of 4, 4 control nodes, a target specifying the state of 4 variables, and a time of 32. These results suggests that a model checking approach to policy design scales well to randomly generated BNs.

6.2. *Application To* D. Melanogaster *Embryo Development*

To test our approach on a BN for a real biological process, we applied it to the task of finding control policies to an existing model of fruit fly embryo development.[3] Briefly, Albert and Othmer have developed a BN model of the segment polarity gene network in *D. Melanogaster* (Fig. 6-left). The model comprises 5 RNAs: (*wingless (wg)*; *engrailed (en)*; *hedgehog (hh)*; *patched (ptc)*; and *cubitus interruptus (ci)*), and 10 proteins: (*WG*; *EN*; *HH*; *PTC*; *CI*; *smoothened (SMO)*; *sloppy-paired (SLP)*; a transcriptional repressor, *(CIR)*, for *wg*, *ptc*, and *hh*; a transcriptional activator, *(CIA)* for *wg* and *ptc*; and the *PTC-HH* complex, (*PH*)). Each molecule is modeled as a Boolean variable and the update rules are Boolean formulas that take into account both intra-cellular state, and inter-cellular communication. The Albert and Othmer research did not consider the question of control policy design.

Albert and Othmer have demonstrated that the Boolean model accurately reproduces both wild-type and mutant behaviors. In their experiments, they consider

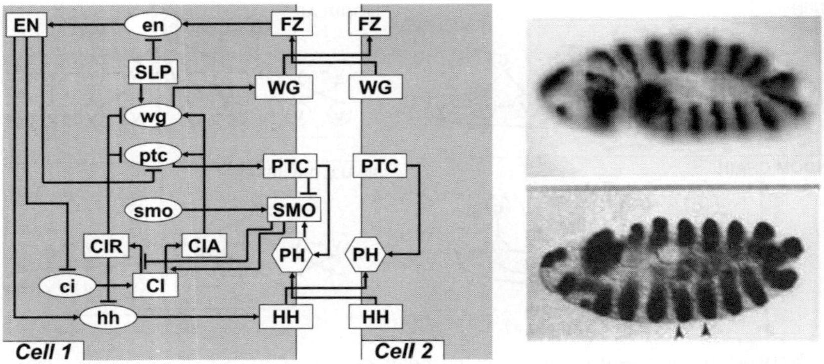

Fig. 6. **(Left)** The drosophila segment polarity BN from Albert and Othmer. The figure shows one cell in detail (large grey box), and the inter-cellular signals (*WG* and *HH*) between two adjacent cells. See text for more details. **(Right)** Expression pattern of *wg* in wild-type (top) and a "broad-stripe" mutant embryo (bottom).

a 1-dimensional array of cells initialized to the experimentally characterized cellular blastoderm phase of *Drosophila* development, which immediately precedes the activation of the segment-polarity network. The purpose of the segment-polarity network is to maintain a pattern of expression throughout the life of the fly that defines the boundaries between *parasegments*, small linear groupings of adjacent cells. Two possible parasegment boundary expression patterns are shown in Figure 6-(right)[b]. In the Albert and Othmer work, the parasegments are four cells wide. We note that the steady-state expression patterns of different sub-populations of cells differ due to inter-cellular communication — this is precisely the mechanism by which the parasegment boundaries are maintained. That is, the fate of every cell is *not* the same, even though each cell is running the same regulatory network.

In our experiment, we modified the Albert and Othmer BN in two ways. *First*, we considered a 32x32, two-dimensional array of cells of dimension, instead of the 1x12 one-dimensional array of cells considered in [3]. We believe that this extension to a two-dimensional model is the first of its kind; we also believe that the 15,360 Boolean variables in our model is the largest ever considered for the purpose of control policy design. Topologically, this network most closely resembles the "modular" network in Figure 5. Adjacent cells in the network can communicate, which introduces loops in overall topology of the BN for the 16x16 array of cells. *Second*, we modified the network such that the RNAs *wg* and *hh* becomes a control node in the network. In principle, one could control RNAs through RNA-silencing or micro RNAs. We used our methods to design two control policies for *hh*. The first is designed to drive the system to either the wild-type expression pattern (Fig. 6-right (top)) and the other to a "broad-stripe" pattern (Fig. 6-right (bottom)). Our algorithms

---

[b]The images in Fig. 6-right are taken from http://www.fruitfly.org (top) and [26] (bottom)

successfully found the two control policies in 6.1 and 6.2 minutes, respectively. The computation was dominated by the time to construct the BDDs. We believe these results strongly suggest that our approach can be used to find control signals for biologically relevant BNs of substantial size.

7. Conclusions and Future Work

We have introduced an effective means for automatically discovering control sequences for Boolean networks based on techniques from the field of model checking. Our approach scales to very large BNs, having as many as 15,360 nodes, and runs in seconds to minutes. We note that, due to the inherent computational complexity of finding control policies in BNs,[1] we cannot claim that our approach will scale to *every* BN of large size. Rather, our results suggest that the modular design of "real" biological networks may reduce the possibility of encountering worst-case instances. This is an interesting question and we believe it is related to the phenomenon of canalizing functions and other generic properties of BNs (e.g., [25]).

BNs have been used widely to model a range of biological phenomena. However, the fact that BNs made strong assumptions about the binary nature of each variable (i.e., active or inactive), the synchronous nature of the updates, the assumption that time unfolds in discrete steps, and the assumption that the dynamic are deterministic. Ultimately, these assumptions limit the overall applicability of BNs. We note that our approach to control policy design can be adapted for use to a much broader range of models including those with continuous-valued variables, asynchronous updates between variables, continuous time, and stochastic transitions. We are presently pursuing these goals as part of ongoing research.

Acknowledgments

This research was supported by a U.S. Department of Energy Career Award (DE-FG02-05ER25696), and a Pittsburgh Life-Sciences Greenhouse Young Pioneer Award to C.J.L.

References

1. T. Akutsu, M. Hayashida, W.K. Ching, and M. Ng. On the complexity of finding control strategies for boolean networks. *Proc. 4th Asian Pacific Bioinf. Conf.*, pages 99–108, 2006.
2. T. Akutsu, S. Miyano, and S. Kuhara. Inferring qualitative relations in genetic networks and metabolic pathways. *Bioinformatics*, 16(8):727–734, 2000.
3. R. Albert and H. G. Othmer. The topology of the regulatory interactions predics the expression pattern of the segment polarity genes in drosophila melanogaster. *Journal of Theoretical Biology*, 223:1–18, 2003.
4. M. Antoniotti, A. Policriti, N. Ugel, and B. Mishra. Model building and model checking for biochemical processes. *Cell Biochem Biophys.*, 38(3):271–286, 2003.
5. C. Baier, E. Clarke, V. Hartonas-Garmhausen, M. Kwiatkowska, and M. Ryan. Symbolic model checking for probabilistic processes. *Proc. 24th International Colloquium on Automata, Languages and Programming (ICALP'97)*, 1256:430–440, 1997.

6. R. E. Bryant. Graph-based algorithms for boolean function manipulation. *IEEE Trans. Comput.*, 35(8):677–691, 1986.
7. J.R. Burch, E. M. Clarke, D. E. Long, K. L. McMillan, and D. L. Dill. Symbolic model checking for sequential circuit verification. *IEEE Transactions on Computer-Aided Design of Integrated Circuits and Systems*, 3(4):401–424, 1993.
8. J.R. Burch, E. M. Clarke, K. L. McMillan, D. L. Dill, and L. J. Hwang. Symbolic Model Checking: 10^{20} States and Beyond. *Proc. Fifth Ann. IEEE Symposium on Logic in Computer Science*, pages 428–439, 1990.
9. M. Calder, V. Vyshemirsky, D. Gilbert, and R. Orton. Analysis of signalling pathways using the PRISM model checker. *Proc. Computational Methods in Systems Biology (CMSB'05)*, pages 179–190, 2005.
10. N. Chabrier and F. Fages. Symbolic Model Checking of Biochemical Networks. *Proc 1st Internl Workshop on Computational Methods in Systems Biology*, pages 149–162, 2003.
11. P. C. Chen and J. W. Chen. A markovian approach to the control of genetic regulatory networks. *Biosystems*, 90(2):535–45, 2007.
12. A. Cimatti, E.M. Clarke, E. Giunchiglia, F. Giunchiglia, P. Pistore, M. Roveri, R. Sebastiani, and A. Tacchella. Nusmv 2: An opensource tool for symbolic model checking. *CAV '02: Proceedings of the 14th International Conference on Computer Aided Verification*, pages 359–364, 2002.
13. E.M. Clarke, M. Fujita, P. C. McGeer, J.C.-Y. Yang, and X. Zhao. Multi-terminal binary decision diagrams: An efficient datastructure for matrix representation. *IWLS '93 International Workshop on Logic Synthesis*, 1993.
14. E.M. Clarke, O. Grumberg, and D. A. Peled. *Model Checking.* MIT Press, Cambridge, MA, 1999.
15. A. Datta, A. Choudhary, M. L. Bittner, and E.R. Dougherty. External control in markovian genetic regulatory networks. *Mach. Learn.*, 52(1-2):169–191, 2003.
16. L. de Alfaro, M. Kwiatkowska, G. Norman, D. Parker, and R. Segala. Symbolic model checking of concurrent probabilistic processes using MTBDDs and the Kronecker representation. *Proc. 6th Int. Conf. on Tools and Algorithms for the Construction and Analysis of Systems (TACAS'00)*, 1785:395–410, 2000.
17. S.E. Harris, B.K. Sawhill, A. Wuensche, and S. Kauffman. A model of transcriptional regulatory networks based on biases in the observed regulation rules. *Complex.*, 7(4):23–40, 2002.
18. S. A. Kauffman. *The Origins of Order: Self-Organization and Selection in Evolution.* Oxford University Press, 1993.
19. M. Kwiatkowska, G. Norman, D. Parker, O. Tymchyshyn, J. Heath, and E. Gaffney. Simulation and verification for computational modelling of signalling pathways. *WSC '06: Proceedings of the 38th conference on Winter simulation*, pages 1666–1674, 2006.
20. C.J. Langmead and S. K. Jha. Predicting protein folding kinetics via model checking. *Lecture Notes in Bioinformatics: The 7th Workshop on Algorithms in Bioinformatics (WABI)*, pages 252–264, 2007.
21. C.J. Langmead and S. K. Jha. Using bit vector decision procedures for analysis of protein folding pathways. *Fourth Workshop on Constraints in Formal Verification*, page in press, 2007.
22. C. Piazza, M. Antoniotti, V. Mysore, A. Policriti, F. Winkler, and B. Mishra. Algorithmic Algebraic Model Checking I: Challenges from Systems Biology. *17th Internl Conf. on Comp. Aided Verification (CAV)*, pages 5–19, 2005.
23. A. Pnueli. The temporal logic of programs. *Proceedings of the 18th IEEE. Foundations of Computer Science (FOCS)*, pages 46–57, 1977.

24. P. Ranadip, D. Aniruddha, L. Bittner, and R. Dougherty. Intervention in context-sensitive probabilistic boolean networks. *Bioinformatics*, 21(7):1211–1218, 2005.
25. I. Shmulevich, H. Lhdesmki, E. R. Dougherty, J. Astola, and W. Zhang. The role of certain post classes in boolean network models of genetic networks. *Proc Natl Acad Sci U S A*, 100(19):10734–10739, 2003.
26. T. Tabata, S. Eaton, and T. B. Kornberg. The drosophila hedgehog gene is expressed specifically in posterior compartment cells and is a target of engrailed regulation. *Genes Dev.*, 6(12B):2635–2645, 1992.
27. H. L. S. Younes and R. G. Simmons. Probabilistic verification of discrete event systems using acceptance sampling. *CAV '02: Proceedings of the 14th International Conference on Computer Aided Verification*, pages 223–235, 2002.

ESTIMATION OF POPULATION ALLELE FREQUENCIES FROM SMALL SAMPLES CONTAINING MULTIPLE GENERATIONS

DMITRY A. KONOVALOV

School of Mathematics, Physics and Information Technology, James Cook University, Townsville, Queensland 4811, Australia

DIK HEG[†]

Department of Behavioural Ecology, Zoological Institute, University of Bern, Hinterkappelen, Switzerland

Estimations of population genetic parameters like allele frequencies, heterozygosities, inbreeding coefficients and genetic distances rely on the assumption that all sampled genotypes come from a randomly interbreeding population or sub-population. Here we show that *small* cross-generational samples may severely affect estimates of allele frequencies, when a small number of progenies dominate the next generation or the sample. A new estimator of allele frequencies is developed for such cases when the kin structure of the focal sample is unknown and has to be assessed simultaneously. Using Monte Carlo simulations it was demonstrated that the new estimator delivered significant improvement over the conventional allele-counting estimator.

1 Introduction

The estimation of population frequencies of codominant genetic markers (*e.g.* microsatellites) from samples with unknown kin structures is of paramount importance to the population genetic studies, since they form the foundation for downstream genetic analyses.[1] The frequencies can be used to estimate, for instance, the genetic distance between two populations, or the effective population size. Similarly, deviations from Hardy-Weinberg Equilibrium (HWE) of these alleles can be used to assess past effects on the genetic structure of the population due to, for instance, genetic drift, inbreeding, and genetic bottlenecks. The population frequencies are normally estimated from a large sample of assumed to be unrelated individuals.[2] In practice, it may be difficult to acquire genotypes from free-living individuals fulfilling this basic assumption of sampling population frequencies and often samples contain a mixture of related genotypes from multiple generations.[3] Currently, it is unknown how the population allele frequencies can be reliably estimated when actual pedigrees within data sets are unknown and have to be assessed simultaneously. Although this may not matter for large sample sizes within a randomly interbreeding population, where all individuals contribute equally to the next

[†] Work partially supported by SNF grant 3100A0-108473.

generation, this certainly will matter for small samples from populations wherein some individuals are more productive than others.[1] For example, if a sample of 100 individuals consists of 40 full-sibs and 60 unrelated individuals,[1] it is very likely that the sample will fail the exact test for HWE,[4] *e.g.* calculated via the GENEPOP program.[5] Such a case is the focus of this study, when the null hypothesis of HWE is rejected (*e.g.* $P < 0.05$), but the sample may still contain sufficient information for the estimation of the population allele frequencies in the HWE sense. That is, the 60 unrelated individuals in the considered example is commonly deemed a "large" sample.[6]

Methods for estimating allele frequencies do exist but they are mostly a by-product of sibship reconstruction.[7-13] However, it is not known if such frequencies could be obtained effectively for a multi generational population sample which could contain any kin groups, such as cousins, half and full sibs including or excluding parental genotypes.[3]

In addition, the generic pedigree reconstruction problem[14] is clearly more difficult than the problem of detecting all unrelated individuals (to be used for allele frequency estimates). Hence there is a much higher chance that the allele frequencies obtained this way would be affected by the pedigree reconstruction errors. Moreover, the population allele frequencies must be estimated iteratively during the sibship reconstruction,[9] thus frequencies' errors feeding into the reconstruction procedure. If incorrectly done, they reduce the reconstruction accuracy drastically, *e.g.* when the frequencies are estimated from the population sample containing a large family of full sibs as in data sets with family sizes of 40,5,2,2, and 1.[11,12,15]

It is important to differentiate the problem at hand from the problem of estimating population allele frequencies when the pedigree of the sampled individuals is known or assumed to be known, in which case population allele frequencies can be calculated exactly.[2,16] In this preliminary study we report for the first time that a robust method for estimation of the outbred population allele frequencies may be possible even when sample genotypes contain individuals from multiple generations and when the actual pedigree is assessed simultaneously using the same genetic markers.

The following is the outline of this study: (1) given the difficulty of inferring allele frequencies and kin structure from the same sample simultaneously, a pair-wise relatedness estimator is developed, which does not require allele frequencies; (2) the structure of the pair-wise relatedness matrix is examined when the sample kin structure is known exactly; (3) using the properties of the relatedness matrix, a new approach is proposed for searching for the largest sample subset, which resembles a set of unrelated individuals; (4) and finally the new approach is tested via Monte Carlo simulations on three different data sets.

2 Method

2.1 *Estimation of Pairwise Relatedness*

Following in some respects Broman[2] and McPeek *et al.*,[16] let a diploid population sample consists of n genotype vectors $\{\mathbf{x}_1, \mathbf{x}_2, ..., \mathbf{x}_n\}$ at a single locus with k co-dominant alleles. The i'th genotype is defined via the number of observed alleles:[17] $\mathbf{x}_i = (..., x_{mi}, ..., x_{m'>mi}, ...)^T = (..., 1, ..., 1, ...)^T$ for heterozygotes and $\mathbf{x}_i = (..., x_{mi}, ...)^T = (..., 2, ...)^T$ for homozygotes, where the rest of the values are zero, and where 'T' denotes the transpose. For example, a genotype (A_1, A_2) is encoded as (1,1,0,0) at a locus with four alleles $\{A_1, A_2, A_3, A_4\}$. Each diploid genotype contains exactly two alleles, $(\mathbf{1} \cdot \mathbf{x}_i) = \sum_{m=1}^{k} x_{mi} = 2$, where $\mathbf{1}$ is the vector of 1's of length k and where the dot-product notation is used for summations when the summation index and range is clear by context, *i.e.* $(\mathbf{x} \cdot \mathbf{y}) = \sum_{m=1}^{k} x_m y_m$.

Let an outbred population (or sub-population) be in HWE and described by the population allele frequencies $\mathbf{p} = (p_1, p_2, \ldots, p_k)^T$. Then each observed (sample) genotype $\mathbf{x}_i$ could be represented as a sum of two statistically independent gamete vectors $\mathbf{x}_i = \boldsymbol{\varepsilon}_i + \boldsymbol{\varepsilon}_i'$, *i.e.* $x_{mi} = \varepsilon_{mi} + \varepsilon_{mi}'$, obtaining $E(\varepsilon_{mi}^2) = E(\varepsilon_{mi}) = p_m$, $\mathrm{var}(\varepsilon_{mi}) = p_m(1-p_m)$, $E(x_{mi}) = 2p_m$, $E(x_{mi}^2) = 2p_m(1+p_m)$, $E(\mathbf{x}_i \cdot \mathbf{x}_i) = 2(1+\gamma)$, and $\mathrm{var}(x_{mi}) = 2p_m(1-p_m)$.[2] The pairwise relatedness matrix could be defined in the identity-by-descent (IBD) sense[18] via $\mathbf{x}_j = r_{ij}\mathbf{x}_i + (1-r_{ij})\mathbf{z}_{ij}$, where $r_{ii} = 1$, and $\mathbf{z}_{ij}$ is statistically independent of $\mathbf{x}_i$. Then $\mathrm{cov}(x_{mi}, x_{mj}) = 2r_{ij}p_m(1-p_m)$, $E(x_{mi}x_{mj}) = 2[r_{ij}p_m(1-p_m) + 2p_m^2]$ and $E(\mathbf{x}_i \cdot \mathbf{x}_j) = 2(r_{ij}h + 2\gamma)$, where $\gamma = (\mathbf{p} \cdot \mathbf{p}) = \sum_{m=1}^{k} p_m^2$ and $h = 1-\gamma$ are the population homozygosity and heterozygosity of the given locus, respectively.

In practice, the pedigree of a sample is often not known *a priori* and hence the relatedness matrix must be estimated together with the allele frequencies. This could be done by using the following estimators of heterozygosity and relatedness, which do not require allele frequencies. An estimator h' of heterozygosity at a locus (and hence homozygosity via $\gamma' = 1-h'$) is given by $h' = \sum_{i=1}^{n} u_i h_{ii}$, where the weights $(u_1, u_2, \ldots, u_n)^T$ are normalized by $\sum_{i=1}^{n} u_i = 1$, and where $h_{ii} = 1$ and $h_{ii} = 0$ for heterozygotes and homozygotes, respectively. If the relatedness matrix $\mathbf{r} = \{r_{ij}\}$ were known, the most optimal weights could be found by minimising $\mathrm{var}(h')$. Since $\mathbf{r}$ is not known, the equal weights $u_i = 1/n$ are used, which yield an unbiased, but not necessarily the most efficient, estimator of heterozygosity in the absence of allele frequencies. The estimate at a locus simply equals to the number of observed heterozygotes averaged over the sample size *n*. Assuming unlinked loci, for multilocus genotypes $\mathbf{X}_i = \{\mathbf{x}_i(1), \mathbf{x}_i(2), \ldots, \mathbf{x}_i(L)\}$, the $h' \equiv h'(l)$ estimator is averaged across loci obtaining $H = \sum_{l=1}^{L} h(l)/L$ and $H' = \sum_{l=1}^{L}\sum_{i=1}^{n} h_{ii}(l)/(nL)$, where $E(H') = H$ and $\mathrm{var}(H') = \sum_{l=1}^{L} \mathrm{var}[h'(l)]/L^2$, *i.e.* the estimate equals to the number of observed heterozygotes averaged over the sample size *n* and number of loci *L*. An estimator for relatedness is given by $r_{ij}'(h) = 1 - d_{ij}^2/H'$, where $d_{ij}^2 = \sum_{l=1}^{L} d_{ij}^2(l)/L$ and $d_{ij}^2(l) = [\mathbf{x}_i(l) - \mathbf{x}_j(l)]^2/4$.

2.2 *Estimation of Allele Frequencies from Known Pedigree*

Following McPeek *et al.*[16] the class of best linear unbiased estimators (BLUE) of allele frequencies is given by

$$\mathbf{q} = \tfrac{1}{2}\sum_{i=1}^{n} w_i \mathbf{x}_i , \tag{1}$$

where the weights $\mathbf{w}^T = (w_1, w_2, ..., w_n)^T$ are normalized by $\sum_{i=1}^{n} w_i = 1$ and hence $E(q_m) = p_m$. The *sample* allele frequencies $\mathbf{s} = (s_1, ..., s_k)^T$ are obtained via $w_i = 1/n$,[6]

$$\mathbf{s} = \sum_{i=1}^{n} \mathbf{x}_i /(2n), \tag{2}$$

which specifies the conventional *allele-counting* estimator. In general, the weights are found by minimizing the variance of each resulting frequency q_m, $\mathrm{var}(q_m) = \tfrac{1}{4}\sum_{i=1}^{n}\sum_{j=1}^{n} w_i w_j \,\mathrm{cov}(x_{mi}, x_{mj})$. Treating each allele with equal weight at the locus, the problem is transformed into finding the weights that minimize

$$V = \sum_{m=1}^{k} \mathrm{var}(q_m) = \tfrac{1}{2} h \sum_{i=1}^{n}\sum_{j=1}^{n} w_i r_{ij} w_j , \tag{3}$$

where the same weights minimize both the absolute and the relative variances, $\sum_{m=1}^{k} \mathrm{var}(q_m)/p_m$. If all individuals are unrelated ($r_{ij} = \delta_{ij}$, $w_i = 1/n$ and $V = h/(2n)$), the commonly used heterozygosity estimator is obtained $h_{\mathrm{Nei}} = 2n(1 - \sum_{m=1}^{k} q_m^2)/(2n-1)$,[6] where δ_{ij} is the Kronecker delta defined by $\delta_{ii} = 1$ and $\delta_{i \neq j} = 0$. The estimator is also known as the gene diversity and is bias corrected for the sample size[19] but not for the sample kin structure.

Since the relatedness matrix r_{ij} is symmetric and positive definite $(V > 0)$, its eigenvectors can always be found and defined as orthonormal $(\xi_\alpha \cdot \xi_\beta) = \delta_{\alpha\beta}$ and sorted by the corresponding real positive eigenvalues $\{0 < \lambda_1 \le \lambda_2 \le ... \le \lambda_n\}$, where $\mathbf{r}\xi_\alpha = \lambda_\alpha \xi_\alpha$. The weights vector in its most generic form is then given by $\mathbf{w} = \sum_{\alpha=1}^{n} C_\alpha \xi_\alpha$ obtaining $V = \tfrac{1}{2} h \sum_{\alpha=1}^{n} C_\alpha^2 \lambda_\alpha$, subject to the original normalization of the weights $\sum_{i,\alpha=1}^{n} C_\alpha \xi_{i\alpha} = 1$. The minimum is found via Lagrange multiplier obtaining $C_\alpha = \zeta_\alpha /(\eta \lambda_\alpha)$ and $\min(V) = h/(2\eta)$, where $\eta = \sum_{\alpha=1}^{n} \zeta_\alpha^2 / \lambda_\alpha$ and $\zeta_\alpha = \sum_{i=1}^{n} \xi_{i\alpha}$. Observing that the inverse matrix of r_{ij} can be written as $(r^{-1})_{ij} = \sum_{\alpha=1}^{n} \xi_{i\alpha}\xi_{j\alpha} / \lambda_\alpha$, the solution can also be expressed via $w_i = \sum_{j=1}^{n} (r^{-1})_{ij} / \eta$, $\eta = \sum_{i,j=1}^{n} (r^{-1})_{ij}$. For multiple loci the resulting formulas for the weights are locus independent, hence the same weights are used to estimate allele frequencies at all loci. The obtained weights (and hence frequencies) provide the *exact* solution to the problem of finding an unbiased estimator of frequencies, which is the most efficient in terms of achieving the smallest possible (absolute and relative) variance of the frequencies in Eq. (3).

When the above formulas are applied (results not shown) to the samples from the *unrelated* data set (see Results section, below) a solution is normally found in the form $w_{i\in U} = 1/u$ and $w_{i\notin U} = 0$ (ignoring rounding errors), where u is the number of elements in the subset U of all unrelated parents in the sample. Note that the weights represent the *theoretical limit* of the allele frequency inference from a single sample, *i.e.* a biologist would select the same weights if he or she knew which individuals are unrelated parents and which are offspring.

2.3 *Unknown Pedigree*

The population allele frequencies could be calculated exactly from a given relatedness matrix but only if the matrix is positive definite. A sample instance of the r'_{ij} matrix may not be positive definite (regardless of which estimator of r is used) and hence it cannot be used directly to infer frequencies. If used, it yields meaningless weights and frequencies essentially amplifying its eigenvectors with near zero eigenvalues (some of them could even be negative; results not shown). This could explain why it was reported that an iterative procedure for estimating relatedness and frequencies yielded worse estimates of relatedness values (and hence the frequencies).[20,21]

This study proposes a new approach where the weights $\{w_i\}$ in Eq. (1) are found by searching for a subset U of unrelated individuals in the sample, $\mathbf{q}(U) = \sum_{i\in U} w_i \mathbf{x}_i /(2u)$, where $w_{i\in U} = 1$ and $w_{i\notin U} = 0$, and where u is the number of elements in the subset U. As indicated earlier, a subset of all unrelated individuals (including unrelated parents) in the sample would give the best theoretically possible estimation of population allele frequencies. If a parent or parents of one or more offspring are missing from the sample, the best one or more representatives of the sibship genotypes should be selected.

The following criterion for selecting U is proposed. The weights could be used to estimate average (over loci) heterozygosity via the standard formula $H(U) = 2u[1 - \frac{1}{L}\sum_{l=1}^{L} q_m^2(U)]/(2u-1)$, which is bias corrected for sample size but not for the sample kin structure. The expected value of the estimate is given by $E[H(U)] = H - R(U)$, where $R(U) = \sum_{i\in U}\sum_{j\neq i\in U} R_{ij}/[u(2u-1)]$, $R_{ij} = Hr_{ij}$, and where $E[H(U)] = H$ if U consists of only unrelated individuals, *i.e.* $r_{ij} = \delta_{ij}$ and $R(U) = 0$. Using the unbiased estimator $R'_{ij} = r'_{ij}H'$, the problem is reduced to finding the minimum of

$$R'(U) = \sum_{i\neq j\in U} R'_{ij}/[u(2u-1)]. \tag{4}$$

The new approach searches for the largest subset of the sample which best resembles a group of mutually unrelated genotypes. In the above analysis, it is implied that U with the largest size u should be preferred. This condition is specified by the denominator

$u(2u-1)$ in Eq. (4). However, a large subset would only be preferred if the resulting R increases slower than $u(2u-1)$, *e.g.* if the sample consists of only full siblings, R becomes $R(u)=0.5(u-1)/(2u-1)$ and $R(u-1)<R(u)$ hence the number of selected full sibs will be minimized (subject to the observed R'_{ij}). While the proposed approach minimizes the number of full-sibs, the approach should also maximize the number of mutually unrelated individuals. This is achieved by using $|r'_{ij}|$ instead of r'_{ij}, which prevents the algorithm from achieving zero in Eq. (4) on not the largest subset U. If r'_{ij} is used, potentially a small number of negative r estimates[1] could cancel out contributions from an equally small number of positive r estimates.

Once a solution is obtained, an exact test[4] for HWE could be used via available software programs[5,10] to assess the solution by verifying that the P value does not reject the HWE null hypothesis. If the original sample does not pass the test for some or all of the loci (*e.g.* $P \leq 0.05$), the new approach offers a practical alternative if it obtains the subset U that passes the test (*e.g.* $P > 0.05$). Note that the proposed solution could be viewed as an approximation for a more general formulation of the problem: "Find the largest subset U that passes such an exact HWE test as the test of Guo & Thompson",[4] where it is assumed that complete sample does not pass the test.

2.4 *Algorithm*

The above approach, when the kin structure of the sample is not known, could be viewed as partitioning of the given sample into two groups: the group of putative unrelated individuals (the subset U) and the rest of the sample. A set of n elements could be partitioned into the two groups 2^n-1 ways, where the single case when all individuals are excluded from U is omitted from consideration. Even though the search space for this problem is "smaller" than the space of the sibship reconstruction problem,[12] it is still non-polynomial and the exhaustive search is possible only for trivially small samples. Moreover, if the relatedness matrix R'_{ij} is viewed as a complete undirected graph (omitting the additional complexity of the dependency on u), the problem of finding a complete sub-graph (*clique*) with the minimum (equivalent to maximum) sum of weights is known to be *NP*-hard,[22] *i.e.* an exact algorithm with polynomial complexity $O(n^{\alpha<\infty})$ does not exist.

Since an exact solution may not be possible, a heuristic approximation is required. One such heuristic for traversing the search space is the simulated annealing technique[23] which was shown to be effective for such related (and more difficult) problems as the sibship[9] and pedigree[14] reconstruction problems. The following algorithm is proposed, where the issue of rare alleles[21] is addressed by ensuring that each putative set U contains at least one instance of every allele observed in the sample. We recognise that the R'_{ij} matrix has a special structure and further study could be done to investigate if a more efficient algorithm exists.

Regarding the design of the algorithm: the main purpose of this study is to develop an algorithm that is implemented in a readily available software program (KINGROUP[10] in our case) so that it could be used by biologists. A typical geneticist/biologist is neither an expert programmer nor a computer scientist, hence we totally agree with the comment of Pearse and Crandall[24] who emphasised that "*improving software usability is essential*". Even though *usability* is often a personal preference, we believe that an algorithm should have as few "magic" numbers controlling the algorithm as possible. Hence the proposed algorithm is controlled by a single parameter, the number of iterations *N*. The number is set to $N = 100 \times n$, *i.e.* each sample genotype is considered 100 times for inclusion or exclusion (on average). User's access to the computing power controls the quality of the solution, *i.e.* the higher the number *N* the higher is the probability of finding the optimal solution. When working with a real sample, the algorithm should be run a number of times with larger *N* each time to verify that the obtained solution is convergent in *N*.

The following algorithm was implemented:

1. Generate an initial configuration by placing all available individuals into the group of putative unrelated individuals, $U_{\text{curr}} = \{1, 2, ..., n\}$. Calculate the current cost function $Z_{\text{curr}} = R'(U_{\text{curr}})$, which is always positive due to the use of $|r'_{ij}|$ and plus H' being non-negative by definition.
2. Generate a new configuration by randomly selecting an individual $1 \le i \le n$. If $i \in U$ and the individual can be taken out of the group, *i.e.* each observed allele at each locus appears at least once in *U*, the individual is removed $(U_{\text{new}} = U_{\text{curr}} - i)$. If $i \in U$ and the individual can not be taken out of the group, another individual is randomly selected. If $i \notin U$, the individual is added $(U_{\text{new}} = U_{\text{curr}} + i)$. Calculate Z_{new} from U_{new}.
3. Calculate relative change via $\Delta Z = (Z_{\text{new}} - Z_{\text{curr}}) / Z_{\text{new}}$. If $\Delta Z \le 0$, the new configuration is accepted becoming "current". If $\Delta Z > 0$, accept the new configuration with the probability $\Pr(\Delta Z) = \exp(-\Delta Z / (k_{\text{B}} T_\alpha))$, where T_α is the annealing temperature, k_{B} is originally the Boltzmann's constant which becomes just a scaling constant, and where the original Boltzmann distribution is used as per Kirkpatrick *et al.*[23]
4. Repeat steps 2 and 3 with $T_\alpha = (N - \alpha + 1)/N$, where α is the iteration count. Since $0 < \Delta Z \le 1$, Boltzmann's constant $k_{\text{B}} = 1/\ln 2 = 1.4427$ is selected to achieve $\Pr(\Delta L = 1) = \exp(-1/k_{\text{B}}) = 0.5$, *i.e.* there is at least 50% chance in accepting the new configuration with larger cost value at the beginning of the annealing process.

3 Results and Discussion

Following Wang[21] a triangular population allele frequency distribution was considered, $p_m(l) = 2m/[(1+k)k]$, yielding the locus heterozygosity of $h = 1 - 2(2k+1)/[3(k+1)k]$. The effect of multiple generations was studied by Monte Carlo simulation using *f* full-sibs in a sample of *n* individuals. A population sample of *n* individuals was generated by firstly generating $n - f$ unrelated individuals based on the given population allele frequencies, **p**. Then, two of the individuals were randomly selected and used to generate *f* full-sibs according to the Mendelian rules of inheritance. The generated set of samples was labelled the *single-family* data set. The theoretically *best* possible estimation of allele frequencies was calculated using only the $n-f$ unrelated individuals, $\mathbf{b} = \sum_{i=1}^{n-f} \mathbf{x}_i / [2(n-f)]$, where, without loss of generality, the unrelated genotypes where labelled from $\mathbf{x}_1$ to $\mathbf{x}_{n-f}$. Assuming the absence of the pedigree information, the frequencies were estimated via the proposed algorithm obtaining the $\mathbf{q}$ frequencies. The mean squared error (MSE) was used to measure the estimation error, where MSE was averaged across loci, $\mathrm{MSE}(\mathbf{q}) = \sum_{m=1}^{k} \sum_{l=1}^{L} [p_m(l) - q_m(l)]^2 / (kL)$.

The second data set was chosen to contain n unrelated individuals. For this *unrelated* data set the best possible estimator of allele frequencies is identical to the allele-counting estimator, *i.e.* $\mathbf{b} \equiv \mathbf{s}$.

The third data set was based on the experimentally observed allele frequencies from a real biological sample of a cooperatively breeding Lake Tanganyika cichlid (*Neolamprologus pulcher*).[3] The cichlid frequencies are specified at $L = 5$ loci with $\{k_1, ..., k_5\} = \{39, 34, 28, 17, 10\}$ alleles and corresponding locus heterozygosities $\{h^{(1)}, ..., h^{(5)}\} = \{0.929, 0.937, 0.847, 0.478, 0.537\}$. This *cichlid* data set is denoted by $G(u, g, s)$, where u is the number of unrelated individuals, g is the number of parental pairs (*i.e.* families), s is the number of full-sibs in the first family. The set is obtained by generating $u + g + 1$ *unrelated* genotypes $\{\mathbf{X}_1, \mathbf{X}_2, ..., \mathbf{X}_{u+g+1}\}$ according to the specified allele frequencies. Then the $s + i - 1$ full-sibs of the i'th group are generated from the $(\mathbf{X}_i, \mathbf{X}_{i+1})$ parental pair.

Fig. 1 presents the root mean square error (RMSE) simulation results: RMSE(**b**), RMSE(**q**) and RMSE(**s**). Fig. 1(a) displays the results for the single-family data set, where $n = 50$ individuals were genotyped with $L = 10$, $k = 10$ ($h = 0.8727$) and variable number of full-sibs *f*. The results for the unrelated data set are displayed in Fig. 1(b), where each sample contained a variable number of individuals genotyped with $L = 5$ and $k = 20$ ($h = 0.9349$). The cichlid data set was generated as $G(u = 10, g, s = 5)$ with a variable number of families. Each point in Fig.1 was obtained by averaging MSE obtained from 100 independent simulation trials and displaying the square root of the average MSE (RMSE).

The results in Fig. 1 are very encouraging as they clearly demonstrate that the new estimator is more accurate than the conventional allele-counting estimator for "dirty" samples with high level of cross-generational contamination, *e.g.* when 20 or more individuals belong to the next generation. Interesting questions still remain for future studies: (1) How much of the RMSE is due to simulated annealing not being able to find the global optimum, and how much is due to the inaccuracy of the relatedness estimates? (2) How robust is the new frequency estimator to the presence of genotyping errors and/or inbreeding? Note that the new estimator is comparable to or even less accurate than the allele-counting estimator for "clean" population samples (Fig. 1(b)) where the level of cross-generational contamination is small. However such clean samples are likely to pass the HWE test anyway and hence the question of a "better" estimation of population allele frequencies would not arise.

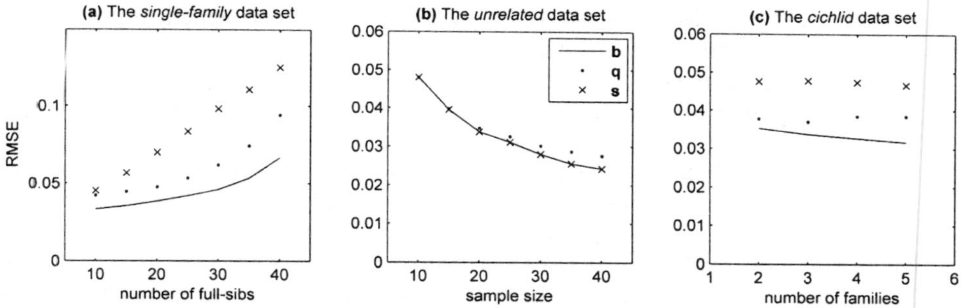

Figure 1. Root mean square error of population allele frequency estimates, where **b** denotes the best possible estimates due to the limited sample size; **q** denotes this study; **s** denotes the allele-counting estimates.

And finally, since the exact HWE test of Guo and Thompson[4] played such an important conceptual role in this study, we would like to comment on the two versions of the HWE test. The first HWE test uses the conventional Monte Carlo (CMC) method and is relatively easy to implement (implemented in KINGROUP[10] and used in this study). This method guarantees P values to within 0.01 with 99% confidence by selecting 17000 simulations regardless of the sample size or the number of observed alleles, hence no "guessing" is required from a software user. Moreover, even Guo and Thompson[4] themselves remarked that the "*method is most suitable for data with a large number of alleles but small sample size*", which is the focus of this study. The second method uses the Markov Chain (MC) estimation. The main argument in favour of the MC method was that it is faster than CMC when the sample size is moderate or large. This argument does not hold in practice since a diligent user would have to run MC a number of times to ensure that the obtained P values are converged, *i.e.* stable to the variations in the three input parameters (dememorization number, number of batches and iterations per batch). In fact, the first method should always be preferred to the second MC method, which is controlled by the three input parameters, which input values are, arguably, meaningless for a typical biologist and can not be deduced easily.

Acknowledgments

This study was partly undertaken when D.K. was on sabbatical leave at the University of Bern. We thank: the University of Bern and James Cook University and, in particular, Michael Taborsky and Bruce Litow for supporting this collaborative project; Peter Stettler for his hospitality; Ross Crozier and Dean Jerry for helpful comments and discussions; and three anonymous reviewers for the thorough review of the earlier version of this manuscript.

References

1. D. A. Konovalov and D. Heg. A maximum-likelihood relatedness estimator allowing for negative relatedness values *Molecular Ecology Notes*, in press, 2007.
2. K. W. Broman. Estimation of allele frequencies with data on sibships. *Genetic Epidemiology*, 20:307-315, 2001.

3. P. Dierkes, D. Heg, M. Taborsky, E. Skubic and R. Achmann. Genetic relatedness in groups is sex-specific and declines with age of helpers in a cooperatively breeding cichlid. *Ecology Letters*, 8:968-975, 2005.
4. S. W. Guo and E. A. Thompson. Performing the Exact Test of Hardy-Weinberg Proportion for Multiple Alleles. *Biometrics*, 48:361-372, 1992.
5. M. Raymond and F. Rousset. Genepop (Version-1.2) - Population-Genetics Software for Exact Tests and Ecumenicism. *Journal of Heredity*, 86:248-249, 1995.
6. M. Nei. Estimation of Average Heterozygosity and Genetic Distance from a Small Number of Individuals. *Genetics*, 89:583-590, 1978.
7. S. C. Thomas and W. G. Hill. Estimating quantitative genetic parameters using sibships reconstructed from marker data. *Genetics*, 155:1961-1972, 2000.
8. B. R. Smith, C. M. Herbinger and H. R. Merry. Accurate partition of individuals into full-sib families from genetic data without parental information. *Genetics*, 158:1329-1338, 2001.
9. J. Wang. Sibship reconstruction from genetic data with typing errors. *Genetics*, 166:1963-1979, 2004.
10. D. A. Konovalov, C. Manning and M. T. Henshaw. KINGROUP: a program for pedigree relationship reconstruction and kin group assignments using genetic markers. *Molecular Ecology Notes*, 4:779-782, 2004.
11. D. A. Konovalov. Accuracy of four heuristics for the full sibship reconstruction problem in the presence of genotype errors. *Series on Advances in Bioinformatics and Computational Biology*, 3:7-16, 2006.
12. D. A. Konovalov, N. Bajema and B. Litow. Modified SIMPSON $O(n^3)$ algorithm for the full sibship reconstruction problem. *Bioinformatics*, 21:3912-3917, 2005.
13. D. A. Konovalov, B. Litow and N. Bajema. Partition-distance via the assignment problem. *Bioinformatics*, 21:2463-2468, 2005.
14. A. Almudevar. A simulated annealing algorithm for maximum likelihood pedigree reconstruction. *Theoretical Population Biology*, 63:63-75, 2003.
15. J. Beyer and B. May. A graph-theoretic approach to the partition of individuals into full-sib families. *Molecular Ecology*, 12:2243-2250, 2003.
16. M. S. McPeek, X. D. Wu and C. Ober. Best linear unbiased allele-frequency estimation in complex pedigrees. *Biometrics*, 60:359-367, 2004.
17. J. M. Olson. Robust Estimation of Gene-Frequency and Association Parameters. *Biometrics*, 50:665-674, 1994.
18. K. F. Goodnight and D. C. Queller. Computer software for performing likelihood tests of pedigree relationship using genetic markers. *Molecular Ecology*, 8:1231-1234, 1999.
19. M. Nei. Analysis of Gene Diversity in Subdivided Populations. *Proceedings of the National Academy of Sciences of the United States of America*, 70:3321-3323, 1973.
20. K. Ritland. Estimators for pairwise relatedness and individual inbreeding coefficients. *Genetical Research*, 67:175-185, 1996.
21. J. Wang. An estimator for pairwise relatedness using molecular markers. *Genetics*, 160:1203-1215, 2002.
22. M. Locatelli, I. M. Bomze and M. Pelillo. The combinatorics of pivoting for the maximum weight clique. *Operations Research Letters*, 32:523-529, 2004.

23. S. Kirkpatrick, C. D. Gelatt and M. P. Vecchi. Optimization by Simulated Annealing. *Science*, 220:671-680, 1983.
24. D. E. Pearse and K. A. Crandall. Beyond F-ST: Analysis of population genetic data for conservation. *Conservation Genetics*, 5:585-602, 2004.

LINEAR TIME PROBABILISTIC ALGORITHMS FOR THE SINGULAR HAPLOTYPE RECONSTRUCTION PROBLEM FROM SNP FRAGMENTS

ZHIXIANG CHEN, BIN FU and ROBERT SCHWELLER
Department of Computer Science, University of Texas-Pan American
Edinburg, TX 78539, USA. E-mail: {chen, binfu, schwellerr}@cs.panam.edu

BOTING YANG
Department of Computer Science, University of Regina
Saskatchewan, S4S 0A2, Canada. E-mail: boting@cs.uregina.ca

ZHIYU ZHAO
Department of Computer Science, University of New Orleans
New Orleans, LA 70148, USA. E-mail: zzha2@cs.uno.edu

BINHAI ZHU
Department of Computer Science, Montana State University
Bozeman, MT 59717, USA. E-mail: bhz@cs.montana.edu.

In this paper, we develop a probabilistic model to approach two scenarios in reality about the singular haplotype reconstruction problem - the incompleteness and inconsistency occurred in the DNA sequencing process to generate the input haplotype fragments and the common practice used to generate synthetic data in experimental algorithm studies. We design three algorithms in the model that can reconstruct the two unknown haplotypes from the given matrix of haplotype fragments with provable high probability and in time linear in the size of the input matrix. We also present experimental results that conform with the theoretical efficient performance of those algorithms. The software of our algorithms is available for public access and for real-time on-line demonstration.

Keywords: Haplotype reconstruction; SNP fragments; Probabilistic algorithms; Inconsistency errors; Incompleteness errors.

1. Introduction

Most part of genomes between two humans are identical. The sites of genomes that make differences among human population are Single Nucleotide Polymorphisms (SNPs). The values of a set of SNPs on a particular chromosome copy define a *haplotype*. Haplotyping an individual involves determining a pair of haplotypes, one for each copy of a given chromosome according to some optimal objective functions.

In recent years, the haplotyping problem has been extensively studied.[1–11] There are several versions of the haplotyping problem. In this paper, we consider the singular haplotype reconstruction problem that asks to reconstruct two unknown haplotypes from the input matrix of fragments as accurately as possible. Like other versions of the problem, this has also been extensively studied.[1,3–5] Because both incompleteness and inconsistency are involved in the fragments, it is not surprising that various versions of the haplotyping problem are NP-hard or even hard to approximate,[1,4,5] and many elegant and powerful methods such as those in (Li, Ma and Wang)[12] cannot be used to deal with incompleteness and inconsistency at the same time. In this paper, we develop a probabilistic approach to overcome some of the difficulties caused by the incompleteness and inconsistency occurred in the input fragments.

2. A Probabilistic Model

Assume that we have two haplotypes H_1, H_2, denoted as $H_1 = a_1a_2\cdots a_m$ and $H_2 = b_1b_2\cdots b_m$. Let $\Gamma = \{S_1, S_2, \ldots, S_n\}$ be a set of n fragments obtained from the DNA sequencing process with respect to the two haplotypes H_1 and H_2. In this case, each $S_i = c_1c_2\cdots c_m$ is either a fragment of H_1 or H_2. Because we lose the information concerning the DNA strand to which a fragment belongs, we do not know whether S_i is a fragment of H_1 or H_2. Suppose that S_i is a fragment of H_1. Because of reading errors or corruptions that may occur during the sequencing process, there is a small chance that either $c_j \neq$ - but $c_j \neq a_j$, or $c_j =$ -, for $1 \leq j \leq m$, where the symbol - denotes a hole or missing value. For the former, the information of the fragment S_i at the j-th SNP site is inconsistent, and we use α_1 to denote the rate of this type of inconsistency error. For the latter, the information of S_i at the j-th SNP is incomplete, and we use α_2 to denote the rate of this type of incompleteness error. It is known[1,3,4] that α_1 and α_2 are in practice between 3% to 5%. Also, it is realistically reasonable to believe that the dissimilarity, denoted by β, between the two haplotypes H_1 and H_2 is big. Often, β is measured using the Hamming distance between H_1 and H_2 divided by the length m of H_1 and H_2, and is assumed to be large, say, $\beta \geq 0.2$. It is also often assumed that roughly half of the fragments in Γ are from each of the two haplotypes H_1 and H_2.

In the experimental studies of algorithmic solutions to the singular haplotype reconstruction problem, we often need to generate synthetic data to evaluate the performance and accuracy of a given algorithm. One common practice[1,3,4] is as follows: First, choose two haplotypes H_1 and H_2 such that the dissimilarity between H_1 and H_2 is at least β. Second, make n_i copies of H_i, $i = 1, 2$. Third, for each copy $H = a_1a_2\cdots a_m$ of H_i, for each $i = 1, 2, \ldots, m$, with probability α_1, flip a_i to a'_i so that they are inconsistent. Also, independently, a_i has probability α_2 to be a hole -. A synthetic data set is then generated by setting parameters $m, n_1, n_2, \beta, \alpha_1$ and α_2. Usually, n_1 is roughly the same as n_2, and $\beta \approx 0.2$, $\alpha_1 \in [0.01, 0.05]$, and $\alpha_2 \in [0.1, 0.3]$.

Motivated by the above reality of the sequencing process and the common practice in experimental algorithm studies, we will present a probabilistic model for the singular haplotype reconstruction problem. But first we need to introduce some necessary notations and definitions.

Let $\Sigma_1 = \{A, B\}$ and $\Sigma_2 = \{A, B, \text{-}\}$. For a set C, $|C|$ denotes the number of elements in C. For a fragment (or a sequence) $S = a_1 a_2 \cdots a_m \in \Sigma_2^m$, $S[i]$ denotes the character a_i, and $S[i,j]$ denotes the substring $a_i \cdots a_j$ for $1 \leq i \leq j \leq m$. $|S|$ denotes the length m of S. When no confusion arises, we alternatively use the terms fragment and sequence.

Let $G = g_1 g_2 \cdots g_m \in \Sigma_1^m$ be a fixed sequence of m characters. For any sequence $S = a_1 \cdots a_m \in \Sigma_2^m$, S is called a $\mathcal{F}_{\alpha_1,\alpha_2}(m, G)$ sequence if for each a_i, with probability at most α_1, a_i is not equal to g_i and $a_i \neq$ -; and with probability at most α_2, $a_i =$ -.

For a sequence S, define *holes*(S) to be the number of holes in the sequence S. If A is a subset of $\{1, \cdots, m\}$ and S is a sequence of length m, $holes_A(S)$ is the number of $i \in A$ such that $S[i]$ is a hole.

For two sequences $S_1 = a_1 \cdots a_m$ and $S_2 = b_1 \cdots b_m$ of the same length m, for any $A \subseteq \{1, \cdots, m\}$, define

$$diff(S_1, S_2) = \frac{|\{i \in \{1, 2, \cdots, m\} | a_i \neq \text{ - and } b_i \neq \text{ - and } a_i \neq b_i\}|}{m}$$

$$diff_A(S_1, S_2) = \frac{|\{i \in A | a_i \neq \text{ - and } b_i \neq \text{ - and } a_i \neq b_i\}|}{|A|}.$$

For a set of sequences $\Gamma = \{S_1, S_2, \cdots, S_k\}$ of length m, define *vote*(Γ) to be the sequence H of the same length m such that $H[i]$ is the most frequent character among $S_1[i], S_2[i], \cdots, S_k[i]$ for $i = 1, 2, \cdots, m$.

We often use an $n \times m$ matrix M to represent a list of n fragments from Σ_2^m and call M an SNP fragment matrix. For $1 \leq i \leq n$, let $M[i]$ represent the i-th row of M, i.e., $M[i]$ is a fragment in Σ_2^m.

We now define our probabilistic model:

The Probabilistic Singular Haplotype Reconstruction Problem: Let β, α_1 and α_2 be small positive constants. Let $G_1, G_2 \in \Sigma_1^m$ be two haplotypes with $diff(G_1, G_2) \geq \beta$. For any given $n \times m$ matrix M of SNP fragments such that n_i rows of M are $\mathcal{F}_{\alpha_1,\alpha_2}(m, G_i)$ sequences, $i = 1, 2$, $n_1 + n_2 = n$, reconstruct the two haplotypes G_1 and G_2, which are unknown to the users, from M as accurately as possible with high probability. We call β (resp., α_1, α_2) dissimilarity rate (resp., inconsistency error rate, incompleteness error rate).

3. Technical Lemmas

For probabilistic analysis we need the following two Chernoff bounds.

Lemma 3.1.[12] *Let $X_1, \cdots, X_n$ be n independent random $0, 1$ variables, where X_i takes 1 with probability at most p. Let $X = \sum_{i=1}^n X_i$. Then for any $1 \geq \epsilon > 0$,* $\Pr(X > pn + \epsilon n) < e^{-\frac{1}{3}n\epsilon^2}$.

Lemma 3.2.[12] *Let $X_1, \cdots, X_n$ be n independent random $0, 1$ variables, where X_i takes 1 with probability at least p. Let $X = \sum_{i=1}^n X_i$. Then for any $1 \geq \epsilon > 0$,*

$\Pr(X < pn - \epsilon n) < e^{-\frac{1}{2}n\epsilon^2}$.

Lemma 3.3. *Let S be a $\mathcal{F}_{\alpha_1,\alpha_2}(m,G)$ sequence. Then, for any $0 < \epsilon \le 1$, with probability at most $2e^{-\frac{\epsilon^2 m}{3}}$, diff$(G_i,S) > \alpha_1 + \epsilon$ or holes$(S) > (\alpha_2+\epsilon)m$.*

Proof. Let X_k, $k = 1\cdots,m$, be random variables such that $X_k = 1$ if $S[k] \neq G_i[k]$ and $S[k] \neq -$, or 0 otherwise. By the definition of the $\mathcal{F}_{\alpha_1,\alpha_2}(m,G)$ sequences, X_k are independent and $Pr(X_k = 1) \le \alpha_1$. So, by Lemma 3.1, with probability at most $e^{-\frac{\epsilon^2 m}{3}}$, $X_1+\cdots+X_m > (\alpha_1+\epsilon)m$. Thus, we have *diff*$(G,S) > \alpha_1+\epsilon$ with probability at most $e^{-\frac{\epsilon^2 m}{3}}$. Similarly, with probability at most $e^{-\frac{\epsilon^2 m}{3}}$, *holes*$(S) > (\alpha_2+\epsilon)m$.□

Lemma 3.4. *Assume that A is a fixed subset of $\{1,2,\cdots,m\}$. Let S be a $\mathcal{F}_{\alpha_1,\alpha_2}(m,G)$ sequence. Then, for any $0<\epsilon\le 1$, with probability at most $2e^{-\frac{\epsilon^2|A|}{3}}$, diff$_A(G_i,S) > \alpha_1+\epsilon$ or* holes$_A(S) > (\alpha_2+\epsilon)|A|$.

Proof. Let S' (resp. G') be the subsequence consisting of all the characters $S[i]$ (resp. $G[i]$), $i \in A$, with the same order as in S (resp. G). Then, *diff*$_A(S,G_i) =$ *diff*(S',G'). The lemma follows from a similar proof for Lemma 3.3. □

Lemma 3.5. *Let N_i be a set of n_i many $\mathcal{F}_{\alpha_1,\alpha_2}(m,G_i)$ sequences, $i=1,2$. Let β and ϵ be two positive constants such that $2\alpha_1+2\alpha_2+2\epsilon<1$, and diff$(G_1,G_2)\ge\beta$. Then, with probability at most $2(n_1+n_2)e^{-\frac{\epsilon^2\beta m}{3}}$, diff$(S_i,S_j) \le \beta(1-2\alpha_1-2\alpha_2-2\epsilon)$ for some $S_i\in N_i$ and some $S_j \in N_j$ with $i\neq j$.*

Proof. For each G_i, let A_i be the set of indexes $\{k \in \{1,2,\cdots,m\} | G_i[k] \neq G_j[k]\}$, where $i \neq j$. Since *diff*$(G_i,G_j)\ge\beta$ and $|G_i| = |G_j| = m$, we have $|A_i| \ge \beta m$. For any $\mathcal{F}_{\alpha_1,\alpha_2}(m,G_i)$ sequence S, by Lemma 3.4, with probability at most $2e^{-\frac{\epsilon^2|A_i|}{3}} \le 2e^{-\frac{\epsilon^2\beta m}{3}}$, *diff*$_{A_i}(S,G_i) > \alpha_1+\epsilon$ or *holes*$_{A_i}(S) > (\alpha_2+\epsilon)|A_i|$. Hence, with probability at most $2n_ie^{-\frac{\epsilon^2\beta m}{3}}$, *diff*$_{A_i}(S,G_i) > \alpha_1+\epsilon$ or *holes*$_{A_i}(S) > (\alpha_2+\epsilon)|A_i|$, for some $S \in N_i$. Therefore, with probability at most $2(n_1+n_2)e^{-\frac{\epsilon^2\beta m}{3}}$, we have *diff*$_{A_i}(S,G_i) > \alpha_1+\epsilon$ or *holes*$_{A_i}(S) > (\alpha_2+\epsilon)|A_{i,j}|$, for some $S\in N_i$, for some $i = 1$ or 2. In other words, with probability at least $1-2(n_1+n_2)e^{-\frac{\epsilon^2\beta m}{3}}$, we have *diff*$_{A_i}(S,G_i)\le\alpha_1+\epsilon$ and *holes*$_{A_i}(S) \le (\alpha_2+\epsilon)|A_{i,j}|$, for all $S\in N_i$ and for $i=1$ and 2.

For any $\mathcal{F}_{\alpha_1,\alpha_2}(m,G_i)$ sequence S_i, $i = 1,2$, if *diff*$_{A_i}(S_i,G_i) \le \alpha_1+\epsilon$ and *holes*$_{A_i}(S_i) \le (\alpha_2+\epsilon)|A_i|$, then *diff*$(S_1,S_2) \ge$ *diff*$_{A_i}(S_1,S_2) \ge \beta(1-2\alpha_1-2\alpha_2-2\epsilon)$. Thus, with probability at least $1-2(n_1+n_2)e^{-\frac{\epsilon^2\beta m}{3}}$, we have *diff*$(S_1,S_2)\ge\beta(1-2\alpha_1-2\alpha_2-2\epsilon)$, for every $S_1\in N_1$ and every $S_2\in N_2$. In words, with probability at most $2(n_1+n_2)e^{-\frac{\epsilon^2\beta m}{3}}$, we have *diff*$(S_1,S_2) < \beta(1-2\alpha_1-2\alpha_2-2\epsilon)$, for some $S_1\in N_1$ and some $S_2 \in N_2$. □

Lemma 3.6. *Let α_1, α_2 and ϵ be three small positive constants that satisfy $0 < 2\alpha_1+\alpha_2-\epsilon < 1$. Assume that $N = \{S_1,\cdots,S_n\}$ is a set of $\mathcal{F}_{\alpha_1,\alpha_2}(m,G)$ sequences.*

Let $H = vote(N)$. Then, with probability at most $2m(e^{-\frac{\epsilon^2 n}{2}})$, $G \neq H$.

Proof. Given any $1 \leq j \leq m$, for any $1 \leq i \leq n$, let X_i be random variables such that $X_i = 1$ if $S_i[j] \neq G[j]$ and $S_i[j] \neq -$, or 0 otherwise. By the definition of the $\mathcal{F}_{\alpha_1,\alpha_2}(m, G)$ sequences, X_i are independent and $Pr(X_i = 1) \leq \alpha_1$. So, by Lemma 3.2, with probability at most $e^{-\frac{\epsilon^2 n}{2}}$, $X_1 + \cdots + X_n < (\alpha - \epsilon)n$. That is, with probability at most $e^{-\frac{\epsilon^2 n}{2}}$, there are fewer than $(\alpha_1 - \epsilon)n$ characters $S_i[j]$ such that $S_i[j] \neq G[j]$ and $S_i[j] \neq$ -. Similarly, with probability at most $e^{-\frac{\epsilon^2 n}{2}}$, there are fewer than $(\alpha_2 - \epsilon)n$ characters $S_i[j]$ such that $S_i[j] =$ -. Thus, with probability at most $2me^{-\frac{\epsilon^2 n}{2}}$, there are fewer than $(\alpha_1 + \alpha_2 - 2\epsilon)n$ characters $S_i[j]$ such that $S_i[j] \neq G[j]$ for some $1 \leq j \leq m$. This implies that, with probability at least $1 - 2me^{-\frac{\epsilon^2 n}{2}}$, there are more than $(1 - \alpha_1 - \alpha_2 + 2\epsilon)n$ characters $S_i[j]$ such that $S_i[j] = G[j]$ for any $1 \leq j \leq m$. Since $0 < 2\alpha_1 + \alpha_2 - \epsilon < 1$ by the assumption of the theorem, we have $(\alpha_1 + \epsilon)n < (1 - \alpha_1 - \alpha_2 + 2\epsilon)n$. This further implies that with probability at least $1 - 2me^{-\frac{\epsilon^2 n}{2}}$, $vote(N)[j] = G[j]$ for any $1 \leq j \leq m$, i.e., $vote(N) = G$. □

4. When the Inconsistency Error Parameter Is Known

Theorem 4.1. *Assume that $\alpha_1, \alpha_2, \beta$, and $\epsilon > 0$ are small positive constants that satisfy $4(\alpha_1 + \epsilon) < \beta$ and $0 < 2\alpha_1 + \alpha_2 - \epsilon < 1$. Let $G_1, G_2 \in \Sigma_1^m$ be the two unknown haplotypes such that diff$(G_1, G_2) \geq \beta$. Let M be any given $n \times m$ matrix of SNP fragments such that M has n_i fragments that are $\mathcal{F}_{\alpha_1,\alpha_2}(m, G_i)$ sequences, $i = 1, 2$, and $n_1 + n_2 = n$. There exists an $O(nm)$ time algorithm that can find two haplotypes H_1 and H_2 with probability at least $1 - 2ne^{-\frac{\epsilon^2 m}{3}} - 2me^{-\frac{\epsilon^2 n_1}{2}} - 2me^{-\frac{\epsilon^2 n_2}{2}}$ such that either $H_1 = G_1$ and $H_2 = G_2$, or $H_1 = G_2$ and $H_2 = G_1$.*

Proof. The algorithm, denoted as SHR-One, is described as follows.

Algorithm SHR-One

Input: M, an $n \times m$ matrix of SNP fragments.
  Parameters α_1 and ϵ.
Output: Two haplotypes H_1 and H_2.
  Set $\Gamma_1 = \Gamma_2 = \emptyset$.
  Randomly select a fragment $r = M[j]$ for some $1 \leq j \leq n$.
  For every fragment r' from M do
    If $(diff(r, r') \leq 2(\alpha_1 + \epsilon))$ then put r' into Γ_1
  Let $\Gamma_2 = M - \Gamma_1$.
  Let $H_1 = vote(\Gamma_1)$ and $H_2 = vote(\Gamma_2)$.
  return H_1 and H_2.

End of Algorithm

Claim 1. With probability at most $ne^{-\frac{\epsilon^2 m}{3}}$, we have either $diff(f, G_1) > \alpha_1 + \epsilon$ for some $\mathcal{F}_{\alpha_1,\alpha_2}(m, G_1)$ sequence f in M, or $diff(g, G_1) > \alpha_1 + \epsilon$ for some $\mathcal{F}_{\alpha_1,\alpha_2}(m, G_2)$ sequence g in M.

By Lemma 3.4, for any fragment $f = M[k]$ such that f is a $\mathcal{F}_{\alpha_1,\alpha_2}(m, G_1)$ sequence, with probability at most $e^{-\frac{\epsilon^2 m}{3}}$ we have $diff(f, G_1) > \alpha_1 + \epsilon$. Since there are n_1 many $\mathcal{F}_{\alpha_1,\alpha_2}(m, G_1)$ sequences in M, with probability at most $n_1 e^{-\frac{\epsilon^2 m}{3}}$, we have $diff(f, G_1) > \alpha_1 + \epsilon$ for some $\mathcal{F}_{\alpha_1,\alpha_2}(m, G_1)$ sequence f in M. Similarly, with probability at most $n_2 e^{-\frac{\epsilon^2 m}{3}}$, we have $diff(g, G_2) > \alpha_1 + \epsilon$ for some $\mathcal{F}_{\alpha_1,\alpha_2}(m, G_2)$ sequence g in M. Combining the above completes the proof for Claim 1.

Claim 2. Let M_i be the set of all the $\mathcal{F}_{\alpha_1,\alpha_2}(m, G_i)$ sequences in M, $i = 1, 2$. With probability at least $1 - ne^{-\frac{\epsilon^2 m}{3}}$, Γ_1 and Γ_2 is a permutation of M_1 and M_2.

By the assumption of the theorem, the fragment r of M is either a $\mathcal{F}_{\alpha_1,\alpha_2}(m, G_1)$ sequence or a $\mathcal{F}_{\alpha_1,\alpha_2}(m, G_2)$ sequence. We assume that the former is true. By Claim 1, with probability at least $1 - ne^{-\frac{\epsilon^2 m}{3}}$, we have $diff(f, G_1) \leq \alpha_1 + \epsilon$ for all $\mathcal{F}_{\alpha_1,\alpha_2}(m, G_1)$ sequences f in M, and $diff(g, G_1) \leq \alpha_1 + \epsilon$ for all $\mathcal{F}_{\alpha_1,\alpha_2}(m, G_2)$ sequences g in M. Hence, for any fragment r' in M, if r' is a $\mathcal{F}_{\alpha_1,\alpha_2}(m, G_1)$ sequence, then with probability at least $1 - ne^{-\frac{\epsilon^2 m}{3}}$, we have $diff(r, r') \leq diff(r, G_1) + diff(r', G_1) \leq 2(\alpha_1 + \epsilon)$. This means that, with probability at least $1 - ne^{-\frac{\epsilon^2 m}{3}}$, all $\mathcal{F}_{\alpha_1,\alpha_2}(m, G_1)$ sequences in M will be included in Γ_1. Now, consider that r' is a $\mathcal{F}_{\alpha_1,\alpha_2}(m, G_2)$ sequence in M. Since $diff(G_1, G_2) \leq diff(G_1, r) + diff(r, G_2) \leq diff(G_1, r) + diff(r, r') + diff(r', G_2)$, we have $diff(r, r') \geq diff(G_1, G_2) - diff(G_1, r) - diff(G_2, r')$. By the given condition of $diff(G_1, G_2) \geq \beta$ and $4(\alpha_1 + \epsilon) < \beta$, with probability at least $1 - ne^{-\frac{\epsilon^2 m}{3}}$, we have $diff(r, r') \geq \beta - diff(G_1, r) - diff(G_2, r') \geq \beta - 2(\alpha_1 + \epsilon) > 2(\alpha_1 + \epsilon)$, i.e., r' will not be added to Γ_1. Therefore, with probability at least $1 - ne^{-\frac{\epsilon^2 m}{3}}$, $\Gamma_1 = M_1$ and $\Gamma_2 = M - \Gamma_1 = M_2$. Similarly, if r is a $\mathcal{F}_{\alpha_1,\alpha_2}(m, G_2)$ sequence, with probability at least $1 - ne^{-\frac{\epsilon^2 m}{3}}$, $\Gamma_1 = M_2$ and $\Gamma_2 = M - \Gamma_1 = M_1$. This completes the proof of Claim 2.

Suppose that Γ_1 and Γ_2 is a permutation of M_1 and M_2. Say, without loss of generality, $\Gamma_1 = M_1$ and $\Gamma_2 = M_2$. By Lemma 3.6, with probability at most $2me^{-\frac{\epsilon^2 n_1}{2}} + 2me^{-\frac{\epsilon^2 n_2}{2}}$, $vote(\Gamma_1) \neq G_1$ or $vote(\Gamma_2) \neq G_2$. Hence, by Claim 2, with probability at most $2ne^{-\frac{\epsilon^2 m}{3}} + 2me^{-\frac{\epsilon^2 n_1}{2}} + 2me^{-\frac{\epsilon^2 n_2}{2}}$, $vote(\Gamma_1) \neq G_1$ or $vote(\Gamma_2) \neq G_2$.

Concerning the computational time of the algorithm, we need to compute the difference between the selected fragment r and each of the rest $n - 1$ fragments in the matrix M. Finding the difference between r and r' takes $O(m)$ steps. So, the total computational time is $O(nm)$, which is linear in the size of the input matrix M. □

5. When Parameters Are Not Known

In this section, we consider the case that the parameters α_1, α_2 and β are unknown. However, we assume the existence of those parameters for the input matrix M of SNP fragments. We will show that in this case we can still reconstruct the two unknown haplotypes from M with high probability.

Theorem 5.1. *Assume that $\alpha_1, \alpha_2, \beta$, and $\epsilon > 0$ are small positive constants that satisfy $2\alpha_1+2\alpha_2+2\epsilon < 1$, $0 < 2\alpha_1+\alpha_2-\epsilon < 1$, and $\beta(1-2\alpha_1-2\alpha_2-2\epsilon) > 2(\alpha_1+\epsilon)$. Let $G_1, G_2 \in \Sigma_1^m$ be the two unknown haplotypes such that $diff(G_1, G_2) \geq \beta$. Let M be any given $n \times m$ matrix of SNP fragments such that M has n_i fragments that are $\mathcal{F}_{\alpha_1,\alpha_2}(m, G_i)$ sequences, $i = 1, 2$, and $n_1 + n_2 = n$. Then, there exists an $O(umn)$ time algorithm that can find two haplotypes H_1 and H_2 with probability at least $1 - (1-\gamma)^u - 4ne^{-\frac{\epsilon^2 \beta m}{3}} - 2me^{-\frac{\epsilon^2 n_1}{2}} - 2me^{-\frac{\epsilon^2 n_2}{2}}$ such that H_1, H_2 is a permutation of G_1, G_2, where $\gamma = \frac{n_1 n_2}{n(n-1)}$ and u is an integer parameter.*

Proof. The algorithm, denoted as SHR-Two, is described as follows.

Algorithm SHR-Two

Input: M, an $n \times m$ matrix M of SNP fragments.
    u, a parameter to control the loop.
Output: two haplotypes H_1 and H_2.
    Let $d_{\min} = \infty$ and $\mathcal{M} = \emptyset$.
    For ($k = 1$ to u) do   //**the k-loop**
        Let $M_1 = M_2 = \emptyset$ and $d_1 = d_2 = 0$.
        Randomly select two fragments $r_1 = M[i_1], r_2 = M[i_2]$ from M
        For every fragment r' from M do
            If ($diff(r_i, r') = min\{diff(r_1, r'), diff(r_2, r')\}$ for $i = 1$ or 2
                then put r' into M_i.
        Let $d_i = \max\{diff(r_i, r') | r' \in M_i\}$ for $i = 1, 2$.
        Let $d = \max\{d_1, d_2\}$.
        If ($d < d_{\min}$) then let $\mathcal{M} = \{M_1, M_2\}$ and $d_{min} = d$.
    return $H_1 = vote(M_1)$ and $H_2 = vote(M_2)$.

End of Algorithm

Claim 3. With probability at most $(1-\gamma)^u$, r_1, r_2 is not a permutation of a $\mathcal{F}_{\alpha,\beta}(m, G_1)$ sequence and a $\mathcal{F}_{\alpha,\beta}(m, G_2)$ sequence in all of the k-loop iterations.

Let N_i be the set of the n_i fragments in M that are $\mathcal{F}_{\alpha_1,\alpha_2}(m, G_i)$ sequences, $i = 1, 2$.

Claim 4. With probability at most $4ne^{-\frac{\epsilon^2 \beta m}{3}}$, $diff(G_i, S) > \alpha_1 + \epsilon$ or $holes(S) > (\alpha_2 + \epsilon)m$ for some S from N_i for some $i = 1$ or 2; or $diff(S_1, S_2) \leq \beta(1 - 2\alpha_1 - 2\alpha_2 - 2\epsilon)$ for some $S_1 \in N_1$ and some $S_2 \in N_2$.

Claim 3 follows from simple counting. Claim 4 follows from Lemmas 3.3 and 3.5.

Claim 5. Let $H_1 = vote(M_1)$ and $H_2 = vote(M_2)$ be the two haplotypes returned by the algorithm. With probability at most $(1-\gamma)^u + 4ne^{-\frac{\epsilon^2 \beta m}{3}}$, M_1, M_2 is not a permutation of N_1, N_2.

We assume that (1) $diff(S_1, S_2) > \beta(1 - 2\alpha_1 - 2\alpha_2 - 2\epsilon)$ for every S_1 from N_1 and every S_2 from N_2; and (2) $diff(G_i, S) \leq \alpha_1 + \epsilon$ and $holes(S) \leq (\alpha_2 + \epsilon)m$ for all $S \in N_i$ for $i = 1, 2$. We consider possible choices of the two random fragments r_1 and r_2 in the following.

At any iteration of the k-loop, if $r_1 \in N_1$ and $r_2 \in N_2$, then by (2) we have $\mathit{diff}(r_1, r') \leq \mathit{diff}(r_1, G_1) + \mathit{diff}(r', G_1) \leq 2(\alpha_1 + \epsilon)$ for any $r' \in N_1$; and $\mathit{diff}(r_2, r') \leq \mathit{diff}(r_2, G_2) + \mathit{diff}(r', G_2) \leq 2(\alpha_1 + \epsilon)$ for any $r' \in N_2$. By (1) and the given condition of the theorem, we have, $\mathit{diff}(r_1, r') > \beta(1 - 2\alpha_1 - 2\alpha_2 - 2\epsilon) > 2(\alpha_1 + \epsilon)$ for any $r' \in N_2$; and $\mathit{diff}(r_2, r') > \beta(1 - 2\alpha_1 - 2\alpha_2 - 2\epsilon) > 2(\alpha_1 + \epsilon)$ for any $r' \in N_1$. This implies that at this loop iteration we have $M_1 = N_1, M_2 = N_2$ and $d \leq 2(\alpha_1 + \epsilon)$. Similarly, if at this iteration $r_1 \in N_2$ and $r_2 \in N_1$, then $M_1 = N_2, M_2 = N_1$ and $d \leq 2(\alpha_1 + \epsilon)$.

If $r_1, r_2 \in N_1$ at some iteration of the k-loop, then for any $r' \in N_2$, either $r' \in M_1$ or $r' \in M_2$. In either case, by (1) of our assumption and the given condition of the theorem, we have $d \geq \beta(1 - 2\alpha_1 - 2\alpha_2 - 2\epsilon) > 2(\alpha_1 + \epsilon)$ at this iteration. Similarly, if $r_1, r_2 \in N_2$ at some iteration of the k-loop, then we also have $d > 2(\alpha_1 + \epsilon)$ at this iteration.

It follows from the above analysis that, under the assumption of (1) and (2), once we have $r_1 \in N_1$ and $r_2 \in N_2$ or $r_1 \in N_2$ and $r_2 \in N_1$ at some iteration of the k-loop, then M_1, M_2 is a permutation of N_1, N_2 at the end of this iteration. Furthermore, if M_1 and M_2 are replaced by M_1' and M_2' after this iteration, then M_1', M_2' must also be a permutation of N_1, N_2. By Claims 3 and 4, with probability at most $(1-\gamma)^u + 4ne^{-\frac{\epsilon^2 \beta m}{3}}$, the assumption of (1) and (2) is not true, or $r_1 \in N_1$ and $r_2 \in N_2$ (or $r_1 \in N_2$ and $r_2 \in N_1$) is not true at all the iterations of the k-loop. Hence, with probability at most $(1-\gamma)^u + 4ne^{-\frac{\epsilon^2 \beta m}{3}}$, the final list of M_1 and M_2 returned by the algorithm is not a permutation of N_1, N_2, so the claim is proved.

For M_1 and M_2 returned by the algorithm, we assume without loss of generality $M_i = N_i$, $i = 1, 2$. By Lemma 3.6 and the given condition of the theorem, with probability at most $2me^{-\frac{\epsilon^2 n_1}{2}} + 2me^{-\frac{\epsilon^2 n_2}{2}}$, we have $H_1 = \mathit{vote}(M_1) \neq G_1$ or $H_2 = \mathit{vote}(M_2) \neq G_2$. Thus, by Claim 6, with probability at most $(1-\gamma)^u + 4ne^{-\frac{\epsilon^2 \beta m}{3}} + 4me^{-\frac{\epsilon^2 n}{2}}$, we have $H_1 \neq G_1$ or $H_2 \neq G_2$.

It is easy to see that the time complexity of the algorithm is $O(umn)$, which is linear in the size of M. □

6. Tuning the Dissimilarity Measure

In this section, we consider a different dissimilarity measure in algorithm SHR-TWO to improve its ability to tolerate errors. We use the sum of the differences between r_i and every fragment $r' \in M_i$, $i = 1, 2$, to measure the dissimilarity of the fragments in M_i with r_i. The new algorithm SHR-Three is given in the following. We will present experimental results in Section 7 to show that algorithm SHR-Three is more reliable and robust in dealing with possible outliers in the data sets.

Algorithm SHR-Three

Input: M, an $n \times m$ matrix of SNP fragments.

u, a parameter to control the loop.

Output: two haplotypes H_1 and H_2.

```
Let d_min = ∞ and 𝓜 = ∅.
For (k = 1 to u) do        //the k-loop
    Let M_1 = M_2 = ∅ and d_1 = d_2 = 0.
    Randomly select two fragments r_1 = M[i_1], r_2 = M[i_2] from M
    For every fragment r' from M do
        If (diff(r_i, r') = min{diff(r_1, r'), diff(r_2, r')} for i = 1 or 2
            then put r' into M_i.
    Let d_i = Σ_{r'∈M_i} diff(r_i, r') for i = 1, 2.
    Let d = max{d_1, d_2}.
    If (d < d_min) then let 𝓜 = {M_1, M_2} and d_min = d.
return H_1 = vote(M_1) and H_2 = vote(M_2).
```

End of Algorithm

Theorem 6.1. *Assume that $\alpha_1, \alpha_2, \beta$, and $\epsilon > 0$ are small positive constants that satisfy $2\alpha_1 + 2\alpha_2 + 2\epsilon < 1$, $0 < 2\alpha_1 + \alpha_2 - \epsilon < 1$, $\eta > \frac{2(\alpha_1+\epsilon)}{\beta(1-2\alpha_1-2\alpha_2-2\epsilon)}$ with $\eta = \frac{\min(n_1,n_2)}{2n}$, and $\beta(1-2\alpha_1-2\alpha_2-2\epsilon) > 2(\alpha_1+\epsilon)$. Let $G_1, G_2 \in \Sigma_1^m$ be the two unknown haplotypes such that $diff(G_1, G_2) \geq \beta$. Let M be any given $n \times m$ matrix of SNP fragments such that M has n_i fragments that are $\mathcal{F}_{\alpha_1,\alpha_2}(m, G_i)$ sequences, $i = 1, 2$, and $n_1 + n_2 = n$. Then, there exists an $O(umn)$ time algorithm that can find two haplotypes H_1 and H_2 with probability at least $1-(1-\gamma)^u - 4ne^{-\frac{\epsilon^2\beta m}{3}} - 2me^{-\frac{\epsilon^2 n_1}{2}} - 2me^{-\frac{\epsilon^2 n_2}{2}}$ such that H_1, H_2 is a permutation of G_1, G_2, where $\gamma = \frac{n_1 n_2}{n(n-1)}$ and u is an integer parameter.*

The proof of Theorem 6.1 is omitted due to space limit.

7. Experimental Results

We have tested both the accuracy and the speed of algorithm SHR-Three. Due to the difficulty of getting real data from the public domain,[8] our experiment data is created following the common practice in literature.[3,8] A random matrix of SNP fragments is created as follows: (1) Haplotype 1 is generated at random with length m ($m \in \{50, 100, 150\}$). (2) Haplotype 2 is generated by copying all the bits from haplotype 1 and flipping each bit with probability β ($\beta \in \{0.1, 0.2, 0.3\}$). This simulates the dissimilarity rate β between two haplotypes. (3) Each haplotype is copied $\frac{n}{2}$ times so that the matrix has m columns and $n (n \in \{10, 20, 30\})$ rows. (4) Set each bit in the matrix to - with probability α_2 ($\alpha_2 \in \{0.1, 0.2, 0.3\}$). This simulates the incompleteness error rate α_2 in the matrix. (5) Flip each non-empty bit with probability $\alpha_1 (\alpha_1 \in \{0.01, 0.02, ..., 0.1\})$. This is the simulation of the inconsistency error rate of α_1.

Due to space limit, we present only one table to show the performance of algorithm SHR-Three with different parameter settings in accordance with those in (Panconesi and Sozio).[8] The typical parameters used there are $m = 100, n = 20, \beta = 0.2, \alpha_2 = 0.2$ and $0.01 \leq \alpha_1 \leq 0.05$. These are considered to be close to the real situations. The computing environment is a PC machine with a typical configuration of 1.6GHz AMD Turion 64X2 CPUs and 1GB memory.

Table 1. Results for $m = 100, \beta = 20\%, \alpha_2 = 20\%$

	$n = 10$		$n = 30$	
α_1 (%)	Time (ms)	Reconstruction Rate (%)	Time (ms)	Reconstruction Rate (%)
1	2.444	99.91	4.744	100.00
2	2.568	99.78	5.046	100.00
3	2.674	99.58	5.261	100.00
4	2.774	99.36	5.605	99.99
5	2.851	99.01	6.045	100.00
6	2.925	98.60	6.302	99.97
7	3.028	98.03	6.567	99.96
8	3.121	97.54	6.870	99.85
9	3.213	96.81	7.307	99.70
10	3.314	95.85	7.635	99.56

The software of our algorithms is available for public access and for real-time on-line demonstration at http://fpsa.cs.uno.edu/HapRec/HapRec.html. We thank Liqiang Wang for implementing the programs in Java and setting up this web site.

References

1. V. Bafna, S. Istrail, G. Lancia and R. Rizzi, *Theoretical Computer Science* **335**, 109 (2005).
2. R. Rizzi, V. Bafna, S. Istrail and G. Lancia, Practical algorithms and fixed-parameter tractability for the single individual SNP haplotyping problem, in *Algorithms in Bioinformatics: Second International Workshop, WABI'02*, 2002.
3. R.-S. Wang, L.-Y. Wu, Z.-P. Li and X.-S. Zhang, *BioInformatics* **21**, 2456 (2005).
4. R. Lippert, R. Schwartz, G. Lancia and S. Istrail, *Briefings in bioinformatics* **3**, 23 (2002).
5. R. Cilibrasi, L. van Iersel, S. Kelk and J. Tromp, On the complexity of several haplotyping problems, in *Algorithms in Bioinformatics, 5th International Workshop, WABI'05, Lecture Notes in Computer Science*, 2005.
6. G. Lancia, M. C. Pinotti and R. Rizzi, *INFORMS Journal on computing* **16**, 348 (2004).
7. G. Lancia and R. Rizzi, *Operations Research letters* **34**, 289 (2006).
8. A. Panconesi and M. Sozio, Fast Hare: A fast heuristic for single individual snp haplotype reconstruction, in *Algorithms in Bioinformatics, 4th International Workshop, WABI'04, Lecture Notes in Computer Science 3240*, 2004.
9. A. Clark, *Molecular Biology Evolution* **7**, 111 (1990).
10. D. Gusfield, A practical algorithm for optimal inference of haplotype from diploid populations, in *The Eighth International Conference on Intelligence Systems for Molecular Biology*, 2000.
11. D. Gusfield, Haplotyping as perfect phylogeny: Conceptual framework and efficient solutions, in *the Sixth Annual International Conference on Computational Biology*, 2002.
12. M. Li, B. Ma and L. Wang, *Journal of the ACM* **49(2)**, 157 (2002).

OPTIMAL ALGORITHM FOR FINDING DNA MOTIFS WITH NUCLEOTIDE ADJACENT DEPENDENCY*

FRANCIS Y. L. CHIN, HENRY CHI MING LEUNG, M. H. SIU and S. M. YIU

Department of Computer Science, University of Hong Kong
Pokfulam, Hong Kong

Abstract: Finding motifs and the corresponding binding sites is a critical and challenging problem in studying the process of gene expression. String and matrix representations are two popular models to represent a motif. However, both representations share an important weakness by assuming that the occurrence of a nucleotide in a binding site is independent of other nucleotides. More complicated representations, such as HMM or regular expression, exist that can capture the nucleotide dependency. Unfortunately, these models are not practical (with too many parameters and require many known binding sites). Recently, Chin and Leung introduced the SPSP representation which overcomes the limitations of these complicated models. However, discovering novel motifs in SPSP representation is still a NP-hard problem. In this paper, based on our observations in real binding sites, we propose a simpler model, the Dependency Pattern Sets (DPS) representation, which is simpler than the SPSP model but can still capture the nucleotide dependency. We develop a branch and bound algorithm (DPS-Finder) for finding optimal DPS motifs. Experimental results show that DPS-Finder can discover a length-10 motif from 22 length-500 DNA sequences within a few minutes and the DPS representation has a similar performance as SPSP representation.

1 Introduction

A *gene* is a segment of DNA that can be decoded to produce functional products like protein. To trigger the decoding process, a molecule, called *transcription factor*, will bind to a short region (*binding site*) preceding the gene. One kind of transcription factor can bind to more than one binding site. These binding sites usually have similar patterns and are collectively represented by a *motif*. Finding motifs and the corresponding binding sites from a set of DNA sequences is a critical step for understanding how genes work.

There are two popular models to represent a motif, string representation [4,6,10,11,16,17,19-22] and matrix representation [2,8,12-14]. String representation uses a length-l string of symbols (or nucleotides) 'A', 'C', 'G' and 'T' to represent a motif of length l. To improve the descriptive power of the representation, IUPAC symbols [6,20,22] can be introduced into the string to represent choices of symbols at a particular position (e.g. 'K' denotes 'G' or 'T'). Matrix representation further improves the descriptive power by using position weight matrices (PWMs) or position specific scoring matrices (PSSMs) to represent a motif. PWMs and PSSMs are matrices of size $4 \times l$ with the j-th column, which has four elements corresponding to the four nucleotides, effectively giving the occurrence probability of each of the four nucleotides at position j. While the matrix representation model appears superior, the solution space for PWMs and

* The research was supported in parts by the RGC grant HKU 7120/06E.

PSSMs is huge, which consists of $4l$ real numbers, and thus, algorithms generally either produce a sub-optimal motif matrix [2,8,12,13] or take too long to run when the motif is longer than 10 [15].

However, both the string and the matrix representations share an important common weakness: they assume that the occurrence of each nucleotide at a particular position of a binding site is independent of the occurrence of nucleotides at other positions. This assumption may not represent the actual situation. According to the analysis of wild-type and mutant Zif268 (Egr1) zinc fingers by Bulyk et al [5], it gives compelling evidence that nucleotides of transcription factor binding sites should not be treated independently, and a more realistic motif model should be able to describe nucleotide interdependence. Man and Stormo [18] have arrived at a similar conclusion in their analysis of Salmonella bacteriophage repressor Mnt: they found that interactions of Mnt with nucleotides at positions 16 and 17 of the 21 bp binding site are in fact not independent.

When there are sufficient number of *known* binding sites of a transcription factor, people can use some complex representations, e.g. the hidden Markov model (HMM) [24], Bayesian network [3] or enhanced PWM [9], to represent nucleotide interdependence. However, when we want to discover novel motif or describe a motif with only a few known binding sites, the input data may not contain enough information for deriving the hidden motif. Chin and Leung overcame the problem by introducing the SPSP representation [7], a generalized model of string representation and matrix representation, that can model the adjacent dependency of nucleotides with much less parameters than HMM and regular expression. Since the SPSP representation is simple, it can be used to discover novel motifs even if there are only five DNA sequences containing the binding sites of the transcription factor. However, like other models, discovering novel motifs in SPSP representation is a NP-hard problem. No efficient algorithm exists that can guarantee finding the hidden motif in reasonable amount of time.

After studying the binding sites of real biological data, we found that many motifs can be described by a simpler model. In this paper, we further simplify the SPSP representation to the Dependency Pattern Sets (DPS) representation. DPS representation is a generalized model of string representation, which can model adjacent nucleotide dependency. Although it has a lower descriptive power than SPSP representation, experimental results on real biological data showed that it has almost the same performance as SPSP representation. Besides, since DPS representation uses fewer parameters to describe a motif, it is possible to find the "optimal" motif in reasonable amount of time. We have introduced a branch and bound algorithm DPS-Finder that guarantees finding the "optimal" motif. In practice, DPS-Finder takes only a few minutes to discover a length-10 motif from 20 length-600 DNA sequences. For other approaches such as HMM, it may take hours or even days for a dataset of similar size.

This paper is organized as follows. In Section 2, we describe the DPS representation and the scoring function for determine the "optimal" motif in a set of DNA sequences. We introduce the branch and bound algorithm DPS-Finder in Section 3. Experimental results on real biological data comparing DPS-Finder with some popular software are given in Section 4, followed by concluding remarks in Section 5.

2 Problem Definition

2.1 DPS Representation

Motif is an abstract model for a set of binding sites with similar patterns. For example, the transcription factor CSRE [25], which activates the gluconeogenic structural genes, can bind to the following binding sites.

CGGA**T**G**A**A**T**GG
CGGA**T**G**A**A**T**GG
CGGA**T**G**A**A**A**GG
CGGA**C**G**G**A**T**GG
CGGA**C**G**G**A**T**GG

Note that there is dependence between the fifth and the seventh symbols, and the binding sites "CGGATGAATGG" occurs twice in the DNA sequences. The string representation models these binding sites by the length-11 string "CGGA**Y**G**R**A**W**GG" where 'Y' denotes 'T' or 'C', 'R' denotes 'A' or 'G' and 'W' denotes 'A' or 'T'. However, this representation has a problem that the strings "CGGA**T**G**G**A**T**GG", "CGGA**T**G**G**A**A**GG", "CGGA**C**G**A**A**T**GG", "CGGA**C**G**A**A**A**GG" and "CGGA**C**G**G**A**A**GG" are also considered as binding sites (false positives). Instead of modeling the CSRE motif by one string, the SPSP representation uses a pattern *P* and a set of score *S* (negative of logarithm of the occurrence probability) to represent the CSRE motif as follows.

$$P = (\mathrm{CGGA})\begin{pmatrix}\mathrm{TGA}\\ \mathrm{CGG}\end{pmatrix}(\mathrm{A})\begin{pmatrix}\mathrm{A}\\ \mathrm{T}\end{pmatrix}(\mathrm{GG}) \text{ and}$$

$$S = (-\log(1))\begin{pmatrix}-\log(0.6)\\ -\log(0.4)\end{pmatrix}(-\log(1))\begin{pmatrix}-\log(0.2)\\ -\log(0.8)\end{pmatrix}(-\log(1))$$

A length-11 string is considered as a binding site of CSRE if it matches with *P* and its score (sum of corresponding entries) is at most some threshold, say 3.1. For example, the score of the binding site "CGGA**TGA**A**T**GG" is -log(1)+ -log(0.6) + -log(1) + -log(0.8) + -log(1) = 1.05 < 3.1. The score of a non-binding site string "CGGA**CGG**A**A**GG" is -log(1)+ -log(0.4) + -log(1) + -log(0.2) + -log(1) = 3.6 > 3.1. The string "**T**GGA**TGA**A**T**GG" does not match with *P*, so it is not a binding site. In this example, the SPSP representation can model the motif with no false positive.

Although SPSP representation can describe the motif well, it is difficult to determine the score *S* for novel motifs (motifs with no known binding site) in real biological data. A challenge is to have a simpler model, which describes real motifs using fewer parameters than the SPSP representation while having fewer false positives than string representation.

We observed that using only the pattern P without S, we already can describe most real motifs. For example, if we consider those strings matching with P as binding sites, we only have one false positive "CGGA**CGGA**AGG" (instead of five for the string representation).

Apart from this, SPSP representation allows a motif having any number of wildcard pattern sets (positions with more than one possible choice of patterns, i.e. brackets with more than one pattern in it). For example, the following pattern P is allowed.

$$P = (\mathrm{C})\begin{pmatrix}\mathrm{GG}\\ \mathrm{GT}\\ \mathrm{TT}\end{pmatrix}(\mathrm{A})\begin{pmatrix}\mathrm{T}\\ \mathrm{C}\end{pmatrix}(\mathrm{G})\begin{pmatrix}\mathrm{A}\\ \mathrm{G}\end{pmatrix}(\mathrm{A})\begin{pmatrix}\mathrm{A}\\ \mathrm{T}\end{pmatrix}(\mathrm{GG})$$

Since the binding sites of a motif should be conserved in most positions, the number of wildcard pattern sets should be small. We found that allowing at most two wildcard pattern sets is enough for describing most motifs. Based on the above observations, we define the *Dependency Pattern Sets* (DPS) representation as follows.

> A DPS representation P contains a list of patterns sets P_i, $1 \le i \le L$, where at most **two** are wildcard pattern sets P_i containing 2 to k length-l_i patterns $P_{i,j}$ of symbols 'A', 'C', 'G' and 'T', $l_i \le l_{max}$ where the Hamming distance between these patterns is at most $d_{\max}$. Each of the other pattern set P_i contains **exactly one** length-l_i pattern $P_{i,1}$ and $\sum_i l_i = l$. A length-l string $\sigma = \sigma_1\sigma_2\ldots\sigma_L$ where $|\sigma_i| = l_i$ is considered as a binding site of P if $\sigma_i \in P_i$, $1 \le i \le L$.

2.2 *Scoring Function and Problem Definition*

Given a set of DNA sequences T with X length-l substrings bound by the same transcription factor, we should find many candidate motifs having different number of binding sites in T. In order to discover the hidden motif, we should have a scoring function for comparing different motifs. Given two motifs P_1 and P_2, a naive scoring function is to count the number of binding sites represented by the motifs, that is, P_1 is more likely to be the hidden motif if P_1 have more binding sites than P_2 in the set of sequences T. However, this scoring function has a weakness that it has not considered the number of possible binding sites for P_1 and P_2. Consider the following motifs.

$$P_1 = (\mathrm{C})\begin{pmatrix}\mathrm{AT}\\ \mathrm{CG}\\ \mathrm{GT}\end{pmatrix}(\mathrm{CC})\begin{pmatrix}\mathrm{A}\\ \mathrm{T}\end{pmatrix}(\mathrm{TC}) \text{ and } P_2 = (\mathrm{ACG})\begin{pmatrix}\mathrm{AG}\\ \mathrm{CT}\end{pmatrix}(\mathrm{AAA})$$

Even P_1 has slightly more binding sites than P_2, we cannot conclude that P_1 is more likely to be the hidden motif because P_1 has more possible binding site patterns ($3 \times 2 = 6$ patterns) than P_2 (2 patterns). In order to have a fair comparison, given a motif P with b binding sites in T, we calculate the probability (p-value) that P has b or more binding sites in T by chance based on a background model. Under the assumption that the hidden motif should have an unexpectedly large number of binding sites, a motif P with small p-value is likely to be the hidden motif. The p-value of a motif can be calculated as follows [7].

Let B be the background model for the non-binding region of the DNA sequences T and $B(\sigma)$ be the probability that a length-l string σ occurs in a particular position in T. B can be a Markov Chain or an uniform distribution etc. Given a DPS motif P with w possible binding sites $s_1, s_2, \ldots, s_w$, the probability that P has a binding site at a particular position in T is $\sum_{i=1}^{w} B(s_i)$. Assuming the probability that motif P has a binding site at any positions in T are independent, the probability that P has b or more binding sites in T is

$$p-\text{value}(P)=\sum_{j=b}^{X}\left[\binom{X}{j}\left(\sum_{i=1}^{w}B(s_i)\right)^{j}\left(1-\sum_{i=1}^{w}B(s_i)\right)^{X-j}\right] \tag{1}$$

Based on the scoring function in Eq(1), we define the motif discovering problem as follows.

> Given a set of DNA sequences T, the background model B and the motif length l, we want to discover a length-l DSP motif P with the minimum p-value.

3 DPS-Finder Algorithm

In this section, we introduce the DPS-Finder Algorithm for solving the motif discovery problem described in Section 2. DPS-Finder Algorithm first constructs a l-factor tree [1], a suffix tree with all nodes of depth $> l$ being removed, to represent all possible motifs in the input sequences T with different positions of the wildcard pattern sets. For each possible motif P, it finds the set of patterns in each wildcard pattern set that minimizes p-value(P) using a branch and bound approach. Experiments showed that DPS-Finder Algorithm has to deal in the best case only 25% of the number of cases to be dealt by the brute force algorithm.

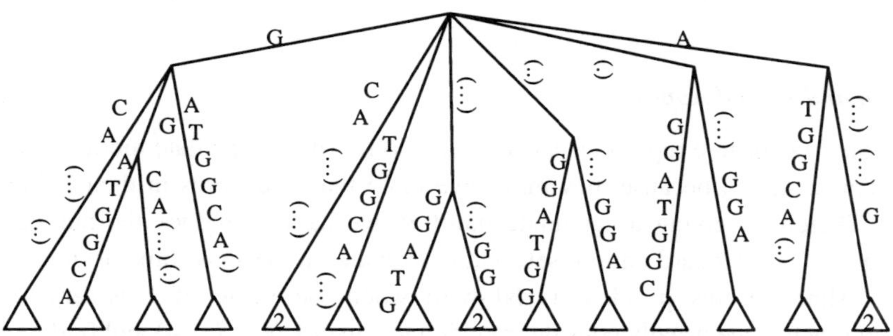

Figure 1. The 8-factor tree of the sequences "CA(...)(...)GGATGGCA(...)(...)GG". For examples, the pattern "(CA)(...)(...)", "(A)(...)(...)(G)" and "(...)(...)(GG)" occur twice in the sequences.

3.1 Factor Tree Representation

In order to discover the optimal motif, we should consider all possible positions ($C(2 + l - 2l_{max}, 2) = O(l^2)$) of the wildcard pattern sets. For example, when the motif length l is 8 and the maximum wildcard pattern length l_{max} is 3, the length-8 substring "CGCAGGTG" (binding site of the AC transcription factor) can be a binding site of motifs in the following formats, (...)(...)(TG), (...)(A)(...)(G), (...)(AG)(...), (C)(...)(...)(G), (C)(...)(G)(...) and (CG)(...)(...), where (...) represents a wildcard pattern set of length 3. Note that motifs with wildcard pattern shorter than 3 or with one wildcard pattern set only have also been considered in the above formats. For example, (...)(AGG)(..) and (...)AGGTG are special case of the motif (...)(AG)(...). When we find the optimal motif in form of (...)(AG)(...), we have also considered motifs in form of (...)(AGG)(..) and (...)AGGTG. Since it takes $O(l)$ time to convert a length-l substring to a motif and there are X length-l substrings in T, brute force method takes $O(Xl^3)$ time to get the list of $O(Xl^2)$ possible forms of motif.

However, when a motif of a substring is considered, we can easily get another motif for the adjacent substring by shifting one symbol. For example, when the motif (CG)(...)(...) of the substring "CGCAGGTG" in the input sequence "...CA**CGCAGGTG**GG..." is considered, by shifting one symbol, we will get another motif (G)(...)(...)(G) for the substring "GCAGGTGG". When we represent the input sequence in the form of "...CA**CGCAGGTG**GG...", each length-8 sliding window containing the two length-3 brackets represents one possible motif. Based on this observation, DPS-Finder Algorithm constructs a generalized l-factor tree [1] of $O(l^2)$ (represent the $O(l^2)$ motifs for a length-l substring) length-$O(X)$ sequences (input sequences with some positions represented by brackets) to represent the $O(Xl^2)$ possible motifs. A l-factor tree is a suffix tree [23] of height l where each path from the root to a leaf represents a length-l substring occurring in the input sequence. Figure 1 shows a factor tree of height-8 for the sequence "CA(...)(...)GGATGGCA(...)(...)GG". Since constructing the generalized l-factor tree takes $O(Xl^2)$ time [1] only, DPS-Finder Algorithm speeds up the process by a factor of $O(l)$ when compares with the brute force algorithm.

3.2 Branch and Bound Approach

Each leaf of the l-factor tree represents a candidate motif. These candidate motifs may not be in DPS representation because they may have more than k patterns in their wildcard pattern sets. Therefore, giving a candidate motif P, we have to reduce the number of patterns in each of its wildcard pattern set to at most k and at the same time, to minimize the p-value. Although this problem is NP-hard when the value of k is large (see Appendix), in practice we usually consider motifs with small k (e.g. $k = 4$) and finding the optimal motif is still feasible.

When refining a candidate motif P to a motif P' in DSP representation with the minimum p-value(P'), we perform a depth-first-search to check all possible combinations of patterns in the two wildcard pattern sets of P. We first pick two patterns, each forms a wildcard pattern set of P. Then we pick more patterns for P' until k patterns have been selected for each wildcard pattern set. In the selection process, we consider patterns with increasing order of p-values. After picking a new pattern P_i, the additional number of binding sites covered by P' is upper bounded by the number of binding sites covered by P_i. Therefore, in many cases, we can stop picking new patterns because the p-value of the refined motif P' must not be smaller than the suboptimal motif we have already found.

Apart from applying a branch and bound approach on refining each candidate motif P, we also apply similar approach on checking the $O(Xl^2)$ candidate motifs. We first refine those candidate motifs with two patterns, each forms a wildcard pattern set, covering the largest number of binding sites. Since the number of binding sites covered by a candidate motif is upper bounded by the total number of binding sites covered by the top-k patterns in its wildcard pattern sets, many candidate motifs can be pruned out.

4 Experimental Results

We compared the performance of some popular motif discovering algorithms, i.e. Weeder [19], MEME [13] and SPSP-Finder [7], with DSP-Finder on the yeast data set in SCPD [25]. SCPD contains information of the motif patterns and the binding sites for a set of transcription factors of yeast. For each transcription factor, we chose the 600 base pairs in the upstream of those genes bound by the transcription factor as the input sequences T. Given the motif length, the four algorithms were used to discover the hidden motif in T.

Weeder and MEME used string representation and matrix representation to model a motif respectively. Both of them could not model the nucleotide dependency in motifs. SPSP-Finder, used the SPSP representation, can model the nucleotide dependency in motifs. However, all these algorithms applied a heuristic approach which cannot guarantee finding the "optimal" motifs.

In the experiments, DSP-finder used an order-0 Markov chain calculated based on the input sequence to model the non-binding regions. The width of a wildcard pattern set was at most 3 ($l_{max} = 3$), the Hamming distance between patterns in a wildcard pattern set was at most 1 ($d = 1$) and there were at most 4 patterns in a wildcard pattern set. The experimental results were shown in Table 1. All algorithms finished in 10 minutes for each dataset. Note that we have not listed out those motifs which could not be discovered by any of the algorithms.

In general, SPSP-Finder and DSP-Finder has better performance than the other algorithms because they can model nucleotide dependency. DSP-Finder performs better than SPSP-Finder when finding motif of MCM1 because DSP-Finder guarantees finding the motif with the lowest p-value while SPSP-Finder is trapped in local minimum.

DSP-Finder performs worse than MEME and SPSP-Finder in two cases, the HAP2/3/4 and SFF datasets. For the HAP2/3/4 dataset, there was nucleotide dependency between the fifth and the sixth nucleotides. However, since the Hamming distance between the possible patterns is 2, DSP-Finder could not discover the motif in our setting ($d = 1$). DSP-Finder could not discover the motif of SFF while MEME was successful because there were no strong bias at most positions of this motif. In these cases, a matrix representation can model the motif better than a string representation, i.e. Weeder also fails in this case.

Table 1. Experimental results on yeast data.

Name	Pattern	Weeder	MEME	SPSPFinder	DPSFinder
13nt	ACGAGGCTTACCG	-	-	(ACGA)(GGCT)(TACC)(G)	(A)(CGA)(GGC)(TTACCG)
ACE2	GCTGGT	-	-	(GCTG)(GT)	(GCA / GCT)(GGT / CGT)
ADR1	TCTCC	-	TCTCC	(TCTC)(C)	(TC)(TCC / TTC / TGC)
AP1	TTANTAA	-	-	(TTA)(G / C)(TAA)	(TTA)(GTA / CTA)(A)
CCBF	CNCGAAA	CACGAAA	-	(C)(A / C / G / T)(CGAA)(A)	(CAC / CGC)(GAA / CAA)(A)
CPF1	TCACGTG	CACGTG	TCACGTG	(CACG)(TA)	(TCA / CCA / GCA)(C)(GTG / GAG)
CSRE	CGGAYRRAWGG	-	-	(CGGA)(TGA / TAA / CGG)(A)(A / T)(GG)	(CGG)(ATG / ATA)(AAT / AAA / AAG)(GG)
CuRE	TTTGCTC	TTTGCTCA	-	(TTT)(GCT / TCG)(C)	(T)(TTG / GTG)(CTC)
GATA	CTTATC	CTTATC	-	(CTTA)(TC)	(CTT / CTA)(ATC)
HAP2/3/4	CCAATCA	-	-	(CCAA)(TC / TG / CC)(A)	-
LEU	CCGNNNNCGG	CCGGGACCGG	CCGGAACCGG	(CGG)(A / G)(ACCG)(G)	(CC)(GGT / GGG)(ACC / CCC)(GG)
MAT2	CRTGTWWWW	CATGTAATTA	-	(GA / GT / TA)(AATT)(AC / TA / TC / TG)(A / C)	(CAT)(GTA / GTT)(ATT / AAT)
MCM1	CCNNNWWRGG	CCCGTTTAGG	CCTAATTAGG	-	(CCTA)(AAA / AAC / AAT)(GGG / TGG / AGG)
SFF	GTMAACAA	-	GTCAACAA	-	-
UASCAR	TTTCCATTAGG	-	-	(T)(GCCC / TCAC / TCCA)(TT)(AGCG / AGGA)	(AT)(TTC / TGC)(CAT)(TAG)

Motifs of transcription factors that cannot be found by any algorithms were not shown in this table. 'M' stands for 'A' or 'C', 'N' stands for any nucleotide. 'R' stands for 'A' or 'G', 'W' stands for 'A' or 'T', 'Y' stands for 'C' or 'T'.

5 Conclusion

In this paper, we introduced the DPS representation to capture the nucleotide dependency in a motif, which is simpler than the SPSP representation. We also developed a branch and bound algorithm DPS-Finder to locate the optimal DPS motif. Experimental results on real biological datasets show that DPS-Finder is efficiency and the DPS representation is powerful enough to capture most of the real motifs. Further directions include extending the model and the algorithm to local motif pairs or non-linear motifs.

References

1. J. Allali and M.F. Sagot. The at most k-deep factor tree. *Internal Report Institut Gaspard Monge, University of Marne-la-Vallee*, IGM 2004-03, Juillet 2004.
2. T. Bailey and C. Elkan. Unsupervised learning of multiple motifs in biopolymers using expectation maximization. *Machine Learning*, 21:51—80, 1995.
3. Y. Barash, G. Elidan, N. Friedman and T. Kaplan. Modeling Dependencies in Protein-DNA Binding Sites. *RECOMB*, 28—37, 2003.
4. J. Buhler and M. Tompa. Finding motifs using random projections. *RECOMB*, 69—76, 2001.
5. M.L. Bulyk, P.L.F. Johnson and G.M. Church. Nucleotides of transcription factor binding sites exert interdependent effects on the binding affinities of transcription factors. *Nuc. Acids Res.*, 30:1255—1261, 2002.
6. F. Chin and H. Leung. An Efficient Algorithm for String Motif Discovery. *APBC*, 79—88, 2006.
7. F. Chin and H. Leung. DNA Motif Representation with Nucleotide Dependency. *TCBB* (to appear)
8. F. Chin, H. Leung, S.M. Yiu, T.W. Lam, R. Rosenfeld, W.W. Tsang, D. Smith and Y. Jiang. Finding Motifs for Insufficient Number of Sequences with Strong Binding to Transcription Factor. *RECOMB*, 125—132, 2004.
9. S. Hannenhalli and L.S. Wang. Enhanced Position Weight Matrices Using Mixture Models. *Bioinformatics*, 21(Supp 1):i204—i212, 2005.
10. U. Keich and P. Pevzner. Finding motifs in the twilight zone. *RECOMB*, 195—204, 2002.
11. S. Kielbasa, J. Korbel, D. Beule, J. Schuchhardt and H. Herzel. Combining frequency and positional information to predict transcription factor binding sites. *Bioinformatics*, 17:1019—1026, 2001.
12. C. Lawrence, S. Altschul, M. Boguski, J. Liu, A. Neuwald and J. Wootton . Detecting subtule sequence signals: a Gibbs sampling strategy for multiple alignment. *Science* 262:208—214, 1993.
13. C. Lawrence and A. Reilly. An expectation maximization (em) algorithm for the identification and characterization of common sites in unaligned biopolymer sequences. *Proteins: Structure, Function and Genetics*, 7:41—51, 1990.
14. H. Leung and F. Chin. Discovering Motifs with Transcription Factor Domain Knowledge. *PSB*, 472—483, 2007.
15. H. Leung and F. Chin. Finding Exact Optimal Motif in Matrix Representation by Partitioning. *Bioinformatics*, 22:ii86—ii92, 2005.
16. M. Li, B. Ma, L. Wang. Finding similar regions in many strings. *Journal of Computer and System Sciences*, 65:73—96, 2002.
17. S. Liang. cWINNOWER Algorithm for Finding Fuzzy DNA Motifs. *Computer Society Bioinformatics Conference*, 260—265, 2003.

18. T.K. Man and G.D. Stormo. Non-independence of Mnt repressor-operator interaction determined by a new quantitative multiple fluorescence relative affinity (QuMFRA) assay. *Nuc. Acids Res.*, 29:2471—2478, 2001.
19. G. Pavesi, P. Mereghetti, F. Zambelli, M. Stefani, G. Mauri and G. Pesole. MoD Tools: regulatory motif discovery in nucleotide sequences from co-regulated or homologous genes. *Nuc. Acids Res.*, 34:566—570, 2006.
20. G. Pesole, N. Prunella, S. Liuni, M. Attimonelli and C. Saccone. Wordup: an efficient algorithm for discovering statistically significant patterns in dna sequences. *Nucl. Acids Res.*, 20(11):2871—2875, 1992.
21. P. Pevzner and S.H. Sze. Combinatorial approaches to finding subtle signals in dna sequences. *ISMB*, 269—278, 2000.
22. S. Sinha S and M. Tompa. A statistical method for finding transcription factor binding sites. *ISMB*, 344—354, 2000.
23. E. Ukkonen. On-line construction of suffix trees. *Algorithmica*, 14(3):249—260, 1995.
24. X. Zhao, H. Huang and T.P. Speed. Finding Short DNA Motifs Using Permuted Markov Models. *RECOMB*, 68—75, 2004.
25. J. Zhu and M. Zhang. SCPD: a promoter database of the yeast Saccharomyces cerevisiae. *Bioinformatics*, 15:563—577, 1999. http://cgsigma.cshl.org/jian/

Appendix

In this section, we prove that the Candidate Motif Refinement Problem is NP-hard.

> **Candidate Motif Refinement (CMR) Problem:** given a motif P, reducing the size of P's wildcard pattern sets to at most k with the minimum p-value.

We prove it by reducing the Balanced Complete Bipartite Subgraph problem, which is NP-hard, to this problem.

> **Balanced Complete Bipartite Subgraph (BCBS) Problem:** given a bipartite graph $G = (V,E)$ and a positive integer k, we want to determine if there are two disjoint subsets V_1, $V_2 \subseteq V$ such that $|V_1| = |V_2| = k$ and $u \in V_1$, $v \in V_2$ implies that $\{u,v\} \in E$.

Given a BCBS Problem, we construct a motif P as follows: Let l_{max} be the smallest integer such that $4^{l_{max}} \geq k|V|$. Each vertex v_i of G is represented by a unique length-l_{max} string $s(v_i)$. The candidate motif P is a length-$2l_{max}$ pattern with exactly two wildcard pattern sets, each contains length-l_{max} string $s(v_i)$, representing the vertices in one partite of G. There are $|E|$ length-$2l_{max}$ input DNA sequences T. $s(v_i)s(v_j)$ is an input DNA sequence if and only if $\{v_i, v_j\} \in E$.

Under the restriction that the size of the wildcard pattern sets is at most k, the refined motif P' has the minimum p-value when the concatenation of each pair of patterns in the two wildcard pattern sets of size k exists in the input DNA sequences T (i.e. P' has exactly k^2 binding sites). Therefore, the BCBS problem can be solved by solving the CMR problem and check if refined motif P' has exactly k^2 binding sites. ∎

PRIMER SELECTION METHODS FOR DETECTION OF GENOMIC INVERSIONS AND DELETIONS VIA PAMP

BHASKAR DASGUPTA*

Department of Computer Science, University of Illinois at Chicago, Chicago, IL 60607-7053
E-mail: dasgupta@cs.uic.edu

JIN JUN† and ION I. MĂNDOIU†

Computer Science & Engineering Department, University of Connecticut, Storrs, CT 06269-2155
E-mail: {jinjun,ion}@engr.uconn.edu

Primer Approximation Multiplex PCR (PAMP) is a recently introduced experimental technique for detecting large-scale cancer genome lesions such as inversions and deletions from heterogeneous samples containing a mixture of cancer and normal cells. In this paper we give integer linear programming formulations for the problem of selecting sets of PAMP primers that minimize detection failure probability. We also show that PAMP primer selection for detection of anchored deletions cannot be approximated within a factor of $2 - \varepsilon$, and give a 2-approximation algorithm for a special case of the problem. Experimental results show that our ILP formulations can be used to optimally solve medium size instances of the inversion detection problem, and that heuristics based on iteratively solving ILP formulations for a one-sided version of the problem give near-optimal solutions for anchored deletion detection with highly scalable runtime.

Keywords: Genomic structural variation detection; PAMP primer selection; Integer linear programming.

1. Introduction

As described by Liu and Carson,[1] PAMP requires the selection of a large number of multiplex PCR primers from the genomic region of interest. Exploiting the fact that the efficiency of PCR amplification falls off exponentially beyond a certain product length, PAMP primers are selected such that (1) no PCR amplification results in the absence of genomic lesions, and (2) with high probability, a genomic lesion brings one or more pairs of primers in the proximity of each other, resulting in PCR amplification. Multiplex PCR amplification products are then hybridized to a microarray to identify the pair(s) of primers that yield amplification. This gives an approximate location for the breakpoints of the genomic lesion; precise breakpoint coordinates can be determined by sequencing PCR products.

As in previous multiplex PCR primer set selection formulations,[2–4] PAMP primers must satisfy standard selection criteria such as hybridizing to a unique site in the genomic

*Supported in part by NSF grants IIS-0346973, IIS-0612044 and DBI-0543365.

†Supported in part by NSF grants IIS-0546457 and DBI-0543365.

region of interest, having melting temperature in a pre-specified range, and lacking secondary structures such as hairpins. Candidate primers meeting these criteria can be found using robust software tools for primer selection, such as the Primer3 package.[5] Similar to some previous works on multiplex PCR primer set selection,[2,4] PAMP also requires subsets of non-dimerizing primers. Indeed, as observed in Bashir et al.,[6] even a single pair of dimerizing primers can lead to complete loss of amplification signal. However, unlike existing works on multiplex PCR primer set selection[2–4] which focus on minimizing the number of primers and/or multiplex PCR reactions needed to amplify a *given* set of discrete amplification targets, the objective in PAMP primer selection is to minimize the probability that an *unknown* genomic lesion fails to be detected by the assay. The only work we are aware on this novel problem is that of Bashir et al.,[6] who proposed integer linear programming (ILP) formulations and simulated annealing algorithms for PAMP primer selection when the goal is to detect genomic deletions known to include a given anchor locus.

In this paper we show that the optimization objective used in the ILP formulation of Bashir et al.[6] is not equivalent to minimization of failure probability, and propose new ILP formulations capturing the later objective in PAMP primer selection for detection of genomic inversions (Section 2) and anchored deletions (Section 3). We also show that PAMP primer selection for detection of anchored deletions cannot be approximated within a factor of $2 - \varepsilon$ (Lemma 3.1), and give a 2-approximation algorithm for a special case of the problem (Lemma 3.2). Experimental results presented in Section 4 show that our ILP formulations can be used to optimally solve medium size instances of the inversion detection problem, and that heuristics based on iteratively solving ILP formulations for a *one-sided* version of the anchored deletion detection problem give near-optimal solutions with highly scalable runtime.

2. Inversion Detection

Throughout the paper, PCR amplification is assumed to occur if and only if there is at least one pair of primers hybridizing to opposite strands at two sites that are at most L bases apart and such that the primers' $3'$ ends face each other. This model assumes that PCR amplification success probability is a simple 1-0 step function of product length, with the transition from fully efficient amplification to no amplification taking place between product lengths L and $L + 1$. Our methods can be easily modified to handle arbitrary amplification success probability functions.

Let $\mathcal{G}$ be a genomic region indexed along the forward strand in $5' - 3'$ orientation. We seek a set of non-dimerizing multiplex PCR primers that does not yield PCR amplification when a specified interval $[x_{min}, x_{max}]$ of $\mathcal{G}$ contains no inversion, and, subject to this condition, minimizes the probability of not getting amplification when an inversion is present in the sample. In order to formalize the optimization objective, we assume a known probability distribution for the pairs of endpoints of inversions within $[x_{min}, x_{max}]$, i.e., we assume that, for every pair (l, r) of endpoints with $x_{min} \leq l < r \leq x_{max}$, we are given the (conditional) probability $p_{l,r} \geq 0$ of encountering an inversion with endpoints l and r, where $\sum_{x_{min} \leq l < r \leq x_{max}} p_{l,r} = 1$. This probability distribution may be as simple as the

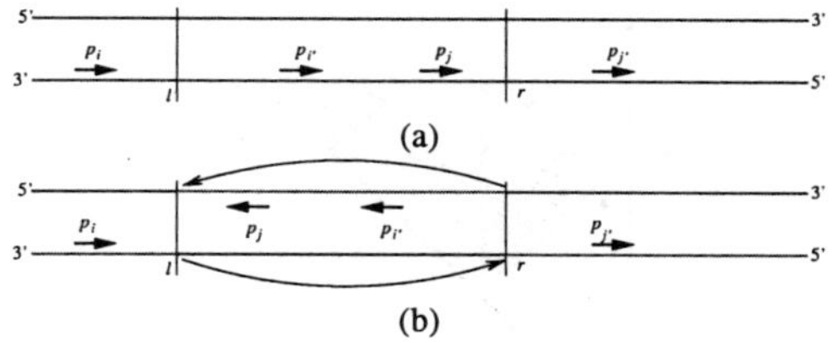

Fig. 1. Hybridization loci for 4 PAMP primers without (a) and with (b) an inversion with endpoints (l, r). DNA strands are color-coded blue or red according to their forward/reverse orientation in the reference genome. Multiplex PCR yields no amplicons when the sample contains no genomic inversion, but yields at least one amplicon if an inversion brings binding sites of primers p_i and p_j within L bases of each other.

uniform distribution (under which every pair of endpoints is equally likely), or can incorporate existing biological knowledge on the distribution of recombination hotspots and/or biases in inversion segment lengths.

In the pre-processing stages of the primer selection process, we collect a large number of candidate primers satisfying appropriate biochemical constraints on melting temperature, lack of hairpin secondary structures, etc. Each candidate primer must also hybridize to the reverse strand of the reference genome at a unique location within $\mathcal{G}$ (see Figure 1(a)). Clearly, multiplex PCR with any subset of the candidate primers should not yield PCR amplification when the genomic sample contains no inversion within $\mathcal{G}$.

Let $\mathcal{P} = \{p_1, p_2, \ldots, p_n\}$ denote the set of candidate primers, and let $x_1 < x_2 < \ldots < x_n$ be the positions of their $3'$ ends when hybridized to the reverse strand of $\mathcal{G}$. Furthermore, let $\mathcal{E}$ denote the set of pairs of primers in $\mathcal{P}$ that form dimers. The *PAMP primer selection problem for inversion detection (PAMP-INV)* can then be formulated as follows:

Given: set $\mathcal{P}$ of candidate primers hybridizing at unique loci of the reverse strand of $\mathcal{G}$, set $\mathcal{E}$ of dimerizing candidate primer pairs, maximum multiplexing degree N, and amplification length upper-bound L
Find: a subset $\mathcal{P}'$ of $\mathcal{P}$ such that

(1) $|\mathcal{P}'| \leq N$
(2) $\mathcal{P}'$ does not include any pair of primers in $\mathcal{E}$, and
(3) The probability that multiplex PCR using the primers of $\mathcal{P}'$ fails to yield amplification, given that $[x_{min}, x_{max}]$ contains an inversion, is minimized. In other words, $\mathcal{P}'$ minimizes

$$\sum_{x_{min} \leq l < r \leq x_{max}} f(\mathcal{P}'; l, r) p_{l,r} \tag{1}$$

where $f(\mathcal{P}'; l, r) = 1$ if $\mathcal{P}'$ fails to yield a PCR product when the inversion with endpoints (l, r) is present in the sample, and $f(\mathcal{P}'; l, r) = 0$ otherwise.

We next formulate PAMP-INV as an integer linear program (ILP). For convenience, we add to $\mathcal{P}$ "dummy" primers p_0 and p_{n+1}, assumed to uniquely hybridize to $\mathcal{G}$ at locations $x_0 = x_{min} - L$ and and $x_{n+1} = x_{max} + L$, respectively. Dummy primers are assumed not

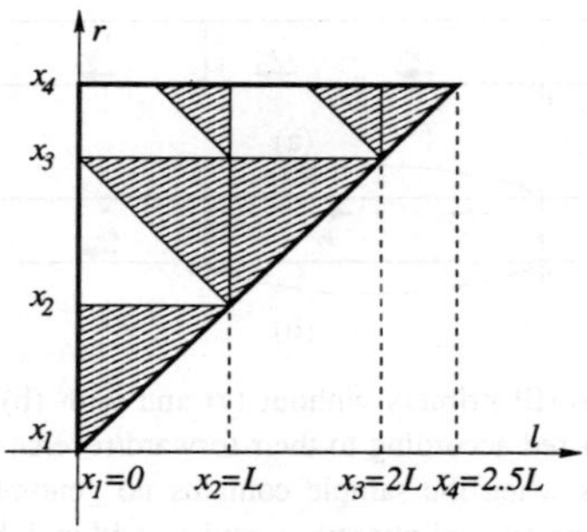

Fig. 2. Graphical representation of the space of endpoint pairs (l, r) (area within thick triangle) for a PAMP-INV instance with $x_{min} = 0, x_{max} = 2.5L$. If primer set $\mathcal{P}'$ consists of 4 primers hybridizing to the reference genome at positions 0, L, $2L$, and $2.5L$, respectively, inversions (l, r) corresponding to the shaded regions fail to yield PCR amplifi cation.

to dimerize with each other or with other primers in $\mathcal{P}$, and thus they can always be included in $\mathcal{P}'$. Without loss of generality we will assume that the location of all candidate primers is between x_0 and x_{n+1}, since primers that hybridize outside the interval $[x_{min}-L, x_{max}+L]$ cannot help in detecting inversions located within $[x_{min}, x_{max}]$.

Consider an inversion with endpoints (l, r) and a set of non-dimerizing primers $\mathcal{P}' \subseteq \mathcal{P}$ with $p_0, p_{n+1} \in \mathcal{P}'$. Let $i = \max\{k : p_k \in \mathcal{P}', x_k < l\}$ and $j = \max\{k : p_k \in \mathcal{P}', x_k < r\}$. Note that if both endpoints of the inversion occur between two consecutive primers of $\mathcal{P}'$ (i.e., $i = j$), then $\mathcal{P}'$ fails to yield any amplification and the inversion remains undetected. When $i < j$, $\mathcal{P}'$ still fails to yield any amplification if $(l-1-x_i)+(r-x_j) > L$. On the other hand, when $i < j$ and $(l - 1 - x_i) + (r - x_j) \leq L$, the multiplex PCR reaction using the primers of $\mathcal{P}'$ yields at least one amplification product given by p_i and p_j.

For every quadruple (i, i', j, j') with $x_i < x_{i'}, x_j < x_{j'}, x_i \leq x_j$, we let $C_{i,i',j,j'} = \sum p_{l,r}$, where the sum is over all inversion endpoint pairs (l, r) such that $\max\{x_i, x_{min}\} < l \leq \min\{x_{i'}, x_{max}\}$, $\max\{x_j, x_{min}\} < r \leq \min\{x_{j'}, x_{max}\}$, $(l-1-x_i)+(r-x_j) > L$. If $(p_i, p_{i'})$ and $(p_j, p_{j'})$ are pairs of consecutive primers of $\mathcal{P}'$, then $C_{i,i',j,j'}$ gives the cumulative probability that an inversion with endpoints $l \in (x_i, x_{i'}] \cap [x_{min}, x_{max}]$ and $r \in (x_j, x_{j'}] \cap [x_{min}, x_{max}]$ fails to yield any amplification product under multiplex PCR with the primers of $\mathcal{P}'$.

To express PAMP-INV as an ILP we use three types of 0/1 variables:

- e_i, which are set to 1 if and only if $p_i \in \mathcal{P}'$,
- $e_{i,i'}$, which are set to 1 if and only if p_i and $p_{i'}$ are consecutive primers in $\mathcal{P}'$, and
- $e_{i,i',j,j'}$, which are set to 1 if and only if $(p_i, p_{i'})$ and $(p_j, p_{j'})$, are consecutive primers in $\mathcal{P}'$ and $i \leq j$.

Variables of last type allow expressing the total failure probability (1) as a sum of appropriate $C_{i,i',j,j'}$'s. The complete PAMP-INV ILP is given below. Constraints (3) and (4) ensure that a variable $e_{i,i',j,j'}$ is set to 1 if and only if both $e_{i,i'}$ and $e_{j,j'}$ are set to 1. Similarly, constraints (5) ensure that a variable $e_{i,j}$ is set to 1 only if both e_i and e_j are set to 1. Variables $e_{i,j}$ which are set to 1 can be viewed as defining a path connecting p_0 to p_{n+1} via a subset of intermediate primers visited in left-to-right order, and this is captured in constraints (6) and (7). Constraint (8) can handle a limitation on the number of allowed primers (N). Finally, constraint (9) is used to ensure that no pair of dimerizing candidate primers is added to the selected set $\mathcal{P}'$.

$$\text{minimize} \quad \sum_{\{(i,i',j,j')\,:\,i<i',j<j',i\leq j\}} C_{i,i',j,j'}\, e_{i,i',j,j'} \tag{2}$$

$$\text{s.t.} \quad e_{i,i'} + e_{j,j'} \geq 2e_{i,i',j,j'},\ i < i',\ j < j', \text{ and } i \leq j \tag{3}$$

$$e_{i,i',j,j'} \geq e_{i,i'} + e_{j,j'} - 1,\ i < i',\ j < j', \text{ and } i \leq j \tag{4}$$

$$e_i + e_j \geq 2e_{ij},\ 1 \leq i < j \leq n \tag{5}$$

$$\sum_{j=1}^{n+1} e_{0j} = \sum_{i=0}^{n} e_{i,n+1} = 1 \tag{6}$$

$$\sum_{i=0}^{j-1} e_{ij} = \sum_{k=j+1}^{n+1} e_{jk},\ 1 \leq j \leq n \tag{7}$$

$$\sum_{1\leq i\leq n} e_i \leq N \tag{8}$$

$$e_i + e_j \leq 1, \text{ for all } (p_i, p_j) \in \mathcal{E} \tag{9}$$

$$e_{i,i',j,j'} \in \{0,1\},\ e_{i,j} \in \{0,1\},\ e_i \in \{0,1\}$$

3. Anchored Deletion Detection

Bashir et al.[6] recently studied the PAMP primer selection problem for deletion detection, which we will refer to as PAMP-DEL. As in their work, we assume that the deletion spans a known genomic location, i.e., we consider detection of *anchored* deletions only. Let $\{p_1, \ldots, p_m\}$ and $\{q_1, \ldots, q_n\}$ be the two sets of forward and reverse candidate primers, indexed by increasing distance from the anchor. Given a set $\mathcal{E}$ of primer pairs that form dimers, the goal is to pick a set $\mathcal{P}'$ of at most N_f forward and at most N_r reverse primers such that no two of the selected primers dimerize, and, subject to this constraint, the probability that the selected primers fail to produce a PCR product when the sample contains a deletion is minimized. The latter probability is computed assuming given PCR amplification threshold L and probability distribution for the pairs of endpoints of the deletion.

PAMP-DEL can be formulated as an ILP using an idea similar to that in previous section. For every quadruple (i, i', j, j'), $i \leq i'$, $j \leq j'$, let $C_{i,i',j,j'}$ denote the total probability that a deletion with ends between the hybridization sites of p_i and $p_{i'}$, respectively q_j and $q_{j'}$, does not result in PCR amplification when $(p_i, p_{i'})$ and $(q_j, q_{j'})$ are consecutive sets of forward, respectively reverse primers of $\mathcal{P}'$. Using 0/1 variables f_i (r_i) to indicate when p_i (respectively q_i) is selected in $\mathcal{P}'$, $f_{i,j}$ ($r_{i,j}$) to indicate that p_i and p_j (respectively q_i and q_j) are consecutive primers in $\mathcal{P}'$, and $e_{i,i',j,j'}$ to indicate that both $(p_i, p_{i'})$ and $(q_j, q_{j'})$ are pairs of are consecutive primers in $\mathcal{P}'$, we obtain the following formulation:

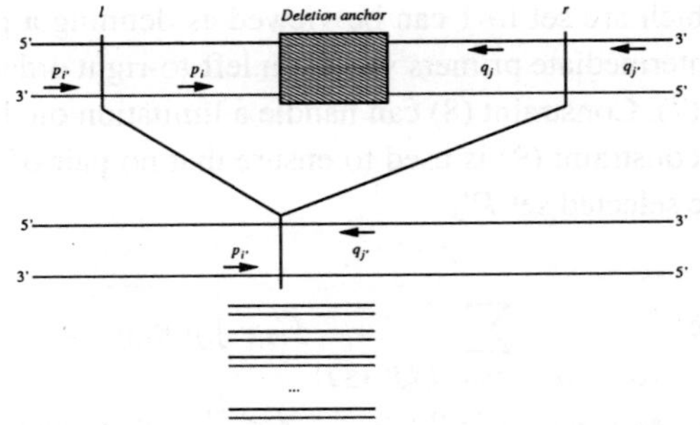

(a)

Fig. 3. Deletion detection using PAMP. If a deletion with endpoints l and r brings the hybridization loci of forward primer $p_{i'}$ and reverse primer $q_{j'}$ within L bases of each other the PAMP assay results in PCR amplifi cation.

$$
\begin{aligned}
\text{minimize} \quad & \sum_{\{(i,i',j,j')\,:\, i<i',j<j'\}} C_{i,i',j,j'}\, e_{i,i',j,j'} \qquad (10)\\
\text{s.t.} \quad & f_{i,i'} + r_{j,j'} \geq 2e_{i,i',j,j'},\ i < i' \text{ and } j < j' \\
& e_{i,i',j,j'} \geq f_{i,i'} + r_{j,j'} - 1,\ i < i' \text{ and } j < j' \\
& f_i + f_j \geq 2f_{i,j},\ 1 \leq i < j \leq m \\
& r_i + r_j \geq 2r_{i,j},\ 1 \leq i < j \leq n \\
& \sum_{j=1}^{m+1} f_{0,j} = \sum_{i=0}^{m} f_{i,m+1} = \sum_{j=1}^{n+1} r_{0,j} = \sum_{i=0}^{n} r_{i,n+1} = 1 \\
& \sum_{i=0}^{j-1} f_{i,j} = \sum_{k=j+1}^{m+1} f_{j,k},\ 1 \leq j \leq m \\
& \sum_{i=0}^{j-1} r_{i,j} = \sum_{k=j+1}^{n+1} r_{j,k},\ 1 \leq j \leq n \\
& \sum_{1 \leq i \leq m} f_i \leq N_f,\ \sum_{1 \leq i \leq n} r_i \leq N_r \\
& f_i + f_j \leq 1, \text{ for all } (p_i, p_j) \in \mathcal{E} \\
& r_i + r_j \leq 1, \text{ for all } (q_i, q_j) \in \mathcal{E} \\
& f_i + r_j \leq 1, \text{ for all } (p_i, q_j) \in \mathcal{E} \\
& e_{i,i',j,j'} \in \{0,1\},\ f_{i,j}, r_{i,j} \in \{0,1\},\ f_i, r_i \in \{0,1\}
\end{aligned}
$$

Bashir et al.[6] also introduced an *one-sided* version of PAMP-DEL, referred to as PAMP-1SDEL, in which one of the deletion endpoints is known in advance. For this version of the problem our ILP formulation can be simplified substantially. Let $x_1 < x_2 < \ldots < x_n$ be the hybridization positions for the reverse candidate primers $q_1, \ldots, q_n$. We introduce two dummy reverse primers that hybridize right after the location x_0 of the anchor, and at position $x_{n+1} = x_{max} + L$, respectively (as usual, dummy primers are assumed not to dimerize). Denoting by $C_{i,j}$ the probability that a deletion whose right endpoint falls between x_i and x_j does not result in PCR amplification, and using 0/1 variables r_i and $r_{i,j}$ as in the PAMP-DEL ILP, we obtain the following formulation for PAMP-1SDEL:

$$\begin{aligned}
\text{minimize } & \sum_{i<j} C_{i,j}\, r_{i,j} && (11)\\
\text{s.t. } & r_i + r_j \geq 2r_{i,j},\ 1 \leq i < j \leq n \\
& \sum_{j=1}^{n+1} r_{0,j} = \sum_{i=0}^{n} r_{i,n+1} = 1 \\
& \sum_{i=0}^{j-1} r_{i,j} = \sum_{k=j+1}^{n+1} r_{j,k},\ 1 \leq j \leq n \\
& \sum_{1 \leq i \leq n} r_i \leq N, \\
& r_i + r_j \leq 1,\ \text{for all } (q_i, q_j) \in \mathcal{E} \\
& r_{i,j} \in \{0,1\},\ r_i \in \{0,1\}
\end{aligned}$$

Discussion. The PAMP-DEL formulation in Bashir et al.[6] does not actually make explicit the underlying probabilistic distribution for the endpoints of the deletion. The ILP proposed by Bashir et al. for PAMP-DEL uses an objective similar to (11) with

$$C_{i,j} = \max\{(x_j - x_i - L/2), 0\} \tag{12}$$

which is measuring the so called "uncovered area." It is not difficult to see that minimizing uncovered area as proposed by Bashir et al. may *not* result in minimizing the probability of failure, even assuming a uniform probability distribution for the deletion endpoints as suggested by (12). An example is as follows. Consider a PAMP-DEL instance in which possible deletions have left endpoint in the interval $(0, L]$ and right endpoint in the interval $(2L, 3L]$, with each endpoint position equally likely. There are non-dimerizing forward primers at *every* position between 0 and L, and three reverse primers at positions $2L$, $2.5L$, and $3L$, with the last two of these primers forming a dimer. The minimum failure probability is in this case zero, and is achieved by selecting all forward primers and the reverse primers at $2L$ and $3L$. However, the minimum uncovered area is $L/2$, since one of the primers at $2.5L$ and $3L$ cannot be selected. The ILP proposed in Bashir et al.[6] may select all forward primers and the reverse primers at $2L$ and $2.5L$, which has optimal uncovered area but fails to detect deletions with probability 1/2.

Lemma 3.1. *Assuming the UNIQUE GAMES conjecture, PAMP-1SDEL (and hence, PAMP-DEL) cannot be approximated to within a factor of* $2 - \varepsilon$ *for any constant* $\varepsilon > 0$.

Sketch of Proof. We reduce the vertex cover problem to PAMP-1SDEL. It is known[7] that, assuming that the UNIQUE GAMES conjecture holds, the vertex cover problem cannot be approximated to within a factor of $2 - \varepsilon$ for any constant $\varepsilon > 0$. Consider an instance

Table 1. Detection probability and ILP runtime for PAMP-INV instances with $x_{max} - x_{min} = 100Kb$ and $L = 20Kb$ (averages over 5 random instances).

Dimerization Rate (%)	n=20(ρ=3.33) N=20	15	10	n=30 (ρ=5) N=20	15	10
	Detection probability(%)					
0	93.91	93.83	91.17	99.25	99.20	96.79
1	93.57	93.54	91.11	98.79	98.69	96.11
2	92.68	92.68	90.55	98.69	98.60	96.06
5	89.78	89.78	88.28	97.84	97.78	95.68
10	84.41	84.41	83.57	94.99	94.98	92.95
20	71.53	71.53	71.53	81.70	81.70	81.64
	Runtime (seconds)					
0	175.01	379.87	994.76	2160.45	5238.17	86115.50
1	211.54	337.44	956.34	2461.93	4919.25	57229.18
2	259.77	260.20	913.67	2081.81	5864.61	31655.12
5	667.87	618.33	868.28	3903.71	6660.55	14266.41
10	535.20	496.97	495.14	6405.27	7081.30	18284.68
20	520.96	470.19	558.82	15506.87	14893.29	14847.14

$G = (V, E)$ of vertex cover with $V = \{1, 2, \ldots, n\}$ and $\{1, n\} \notin E$. We define an instance of PAMP-1SDEL with reverse primers $q_1, \ldots, q_n$ at positions $x_i = iL$, $i = 1, \ldots, n$, where the pairs of dimerizing primers correspond to the edges of G. Further, assume that the position of the right endpoint of the deletion is uniformly distributed in the interval $[0, nL]$.

Let V' be a vertex cover of G containing k vertices. Then, $V \setminus V'$ is an independent set of G, and the set of primers $\{q_i \; : \; i \in V \setminus V'\}$ is a feasible PAMP-1SDEL solution whose failure probability is k/n. Conversely, consider a solution $\mathcal{P}'$ of PAMP-1SDEL with failure probability k/n. Under the uniform probability distribution, it follows that $|\mathcal{P}'| = n - k$. Clearly $\{i \; : \; q_i \in \mathcal{P}'\}$ is an independent set of G, and so $\{i \; : \; q_i \notin \mathcal{P}'\}$ is a vertex cover of size k of G. ∎

On the positive side we have the following result, whose proof we omit due to space the limitation.

Lemma 3.2. *There is a 2-approximation algorithm for the special case of PAMP-1SDEL in which candidate primers are spaced at least L bases apart and the deletion endpoint is distributed uniformly within a fixed interval* $(x_0, x_{max}]$.

4. Experimental Results

We used the Cplex 10.1 solver to solve ILP formulations given in Sections 2 and 3. All reported runtimes are for a Dell PowerEdge 6800 server with four 2.66GHz Intel Xeon dual-core processors (only one of which is used by Cplex).

Table 1 gives the detection probability (one minus failure probability) and runtimes for the ILP from Section 2 for randomly generated PAMP-INV instances with $x_{max} - x_{min}$=100Kb, L=20Kb (which is representative of long-range PCR), number of candidate primers n between 20 and 30 (candidate primer density $\rho = nL/(x_{max} - x_{min} + L)$ be-

Table 2. Comparison of PAMP-DEL ILP, ITERATED-1SDEL, and INCREMENTAL-1SDEL for instances with $m = n = N_f = N_r = 15$, $x_{max} - x_{min} = 5Kb$, and $L = 2Kb$ (averages over 5 random instances for each dimerization rate between 0 and 20%).

Dimerization Rate (%)	PAMP-DEL ILP		ITERATED-1SDEL		INCREMENTAL-1SDEL	
	Detection Prob. (%)	#Primers	Detection Prob. (%)	#Primers	Detection Prob. (%)	#Primers
0	97.29	(15.0,15.0)	97.29	(15.0,15.0)	97.29	(10.4, 8.8)
1	96.81	(14.2,12.6)	96.81	(14.4,12.6)	96.81	(11.4, 9.6)
2	96.73	(13.4,11.6)	96.70	(13.6,11.4)	96.73	(11.6,10.0)
5	93.13	(10.8, 8.0)	88.91	(10.4, 7.4)	91.60	(10.0, 7.8)
10	87.58	(8.2, 6.2)	84.34	(8.4, 6.4)	83.19	(7.0, 5.8)
20	72.95	(6.0, 4.8)	56.03	(6.4, 3.8)	68.89	(5.4, 4.0)

tween 3.33 and 5), maximum multiplexing degree N between 10 and 20, and primer dimerization rate between 0 and 20%. Both the hybridization locations for candidate primers and the pairs of candidate primers that dimerize were selected uniformly at random. In this experiment all inversions longer than 10Kb were assumed to be equally likely. The PAMP-INV ILP can usually be solved to optimality within a few hours, and the runtime is relatively robust to changes in dimerization rate, candidate primer density, and constraints on multiplexing degree. The detection probability varies from 75% to over 99% depending on instance parameters.

Unfortunately the runtime for solving the PAMP-DEL ILP in Section 3 is impractical for all but very small problem instances. In contrast, the PAMP-1SDEL ILP can be solved efficiently for very large instances. Therefore, we considered a practical PAMP-DEL heuristic which relies on iteratively solving simpler PAMP-1SDEL instances, as follows. First, we solve a PAMP-1SDEL for one side – say, for reverse primers – assuming that the position of the other deletion endpoint is right next to the anchor. Then we solve a PAMP-1SDEL for selecting a set of forward primers from candidates that do not dimerize with the already selected reverse primers. The PAMP-1SDEL ILP in this second step is as in Section 3, however, coefficients $C_{i,j}$ in (11) represent the two-sided failure probability reflecting the fixed set of reverse primers. The process is repeated until there is no further decrease in failure probability.

The above iterative heuristic is referred to as ITERATED-1SDEL. One drawback of ITERATED-1SDEL is that it may result in unbalanced sets of primers for high dimerization rates. This happens since the first step will typically select the maximum possible number of reverse primers, and this may leave very few non-dimerizing forward primers. To avoid this drawback, we have also implemented a version of ITERATED-1SDEL, referred to as INCREMENTAL-1SDEL, which in the first iteration limits the number of selected reverse and forward primers to some proportional number of the given bounds N_r and N_f, for example, half of the given bounds, then increments these limits by a fixed factor in each of the subsequent iterations.

Table 2 compares the detection probability and average number of forward and reverse primers selected using the PAMP-DEL ILP, ITERATED-1SDEL, and INCREMENTAL-1SDEL on a set of small randomly generated instances for which the PAMP-DEL ILP

Table 3. Detection probability and runtime (in seconds) of INCREMENTAL-1SDEL.

Instance size	Dimer. Rate (%)	N=55	N=44	N=33	N=22
	0	93.24 (3902.16)	93.23 (3901.92)	93.02 (3901.68)	91.73 (3900.54)
	1	91.91 (93.80)	91.91 (93.70)	91.89 (93.60)	90.86 (93.40)
$2 \times 200Kb$	2	90.54 (12.24)	90.54 (12.14)	90.54 (12.04)	89.90 (11.94)
$n = 55$	3	86.40 (5.58)	86.40 (5.50)	86.40 (5.42)	86.05 (5.34)
	4	82.68 (5.36)	82.68 (5.20)	82.68 (5.04)	82.56 (4.88)
	5	76.09 (2.46)	76.09 (2.40)	76.09 (2.34)	76.09 (2.28)
		N=105	N=84	N=63	N=42
	1	91.04 (1258.70)	91.04 (1258.22)	91.04 (1257.74)	90.13 (1257.24)
$2 \times 400Kb$	2	78.28 (56.48)	78.28 (55.90)	78.28 (55.32)	77.30 (54.74)
$n = 105^*$	3	65.88 (29.31)	65.88 (28.03)	65.88 (26.75)	65.86 (25.45)
	4	54.12 (89.33)	54.12 (85.39)	54.12 (81.45)	54.12 (76.43)
	5	54.66 (276.93)	54.66 (272.19)	41.87 (267.45)	41.22 (257.21)

Note: * runtime for 0 dimerization rate exceeded 48 hours.

can be solved in practical runtime. The results show that both ITERATED-1SDEL and INCREMENTAL-1SDEL solutions are very close to optimal for low dimerization rates. For larger dimerization rates INCREMENTAL-1SDEL detection probability is still close to optimal, while ITERATED-1SDEL detection probability degrades substantially. As shown in Table 3, the runtimes of INCREMENTAL-1SDEL remain practical for large random instances except for the largest instance with no dimerization rate.

5. Conclusions

In this paper we propose ILP formulations for selecting sets of PAMP primers with high probability of detecting genomic inversions and anchored deletions in cancer tumors. In ongoing work we are assessing the performance of our methods on real biological datasets[6] and exploring scalable heuristics and approximation algorithms for un-anchored deletion detection via PAMP.

References

1. Y.-T. Liu and D. Carson, *PLoS ONE* **2**, p. e380 (2007).
2. K. Doi and H. Imai, *Genome Informatics* **10**, 73 (1999).
3. K. Konwar, I. Măndoiu, A. Russell and A. Shvartsman, Improved algorithms for multiplex PCR primer set selection with amplifi cation length constraints, in *Proc. 3rd Asia-Pacific Bioinformatics Conference (APBC)*, 2005.
4. P. Nicodème and J.-M. Steyaert, Selecting optimal oligonucleotide primers for multiplex PCR, in *Proc. 5th Intl. Conference on Intelligent Systems for Molecular Biology*, 1997.
5. S. Rozen and H. Skaletsky, Primer3 on the WWW for general users and for biologist programmers, in *Bioinformatics Methods and Protocols: Methods in Molecular Biology*, eds. S. Krawetz and S. Misener (Humana Press, Totowa, NJ, 2000).
6. A. Bashir, Y.-T. Liu, B. Raphael, D. Carson and V. Bafna, *Bioinformatics* (Advance Access published online on August 30, 2007).
7. S. Khot and O. Regev, Vertex cover might be hard to approximate to within 2-ε, in *Proc. 18th IEEE Annual Conference on Computational Complexity*, 2003.

GenePC AND ASPIC INTEGRATE GENE PREDICTIONS WITH EXPRESSED SEQUENCE ALIGNMENTS TO PREDICT ALTERNATIVE TRANSCRIPTS

TYLER S. ALIOTO* and RODERIC GUIGÓ

Center for Genomic Regulation, c/ Dr. Aiguader, 88
08003, Barcelona, Spain
**E-mail: tyler.alioto@crg.es*

ERNESTO PICARDI and GRAZIANO PESOLE

Dipartimento di Biochimica e Biologia Molecolare
University of Bari
and Istituto Tecnologie Biomediche del C.N.R. (sede di Bari)
via Orabona 4
70126 Bari, Italy

We have developed a generic framework for combining introns from genomicly aligned expressed–sequence–tag clusters with a set of exon predictions to produce alternative transcript predictions. Our current implementation uses ASPIC to generate alternative transcripts from EST mappings. Introns from ASPIC and a set of gene predictions from many diverse gene prediction programs are given to the gene prediction combiner GenePC which then generates alternative consensus splice forms. We evaluated our method on the ENCODE regions of the human genome. In general we see a marked improvement in transcript-level sensitivity due to the fact that more than one transcript per gene may now be predicted. GenePC, which alone is highly specific at the transcript level, balances the lower specificity of ASPIC.

Keywords: Alternative Splicing; Gene Prediction; Combiner.

1. Introduction

1.1. *Gene Prediction Combiners*

The computational integration of multiple gene predictions into one unified or consensus prediction has increasingly become the first-pass annotation of choice for newly sequenced genomes. Examples of such programs are EuGène,[1] GAZE,[2] GLEAN, JIGSAW,[3] Genomix[4] and even the Ensembl pipeline.[5] These programs are popular because they perform a task that human annotators usually begin with — examining multiple predictions and looking for consensus. Of course, human annotators have the ability to look at all sources of information including, but not limited to, CpG islands, putative promoter elements, protein domain integrity, protein family conservation, genome integrity, and literature references. Nevertheless, the primary tools continue to be a combination of *ab initio*, homology-based, and expressed sequence tag (EST)/cDNA-based gene prediction. In this area, gene combiners are already demonstrating their utility. In the future perhaps combiners will be able to incorporate even more diverse sources of information such as those mentioned above.

1.2. *Diversity Combiners and Wise Crowds*

Because gene prediction combiners take advantage of the diversity of input, they can be considered a class of diversity combiner.[6] Diversity combiners used in the wireless telecommunication arena are effective at canceling out noise when receiving multiple independent inputs. The most effective of these are called maximal-ratio combiners; these combiners use a weighted sum of their inputs where the weight is based on their signal–to–noise ratio. We reasoned that a similar approach might help in integrating multiple gene predictions. Additionally, we tried to enforce the following four principles of collective wisdom:[7]

(1) *Diversity of opinion* — Input gene predictors should represent a diversity of algorithms.
(2) *Independence* — Input gene predictors should not be influenced by each others' predictions.
(3) *Decentralization* — Predictors should be able to specialize and draw upon local private knowledge.
(4) *Aggregation* — Some mechanism must exist to combine the predictions.

Our gene prediction combiner, GenePC, constructs a consensus prediction using the principle that exons co-predicted by diverse input prediction algorithms are more likely to be real. By explicitly correcting for correlated input, GenePC corrects for lack of diversity and interdependence. This procedure also takes the guesswork out of choosing which gene predictions to combine.

Decentralization is normally not an issue. However, when one organization is in charge of annotating a genome, it tends to run a set of gene prediction programs using the same set of aligned expressed sequence or proteins or, if using comparative genomics, they tend to use the same genomes for comparison. Likewise, when a gene annotation assessment project is undertaken (GASP,[8] EGASP,[9] nGASP), the allowed input and allowed training sets are usually fixed. Of course, this allows a fair assessment of individual gene predictors; however, in principle it should have a negative effect on the performance of gene prediction combiners.

The method we present here is an aggregation method that explicitly corrects for lack of diversity and independence among all pairs of inputs. Additionally, we correct for over-aggregation. A problem for consensus building is the special case in which more than one solution is correct. This is particularly relevant to gene prediction, where a gene can code for more than one alternative transcript. Our method reintroduces high-quality transcript evidence during the last step of aggregation to guide the division of a single consensus into multiple likely consensus transcripts.

1.3. *Alternative Splicing Prediction*

The annotation of multiple alternative splice forms has been a stumbling block for both the standard class of gene predictors and for combiners. Programs that are successful at alternative transcript prediction are in general those that use expressed sequence tag (EST) or mRNA evidence (a notable exception is the latest version of Augustus,[10] which is able to output multiple transcripts per loci in the absence of expressed sequence evidence.) Many such programs, including ASPIC, rely only on transcript mapping and often output incomplete transcripts. EuGène is one of the first gene predictors/combiners to output full-length alternative transcripts using often incomplete transcript evidence. Our approach is similar in spirit to that of EuGène, but in implementation there are many differences. The most obvious of which is our use of ASPIC,[11] a software tool for Alternative Splicing PredICtion, to drive the construction of alternative gene models by GenePC.

2. Methods

Before we begin we must mention one caveat. Although more and more programs are predicting UTR segments of genes by using mRNA and EST evidence, we have decided to restrict ourselves to only the the coding segments (CDS) of genes as these are the most reliably predicted features of protein-coding genes. As such we will naturally not be able to annotate the variation in splicing of UTR sequence which actually represents a significant proportion of alternative splicing.

2.1. *Approach*

Our aggregation method combines ASPIC CDS predictions with GenePC, an extended version of GeneID[12] which combines gene predictions at the level of exact exons based on diversity and performance. Such a combination allows multiple consensus transcripts to be predicted per locus, depending on the number of mutually incompatible sets of introns inferred from the EST alignments.

Our method involves three steps. In the first step, we run GenePC using all available gene predictions on a genome or subset of the genome. GenePC scores every uniquely predicted exon as a function of the set of gene predictors that predicted it. We calculate an Exon Score S which is given by

$$S = \alpha \sum_{i=1}^{n} S_{i_{norm}} + \beta \sum_{i=1}^{n} \sum_{\substack{j=1 \\ j \neq i}}^{n} \left(1 - \frac{SNet_{ij} + SPet_{ij}}{2}\right) + \gamma \sum_{i=1}^{n} (x \cdot SNet_{ia} + y \cdot SPet_{ia}) \tag{1}$$

where $S_{i_{norm}}$ is the Z-score normalized exon score provided by program i and $SNet$ and $SPet$ are the exon-level sensitivity and specificity, respectively, of program i against program j or program i against the annotation a. The weighting parameters α, β and γ can be optimized for a particular combination of gene predictions and genomes. The coefficients x and y are adjusted to give more weight to sensitivity or specificity of a gene prediction program. We generally set $x = 0.25$ and $y = 0.75$. The parameters for First, Internal, Terminal and Single exon types are optimized independently using a training set of confirmed genes within the prediction region or on a separate set. If no annotation is available, performance values can either be estimated from the literature or γ may be set to zero. A graphical overview of this procedure is shown in Figure 1.

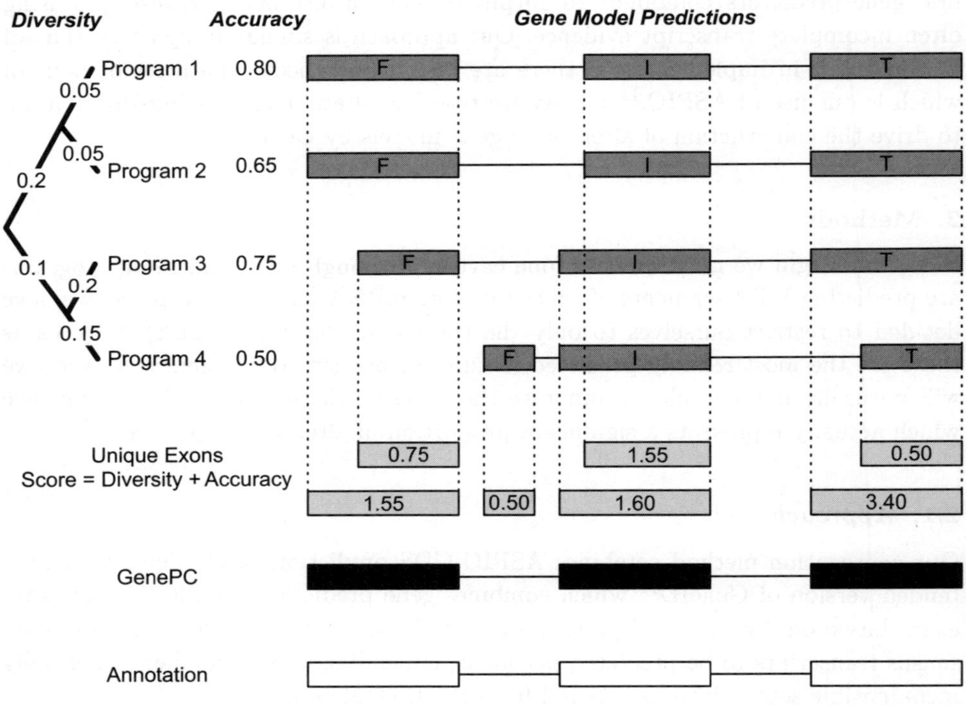

Fig. 1. Schematic of GenePC prediction aggregation using Diversity (distance between programs) and Performance (Accuracy against the annotation). Self-reported exon scores are not used ($\alpha = 0$.) In practice, we use a pairwise distance matrix rather than the phylogram shown here.

In the second step, we run the alternative splicing prediction program ASPIC, which is freely available at http://t.caspur.it/ASPIC/, with a set of ESTs aligned to the genome. We perform an initial alignment of ESTs to the genome with the program GMAP.[13] The ESTs are then clustered based on the sharing of at least one splice site (a fuzzy clustering methods that we developed allows for small shifts in intron position due to spliced alignment error) and then each cluster is given to ASPIC. ASPIC then performs a multiple genome-EST alignment optimized for producing a set of transcripts with a minimal number of exons.[11,14] The longest open reading frame is then determined for each transcript output by ASPIC and overlapping CDS spans are assigned to different compatibility groups or bins. The number of bins is determined by the locus with the largest number of alternative transcripts. For loci with less than the maximum number of alternative transcripts, CDS spans are reassigned to empty bins so that we maximally cover the genome with the available evidence. Finally, for use in the next step, we convert the transcript data into sets of introns in GFF format for input into GenePC.

In the final step, we use the dynamic programming algorithm `genamic`[15] as implemented in GeneID to chain together the highest scoring set of exons from each intron compatibility set generated in the previous step. Only the introns from ASPIC are provided to GenePC so that exons not predicted by any gene predictor will not contribute to the final gene models. This feature is made possible in version 1.3 of GeneID, freely available at http://genome.imim.es/software/geneid/, which has been modified to use introns as evidence but will not output a gene model if no exons are present. In the last step, redundant transcripts are removed from the combined predictions of each run of GeneID.

3. Results

3.1. *Training and Test Sets*

We evaluated our method on the ENCODE[16] regions using 14 sets of gene predictions submitted to the EGASP competition.[9] The EGASP competition solicited gene predictions on the 44 ENCODE regions encompassing 1% of the human genome. Training was allowed on 13 of these regions. We trained GenePC on these same 13 regions, obtaining performance (prediction vs. annotation) and distance (prediction vs. prediction) values separately for First, Internal, Terminal and Single exons. Self-reported exon scores were not used since this information was not provided for all inputs. Input predictions included all non-combiner programs that predicted complete gene structures, thus excluding Spida and JIGSAW.

The set of ESTs used by the HAVANA team to annotate the ENCODE regions were downloaded from Genbank and aligned to the genome with GMAP. They were clustered based on shared splice junctions and given as input to ASPIC. The resulting transcripts (some of which are full-length but many of which are partial) were distributed into the minimal number of internally compatible sets of transcripts. In this case, the locus with the largest number of alternative incompatible transcripts was 12. Therefore, we were required to run GeneID twelve times on the ENCODE regions, each time using the same set of GenePC-scored non-redundant exons and a different set of ASPIC introns.

We evaluated ASPIC-GenePC predictions and all input predictions against the March 2007 Gencode[17] human ENCODE gene annotation release using the pro-

Table 1. Transcript-level performance.

Category	Program	SNt	SPt	$(SNt + SPt)/2$
Ab initio	GeneID	0.03	0.04	0.04
	Exonhunter	0.06	0.03	0.05
	Genemark	0.05	0.04	0.05
	AugustusAbinitio	0.13	0.16	0.15
Multi-genome	Sgp2	0.05	0.04	0.05
	Twinscan	0.10	0.08	0.09
	AugustusDual	0.15	0.17	0.16
	N-SCAN	0.21	0.35	0.28
EST/Protein and Pipelines	Ensembl	0.25	0.24	0.25
	AugustusAny	0.27	0.34	0.31
	AugustusEst	0.27	0.36	0.32
	Fgenesh	0.43	0.38	0.41
	Exogean	0.51	0.43	0.47
	Pairagon-N-SCAN	0.51	0.46	0.49
	ASPIC CDS	0.63	0.37	0.50
Combiners	JIGSAW	0.42	0.63	0.53
	GenePC	0.41	**0.65**	0.53
	ASPIC-GenePC	**0.65**	0.51	**0.58**

Note: All predictions except for ASPIC CDS, GenePC and ASPIC-GenePC predictions correspond to EGASP submissions downloaded from the UCSC genome browser.

gram `evaluation.pl` (Eduardo Eyras, personal communication) which takes into account alternative transcripts in both the annotation and predictions. Only full-length known and putative Gencode genes were used for evaluation.

3.2. *Evaluation*

We evaluated ASPIC-GenePC using a large number of measures including sensitivity and specificity at the level of nucleotides, exons, transcripts and genes (see Tbl. 1). JIGSAW and GenePC outperform all other predictors in nearly every category (results not shown). We believe the most significant result is the accuracy at the transcript level. GenePC and JIGSAW already perform quite well in predicting exact transcripts. Further gains in transcript-level sensitivity are realized when EST evidence is used to guide alternative best predictions. We observed an average 5% gain in accuracy. This is a combination of a 24% gain in sensitivity and 14% loss in specificity. ASPIC-GenePC predicts 466 of 719 Gencode-annotated transcripts in 381 of 397 genes. This is in contrast with the 295 transcripts predicted correctly by GenePC alone.

Table 2. Evaluation at gene, transcript, exon, intron and nucleotide levels.

	JIGSAW	GenePC	ASPIC CDS	ASPIC-GenePC
SNg	**0.96**	0.94	0.84	**0.96**
SPg	0.82	**0.83**	0.69	0.78
SNt	0.42	0.41	0.63	**0.65**
SPt	0.63	**0.65**	0.37	0.51
SNe	0.88	0.88	0.87	**0.93**
SPe	0.84	**0.86**	0.75	0.80
SNet	0.93	**0.96**	0.84	0.94
SPet	0.93	**0.97**	0.95	0.92
SNi	0.88	0.86	0.89	**0.95**
SPi	0.85	0.86	**0.87**	0.82
SNit	0.93	**0.95**	0.85	**0.95**
SPit	0.94	**0.96**	**0.96**	0.93
SNn	0.95	0.94	0.81	**0.96**
SPn	0.86	**0.87**	0.85	0.85
SNnt	0.95	**0.97**	0.86	0.96
SPnt	0.96	**0.98**	0.97	0.95

Note: SN indicates sensitivity. SP indicates specificity. Gene (g), transcript(t), exon(e), intron(i) and nucleotide(n) levels were assessed. For each exon, intron and nucleotide, we evaluated against the set of projected unique predicted features to the projected unique annotated features. We also compared the features of a predicted transcript to the features of the annotated transcript with the highest overlap (best hit), denoted by the suffix *t*. Gene sensitivity is defined as any exonic overlap.

For comparison, we also show the performance of JIGSAW, ASPIC, GenePC and ASPIC-GenePC at the exon and nucleotide levels in Table 2. It is interesting to note that with the addition of ASPIC introns, sensitivity increases and specificity decreases in almost every level compared with GenePC alone or with JIGSAW.

4. Discussion

The phenomenon of alternative splicing continues to represent a challenge to *ab initio* gene prediction. Two approaches to the problem have generally been taken in the past: (a) output of suboptimal predictions and (b) output of multiple transcripts supported by mutually incompatible evidence. We have taken the latter approach as it is straightforward and is backed by strong evidence of transcription. However, this does not preclude the use of a combiner to gather a consensus on how many alternative transcripts exist for a particular gene and output that number of transcripts, without relying on another source of evidence. We have not moved in this direction yet, considering most *ab initio* predictors have yet to produce more than one transcript per locus.

Nevertheless, using our straightforward EST-driven approach, we were struck by the 24% gain in transcript sensitivity over GenePC alone that is conferred by ASPIC-guided alternative transcript prediction. Part of this gain was offset by a decrease in specificity. However, the number of additional genes predicted was not dramatic (28 new genes, 9 of which are annotated), meaning that the majority of the 171 new transcripts predicted are alternative transcripts of known loci. While all of them by definition have some level of EST support, many may actually be partial or non-coding transcripts predicted by ASPIC. Since the corresponding transcripts have been filtered out of the annotation, these predictions demonstrate themselves as false positives.

The success of our aggregation method is likely due to the high sensitivity of ASPIC in predicting reliable transcripts. ASPIC in fact has been mainly designed to perform a multiple alignment of ESTs to a genomic sequence based on the combined analysis of all available expressed sequence tags. Moreover, it refines the exon-intron boundaries by an appropriate dynamic programming module and generates the most likely transcripts using a new algorithm based on graph theory. However, where ASPIC fails to find exons or genes, GenePC can fill in the gaps. Furthermore, where ASPIC predicts transcripts with exons not present in the combined set of input gene predictions to GenePC, no transcripts are predicted.

The general framework we have outlined — dividing transcript evidence into compatible sets and providing them as intron evidence for exon structure assembly — should be extensible to nearly any gene prediction method that directly takes EST/cDNA evidence or external constraints such as introns, as we have used here. While substantial gains in accuracy from the incorporation of expression evidence would be expected (and have been demonstrated) for *de novo* or *ab initio* gene predictors, the enhanced performance of a combiner was not anticipated. In theory the expressed sequence evidence is already used by at least several input methods, making its re-introduction redundant. However, combiners such as GenePC are afflicted with the same problem faced by traditional gene finders: they output only the optimal gene model. Thus, the enhancement we observe is acheived by essentially unflattening the consensus gene model determined by GenePC into multiple models that maximally represent the transcript evidence. Incompatibilities among ESTs are not ignored but are, if they are of sufficient quality, utilized to their full diagnostic potential.

In the future, we anticipate an increased focus on alternative transcript prediction. While we currently have to rely on the crutch of EST and cDNA sequences, advances in our understanding of the regulation of alternative splicing may allow the prediction of alternative transcripts based on knowledge of the critical factors present and active, for example, in particular cell-types or at particular developmental stages or in response to particular external stimuli. We suggest that our iterative approach in conjunction with an evidence combiner would be equally applicable in this context.

Acknowledgments

This work was supported by grants from the European Commission FP6 Programme (BioSapiens Network of Excellence) to RG and TA and from the Ministero Università e Ricerca, Italy (FIRB project "Laboratorio Italiano di Bioinformatica") to GP.

References

1. S. Foissac and T. Schiex, *BMC bioinformatics* **6**, p. 25 (2005), 10.1186/1471-2105-6-25.
2. K. Howe, T. Chothia and R. Durbin, *Genome research* **12**, 1418(Sep 2002), 10.1101/gr.149502.
3. J. Allen and S. Salzberg, *Bioinformatics (Oxford, England)* **21**, 3596(Sep 2005), 10.1093/bioinformatics/bti609.
4. A. Coghlan and R. Durbin, *Bioinformatics (Oxford, England)* **23**, 1468(Jun 2007), 10.1093/bioinformatics/btm133.
5. T. Hubbard, B. Aken, K. Beal, B. Ballester, M. Caccamo, Y. Chen, L. Clarke, G. Coates, F. Cunningham, T. Cutts, T. Down, S. Dyer, S. Fitzgerald, J. Fernandez-Banet, S. Graf, S. Haider, M. Hammond, J. Herrero, R. Holland, K. Howe, N. Johnson, A. Kahari, D. Keefe, F. Kokocinski, E. Kulesha, D. Lawson, I. Longden, C. Melsopp, K. Megy, P. Meidl, B. Ouverdin, A. Parker, A. Prlic, S. Rice, D. Rios, M. Schuster, I. Sealy, J. Severin, G. Slater, D. Smedley, G. Spudich, S. Trevanion, A. Vilella, J. Vogel, S. White, M. Wood, T. Cox, V. Curwen, R. Durbin, X. Fernandez-Suarez, P. Flicek, A. Kasprzyk, G. Proctor, S. Searle, J. Smith, A. Ureta-Vidal and E. Birney, *Nucleic acids research* **35**, 610(Jan 2007), 10.1093/nar/gkl996.
6. D. Brennan, *Proc. IRE* **47**(Jun 1959).
7. J. Surowiecki, *The Wisdom of Crowds* (Anchor, August 2005).
8. M. Reese, G. Hartzell, N. Harris, U. Ohler, J. Abril and S. Lewis, *Genome research* **10**, 483(Apr 2000).
9. R. Guigó, P. Flicek, J. Abril, A. Reymond, J. Lagarde, F. Denoeud, S. Antonarakis, M. Ashburner, V. Bajic, E. Birney, R. Castelo, E. Eyras, C. Ucla, T. Gingeras, J. Harrow, T. Hubbard, S. Lewis and M. Reese, *Genome biology* **7 Suppl 1**, 2 (2006), 10.1186/gb-2006-7-s1-s2.
10. M. Stanke, O. Keller, I. Gunduz, A. Hayes, S. Waack and B. Morgenstern, *Nucleic acids research* **34**, W435(Jul 2006), 10.1093/nar/gkl200.
11. P. Bonizzoni, R. Rizzi and G. Pesole, *BMC bioinformatics* **6**, p. 244 (2005), 10.1186/1471-2105-6-244.
12. G. Parra, E. Blanco and R. Guigó, *Genome research* **10**, 511(Apr 2000).
13. T. Wu and C. Watanabe, *Bioinformatics (Oxford, England)* **21**, 1859(May 2005), 10.1093/bioinformatics/bti310.
14. T. Castrignanò, R. Rizzi, I. Talamo, P. D'Onorio, A. Anselmo, P. Bonizzoni and G. Pesole, *Nucleic acids research* **34**, W440(Jul 2006), 10.1093/nar/gkl324.
15. R. Guigó, *Journal of computational biology : a journal of computational molecular cell biology* **5**, 681 (1998).
16. *Nature* **447**, 799(Jun 2007).
17. J. Harrow, F. Denoeud, A. Frankish, A. Reymond, C.-K. Chen, J. Chrast, J. Lagarde, J. Gilbert, R. Storey, D. Swarbreck, C. Rossier, C. Ucla, T. Hubbard, S. Antonarakis and R. Guigo, *Genome biology* **7 Suppl 1**, 4 (2006), 10.1186/gb-2006-7-s1-s4.

COMPARING AND ANALYSING GENE EXPRESSION PATTERNS ACROSS ANIMAL SPECIES USING 4D*XPRESS*

YANNICK HAUDRY

European Molecular Biology Laboratory EMBL, Meyerhofstrasse 1
69117 Heidelberg, Germany

CHUANG KEE ONG

European Molecular Biology Laboratory EMBL, Meyerhofstrasse 1
69117 Heidelberg, Germany

LAURENCE ETTWILLER

European Molecular Biology Laboratory EMBL, Meyerhofstrasse 1
69117 Heidelberg, Germany

HUGO BERUBE

European Bioinformatics Institute,
EMBL-EBI Wellcome Trust Genome Campus Hinxton
Cambridge, CB10 1SD, UK

IVICA LETUNIC

European Molecular Biology Laboratory EMBL, Meyerhofstrasse 1
69117 Heidelberg, Germany

MISHA KAPUSHESKY

European Bioinformatics Institute,
EMBL-EBI Wellcome Trust Genome Campus Hinxton
Cambridge, CB10 1SD, UK

PAUL-DANIEL WEEBER

European Molecular Biology Laboratory EMBL, Meyerhofstrasse 1
69117 Heidelberg, Germany

XI WANG

European Molecular Biology Laboratory EMBL, Meyerhofstrasse 1
69117 Heidelberg, Germany

JULIEN GAGNEUR

*European Molecular Biology Laboratory EMBL, Meyerhofstrasse 1
69117 Heidelberg, Germany*

CHARLES GIRARDOT

*European Molecular Biology Laboratory EMBL, Meyerhofstrasse 1
69117 Heidelberg, Germany*

DETLEV ARENDT

*European Molecular Biology Laboratory EMBL, Meyerhofstrasse 1
69117 Heidelberg, Germany*

PEER BORK

*European Molecular Biology Laboratory EMBL, Meyerhofstrasse 1
69117 Heidelberg, Germany*

ALVIS BRAZMA

*European Bioinformatics Institute,
EMBL-EBI Wellcome Trust Genome Campus Hinxton
Cambridge, CB10 1SD, UK*

EILEEN FURLONG

*European Molecular Biology Laboratory EMBL, Meyerhofstrasse 1
69117 Heidelberg, Germany*

JOACHIM WITTBRODT

*European Molecular Biology Laboratory EMBL, Meyerhofstrasse 1
69117 Heidelberg, Germany*

THORSTEN HENRICH†

*European Molecular Biology Laboratory EMBL, Meyerhofstrasse 1
69117 Heidelberg, Germany*

High-resolution spatial information on gene expression over time can be acquired through whole mount in-situ hybridisation experiments in animal model species such as fish, fly or mouse. Expression patterns of many genes have been studied and data has been integrated into dedicated model organism databases like ZFIN for zebrafish, MEPD for medaka, BDGP for drosophila or MGI for mouse. Nevertheless, a central repository that allows users to query and compare gene expression patterns across different species has not yet been established. For this purpose we have integrated gene expression data for zebrafish, medaka, drosophila and mouse into a central public repository named 4D*Xpress* (http://ani.embl.de/4DXpress). 4D*Xpress* allows to query anatomy ontology based expression annotations across species and quickly jump from one gene to the orthologs in other species based on ensembl-compara relationships. We have set up a linked resource for microarray data at ArrayExpress. In addition we have mapped developmental stages between the species to be able to compare corresponding developmental time phases. We have used clustering algorithms to

†corresponding author

classify genes based on their expression pattern annotations. To illustrate the use of 4D*Xpress* we systematically analysed the relationships between conserved regulatory inputs and spatio-temporal gene expression derived from 4D*Xpress* and found significant correlation between expression patterns of genes predicted to have similar regulatory elements in their promoters.

1. Introduction

Embryonic development is a process in which cells signal to each other and thereby acquire different identities which is necessary to establish the basic body plan of the organism. This process results in amazingly complex gene expression patterns that can be visualised by whole mount in situ hybridisation experiments. Whoever has seen expression patterns of typical developmental regulators like FGF, HOX, or PAX genes will understand that the spatial regulation of such genes can not be analysed by microarray experiments.

To know the exact time and location of gene transcripts is crucial when studying the functions of genes involved in development as well as for trying to decipher the code of cis-regulatory modules. It is important to store images, which lets biologists see and judge expression together with an organized annotation. Ontology based annotations let users query the data and make data accessible to computational analysis.

Expression localisation data has been gathered in the model species databases but a central resource, which allows users to compare gene expressions in different species, has not yet been established until recently. We have set up such a platform for cross species expression pattern comparisons (1), comprising annotations on 16505 genes. This is the largest collection available to date. Our vision is that in a few years time the exact localisation of each single transcript will be known for the major model species. We hope that our resource will help to store them in an organised way, to compare different species expression patterns and to provide tools to analyse this data. We show that the organised storage of expression annotation data is sufficient to classify genes into clusters of similar expressed genes and thereby offers an entry point for cross species comparisons through computational biology.

As an example of such an approach we present an application of deciphering the code of cis-regulatory modules. In this context we take advantage of the information stored in 4D*Xpress* and analyse the correlation between regulatory input and the spatio-temporal expression pattern. More specifically, we systematically investigated whether a significant correlation between expression annotation and the occurrence of at least one common conserved transcription factor (TF) binding site exists. Using binding site information for 309 TFs we found a significant correlation for the predicted target genes of 4 TFs. This demonstrates that in some cases, genes predicted to be the common targets of at least one transcription factor have similar pattern of expressions.

2. 4D*Xpress*

We have integrated gene expression data for drosophila (2), medaka (3), zebrafish (4) and mouse (5) so far. In table 1 we give an overview on the gene expression patterns that have been integrated for 4D*Xpress*. The best-annotated model species are drosophila and zebrafish at the moment with almost 6000 annotated genes each. Mouse has slightly less annotated genes reflecting the difficulty to yield large amounts of specimens when compared to egg-laying species like fish and fly. The annotated mouse genes represent a large set of important developmental regulators and are annotated in great detail sometimes using a 3D virtual embryo (6).

Expression data has been gathered in different ways. For drosophila and medaka the major annotation results from a screen. Expression has been analysed at 3 or 4 distinct time points (table 1, stages per gene), whereas zebrafish expression patterns are additionally annotated from literature by a team of database curators. Annotation is done for continues developmental stages.

Table 1: Content of 4D*Xpress*. Annotation status of gene expression patterns at present time.

	Source	Genes	Stages	Stages per Gene	Anatomy Terms	Anat. Terms per Gene	Anat. Terms per Stage	Distinct Anat. Terms
drosophila	bdgp	5951	21048	3.54	29867	5.02	1.42	288
medaka	mepd	882	2746	3.11	5047	5.72	1.84	338
zebrafish	zfin	5779	102671	17.77	178851	30.95	1.74	694
mouse	mgi	3893	12799	3.29	17291	4.44	1.35	1661
		16505	139264	8.44	231056	14.00	1.66	2981

Anatomy ontologies are often huge, however only a limited fraction of the terms is actually used for expression annotation (table 1, distinct annotations). Again, ZFIN uses a rich vocabulary with almost 700 distinct terms. The values for mouse and medaka need to be treated with care, as the ontologies used for annotation here are the cross product of anatomy and stage ontologies and therefore overestimates vocabulary richness.

The web application is based on a MVC (Model-View-Controller) architecture using the Struts Framework, and enhanced with applets, JavaScript and AJAX (Asynchronous JavaScript and XML) technologies to build a powerful, interactive, user-friendly interface. 4D*Xpress* is available at http://ani.embl.de/4DXpress. The usage of the interface is documented in detail on our home page. Genes can be searched either by a range of external identifiers, symbol, name or by their expression pattern annotation. Ontologies that were used to annotate gene expression can be browsed. Anatomy ontologies are browse-able by a tree-based tool, which allows users to query terms and expand and collapse individual nodes. Developmental stage anatomies can be browsed by species and external links provide more information on stage definitions. Species-specific stage ontologies were mapped onto a common stage list and thereby establish temporal relationships, which can be accessed via web interface. Our annotation tool allows users to annotate gene expression patterns resulting from either whole mount in situ hybridisation experiments, transgenic reporter gene expression or antibody staining. The same tool can be used for all supported species (for now: zebrafish, mouse, medaka, drosophila, platynereis).

Expression data acquired through either in situ hybridisation, antibody or transgenic expression and microarrays can complement each other. The first provides high-resolution data; the latter can quickly give an overview on all genes in a genome. That is why we have decided to set up a data warehouse for microarray data at EBI: 4D ArrayExpress Data Warehouse (4DDW). It is accessible at: http://www.ebi.ac.uk/microarray-as/4DDW_EMBL/. So far we have established 4737 reciprocal links for mouse, drosophila and zebrafish. When querying microarray data at ArrayExpress (7) users can quickly go to 4D*Xpress* and vice versa.

The 4D*Xpress* schema is based on the common MISFISHIE (8) standard allowing a straightforward integration of additional species and data exchange with other databases.

2.1. *Cross Species Relationships*

One of the major goals of our project is to be able to compare gene expression patterns between the different model species. For doing so relationships need to be established between genes (orthology), between time windows (developmental stages) and most challenging between anatomical structures.

2.1.1. *Orthology*

EnsEMBL compara (9) provides a reliable source of sequence homology relationships, which was computed using a tree-based approach. We have chosen to use this and update regularly upon new EnsEMBL releases. We assigned each gene to a cluster of orthologs using the one2one-, one2many and many2many orthology relationships. These clusters are visualised as a network in the gene view and used to sort the gene list retrieved from a query.

2.1.2. *Developmental Stages*

It is very difficult to identify corresponding developmental stages in two species, even when comparing two closely related fish like medaka and zebrafish. It is almost impossible to find an exact corresponding stage for one of the 46 medaka stages in zebrafish, because within the embryo different structures develop with different speed. E.g. the head and brain develops faster, whereas the tail and somites develops slower in medaka. So if one finds a matching stage in zebrafish regarding the number of somites, which is a very popular staging feature, the head would actually correspond to a later zebrafish stage.

However there is a list of 8 embryonic stages that is described in all developmental biology text books and is common to all bilaterian animals: zygote, cleavage, blastula, gastrula, neurula, organogenesis, juvenile and adult. By mapping each of the species stages onto one of the bilaterian stages the link between species stages can be done and combinatorial explosion can be prevented. A new species will only need to be mapped to the common stages (fig. 1) and not against all stages of all other species.

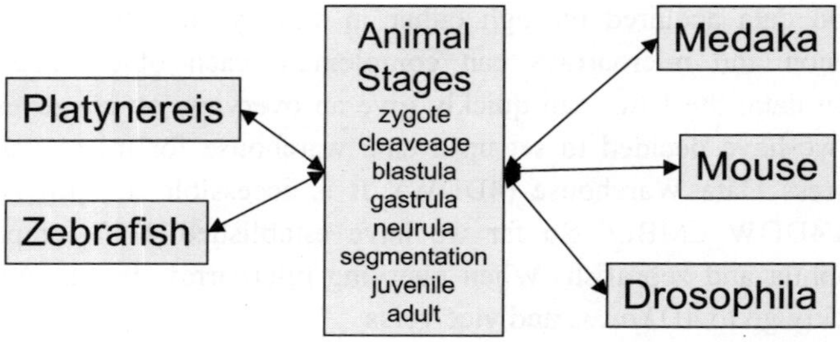

Fig 1. Mapping of developmental species was done through a list of stages common to all bilaterian animals.

2.1.3. *Anatomical Structures*

We have used lexical cues to start a simple anatomical mapping. 50% of all unique terms used for annotations could be mapped to high-level terms common to all species. Anatomy structure and co-expression cues will be used to refine these relationships. For co-expression we will use the integrated expression data of 4D*Xpress* as this data has the best spatial resolution.

The Common Anatomy Reference Ontology (CARO) is being developed to facilitate interoperability between existing anatomy ontologies for different species. It aims to provide a template for building new anatomy ontologies. We want to use CARO to build an anatomy ontology shared by all bilaterians. Similar to the stage mapping we then want to map species-specific anatomy terms onto the common ontology.

3. Co-Expression Analysis

Having gene expression annotation available in an organised way allows us to analyse gene expression annotation computationally. Using the same clustering tools as used for micro array analysis such as the TIGR Multi Experiment Viewer (10) simple hierarchical clustering can identify genes with similar expression patterns (fig 2).

The method can be validated by looking at a few examples of genes that have been clustered together and by examining their in situ images. Indeed simple expression patterns such as cypa and ctss or nanos and h1m would have also been described as similar by researchers.

More complex expression patterns like those of developmental regulators do not cluster well with other genes though. However there is still room for improvements. I.e. the ontology relations have not yet been exploited by the clustering algorithm. Semantic similarities measures are able to account for that. We will apply this method on gene expression annotation.

When comparing genes across species we need maximum overlap. Zebrafish and drosophila are the most complete data sets. We have 964 ortholog groups annotated in the two species and 336 for three species (including mouse). They also overlap temporally in the developmental stages: blastula, gastrula, neurula, organogenesis and juvenile.

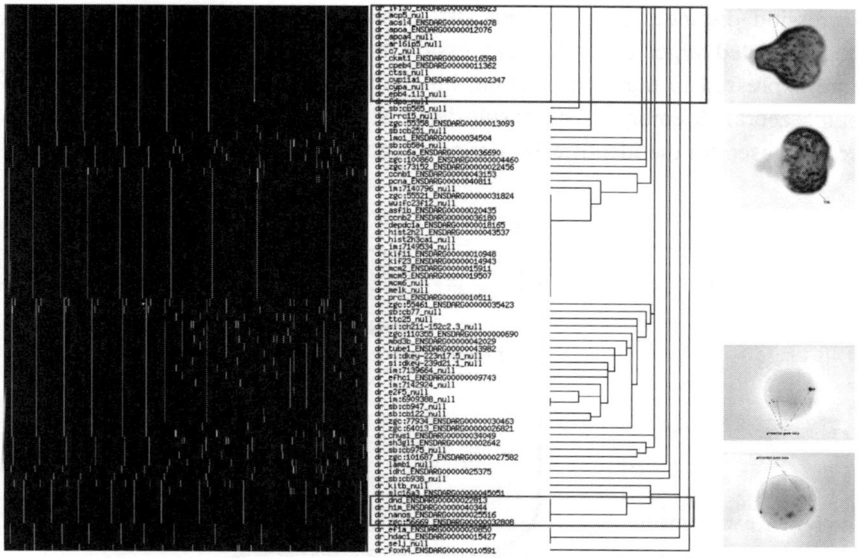

Fig 2. Genes can be classified upon their expression pattern annotation. Genes were clustered using the binary distance and a hierarchical clustering algorithm with the TIGR MEV package.

We will use semantic similarities to generate co-expression networks for each species individually and then use orthology relationships to identify conserved patterns in these networks. Looking at the terms used in different species for annotating conserved patterns we hope to find candidates for the cross species anatomy mapping. In addition it will be interesting to study the regulatory sequences of genes appearing in the same conserved network pattern (see below).

4. Application

To demonstrate the value of 4D*Xpress* for computational approaches, we analysed the correlation between regulatory inputs of genes and their corresponding expression patterns derived from the zebrafish in-situ annotation taken from 4D*Xpress*. In a first step, the human target genes of all transcription factors with known binding sites in TRANSFAC (11) were predicted using a similar approach as described previously (12) (fig. 3). This approach is based on extreme conservation of the binding site from human to fish, only the predicted most conserved target genes for each transcription factor are further selected. The corresponding ortholog genes in zebrafish are then retrieved and the in-situ expression pattern information is mapped to the predicted target genes.

For each target gene set, the average expression distance is computed using the Jaccard metric. In order to evaluate if the average distance is significantly lower than expected under a random model, randomizations of the expression matrix is done by shuffling the gene IDs and the random distances are computed and compare to the real distance. Out of 309 predicted target gene sets, 4 show a significantly lower ($p<0.01$)

average expression distances (fig. 3). This list includes the transcription factor Hes1 where all the predicted target genes with expression information (Her6, Atoh1, ide2a and Neurog1) are expressed in the diencephalon, hindbrain, midbrain and telencephalon of the developing zebrafish embryo. This result demonstrates that in situ data as stored in 4D*Xpress* can be used for the identification of new regulatory sequences in the genome.

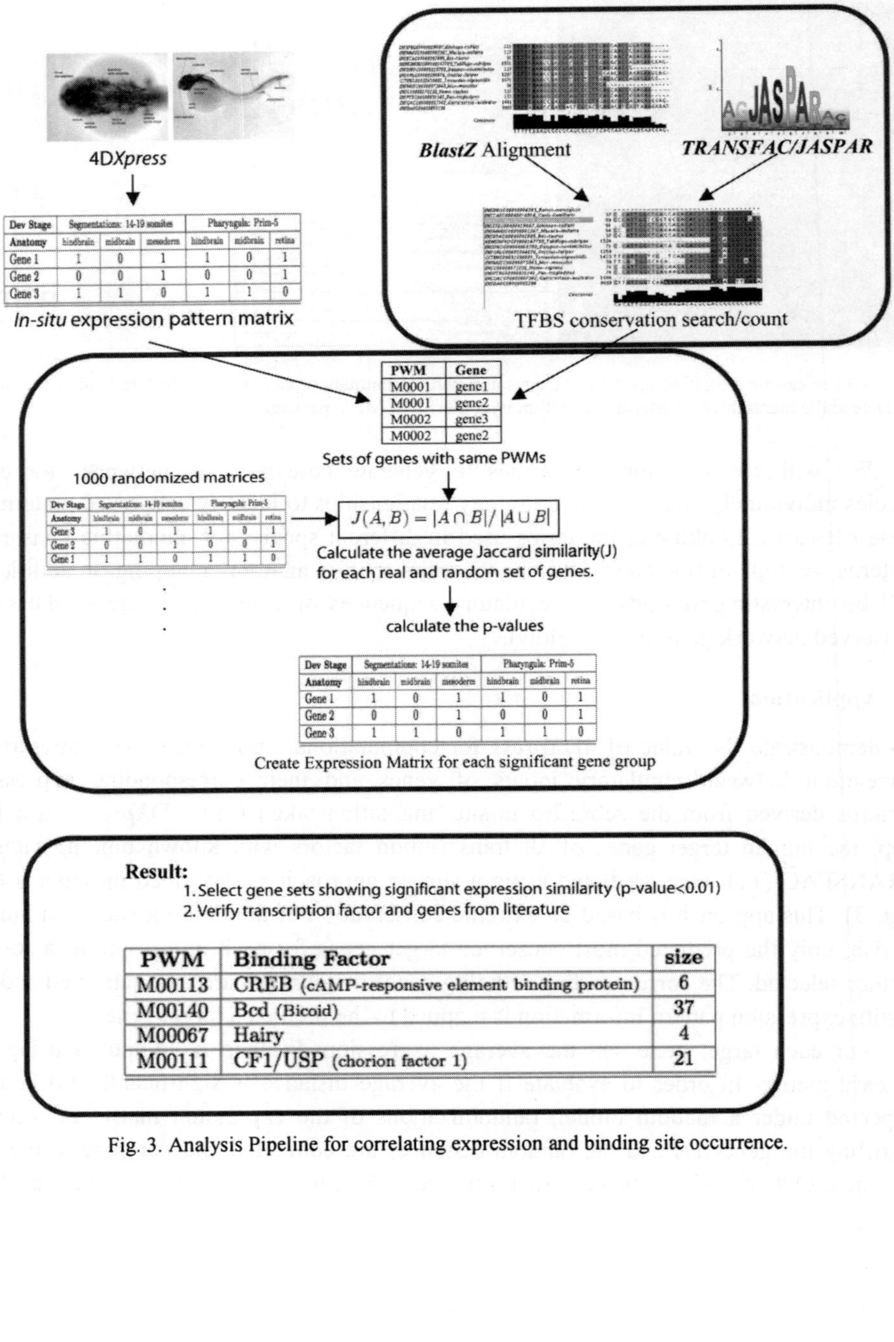

Fig. 3. Analysis Pipeline for correlating expression and binding site occurrence.

5. Future Perspective

Now that we have a stable database schema and a efficient web interface to access data, in the future we will focus on three points:

1. Integrating more data and species
   There is more gene expression pattern annotation available in the public domain. The next species we are aiming for is Xenopus laevis with 17.000 images (Naoto Ueno, NIBB in Okazaki), Ciona intestinalis (7000 genes) and C. elegans.
2. Developing tools for data analysis
   We will calculate distances between genes based on their expression annotation vector. This will enable users to easily find genes with similar expression patterns as the gene of interest. We are planning to establish tools integrated in the web interface, that allows users to cluster expression data an correlate clusters of genes with other sources of data like chromosomal location, GO, KEGG or occurrence of binding sites.
3. Mapping Anatomy Ontologies
   Our strategy is described above. Establishing such a resource is as challenging as it will be valuable when achieved.

Acknowledgement

This work was carried out in the Centre for Computational Biology at EMBL. We are grateful to the model organism database crews for providing us their expression pattern data. In particular we thank Martin Ringwald and Susan McClatchy to give us direct access to the MGI database; thanks to Monte Westerfield and Judy Sprague for helping us with the ZFIN data reports; we thank Pavel Tomancak for helping us to understand the BDGP mySQL schema. We thank Mirana Ramialison for extending the medaka annotation and Francois Spitz for mouse in situ images. We are grateful to the MISFISHIE team for initiating an expression pattern data exchange format.

References

1. Haudry, Y., Berube, H., Letunic, I., Weeber, P., Gaugneur, J., Girardot, C., Kapushesky, M., Arendt, D., Bork, P., Brazma, A. *et al.* (in press NAR 2008) 4DXpress: A database for cross species expression pattern comparisons. *Nucleic Acids Res.*
2. Tomancak, P., Berman, B.P., Beaton, A., Weiszmann, R., Kwan, E., Hartenstein, V., Celniker, S.E. and Rubin, G.M. (2007) Global analysis of patterns of gene expression during Drosophila embryogenesis. *Genome Biol*, **8**, R145.
3. Henrich, T., Ramialison, M., Wittbrodt, B., Assouline, B., Bourrat, F., Berger, A., Himmelbauer, H., Sasaki, T., Shimizu, N., Westerfield, M. *et al.* (2005) MEPD: a resource for medaka gene expression patterns. *Bioinformatics*, **21**, 3195-3197.
4. Sprague, J., Bayraktaroglu, L., Clements, D., Conlin, T., Fashena, D., Frazer, K., Haendel, M., Howe, D.G., Mani, P., Ramachandran, S. *et al.* (2006) The

Zebrafish Information Network: the zebrafish model organism database. *Nucleic Acids Res*, **34**, D581-585.
5. Smith, C.M., Finger, J.H., Hayamizu, T.F., McCright, I.J., Eppig, J.T., Kadin, J.A., Richardson, J.E. and Ringwald, M. (2007) The mouse Gene Expression Database (GXD): 2007 update. *Nucleic Acids Res*, **35**, D618-623.
6. Christiansen, J.H., Yang, Y., Venkataraman, S., Richardson, L., Stevenson, P., Burton, N., Baldock, R.A. and Davidson, D.R. (2006) EMAGE: a spatial database of gene expression patterns during mouse embryo development. *Nucleic Acids Res*, **34**, D637-641.
7. Parkinson, H., Kapushesky, M., Shojatalab, M., Abeygunawardena, N., Coulson, R., Farne, A., Holloway, E., Kolesnykov, N., Lilja, P., Lukk, M. *et al.* (2007) ArrayExpress--a public database of microarray experiments and gene expression profiles. *Nucleic Acids Res*, **35**, D747-750.
8. Deutsch, E.W., Ball, C.A., Bova, G.S., Brazma, A., Bumgarner, R.E., Campbell, D., Causton, H.C., Christiansen, J., Davidson, D., Eichner, L.J. *et al.* (2006) Development of the Minimum Information Specification for In Situ Hybridization and Immunohistochemistry Experiments (MISFISHIE). *Omics*, **10**, 205-208.
9. Hubbard, T.J., Aken, B.L., Beal, K., Ballester, B., Caccamo, M., Chen, Y., Clarke, L., Coates, G., Cunningham, F., Cutts, T. *et al.* (2007) Ensembl 2007. *Nucleic Acids Res*, **35**, D610-617.
10. Saeed, A.I., Sharov, V., White, J., Li, J., Liang, W., Bhagabati, N., Braisted, J., Klapa, M., Currier, T., Thiagarajan, M. *et al.* (2003) TM4: a free, open-source system for microarray data management and analysis. *Biotechniques*, **34**, 374-378.
11. Wingender, E., Chen, X., Fricke, E., Geffers, R., Hehl, R., Liebich, I., Krull, M., Matys, V., Michael, H., Ohnhauser, R. *et al.* (2001) The TRANSFAC system on gene expression regulation. *Nucleic Acids Res*, **29**, 281-283.
12. Del Bene, F., Ettwiller, L., Skowronska-Krawczyk, D., Baier, H., Matter, J.-M., Birney, E. and Wittbrodt, J. (2007) In vivo Validation of a Computationally Predicted Conserved Ath5 Target Gene Set. *Plos Genetics*.

NEAR-SIGMOID MODELING TO SIMULTANEOUSLY PROFILE GENOME-WIDE DNA REPLICATION TIMING AND EFFICIENCY IN SINGLE DNA REPLICATION MICROARRAY STUDIES*

JUNTAO LI [1], MAJID ESHAGHI[2], JIANHUA LIU[3] and
RADHA KRISHNA MURTHY KARUTURI[4†]

[1,4]Computational & Mathematical Biology, [2,3]Systems Biology, Genome Institute of Singapore, #02-01, Genome, 60 Biopolis ST, Republic of Singapore 138672

DNA replication is a key process in cell division cycle. It is initiated in coordinated manner in several species. To understand the DNA replication in a species one needs to measure the half replication timing (or replication timing) and the efficiency of replication which vary across genome in higher eukaryotes. In the previous studies, no direct assessment of replication efficiency on a genomic scale was performed while the replication timing was indirectly assessed using average DNA. In this paper, we present a first-ever-method of directly measuring both half replication timing and efficiency simultaneously from a single DNA microarray time-course data. We achieve it by fitting the so called *near-sigmoid model* to each locus of the DNA. We use this model apply *S. pombe* DNA replication microarray data and show that it is effective for genome-scale replication timing and efficiency profiling studies.

1 Introduction

DNA replication is a very important process in cell cycle progression, takes place within a short cell cycle phase called S-phase. It is initiated at multiple sites or origins at varying times in eukaryotic genomes [1-3] within S-phase. It was shown that [4] some regions of the genome initiate replication early, some in the middle, and the others near the end, attributing to a strict timing and coordination of firing at origins with a few exceptions such as frog embryo [24]. The replication carried out by the fork initiated (also called *replication firing*) at a locus is called *active replication* and the site is called *origin of replication* or *origin*. Whereas the replication carried by the passing forks resulting from the nearby origins is called *passive replication*. The active and passive replications are well defined in S. cerevisiae and fuzzily defined in the other higher eukaryotes when efficiency of replication is relatively low.

---

† Corresponding author (karuturikm@gis.a-star.edu.sg)

* This work was supported by Genome Institute of Singapore and the Agency for Science, Technology and Research, Singapore.

The two genome-wide profiles of interest in DNA replication studies are *half replication timing* (or *replication timing*) and *replication efficiency.* Half replication timing of a locus is the time it takes to complete its replication in half of the cells or the probability that it is replicated is 0.5. This is especially significant if the replication efficiency is less than 100% i.e. replication of the locus takes significant time of the S-phase. The importance of the half replication timing is its estimation stability to identify the origins of replication. Genome-wide DNA microarray analyses have been widely used to determine profiles of half replication timing at the genomic scale [4-7, 14]. It is measured indirectly by average DNA content at the loci and the peaks in the average DNA content profile show the origins or most likely active replication sites.

Flexible timing of firing or inefficient firing at origins was observed in several species such as human [7], *S. pombe* [8,9] and frog embryos [24] unlike in *S. cerevisiae* [4]. Replication efficiency is the measure of strictness of timing of replication of the locus under consideration i.e. 100% efficient locus is the one that is always replicated strictly at the same time in all cell cycles whereas an inefficient locus is replicated at different timings in different cell cycles within a given period within S-phase.

The efficiency of replication has been measured using different techniques such as Single strand DNA (ssDNA) [17] and DNA combing [9]. DNA combing technique analyzes DNA replication/firing only at single origins leaving it to be a tedious and low throughput technique to measure efficiency and half replication time. Hence it cannot be used for genome scale study. In case of ssDNA technique, the amount of nascent DNA accumulated at sites of replication initiation during HU treatment may not be proportional to firing efficiency on a genomic scale resulting in inaccurate assessment of replication efficiency. Moreover, the current approach requires two different technologies and datasets, one to measure half replication timing and the other to measure efficiency while not resulting in any advantage.

Though genome-wide microarray analyses have been widely used to determine profiles of half replication timing at the genomic scale, direct estimation of replication efficiency at various loci of the genome based on the genome-wide replication profiles has not been performed previously. In this paper, we demonstrate that replication efficiency, together with half replication timing, can be estimated using a novel approach - the *near-sigmoid modeling* for the increase in DNA copy number as a function of time at individual loci. The *near-sigmoid modeling* approach permits estimation of replication start timing and replication end timing at various loci of the genome. Based on these measurements, we attain the genome-wide profiles of half replication timing and replication efficiency. The rest of the paper describes near-sigmoid modeling and proceeds to show its efficacy on genome-wide profiling of DNA replication timing and efficiency of S. Pombe.

2 Near-Sigmoid Modeling

In our approach, DNA replication process is described in three steps: *initiation*, *linear progression*, and *completion* of replication. The time period from the replication initiation to the completion is defined as the duplication time *DT*. As one and only one copy of DNA at all loci would be synthesized during the S-phase, $\eta_r \cdot DT = 1$, where η_r is the (average) replication efficiency = *1/DT* i.e. rate of fraction of cells replicate the given locus in a given time upon initiation of replication, higher the *DT* lower the efficiency.

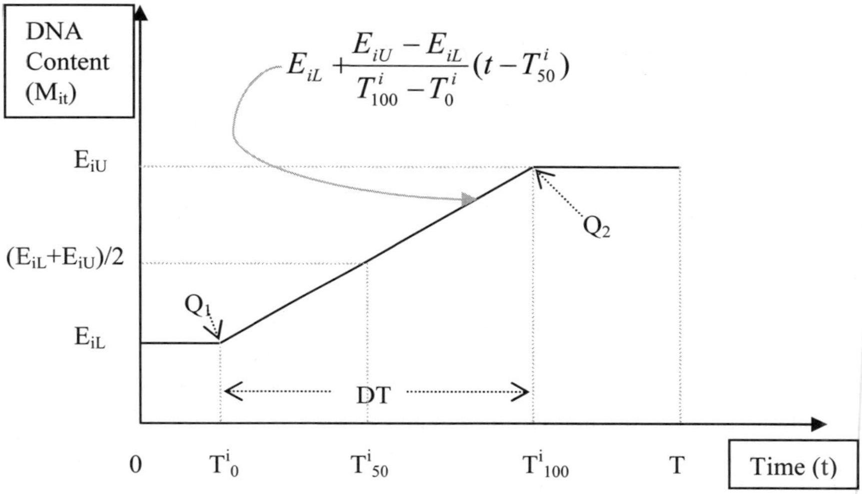

Figure 1. Graphical illustration of near-sigmoid model to measure half replication timing and efficiency. Two point of inflexion Q_1 and Q_2 define the model, a piecewise linear approximation of sigmoid, hence the name.

Near-sigmoid model represents the above replication model as a specialized 3-piecewise linear model as shown in figure 1. This model is called near-sigmoid model because it is a piecewise linear approximation of sigmoid model. The two points of inflexion $Q_1 = (T^i_0, E_{iL})$ and $Q_2 = (T^i_{100}, E_{iU})$ signify the quantitative state of replication of locus l_i just prior to the replication initiation and just after the replication completion respectively. T^i_0 and T^i_{100} indicate replication initiation and replication completion timings of l_i respectively; T^i_{50} is its half replication timing, the average of T^i_0 and T^i_{100} i.e. $T^i_{50} = (T^i_0+T^i_{100})/2$. T is the end of the experiment. Duplication time (DT^i), inverse of replication efficiency (η^i_r), is $T^i_{100} - T^i_0$. E_{iL} is the initial DNA content at l_i which remains constant till T^i_0 and starts increasing linearly at T^i_0 till it reaches E_{iU} at time T^i_{100}. T^i_0 and T^i_{100} obey the conditions $0 \leq T^i_0 < T^i_{100} \leq T$ and $1 \leq E_{iL} \leq E_{iU} \leq 2$. E_{iL} and E_{iU} are ideally expected to be 1 and 2 respectively. We used them in this manner to signify the fact that some loci may have already replicated to varying extent before the release of the arrested cells to progress in which case $E_{iL} > 1$ and the experiment may have stopped before the end of S-phase i.e. $T <$ S-phase period resulting in $E_{iU} < 2$ for some late replicating or highly inefficient loci. The model is mathematically expressed as follows.

Let C_{it} be the relative DNA content in the synchronized S-phase cells with that of the reference genome at locus l_i at time t. Let C_{itk} be the kth repeated measurement of C_{it}. Let M_{it} be the estimation of $log(C_{it})$ as described by the near-sigmoid model in equation (1). In this model we assume that, like in typical microarray studies, $log(C_{itk}) \sim N(log(C_{it}), \sigma_i^2)$.

$$M_{it} = \begin{cases} E_{iL} & for \quad t \in [0, T_0^i] \\ E_{iU} & for \quad t \in [T_{100}^i, T] \\ E_{iL} + \dfrac{E_{iU} - E_{iL}}{T_{100}^i - T_0^i}(t - T_0^i) & for \quad t \in [T_0^i, T_{100}^i] \end{cases} \tag{1}$$

$$E_{iL} = \frac{1}{K_L^i} \sum_{t=0}^{T_0^i} \sum_{k=1}^{n_t^i} \log(C_{itk}) \qquad and \qquad E_{iU} = \frac{1}{K_U^i} \sum_{t=T_{100}^i}^{T} \sum_{k=1}^{n_t^i} \log(C_{itk})$$

$$K_L^i = \sum_{t=0}^{T_0^i} n_t^i \qquad and \qquad K_U^i = \sum_{t=T_{100}^i}^{T} n_t^i$$

Where n_t^i is the number of repeats for l_i at time t, the superscript 'i' signifies the fact that some observations may be missing and their numbers vary from locus to locus at each time point. K^i_L is the number of measurements up to T^i_0 and K^i_U is the number of measurements from time T^i_{100} to T, the end of the time-course.

2.1 *Near-sigmoid model parameter estimation*

As can be seen from the definitions in equation (1), we need to estimate optimal values for T^i_0 and T^i_{100} given $\{C_{itk}\}_i$ which automatically leads to the remaining parameters of the model. They are estimated by exhaustive search by minimizing the mean-squared error between M_{it} and $log(C_{itk})$ as depicted in equation (2).

$$O = \frac{1}{K^i} \sum_{t=0}^{T} \sum_{k=1}^{n_t^i} \left(M_{it} - \log(C_{itk})\right)^2 \qquad K^i = \sum_{t=0}^{T} n_t^i \tag{2}$$

Where K^i is the total number of measurements made for l_i at all time points put together.

2.2 *Statistical significance of near-sigmoid fit*

Upon estimation of the model parameters (Q_1 & Q_2), we have to examine whether the near-sigmoid fit is better than constant or flat fit which we evaluate using hypothesis testing and false discovery rate estimation. The statistical significance (or the p-value) of the fit was calculated under the null hypothesis that $E_{iU} = E_{iL} = E_i$ and the alternative hypothesis is $E_{iU} > E_{iL}$. The following ANOVA table is used to derive the statistic F^i.

Table 1. ANOVA table for near sigmoid model fit. The F^i statistic follows central F distribution with degrees of freedom 3 and (K^i-4) i.e. $F_{3,K^i-4}(x)$. Higher the F^i better the fit.

	Sum of squares, *SS*	Degrees of freedom, *df*	F-statistic, F^i
Regression	$SSR^i = \sum_{t=0}^{T}(M_{it} - \overline{\log(C_i)})^2$ $\overline{\log(C_i)} = \frac{1}{K^i}\sum_{t=0,k=1}^{t=T,k=n_t^i}\log(C_{itk})$	3	$F^i = \frac{\frac{SSR^i}{3}}{\frac{SSE^i}{(K^i-4)}}$
Error	$SSE^i = \sum_{t=0,k=1}^{t=T,k=n_t^i}(\log(C_{itk}) - M_{it})^2$	K^i-4	
Total	$TSS^i = \sum_{t=0,k=1}^{t=T,k=n_t^i}(\log(C_{itk}) - \overline{\log(C_i)})^2$	K^i-1	

P-value of the fit (p_i) is given by area under F_{3,K^i-4} from F^i to ∞. The p-values of all loci were used to obtain false discovery rate (FDR) with monotonicity correction [23]. The loci above an FDR cut-off are declared to be unfit or flat responsive loci.

2.3 *Meaning of flat responsive or unfit loci*

In principle, each and every locus of the DNA has to be replicated before the cell cycle progresses into its next stage i.e. G2 phase. The insignificant FDR (p-value) shows that the DNA content at the locus has not changed from the start of the study to the end which means either the probe is bad or the locus has replicated even before the release of the cells from synchronization block such as HU arrest to progress in the S-phase. We believe that almost all probes in an array were tested for their goodness, leaving only possibility that the loci have replicated even before the release of the cells to progress. Hence the flat responsive or unfit loci (probes) signify that the corresponding loci are early efficient replication regions. We show in the next section (3) that it is indeed the case in *S. pombe*.

3 *S. pombe* DNA Replication Data Analysis using Near-Sigmoid Modeling

To investigate the efficacy of our approach we present its application on genome-wide DNA replication timing and efficiency. Appropriately normalized DNA microarray data on DNA replication was obtained from [15]. It is based on *S. pombe* genome-wide ORF-specific microarray and the increase in DNA copy numbers at individual loci (from all three chromosomes) was studied in cells released after HU block. The microarray has an average resolution of one locus per ~2.4 Kb. Each locus (or ORF) was represented by two different 50-mer oligonucleotide probes whose average ratio was used for profiling. The length and resolution of the time-course are 60min and 5 min respectively, two repeats are available at each time point

We applied the near-sigmoid model to fit DNA copy number increase as a function of time at individual loci for estimation of replication initiation timing T_0 and completion timing T_{100}. To this end, the T_0 and T_{100} at the majority of loci (> 96% loci of the genome) were attained based on the criterion of FDR less than 0.01%.

Our methodological approach is validated by the following genome-wide observations: (1) predicted replication origins are close to A+T islands and the previously predicted origins; (2) the unfit or flat responsive loci are close to A+T islands and other previously predicted origins; (3) chrIII has early half replication timing and lower efficiency relative to the remaining two chromosomes; (4) telomeres on chrI and chrII are late replicating and more efficient. The importance of A+T islands in our observations stems from the fact that the origins of replication in *S. pombe* were shown to be close to A+T islands [21]. The detailed results are described in the following subsections 3.1 through 3.5.

3.1 *Replication origins are close to A+T islands and other predicted origins*

The origins of replication were predicted using Peak finder software [20] on T_{50} profile. Of the 516 origins predicted 285 overlap with A+T islands [21 (A+T)] and 360, 48, 305, 295, 239 and 318 match with the peaks predicted by [17, 21(Validated), 22 (ORC1), 22 (MCM6), 16 (Wt), 16 (ΔCDS1)] respectively.

3.2 *Flat responsive loci coincide with A+T islands and predicted origins*

193 loci were flat responsive which were analyzed to check whether they match closely to A+T islands and the predicted origins of replication. The islands are expected to be close to the origins of replication, especially the early ones. Of the 193 flat responsive loci, 146, 173, 36, 155, 154, 129 and 147 loci match with A+T islands [21 (A+T)] and the other predicted origins [17, 21(Validated), 22 (ORC1), 22 (MCM6), 16 (Wt), 16 (ΔCDS1)] respectively. This shows that these loci are indeed close to the A+T islands and predicted origins of replication which proves the efficacy of our approach and interpretation.

3.3 *Early replication timing of chrIII relative to chrI & chrII*

Further evaluation of the effectiveness of our approach to measure half replication timing was further carried out by comparing the half replication timing of chrIII with that of chrI & chrII. ChrIII was observed to be early half replicating as compared to that of chrI & chrII [17]. This was reinforced in our analysis with P-value $< 2.2\text{x}10^{-16}$. The box plots of the half replication timings of chrI & chrII put together and chrIII are shown in figure 2.

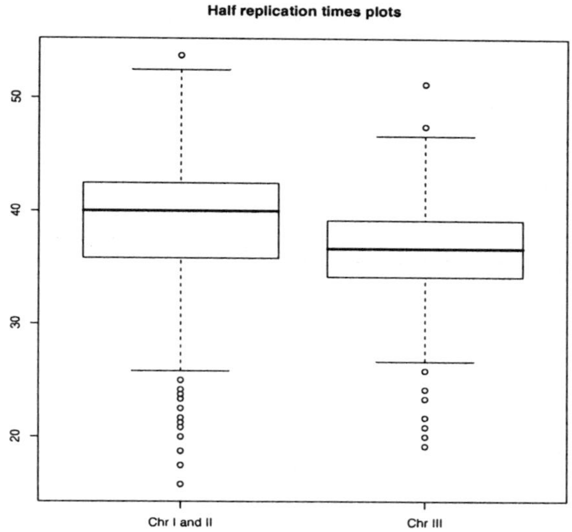

Figure 2. Boxplots of half replication timing (vertical axis) of loci on chrI & chrII and chrIII (horizontal axis). chrIII is systematically has early half replication timing compared to the remaining chromosomes, p-value < 2.2×10^{-16} by Wilcoxon rank-sum test.

3.4 *Low efficiency of replication of chrIII relative to chrI & chrII*

The evaluation of the effectiveness of our model to measure efficiency of replication is carried out by comparing the overall efficiency of chrI & chrII with that of chrIII. ChrIII was observed to be less efficient compared to that of chrI &chrII, with P-value < 2.2×10^{-16} and the box plots of efficiency are shown in figure 3. This observation was supported by the fact that chrIII was shown to have lot more origins [22] with earlier half replication timing compared to the remaining two chromosomes. This is due to the fact that inefficient origins are expected to have systematically early half replication timing in order to complete replication and there should be many such origins owing to their inefficiency.

3.5 *Higher efficiency of telomeres compared to the other regions*

The evaluation of the effectiveness of our model to measure efficiency of replication is carried out by comparing telomere regions with that of the others in chrI & chrII. Telomeres are defined as the regions in chrI for < 0.2Mb and >5.4Mb, for chrII < 0.2Mb and > 4.4Mb. We observed that the telomeres are more efficient (p-value is 0.06) as telomeres are known to be late replicating which is possible only if they are more efficient than the other regions. The box plots of efficiency of all loci in telomeres and the other regions are shown in figure 4.

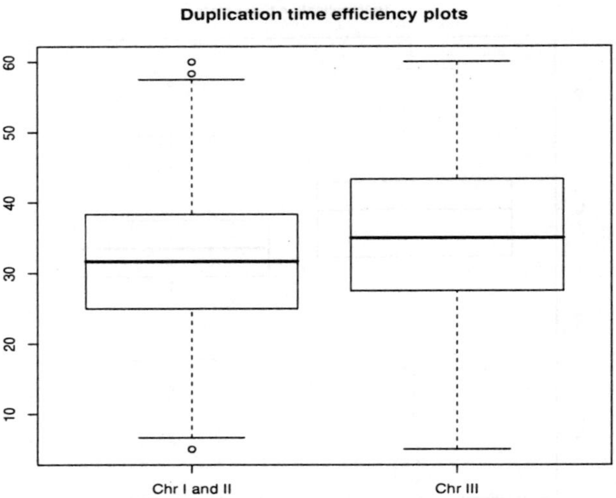

Figure 3. Boxplots of duplication times (vertical axis) of loci on chrI & chrII and chrIII (horizontal axis). chrIII is systematically has higher duplication time or lower efficiency compared to the remaining chromosomes, p-value $< 2.2 \times 10^{-16}$ by Wilcoxon rank-sum test.

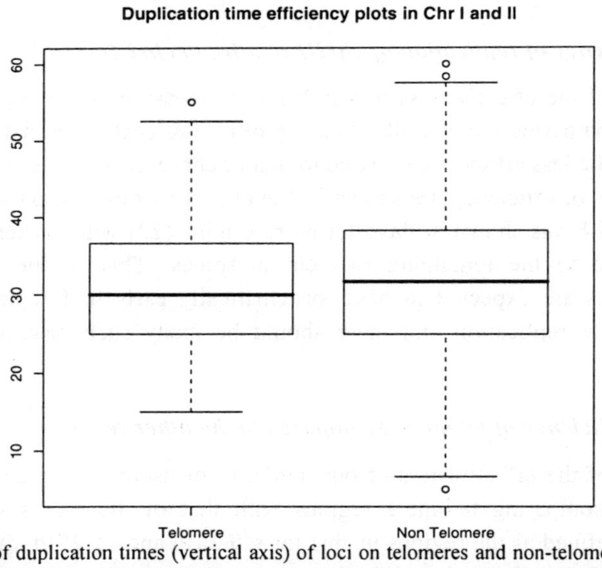

Figure 4. Boxplots of duplication times (vertical axis) of loci on telomeres and non-telomeres (horizontal axis) of chrI & chrII. Telomeres, systematically have lower duplication time or higher efficiency compared to the non-telomere regions, p-value < 0.06 by Wilcoxon rank-sum test.

4 Discussion

We have presented the first-of-its-kind approach that permits direct assessment of genome-wide replication efficiency at individual loci from the same dataset used to estimate the half replication timing. This is a significant step in DNA replication studies since the direct and accurate genomic scale estimation of replication efficiency at various loci of the genome based on the genome-wide replication profiles which has not been performed previously though the genome-wide microarray analyses have been widely used to determine profiles of average DNA replication timing at the genomic scale. In this paper, we demonstrated that replication efficiency, together with average replication timing, can be estimated using a novel approach - the near-sigmoid fitting for the increase in DNA copy number as a function of time at individual loci. Based on their estimates at each locus, we attained the genome-wide profiles of half replication timing and replication efficiency. We have shown the efficacy of our approach by various observations and their concordance with the literature on the analysis of *S. pombe* DNA replication microarray data.

Timing of firing at origins is proximated by the average replication timing of the peak loci. This approach has limitations on identifying (inefficient) late-firing origins closely located to the other (efficient) early-firing origins [2,4]. As firing at origins is relatively inefficient in *S. pombe* [8,9,16], it would not only under-estimate the number of efficient late firing origins, but also would fail to identify most, if not all, inefficient late-firing origins. This is because inefficient late-firing origins are unlikely to be self-sufficient in replication of the origin DNA. Nevertheless, this is still the most effective way to predict origins of replication at the genomic scale [14].

Acknowledgments

We thank Edison T. Liu and Neil D. Clarke for their support during this work.

References

1. M. Barranco and J. R. Buchler. Thermodynamic properties of hot nucleonic matter. *Phys. Rev.,* C22:1729—1737, 1980.
2. H. Müller and B. D. Serot. Phase transition in warm, asymmetric nuclear matter. *Phys. Rev.,* C52:2072—2091, 1995.
3. V. Baran, M. Colonna, M. Di Toro and A. B. Larionov. Spinodal decomposition of low-density asymmetric nuclesr matter. *Nucl. Phys.,* A632:287—303, 1998.
4. V. Baran, M. Colonna, M. Di Toro and V. Greco. Nuclear fragmentation: Sampling the instability of binary systems. *Phys. Rev. Lett.,* 86:4492—4495, 2001.
5. Gilbert DM (2001) Nuclear position leaves its mark on replication timing. J Cell Biol 152: F11-15.
6. MacAlpine DM, Bell SP (2005) A genomic view of eukaryotic DNA replication. Chromosome Res 13: 309-326.
7. Bell SP, Dutta A (2002) DNA replication in eukaryotic cells. Annu Rev Biochem 71: 333-374.
8. Raghuraman MK, Winzeler EA, Collingwood D, Hunt S, Wodicka L, et al. (2001) Replication dynamics of the yeast genome. Science 294: 115-121.

9. Schubeler D, Scalzo D, Kooperberg C, van Steensel B, Delrow J, et al. (2002) Genomewide DNA replication profile for Drosophila melanogaster: a link between transcription and replication timing. Nat Genet 32: 438-442.
10. Yabuki N, Terashima H, Kitada K (2002) Mapping of early firing origins on a replication profile of budding yeast. Genes Cells 7: 781-789.
11. Jeon Y, Bekiranov S, Karnani N, Kapranov P, Ghosh S, et al. (2005) Temporal profile of replication of human chromosomes. Proc Natl Acad Sci U S A 102: 6419-6424.
12. Kim SM, Huberman JA (2001) Regulation of replication timing in fission yeast. Embo J 20: 6115-6126.
13. Patel PK, Arcangioli B, Baker SP, Bensimon A, Rhind N (2006) DNA replication origins fire stochastically in fission yeast. Mol Biol Cell 17: 308-316.
14. Eklund H, Uhlin U, Farnegardh M, Logan DT, Nordlund P (2001) Structure and function of the radical enzyme ribonucleotide reductase. Prog Biophys Mol Biol 77: 177-268.
15. Majid Eshaghi, R. Krishna M. Karuturi, Juntao Li, Zhaoqing Chu, Edison T. Liu and Jianhua Liu. Global Profiling of DNA Replication Timing and Efficiency Reveals that Efficient Replication/ Firing Occurs Late During S-phase in Cells Released after HU-block in S. pombe, PLoS One, 2(8): e722. doi:10.1371/journal.pone.0000722.
16. Feng W, Collingwood D, Boeck ME, Fox LA, Alvino GM, et al. (2006) Genomic mapping of single-stranded DNA in hydroxyurea-challenged yeasts identifies origins of replication. Nat Cell Biol 8: 148-155.
17. Heichinger C, Penkett CJ, Bahler J, Nurse P (2006) Genome-wide characterization of fission yeast DNA replication origins. Embo J 25: 5171-5179.
18. MacNeill SA, Fantes PA (1997) Genetic and physiological analysis of DNA replication in fission yeast. Methods Enzymol 283: 440-459.
19. Rhind N (2006) DNA replication timing: random thoughts about origin firing. Nat Cell Biol 8: 1313-1316.
20. Glynn EF, Megee PC, Yu HG, Mistrot C, Unal E, et al. (2004) Genome-wide mapping of the cohesin complex in the yeast S. cerevisiae. PLoS Biol 2: E259.
21. Segurado M, de Luis A, Antequera F (2003) Genome-wide distribution of DNA replication origins at A+T-rich islands in Schizosaccharomyces pombe. EMBO Rep 4:1048-1053.
22. Hayashi M, Katou Y, Itoh T, Tazumi M, Yamada Y, et al. (2007) Genome-wide localization of pre-RC sites and identification of replication origins in fission yeast. Embo J 26: 1327-1339.
23. Benjamini, Y., and Hochberg, Y. (1995). Controlling the false discovery rate: a practical and powerful approach to multiple testing. J. Roy. Stat. Soc.B *57*, 289–300.
24. Herrick J, Jun S, Bechhoefer J, Bensimon A (2002). Kinetic model of DNA replication in eukaryotic organisms. J Mol Biol. 320(4):741-50.

AUTHOR INDEX